Selected Papers
of
NORBERT WIENER

NORBERT WIENER

1894–1964

Professor Daniel Rosich

Selected Papers of NORBERT WIENER

including

Generalized Harmonic Analysis

and

Tauberian Theorems

with contributions by

Y. W. Lee

N. Levinson

W. T. Martin

Published jointly by

SOCIETY FOR INDUSTRIAL AND APPLIED MATHEMATICS

and

THE M. I. T. PRESS

Massachusetts Institute of Technology

Cambridge, Massachusetts

LIBRARY OF CONGRESS CATALOG CARD NUMBER 64-7694
PRINTED IN THE UNITED STATES OF AMERICA

Professor Norbert Wiener (1894–1964) believed that significant research topics are to be found in the "crack" between two fields. Motivated in this way, he spent much of his life in areas bordering on electrical engineering, physics, and biophysics. His exceptional intuition and profound understanding of mathematics exhibited to him a unity where previously only diversity had been in evidence. It is to his initiative and genius in elucidating these complex structures that the present volume is dedicated.

Society for Industrial and Applied Mathematics
Philadelphia, November 26, 1964

Foreword

W. T. MARTIN

Department of Mathematics
Massachusetts Institute of Technology

This volume is a collection of some of Professor Norbert Wiener's outstanding mathematical and other scientific contributions. Those of us who had the pleasure of knowing him personally also gained something far beyond his contributions to science and mathematics.

In our associations with Professor Wiener we found that he could always be counted upon as a friend of his younger colleagues. He was generous with his ideas and always accessible to give advice and counsel when needed.

In our association with Professor Wiener over a period of twenty-five to thirty years, we learned certain attitudes from him that we found very valuable and helpful. One of these is the high and hard mathematical standards that he demonstrated in every piece of work he did. A second was the importance and dignity of the human individual. A third was his outlook in mathematics and science which emphasized a central unifying theme in all branches of mathematics and in the relation of mathematics to other fields of science and engineering. Finally, we learned from him the importance of judging which factors are basic in any large issue and of trying to base decisions on these basic factors.

As President Julius A. Stratton stated at the time of a special dinner in honor of Professor Wiener in February, 1961:

"We honor Norbert Wiener not solely as an extraordinarily gifted mathematician or simply as a scholarly, productive, creative mind—but equally for his warmth of understanding, for his humanity. We respect him, too, for the principles he defends, for the things for which he stands.

"During the more than forty years of his stay with us—years in which the Institute has grown in stature and in breadth, he has been a symbol of fine scholarship, of the highest goals of the academic world—indeed of all that we have hoped to be.

"M.I.T. is proud to count him as a member of the faculty. We are deeply indebted to him for his accomplishments and for his presence among us—to him and to Mrs. Wiener, who shares in our tribute on this great occasion."

Contents

Introduction and Some Comments on Wiener's Selected Papers

NORMAN LEVINSON

Department of Mathematics
Massachusetts Institute of Technology

The selection of Norbert Wiener's work reproduced here is taken mainly from his earlier work. More than three-quarters of the pages come from papers published in 1932 or earlier. One reason for this is that some of the later work is contained in books which are still available. Another reason is that Wiener's later work is in many ways based on, and a development of, his earlier work. A third reason is that much of Wiener's early work was premature in that an appreciation of its importance and usefulness did not come until years and sometimes decades after his discoveries. Thus these papers are of great current interest. Finally, more than half of this volume is taken up with two of Wiener's early papers, "Generalized Harmonic Analysis" and "Tauberian Theorems," which no selection of his work could omit and which appeared in 1930 and 1932, respectively. Wiener himself believed that the pattern of his subsequent work was laid down in his early work.

> Mathematics is very largely a young man's game. It is the athleticism of the intellect, making demands which can be satisfied to the full only when there is youth and strength. After one or two promising papers, many young mathematicians who have shown signs of ability sink into that very same limbo which surrounds yesterday's sports heroes.
>
> Yet it is not bearable to contemplate a brief distinction and burgeoning of activity which is to be followed by a lifetime of boredom. If the career of a mathematician is to be anything but an anti-climactic one, he must devote this brief springtime of top creative ability to the discovery of new fields and new problems, of such richness and compelling character that he can scarcely exhaust them in his lifetime. It has been my good fortune that the problems which excited me as a youth, and which I did a considerable amount to initiate, still do not seem to have lost their power to make maximum demands on me in my sixtieth year.[1]

A perusal of Wiener's work shows that he was motivated by the applications of mathematics to physics and engineering. As we shall see, this is true even of his work in Tauberian theorems. On the other hand, once Wiener tackled a problem, his treatment was rigorous, general, and aesthetic. Perhaps the framing of his theories in the full generality and abstraction of the Lebesgue integral delayed their

[1] Norbert Wiener, *I Am a Mathematician*, New York, 1956, p. 42.

accessibility to the engineer. Yet he could do it no other way. While his inspiration came from applications, he was and remained a Cambridge-trained analyst.

> Hardy . . . led me through the complicated logic of higher mathematics with such clarity and in such detail that he . . . gave me a real sense of what is necessary for a mathematical proof. He also introduced me to the Lebesgue integral, which was to lead directly to the main achievements of my early career.[2]

In combining the abilities of pure and applied mathematician, Wiener was in the best tradition of Gauss, Riemann, Poincaré, and Hadamard. However, with the inevitable increase of specialization, this breadth of facility is on the wane among contemporary mathematicians.

Random processes are treated in Wiener's early work and recur in one form or another in many of his later papers. The paper entitled "Differential Space" is the first paper in this volume treating this subject. Of his early work in this field Wiener says:

> Now consider the molecules of a fluid, whether gas or liquid. These molecules will not be at rest but will have a random irregular motion like that of the people in the crowd. This motion will become more active as the temperature goes up. Let us suppose that we have introduced into this fluid a small sphere which can be pushed about by the molecules in much the way that the pushball is agitated by the crowd. If this sphere is extremely small we cannot see it, and if it is extremely large and suspended in a fluid, the collisions of the particles of the fluid with the sphere will average out sufficiently well so that no motion is observable. There is an intermediate range in which the sphere is large enough to be visible and small enough to appear under the microscope in a constant irregular motion. This agitation, which indicates the irregular movement of the molecules is known as the Brownian motion. It had first been observed by the microscopists of the eighteenth century as a universal agitation of all sufficiently small particles in the microscopic field.
>
> Here I had a situation in which particles describe not only curves but statistical assemblages of curves. It was an ideal proving ground for my ideas concerning the Lebesgue integral in a space of curves, and it had the abundantly physical texture of the work of Gibbs. I met with a considerable degree of success.
>
> The Brownian motion was nothing new as an object of study by physicists. There were fundamental papers by Einstein and Smoluchowski that covered it, but whereas these papers concerned what was happening to any given particle at a specific time, or the long-time statistics of many particles, they did not concern themselves with the mathematical properties of the curve followed by a single particle.
>
> Here the literature was very scant but it did include a telling comment by the French physicist Perrin in his book *Les Atomes*, where he said in effect that the very irregular curves followed by particles in the Brownian motion led one to

[2] *Ibid.*, p. 22.

think of the supposed continuous non-differentiable curves of the mathematicians. He called the motion continuous because the particles never jump over a gap and non-differentiable because at no time do they seem to have a well-defined direction of movement.

In the physical Brownian motion, it is of course true that the particle is not subject to an absolutely perpetual influence resulting from the collision of the molecules but that there are short intervals of time between one collision and the next. These, however, are far too short to be observed by any ordinary methods. It therefore becomes natural to idealize the Brownian motion as if the molecules were infinitesimal in size and the collisions continuously described. It was this idealized Brownian motion which I studied, and which I found to be an excellent surrogate for the cruder properties of the true Brownian motion.

To my surprise and delight I found that the Brownian motion as thus conceived had a formal theory of a high degree of perfection and elegance. Under this theory I was able to confirm the conjecture of Perrin and to show that, except for a set of cases of probability 0, all the Brownian motions were continuous non-differentiable curves.

The papers which I wrote on this subject were, I believe, the first to disclose anything very new—combining the Lebesgue technique of integration with the physical ideas of Gibbs.[3]

"Differential Space" formulates a theory of integration in a space of functions, the functions being the sample paths of Brownian motion. Differential space occurs again in "Generalized Harmonic Analysis," where a spectral analysis of random processes is made. Random processes occur also in "The Homogeneous Chaos," in the last chapters of the book (with R. E. A. C. Paley) *Fourier Transforms in the Complex Domain,* and in the book *Extrapolation, Interpolation, and Smoothing of Stationary Time Series.*

On the cover of Wiener's book *I Am A Mathematician* appears the formula

$$\lim_{\epsilon \to 0} \frac{1}{2\epsilon} \int_{-\infty}^{\infty} | g(\omega + \epsilon) - g(\omega - \epsilon) |^2 \, d\omega = \frac{2}{\pi} \lim_{A \to \infty} \frac{1}{2A} \int_{-A}^{A} | f(t) |^2 \, dt. \tag{1}$$

This is one of the key formulas of generalized harmonic analysis which is one of Wiener's great creations. To obtain this formula, Wiener was confronted with the necessity of proving the equivalence of

$$\frac{1}{\pi} \lim_{\epsilon \to 0} \int_{-\infty}^{\infty} \frac{\sin^2 \epsilon t}{\epsilon t^2} | f(t) |^2 \, dt = \lim_{A \to \infty} \frac{1}{2A} \int_{-A}^{A} | f(t) |^2 \, dt. \tag{2}$$

[3] *Ibid.,* pp. 38–39.

Note that both integrals are weighted averages of $|f|^2$ over $(-\infty, \infty)$. In the course of establishing (2), Wiener developed a general theory for showing the equivalence of different weighted averages of a function. Wiener saw in this equivalence procedure the clue to the correct way to formulate the theory of Tauberian theorems, and in the course of proving (2), arising as an aspect of generalized harmonic analysis, Wiener developed an entirely novel and extremely powerful theory of Tauberian theorems. Thus Wiener came to Tauberian theorems in the course of developing a spectral theory satisfactory for treating white light, noise, etc. In origin the generalized harmonic analysis and Tauberian theorems are intertwined, and it is not an accident that Wiener's two great memoirs on these subjects are separated by only two years.

In generalized harmonic analysis Wiener set himself the problem of obtaining the harmonic analysis of phenomena more general than either the purely periodic process for which Fourier series suffice or the transient function small at the remote ends of the time scale for which Fourier transform theory is used. Wiener wanted to analyze such persistent phenomena as a continuing source of light, a continuing broadcast signal, a continuing Brownian motion, etc.

The really satisfactory theories of the Fourier series and integral were too new in 1920 to have trickled down to the working electrical engineer. Moreover, the sort of phenomenon in which the engineer is chiefly interested had almost entirely escaped the treatment of the pure mathematicians. The Fourier series, which the pure mathematicians had treated, was useful only for the study of those phenomena which repeat themselves after a fixed time. The standard form of the theory of the Fourier integral, as developed by Plancherel and others, concerns curves which are small in the remote past and are destined to become small in the remote future. In other words, the standard theory of the Fourier integral deals with phenomena which in some sense or other both begin and end, and do not keep running indefinitely at about the same scale. The sort of continuing phenomenon that we find in a noise or a beam of light had been completely neglected by the professional mathematician, and had been left to such mathematically-minded physicists as Sir Arthur Schuster of Manchester.[4]

In mathematical terms the class of functions $\{f(t)\}$ Wiener considered were those for which

$$\phi(x) = \lim_{T\to\infty} \frac{1}{2T} \int_{-T}^{T} f(x+t)\bar{f}(t)\, dt \tag{3}$$

exists for all x. If one applies (3) to the case

$$f(t) = \sum_{1}^{N} a_n \exp(i\lambda_n t),$$

[4] *Ibid.*, p. 77.

then

$$\phi(x) = \sum_{1}^{N} | a_n |^2 \exp (i\lambda_n x).$$

If

$$(4) \qquad S(u) = \frac{1}{2\pi} \int_{-\infty}^{\infty} \phi(x) \frac{\exp(-iux) - 1}{-ix} dx,$$

then the fact that

$$\frac{1}{2\pi} \int_{-\infty}^{\infty} \frac{\sin vx}{x} dx = \begin{cases} \frac{1}{2}, & v > 0, \\ -\frac{1}{2}, & v < 0 \end{cases}$$

shows that in this case $S(u)$ is a step function with jumps of $| a_n |^2$, where $u = \lambda_n$ and constant for all other u. Thus

$$(5) \qquad \phi(x) = \int_{-\infty}^{\infty} \exp(iux)\, dS(u).$$

Clearly if u_1 and u_2 are not equal to any λ_n,

$$S(u_2) - S(u_1) = \sum_{u_1<\lambda_n<u_2} | a_n |^2.$$

Thus $S(u_2) - S(u_1)$ is a measure of the amount of energy of $f(t)$ between frequencies u_1 and u_2. It turns out that even if f is not of the form of a trigonometric polynomial, $S(u)$ is an increasing function, and (4) and (5) are valid. If f is random in character, it is often the case that $S(u)$ becomes a differentiable function and

$$S'(u) = \frac{1}{2\pi} \int_{-\infty}^{\infty} \phi(x) \exp(-iux)\, dx,$$

$$\phi(x) = \int_{-\infty}^{\infty} \exp(iux) S'(u)\, du,$$

at least in the sense of convergence in the mean. In this case, $S'(u)$ is sometimes called the power density spectrum or simply the power spectrum and $S(u)$, the integrated power spectrum. From $S(u)$, or $S'(u)$, one can determine how much of the energy of $f(t)$ is concentrated in a given range of the spectrum, $u_1 < u < u_2$. This is a result of great theoretical and applied interest, and explains the importance of generalized harmonic analysis in contemporary mathematics.

If in (2) one sets

$$F(t) = \tfrac{1}{2}(| f(t) |^2 + | f(-t) |^2),$$

then (2) becomes

$$\frac{2}{\pi} \lim_{\epsilon \to 0} \int_0^\infty \frac{\sin^2 \epsilon t}{\epsilon t^2} F(t)\, dt = \lim_{A \to \infty} \frac{1}{A} \int_0^A F(t)\, dt \tag{6}$$

in the sense that it is required to show that the existence of either limit implies the other if $F(t) \geq 0$.

Wiener recast (6) in new variables by setting $t = e^y$, $A = e^x$, and $\epsilon = e^{-x}$. With $F(e^y) = H(y)$, (6) becomes

$$\frac{2}{\pi} \lim_{x \to \infty} \int_{-\infty}^\infty \exp(x - y) \sin^2 \exp[-(x - y)] H(y)\, dy$$

$$= \lim_{x \to \infty} \int_{-\infty}^x \exp[-(x - y)] H(y)\, dy. \tag{7}$$

These are convolution integrals. Wiener now viewed the problem in the following light. Let $K(x)$ be integrable and satisfy

$$\int_{-\infty}^\infty |K(x)|\, dx < \infty, \qquad \int_{-\infty}^\infty K(x)\, dx = 1.$$

Let $K_1(x)$ satisfy the same conditions. Then under what condition does the existence of

$$\lim_{x \to \infty} \int_{-\infty}^\infty K(x - y) f(y)\, dy = a$$

imply that

$$\lim_{x \to \infty} \int_{-\infty}^\infty K_1(x - y) f(y)\, dy = a?$$

Note that (7) is precisely of this form with

$$K(x) = \frac{2}{\pi} e^x \sin^2 e^{-x}$$

and

$$K_1(x) = \begin{cases} e^{-x}, & x > 0, \\ 0, & x < 0 \end{cases}$$

in going in one direction in (7), and then the roles of K and K_1 are reversed in going in the other direction.

Wiener saw that formulating his particular problem in terms of K and K_1 made it a particular case of a comprehensive theory which included the classical Tauberian theorems and also suggested a general method of solution.

To see how the classical Tauberian theorems can be formulated in terms of a convolution, consider the hypothesis

$$\lim_{t\to 1-0} \sum_{1}^{\infty} a_k t^k = L$$

where the a_k are constants. (The Tauberian theorem of Hardy and Littlewood, for example, states that if $| a_n | \leq K/n$ for some constant K and if L exists as above, then $\lim_{n\to\infty} \sum_{1}^{n} a_k = L$.)

Let

$$s(u) = \sum_{k\leq u} a_k.$$

Then

$$\sum_{1}^{\infty} a_k t^k = \int_0^{\infty} t^u \, ds(u)$$

$$= -\int_0^{\infty} t^u \log t \; s(u) \, du.$$

By setting $t = \exp(-e^{-x})$ and $u = e^y$,

$$\sum_{1}^{\infty} a_k t^k = \int_{-\infty}^{\infty} \exp\{-\exp[-(x-y)]\} \exp[-(x-y)] \, s(e^y) \, dy.$$

Hence the hypothesis becomes

$$\lim_{t\to 1-0} \sum_{1}^{\infty} a_k t^k = \lim_{x\to\infty} \int_{-\infty}^{\infty} K(x-y) f(y) \, dy = L$$

where

$$K(x) = e^{-x} \exp(-e^{-x}), \qquad f(y) = s(e^y),$$

and thus the hypothesis is formulated as a convolution. Note that the convolution integral represents an averaging process on f with weight 1 since

$$\int K(x) \, dx = 1.$$

To see how the reformulation of Tauberian theorems as a convolution helped Wiener solve the problem, let us consider Wiener's early method of proof (prior to his introduction of the reciprocal of an absolutely convergent nonvanishing Fourier series). Here is one of Wiener's general Tauberian theorems. (All functions considered are assumed to be measurable.)

THEOREM. *Let $f(x)$, $-\infty < x < \infty$, be bounded:*

$$|f(x)| \leq M.$$

Let $K(x)$ satisfy

$$\int_{-\infty}^{\infty} |K(x)|\,dx < \infty, \qquad \int_{-\infty}^{\infty} K(x)\,dx = 1, \tag{8}$$

where the second condition will be called normality. Let

$$k(u) = \int_{-\infty}^{\infty} K(x) \exp(iux)\,dx \neq 0, \qquad -\infty < u < \infty. \tag{9}$$

(This is the key assumption.) Let $K_1(x)$ also satisfy (8). Then if

$$\lim_{x\to\infty} \int_{-\infty}^{\infty} K(x-y)f(y)\,dy = a, \tag{10}$$

it follows that

$$\lim_{x\to\infty} \int_{-\infty}^{\infty} K_1(x-y)f(y)\,dy = a. \tag{11}$$

What Wiener's theorem does is allow one to go from an intractable weighting function K on f to a tractable one K_1 from which one can then conclude, with some auxiliary condition on f, that

$$\lim_{x\to\infty} f(x) = a. \tag{12}$$

The major step is that from (10) to (11).

First the elementary converse, due in a special case to Abel, will be proved.

ABEL'S THEOREM. *Let $|f(x)| \leq M$ and let (12) hold. Then for any K satisfying (8), the limit (10) exists.*

Thus if $f(x) \to a$ as $x \to \infty$, the weighted average of f approaches the same limit, a result that is not surprising.

Proof of Abel's Theorem. Since K is normal,

$$J = \int_{-\infty}^{\infty} K(x-y)f(y)\,dy - a$$

$$= \int_{-\infty}^{\infty} K(x-y)[f(y) - a]\,dy = J_1 + J_2$$

where

$$J_1 = \int_{-\infty}^{\frac{1}{2}x} K(x - y)[f(y) - a]\, dy$$

and

$$J_2 = \int_{\frac{1}{2}x}^{\infty} K(x - y)[f(y) - a]\, dy.$$

Since

$$|f(y) - a| \leq |f(y)| + |a| \leq 2M,$$

$$|J_1| \leq 2M \int_{-\infty}^{\frac{1}{2}x} |K(x - y)|\, dy.$$

By setting

$$v = x - y,$$

$$|J_1| \leq 2M \int_{\frac{1}{2}x}^{\infty} |K(v)|\, dv.$$

Hence,

$$\lim_{x\to\infty} J_1 = 0.$$

If $\epsilon > 0$ is given, then for sufficiently large x

$$|f(y) - a| < \epsilon$$

for $y > \frac{1}{2}x$ by (12). Hence,

$$|J_2| \leq \epsilon \int_{\frac{1}{2}x}^{\infty} |K(x - y)|\, dy \leq \epsilon \int_{-\infty}^{\infty} |K(v)|\, dv.$$

Since $J = J_1 + J_2$, it now follows that

$$\limsup_{x\to\infty} |J| \leq \epsilon \int_{-\infty}^{\infty} |K(v)|\, dv.$$

Since ϵ is arbitrary, this implies

$$\lim_{x\to\infty} J = 0,$$

and proves Abel's theorem.

In Wiener's early proof of the Tauberian theorem the additional assumption

$$\int_{-\infty}^{\infty} |xK(x)|\, dx < \infty \tag{13}$$

is made. This condition is met in all applications of the Tauberian theorem familiar to me. It is a testimonial to Wiener's aesthetic feeling that he found it desirable to dispense with (13) by introducing in his later proofs the very beautiful and rich concept associated with absolutely convergent Fourier series which is not really required in applications of the general Tauberian theorem but which has inspired much mathematical activity. Here, in order to reveal the essential simplicity of Wiener's theory, (13) will be used.

Proof of Theorem. Let $R(x)$ satisfy the absolute integrability and normality condition (8). Then because of the absolute convergence of the following repeated integral, the limits can be interchanged to give

$$\int_{-\infty}^{\infty} R(x-\xi)\, d\xi \int_{-\infty}^{\infty} K(\xi-y)f(y)\, dy$$

$$= \int_{-\infty}^{\infty} K_2(x-y)f(u)\, dy \tag{14}$$

where

$$K_2(x) = \int_{-\infty}^{\infty} R(x-\xi)K(\xi)\, d\xi. \tag{15}$$

It follows from (15) that K_2 satisfies (8). On the other hand, by Abel's theorem with R playing the role of K and

$$\int_{-\infty}^{\infty} K(\xi-y)f(y)\, dy$$

playing the role of f, it follows that

$$\lim_{x\to\infty} \int_{-\infty}^{\infty} R(x-\xi)\, d\xi \int_{-\infty}^{\infty} K(\xi-y)f(y)\, dy = a,$$

and hence by (14)

$$\lim_{x\to\infty} \int_{-\infty}^{\infty} K_2(x-y)f(y)\, dy = a. \tag{16}$$

Thus, given that the weighted average of f by K tends to a limit, the same is true for any K_2 given by (15) providing only that R satisfies (8). However, the weighting functions, or kernels, K_2 are not always tractable when one wants to proceed to the conclusion (12) for appropriately restricted f. Wiener saw that K_2 could be made to approximate arbitrarily closely any K_1, and this will be shown later.

Given K and K_2, (15) is an integral equation of the convolution type for R. By taking the Fourier transform of (15),

$$k_2(u) = k(u)r(u)$$

where $k(u)$ is given by (9); k_2 is similarly related to K_2, and

$$r(u) = \int_{-\infty}^{\infty} R(x) \exp(iux)\, dx.$$

Hence,

$$r(u) = \frac{k_2(u)}{k(u)}. \tag{17}$$

It is convenient to choose $k_2(u) = 0$ for large $|u|$, so that if $k(u)$ is very small for large $|u|$, $r(u)$ will not be too large, for large $|u|$, to be a Fourier transform. It is also convenient to take $k_2(u)$ differentiable. For these and other reasons Wiener chose

$$k_2(u) = \begin{cases} 1 - |u|/2N, & |u| \leq 2N, \\ 0, & |u| > 2N \end{cases} \tag{18}$$

for constant N. Thus (17) can now be used as the definition of $r(u)$; $R(x)$ is defined by

$$R(x) = \frac{1}{2\pi} \int_{-\infty}^{\infty} r(u) \exp(-iux)\, du; \tag{19}$$

and now it must be shown that R in fact satisfies (8). From (18) follows

$$K_2(x) = \frac{1}{2\pi} \int_{-2N}^{2N} \left(1 - \frac{|u|}{2N}\right) \exp(-iux)\, du$$

$$= \frac{1}{\pi} \int_{0}^{2N} \left(1 - \frac{u}{2N}\right) \cos ux\, du$$

$$= \frac{1}{\pi} \left(1 - \frac{u}{2N}\right) \frac{\sin ux}{x} \Bigg|_0^{2N} + \frac{1}{2\pi N x} \int_0^{2N} \sin ux\, du$$

$$= \frac{1 - \cos 2Nx}{2\pi N x^2} = \frac{\sin^2 Nx}{\pi N x^2}. \tag{20}$$

Since $k_2(0) = 1$, K_2 is normal.

Since $k(u) \neq 0$, it follows from (17) that

$$\int_{-\infty}^{\infty} |\, r(u) \,|^2 \, du = \int_{-2N}^{2N} \frac{[1 - (|\, u \,|/2N)]^2}{|\, k(u) \,|^2} \, du < \infty. \tag{21}$$

By (13), $k'(u)$ exists and

$$k'(u) = \int_{-\infty}^{\infty} ixK(x) \exp(iux) \, dx.$$

It follows from (13) that $|\, k'(u) \,|$ is bounded. From (17),

$$\begin{aligned} r'(u) &= \frac{1}{2Nk(u)} - \frac{k_2(u)}{k^2(u)} k'(u), & -2N \le u \le 0, \\ &= \frac{-1}{2Nk(u)} - \frac{k_2(u)}{k^2(u)} k'(u), & 0 \le u \le 2N, \\ &= 0, & |\, u \,| > 2N. \end{aligned}$$

Hence

$$\int_{-2N}^{2N} |\, r'(u) \,|^2 \, du = \int_{-\infty}^{\infty} |\, r'(u) \,|^2 \, du < \infty. \tag{22}$$

From Plancherel's theorem, (21) and (22) imply

$$\int_{-\infty}^{\infty} R^2(x) \, dx < \infty, \qquad \int_{-\infty}^{\infty} x^2 R^2(x) \, dx < \infty$$

and hence

$$\int_{-\infty}^{\infty} (1 + x^2) R^2(x) \, dx < \infty.$$

Thus by the Schwarz inequality,

$$\begin{aligned} \int_{-\infty}^{\infty} |\, R(x) \,| \, dx &= \int_{-\infty}^{\infty} (1 + x^2)^{\frac{1}{2}} \,|\, R(x) \,|\, (1 + x^2)^{-\frac{1}{2}} \, dx \\ &\le \left[\int_{-\infty}^{\infty} (1 + x^2) R^2(x) \, dx\right]^{\frac{1}{2}} \left[\int_{-\infty}^{\infty} \frac{dx}{1 + x^2}\right]^{\frac{1}{2}} < \infty. \end{aligned}$$

Hence $R(x)$ is in fact absolutely integrable. That $R(x)$ is normal follows from $r(0) = 1$, which is a consequence of (17) and the fact that $k_2(0) = k(0) = 1$. Thus it has been proved that K_2 in (16) can be taken as $\sin^2 Nx/(\pi N x^2)$ to give

$$\lim_{x\to\infty} \int_{-\infty}^{\infty} \frac{\sin^2 N(x-y)}{\pi N(x-y)^2} f(y)\,dy = a. \tag{23}$$

A further reason for the choice of $K_2(x)$ as $\sin^2 Nx/(\pi N x^2)$ is that with this choice $K_2(x - y)$ tends to the Dirac delta function $\delta(x - y)$ as $N \to \infty$. Why this is useful will now be seen.

Let $K_1(x)$ be absolutely integrable and normal so that it satisfies (8). Then by using Abel's theorem much as was done below (15) but with K_1 now playing the role of R, it follows that

$$\lim_{x\to\infty} \int_{-\infty}^{\infty} K_1(x-\xi)\,d\xi \int_{-\infty}^{\infty} \frac{\sin^2 N(\xi - y)}{\pi N(\xi - y)^2} f(y)\,dy = a$$

or inverting the order of integration, which is allowed by the absolute integrability,

$$\lim_{x\to\infty} \int_{-\infty}^{\infty} f(y)\,dy \int_{-\infty}^{\infty} K_1(x-\xi) \frac{\sin^2 N(\xi - y)}{\pi N(\xi - y)^2}\,d\xi = a.$$

By replacing $\xi - y$ by u,

$$\lim_{x\to\infty} \int_{-\infty}^{\infty} f(y)\,dy \int_{-\infty}^{\infty} K_1(x-u-y) \frac{\sin^2 Nu}{\pi N u^2}\,du = a$$

or by inverting again,

$$\lim_{x\to\infty} \int_{-\infty}^{\infty} \frac{\sin^2 Nu}{\pi N u^2}\,du \int_{-\infty}^{\infty} K_1(x-u-y) f(y)\,dy = a. \tag{24}$$

It will now be shown that this implies (11), which in effect will show that one can let $N \to \infty$ in (24) before letting $x \to \infty$. To begin with the fact that $\sin^2 Nu/(\pi N u^2)$ is normal shows that

$$\int_{-\infty}^{\infty} \frac{\sin^2 Nu^2}{\pi N u^2}\,du \int_{-\infty}^{\infty} K_1(x-y) f(y)\,dy = \int_{-\infty}^{\infty} K_1(x-y) f(y)\,dy.$$

Hence if

$$J = \int_{-\infty}^{\infty} \frac{\sin^2 Nu^2}{\pi N u^2}\,du \int_{-\infty}^{\infty} K_1(x-y-u) f(y)\,dy$$

$$- \int_{-\infty}^{\infty} K_1(x-y) f(y)\,dy, \tag{25}$$

then

$$J = \int_{-\infty}^{\infty} \frac{\sin^2 Nu^2}{\pi Nu^2}\, du \int_{-\infty}^{\infty} [K_1(x-y-u) - K_1(x-y)]f(y)\, dy.$$

Clearly

$$|J| \leq J_1 + J_2 + J_3 \tag{26}$$

where if $\delta > 0$ and since $|f(y)| \leq M$,

$$J_1 = M \int_{-\infty}^{-\delta} \frac{\sin^2 Nu^2}{\pi Nu^2}\, du \int_{-\infty}^{\infty} (|K_1(x-y-u)| + |K_1(x-y)|)\, dy,$$

$$J_2 = M \int_{-\delta}^{\delta} \frac{\sin^2 Nu^2}{\pi Nu^2}\, du \int_{-\infty}^{\infty} |K_1(x-y-u) - K_1(x-y)|\, dy,$$

and J_3 is similar to J_1 but over $\delta \leq u < \infty$. Clearly

$$J_1 \leq \frac{M}{\pi N} \int_{-\infty}^{-\delta} \frac{du}{u^2}\, 2 \int_{-\infty}^{\infty} |K_1(v)|\, dv$$

$$= \frac{2M}{\pi N \delta} \int_{-\infty}^{\infty} |K_1(v)|\, dv. \tag{27}$$

The same holds for J_3. Now

$$J_2 \leq M \max_{|\eta| \leq \delta} \int_{-\infty}^{\infty} |K_1(v+\eta) - K_1(v)|\, dv \int_{-\delta}^{\delta} \frac{\sin^2 Nu}{\pi Nu^2}\, du$$

$$< M \max_{|\eta| \leq \delta} \int_{-\infty}^{\infty} |K_1(v+\eta) - K_1(v)|\, dv \tag{28}$$

since $\sin^2 Nu^2/(\pi Nu^2)$ is normal.

Given $\epsilon_1 > 0$, choose $H_1(x)$ as a continuous function that vanishes outside of a finite interval and such that

$$\int_{-\infty}^{\infty} |H_1(v) - K_1(v)|\, dv < \epsilon_1.$$

By the continuity of H_1, if δ is small enough,

$$\max_{|\eta| \leq \delta} \int_{-\infty}^{\infty} |H_1(v+\eta) - H_1(v)|\, dv < \epsilon_1.$$

Hence,

$$\max_{|\eta| \leq \delta} \int_{-\infty}^{\infty} | K_1(v + \eta) - K_1(v) | \, dv \leq 3\epsilon_1.$$

Choose $\epsilon > 0$ and let $\epsilon_1 = \epsilon/(9M)$. Then by (28),

$$J_2 \leq \tfrac{1}{3}\epsilon$$

for δ small enough. Having chosen δ, one can choose N so large in (27) that

$$| J_1 | \leq \tfrac{1}{3}\epsilon$$

and similarly for J_3. Hence, by (26),

$$| J | \leq \epsilon.$$

Using this with (25) and (24) gives

$$\limsup_{x \to \infty} \left| a - \int_{-\infty}^{\infty} K_1(x - y) f(y) \, dy \right| \leq \epsilon.$$

Since ϵ is arbitrary, this yields (11) and proves Wiener's theorem.

Contributions of Norbert Wiener to Linear Theory and Nonlinear Theory in Engineering

Y. W. LEE

Department of Electrical Engineering and Research Laboratory of Electronics Massachusetts Institute of Technology

A new foundation. Before the introduction of statistical concepts to communication problems, classical communication theory had been established on the basis that a message or a noise could be considered either as a periodic phenomenon or as a transient phenomenon. In the absence of a suitable method of representation for the actual messages and noise, classical theory replaced them by periodic or transient functions depending upon which one was the more "reasonable" substitute.

In 1942, in his classic work "The Extrapolation, Interpolation and Smoothing of Stationary Time Series," Wiener laid a new foundation for communication and control theory by considering messages and noise as statistical phenomena rather than as deterministic ones as in classical theory. He said:*

Let us now turn from the study of time series to that of communication engineering. This is the study of messages and their transmission, whether these messages be sequences of dots and dashes as in the Morse code or the teletypewriter, or sound-wave patterns, as in the telephone or phonograph, or patterns representing visual images, as in telephoto service and television. In all communication engineering—if we do not count such rude expedients as the pigeon post as communication engineering—the message to be transmitted is represented as some sort of array of measurable quantities distributed in time. In other words, by coding or the use of the voice or scanning, the message to be transmitted is developed into a time series. This time series is then subjected to transmission by an apparatus which carries it through a succession of stages, at each of which the time series appears by transformation as a new time series. These operations, although carried out by electrical or mechanical or other such means, are in no way essentially different from the operations computationally carried out by the time series statistician with slide rule and computing machine.

* Norbert Wiener, *The Extrapolation, Interpolation, and Smoothing of Stationary Time Series*, The Technology Press of The Massachusetts Institute of Technology and John Wiley & Sons, Inc., New York, 1949, pp. 2–4. Original work appeared as a report in 1942.

The proper field of communication engineering is far wider than that generally assigned to it. Communication engineering concerns itself with the transmission of messages. For the existence of a message, it is indeed essential that variable information be transmitted. The transmission of a single fixed item of information is of no communicative value. We must have a repertory of possible messages, and over this repertory a measure determining the probability of these messages.

A message need not be the result of a conscious human effort for the transmission of ideas. For example, the records of current and voltage kept on the instruments of an automatic sub-station are as truly messages as a telephone conversation. From this point of view, the record of the thickness of a roll of paper kept by a condenser working an automatic stop on a Fourdrinier machine is also a message, and the servo-mechanism stopping the machine at a flaw belongs to the field of communication engineering, as indeed do all servo-mechanisms.

. . . .

Let us now see what are the fields from which the present-day statistician and the present-day communication engineer draw their techniques. First, let us consider the statistician. Behind all statistical work lies the theory of probabilities. The events which actually happen in a single instance are always referred to a collection of events which might have happened; and to different subcollections of such events, weights or probabilities are assigned . . .. A statistical method, as for example a method of extrapolating a time series into the future, is judged by the probability with which it will yield an answer correct within certain bounds, or by the mean (taken with respect to probability) of some positive function or norm of the error contained in its answer.

. . . .

In other words, the statistical theory of time series does not consider the individual time series by itself, but a distribution or *ensemble* of time series. Thus the mathematical operations to which a time series is subjected are judged, not by their effect in a particular case, but by their average effect. While one does not ordinarily think of communication engineering in the same terms, this statistical point of view is equally valid. No apparatus for conveying information is useful unless it is designed to operate, not on a particular message, but on a set of messages, and its effectiveness is to be judged by the way in which it performs on the average on messages of this set. "On the average" means that we have a way of estimating which messages are frequent and which rare or, in other words, that we have a *measure* or probability of possible messages. The apparatus to be used for a particular purpose is that which gives the best result "on the average" in an appropriate sense of the word "average."

These ideas on communication and control enunciated by Wiener in 1942 are the basic ideas upon which modern communication theory has been developed.

Linear theory. The problem that led Wiener to these ideas was that of the design of fire-control apparatus for anti-aircraft guns on which he worked during World War II. This was the problem of predicting the future position of an airplane based upon the observed past positions of the airplane. How he arrived at these ideas through this problem is interesting history.* Wiener said:

The mathematical processes which suggested themselves to me in the first instance for prediction were, in fact, impossible of execution, for they assumed an already existing knowledge of the future. However, I was able to show there was a certain sense in which these processes might be approximated by processes free from this objection. . . .

I did in fact consider certain possibilities of approximating to non-realizable operations by realizable operators. . . . What was interesting and exciting . . . was that the pieces of apparatus designed for best following a smooth curve were oversensitive and were driven into violent oscillation by a corner. . . . Then the idea suggested itself to me: Perhaps this difficulty is in the order of things, and there is no way in which I can overcome it. Perhaps it belongs to the nature of prediction that an accurate apparatus for smooth curves is an excessively sensitive apparatus for rough curves. Perhaps we have here the example of the same sort of malice of nature which appears in Heisenberg's principle, which forbids us to say precisely and simultaneously both where a particle is and how fast it is going.

The more we studied the problem, the more we became convinced that we were right and that the difficulty was fundamental. . . . If errors of inaccuracy and errors of hypersensitivity always seemed to be in opposite directions, on what basis could we make a compromise between those two errors?

The answer was that we could make such a compromise only on a statistical basis. For the actual distribution of curves which we wanted to predict . . . we might seek a prediction making some quantity a minimum; and the most natural quantity to choose at the start . . . was the mean square error of prediction. . . .

Thus we could set up the prediction problem as a minimization problem . . . [which] leads to . . . an integral equation.

This was lucky for me, for integral equations were well within my field of interest; but the even luckier thing was that the particular integral equation to which the problem leads is a slight extension of the one which had been considered by Eberhard Hopf and myself. The result was that not only was I able to formulate the prediction problem but also to solve it; and what was even luckier was the fact that the solution came out in a simple form.

In solving the prediction problem, Wiener realized that he had, in fact, included in the new formulation and solution a large class of communication problems besides prediction. One particularly important

* Norbert Wiener, *I am a Mathematician*, Doubleday & Co., Inc., New York, 1956, pp. 241–245.

one is that of filtering. In classical communication theory, prediction and filtering are not conceived to be related problems. The unification of these and other problems under one formulation from the statistical point of view is a major contribution of Wiener to communication and control engineering. The class of problems that Wiener solved is generally known as that of the design of optimum linear systems.

In a time-invariant linear system with an input time function $x(t)$ and an output time function $y(t)$, the input-output relation is given by the convolution integral

$$y(t) = \int_{-\infty}^{\infty} h(\tau)x(t - \tau)\, d\tau, \tag{1}$$

in which $h(t)$ is the response of the system to an excitation in the form of an impulse that occurs at $t = 0$ and whose integral is unity. Because in a physical system the effect cannot precede the cause, we must have

$$h(t) = 0, \qquad t < 0. \tag{2}$$

Since we represent the input and output statistically, $x(t)$ is a member function of the ensemble of possible inputs and $y(t)$ is the corresponding member of the ensemble of possible outputs. We assume that the ensembles are ergodic. The random process $x(t)$ is generally a mixture of a message $m(t)$ and a noise $n(t)$. Although this mixture can be in many forms it is frequently additive so that

$$x(t) = m(t) + n(t). \tag{3}$$

Now, for the determination of the linear system characteristic $h(t)$ that fulfills the requirements of a problem, we specify a desired output $z(t)$. For instance, if the problem is pure prediction, we would have $x(t) = m(t)$ and $z(t) = m(t + \alpha)$; that is, the input is a pure message, and the desired output is the same message with an advance in time of α seconds ($\alpha > 0$). If the problem is filtering, we would have $x(t) = m(t) + n(t)$ and $z(t) = m(t - \alpha)$; that is, the input is an additive mixture of message and noise, and the desired output is the message with a time lag of α seconds. In another problem we might have $x(t) = m(t) + n(t)$ and $z(t) = m'(t + \alpha)$. This problem requires the output to be the derivative of the message with a forward displacement of α seconds when the input is an additive mixture of message and noise. This is a problem of filtering, differentiation, and prediction. Clearly there are many other possible situations.

When we specify a desired output in this manner, we assume an ideal result that we do not expect to obtain perfectly from the physical linear system. There will be an error. The measure of error Wiener chose was the mean-square error. Thus the mean-square error is

$$\overline{\epsilon^2(t)} = \lim_{T\to\infty} \frac{1}{2T} \int_{-T}^{T} [y(t) - z(t)]^2 \, dt. \tag{4}$$

By expanding the expression and applying (1), we obtain

$$\overline{\epsilon^2(t)} = \int_{-\infty}^{\infty} h(\tau)\, d\tau \int_{-\infty}^{\infty} h(\sigma) \phi_{xx}(\tau - \sigma)\, d\sigma - 2 \int_{-\infty}^{\infty} h(\tau) \phi_{xz}(\tau)\, d\tau + \phi_{zz}(0), \tag{5}$$

where

$$\phi_{xx}(\tau) = \lim_{T\to\infty} \frac{1}{2T} \int_{-T}^{T} x(t) x(t + \tau)\, d\tau, \tag{6}$$

which is the autocorrelation function of the input,

$$\phi_{xz}(\tau) = \lim_{T\to\infty} \frac{1}{2T} \int_{-T}^{T} x(t) z(t + \tau)\, d\tau, \tag{7}$$

which is the input desired-output crosscorrelation function, and

$$\phi_{zz}(0) = \lim_{T\to\infty} \frac{1}{2T} \int_{-T}^{T} z^2(t)\, dt. \tag{8}$$

The functions $\phi_{xx}(\tau)$ and $\phi_{xz}(\tau)$ upon which the mean-square error depends are expressed here as time averages. Because the input, actual output, and desired output ensembles are ergodic, these time averages are equivalent to their respective ensemble averages.

To find the necessary and sufficient condition under which the linear system characteristic $h(t)$ in expression (5) minimizes the mean-square error, an application of calculus of variations is made. The result is the integral equation

$$\phi_{xz}(\tau) - \int_{-\infty}^{\infty} h_{\text{opt}}(\sigma) \phi_{xx}(\tau - \sigma)\, d\sigma = 0, \qquad \tau \geqq 0, \tag{9}$$

in which $h_{\text{opt}}(t)$ is the unit-impulse response of the optimum linear system, the system that minimizes the mean-square error. Note that (9) holds only for $\tau \geqq 0$. This restriction is the result of imposing the physical realizability condition (2). The integral equation is known as the Wiener-Hopf equation.

The solution of the Wiener-Hopf equation for the optimum system unit-impulse response is based upon an ingenious technique of Wiener and Hopf called spectrum factorization. In the application of this technique, we first obtain the Fourier transform $\Phi_{xx}(\omega)$ of the input autocorrelation $\phi_{xx}(\tau)$. Thus

$$\Phi_{xx}(\omega) = \frac{1}{2\pi}\int_{-\infty}^{\infty} \phi_{xx}(\tau)e^{-j\omega\tau}\,d\tau. \tag{10}$$

According to the Wiener theorem for autocorrelation, this is the power density spectrum of the random process $x(t)$. The function $\Phi_{xx}(\omega)$ is a positive, real, and even function of the angular frequency ω. We now introduce the complex variable λ with $\lambda = \omega + j\sigma$, in which σ is a real variable independent of ω, and $j = \sqrt{-1}$, and consider the power density spectrum as a function of the complex variable; that is, we consider $\Phi_{xx}(\lambda)$. We shall represent $\Phi_{xx}(\lambda)$ by a rational function. The poles and zeros of $\Phi_{xx}(\lambda)$ are located in both the upper half and lower half of the λ-plane. Spectrum factorization consists in writing $\Phi_{xx}(\lambda)$, after having expressed it as a rational function, as the product of two factors, $\Phi_{xx}^{+}(\lambda)$ and $\Phi_{xx}^{-}(\lambda)$; that is,

$$\Phi_{xx}(\lambda) = \Phi_{xx}^{+}(\lambda)\Phi_{xx}^{-}(\lambda) \tag{11}$$

with $\Phi_{xx}^{+}(\lambda)$ containing all the zeros and poles of $\Phi_{xx}(\lambda)$ that are in the upper half-plane, and $\Phi_{xx}^{-}(\lambda)$ containing all the zeros and poles of $\Phi_{xx}(\lambda)$ that are in the lower half-plane. Because of the properties of $\Phi_{xx}(\omega)$ already mentioned, $\Phi_{xx}^{+}(\lambda)$ and $\Phi_{xx}^{-}(\lambda)$ are complex conjugates. Making use of the fact that the complex Fourier transform of a real function vanishing to the left of the origin has no poles in the lower half-plane, and that of a real function vanishing to the right of the origin has no poles in the upper half-plane, Wiener was able to show that the solution of the Wiener-Hopf equation (9) is

$$H_{\text{opt}}(\lambda) = \frac{1}{2\pi\Phi_{xx}^{+}(\lambda)}\int_{0}^{\infty} e^{-j\lambda\tau}\,d\tau \int_{-\infty+jv_1}^{\infty+jv_1} \frac{\Phi_{xz}(w)}{\Phi_{xx}^{-}(w)}\,e^{jw\tau}\,dw, \qquad v_1 > 0, \tag{12}$$

in which the complex variable of integration w is $w = u + jv$,

$$H_{\text{opt}}(\lambda) = \int_{-\infty}^{\infty} h_{\text{opt}}(t)e^{-j\lambda t}\,dt \tag{13}$$

is the system function of the optimum linear system, and

$$\Phi_{xz}(\lambda) = \frac{1}{2\pi}\int_{-\infty}^{\infty} \phi_{xz}(\tau)e^{-j\lambda\tau}\,d\tau \tag{14}$$

is the transform of the input–desired-output crosscorrelation. The construction of the optimum linear system in terms of standard electrical components after having determined the optimum linear system function, or its transform, the optimum linear system unit-impulse response, can be accomplished through well-known synthesis techniques in classical theory.

An important feature of the theory is that the mean-square error of the optimum linear system (the minimum mean-square error) is expressible in a simple form. If we let

$$\psi(\tau) = \int_{-\infty+jv_1}^{\infty+jv_1} \frac{\Phi_{xz}(w)}{\Phi_{xx}^{-}(w)} e^{jw\tau}\, dw, \tag{15}$$

then the minimum mean-square error is

$$\overline{\epsilon^2(t)}\,\Big|_{\min} = \phi_{zz}(0) - \frac{1}{2\pi}\int_0^{\infty} \psi^2(\tau)\, d\tau. \tag{16}$$

In analyzing this expression in a problem of filtering where the desired message is allowed a lag and the error is considered as a function of the lag, we find that the minimum mean-square error decreases as the lag increases. When the lag tends to infinity, the minimum mean-square error tends to an irremovable error.

Wiener's theory of optimum linear systems is a milestone in the development of communication theory. It is totally different from the classical theory of filtering. The message and the noise—the primary quantities with which a communication system is concerned—were conceived by Wiener as statistical quantities. He provided concepts and techniques for their representation in a problem. The problems of filtering, prediction, and other similar operations were given a unity in formulation by the introduction of the idea that they all have in common an input and a desired output. Then the minimization of a measure of error, which is absent in classical theory, was carried out. The solution of the resulting integral equation was an application of his theory of generalized harmonic analysis and his ingenious technique of spectrum factorization. For optimum linear systems Wiener arrived at a general and explicit expression in a closed form. It involves only the input autocorrelation and the input–desired-output crosscorrelation. The entire theory from its inception to the final expressions for the system function and the minimum mean-square error is invaluable to the communication engineer in the understanding of many communication problems in a new light.

Nonlinear theory. When Wiener was working on linear theory in 1942, he also undertook research in nonlinear theory. In particular, he applied the functionals of Volterra to the study of the response of a nonlinear system to noise.* However, his major work in this area came a few years later.

On December 8, 1949, while he was in Mexico with the Instituto Nacional de Cardiología, he sent a letter to Dr. J. B. Wiesner, who was then Associate Director of the Research Laboratory of Electronics, M.I.T., saying: "I am shipping you some dope on what I think of nonlinear circuits and their testing. The instrument can be made, and Lee can design them. Theoretically, there is nothing new to this, but I think as engineering technique, it is red hot." Accompanying the letter was a memorandum of twenty pages entitled "The Characteristic Properties of Linear and Nonlinear Systems." About half of the memorandum dealt with the subject in general, and the Ergodic Theorem. Unfortunately the "hot dope" that was in the last half of the memorandum was not understandable to me, so that the possibility of my designing the circuits, at that time, was not given serious consideration. The document was studied patiently by several of our colleagues and graduate students, but none of us could produce a satisfactory interpretation. Nevertheless, we knew it was an important piece of work that should receive continued attention. Fortunately our combined effort in research in this area, which Wiener started, during the period 1950–1964 produced more than ten doctoral theses with many activities in the possibilities of practical application. The major contributions of Wiener in this field have been recorded in his book *Nonlinear Problems in Random Theory*† in the form of transcribed lectures.

To present Wiener's nonlinear theory in an elementary form, we start with the generalization of the convolution integral for a linear system. Assuming that a nonlinear system could be characterized by a set of kernel functions $\{h_n(\tau_1, \cdots, \tau_n)\}$, $n = 1, 2, \cdots$, Wiener expressed the relation between the input time function $x(t)$ and the corresponding output $y(t)$ of a nonlinear system by the expression

$$y(t) = h_0 + \sum_{n=1}^{\infty} \int_{-\infty}^{\infty} \cdots \int_{-\infty}^{\infty} h_n(\tau_1, \cdots, \tau_n) x(t - \tau_1) \cdots x(t - \tau_n)\, d\tau_1 \cdots d\tau_n, \tag{17}$$

in which h_0 is a constant. To satisfy the condition that the output of a system does not depend upon the future of the input, we have $h_n(\tau_1, \cdots, \tau_n) = 0$ for any $\tau_i < 0$, $i = 1, \cdots, n$. If the system is

* Norbert Wiener, *Response of a Nonlinear Device to Noise*, Report No. 129, Radiation Laboratory, M.I.T., Cambridge, Mass., 1942.

† Published jointly by The Technology Press of M.I.T., and John Wiley & Sons Inc., New York, 1958.

linear, then $h_0 = 0$ and $n = 1$. The integrals in (17) are multidimensional convolution integrals and they are known as Volterra functionals.* Volterra studied these functionals as early as 1913.

The major part of Wiener's work on nonlinear systems is not in the form (17) but rather in the form of an orthogonal series of Volterra functionals with the input being a white Gaussian process. He was interested in the analysis and synthesis of a nonlinear system, and he showed clearly that the proper probe for a nonlinear system was not the sinusoid but rather the white Gaussian process.

Based on a theorem of Paley and Wiener and the orthogonal properties of Hermite polynomials, Cameron and Martin† in 1947 introduced an orthogonal development of nonlinear functionals which they called Fourier-Hermite functionals. By introducing Volterra functionals and the white Gaussian process as the probe for a nonlinear system into the work of Cameron and Martin, Wiener was able to formulate the problem of analysis and synthesis of a nonlinear physical system. This formulation was given in his memorandum that we mentioned earlier.

We shall show how the orthogonal functionals are developed and applied to the study of nonlinear systems. First, let us note the terms *homogeneous functionals* and *nonhomogeneous functionals*. A homogeneous functional of degree n is the functional

$$\int_{-\infty}^{\infty} \cdots \int_{-\infty}^{\infty} h_n(\tau_1, \cdots, \tau_n) x(t - \tau_1) \cdots x(t - \tau_n)\, d\tau_1 \cdots d\tau_n, \tag{18}$$

and a nonhomogeneous functional of degree n is the sum of functionals

$$\begin{aligned} &\int_{-\infty}^{\infty} \cdots \int_{-\infty}^{\infty} h_n(\tau_1, \cdots, \tau_n) x(t - \tau_1) \cdots x(t - \tau_n)\, d\tau_1 \cdots d\tau_n \\ &\qquad + \int_{-\infty}^{\infty} \cdots \int_{-\infty}^{\infty} h_{n-1}(\tau_1, \cdots, \tau_{n-1}) x(t - \tau_1) \\ &\qquad\qquad \cdots x(t - \tau_{n-1})\, d\tau_1 \cdots d\tau_{n-1} + \cdots + h_0. \end{aligned} \tag{19}$$

To develop a set of orthogonal functionals for a white Gaussian process $x(t)$, we start with the first-degree nonhomogeneous functional

$$\int_{-\infty}^{\infty} k_1(\tau) x(t - \tau)\, d\tau + k_0 \tag{20}$$

* V. Volterra, *Theory of Functionals*, Blackie and Son, Glasgow, 1930.

† R. H. Cameron and W. T. Martin, "The Orthogonal Development of Nonlinear Functionals in Series of Fourier-Hermite Functionals," *Annals of Math.*, *48*, 385–392 (1947).

and determine k_0 satisfying the requirement that (20) be orthogonal to any constant c; that is

$$\lim_{T\to\infty} \frac{1}{2T} \int_{-T}^{T} \left[\int_{-\infty}^{\infty} k_1(\tau) x(t-\tau)\, d\tau + k_0 \right] c\, dt = 0. \tag{21}$$

Here we designate the first-order kernel in (20) by k_1 and the constant by k_0 in order to avoid confusion with h_1 and h_0 in (17). We shall call h_n in (17) the Volterra kernel and k_1 in (21) and other k's in expressions involving higher-order kernels the Wiener kernels. The next step is to take a second-degree nonhomogeneous functional

$$\int_{-\infty}^{\infty} \int_{-\infty}^{\infty} k_2(\tau_1, \tau_2) x(t-\tau_1) x(t-\tau_2)\, d\tau_1\, d\tau_2 + \int_{-\infty}^{\infty} k_{1(2)}(\tau_1) x(t-\tau_1)\, d\tau_1 + k_{0(2)} \tag{22}$$

and determine $k_{1(2)}$ and $k_{0(2)}$ in terms of k_2 satisfying the conditions that (22) be orthogonal to any constant, and to any first-degree homogeneous functional. In (22), $k_{1(2)}$ is a first-order Wiener kernel and $k_{0(2)}$ is a constant. The numeral 2 in parentheses indicates that these k's are associated with the second-degree nonhomogeneous functional. Continuing this procedure we take an nth-degree nonhomogeneous functional

$$\int_{-\infty}^{\infty} \cdots \int_{-\infty}^{\infty} k_n(\tau_1, \cdots, \tau_n) x(t-\tau_1) \cdots x(t-\tau_n)\, d\tau_1 \cdots d\tau_n + \int_{-\infty}^{\infty} \cdots \int_{-\infty}^{\infty} k_{n-1(n)}(\tau_1, \cdots, \tau_{n-1}) x(t-\tau_1) \cdots x(t-\tau_{n-1})\, d\tau_1 \cdots d\tau_{n-1} + \cdots + k_{0(n)} \tag{23}$$

and determine $k_{n-1(n)}, \cdots, k_{0(n)}$ in terms of k_n under the conditions that (23) be orthogonal to any constant and all homogeneous functionals of degrees less than n. (Here $k_{n-1(n)}$ denotes the Wiener kernel of order $n-1$ associated with the nonhomogeneous functional of degree n.)

Let the autocorrelation function $\phi_{xx}(\tau)$ of the white Gaussian process be $\phi_{xx}(\tau) = Au(\tau)$, in which A is a constant and $u(\tau)$ represents the unit-impulse function. The power density spectrum of $x(t)$ is $\Phi_{xx}(\omega) = A/2\pi$. The orthogonalization procedure just described would yield, for (20) satisfying (21), the functional (we shall represent it by $G_1[k_1, x(t)]$)

$$G_1[k_1, x(t)] = \int_{-\infty}^{\infty} k_1(\tau) x(t-\tau)\, d\tau \tag{24}$$

and for (22) satisfying the requirement mentioned immediately following (22) the functional

$$G_2[k_2, x(t)] = \int_{-\infty}^{\infty} \int_{-\infty}^{\infty} k_2(\tau_1, \tau_2) x(t - \tau_1) x(t - \tau_2)\, d\tau_1\, d\tau_2$$

$$- A \int_{-\infty}^{\infty} k_2(\tau_1, \tau_1)\, d\tau_1. \tag{25}$$

Similarly for (23) with $n = 3, 4, 5$, the results of orthogonalization are

$$G_3[k_3, x(t)]$$

$$= \int_{-\infty}^{\infty} \cdots \int_{-\infty}^{\infty} k_3(\tau_1, \cdots, \tau_3) x(t - \tau_1) \cdots x(t - \tau_3)\, d\tau_1 \cdots d\tau_3$$

$$- 3A \int_{-\infty}^{\infty} \int_{-\infty}^{\infty} k_3(\tau_1, \tau_2, \tau_2) x(t - \tau_1)\, d\tau_1\, d\tau_2, \tag{26}$$

$$G_4[k_4, x(t)] = \int_{-\infty}^{\infty} \cdots \int_{-\infty}^{\infty} k_4(\tau_1, \cdots, \tau_4) x(t - \tau_1) \cdots x(t - \tau_4)\, d\tau_1 \cdots d\tau_4$$

$$- 6A \int_{-\infty}^{\infty} \cdots \int_{-\infty}^{\infty} k_4(\tau_1, \tau_2, \tau_3, \tau_3) x(t - \tau_1) x(t - \tau_2)\, d\tau_1\, d\tau_2\, d\tau_3$$

$$+ 3A^2 \int_{-\infty}^{\infty} \int_{-\infty}^{\infty} k_4(\tau_1, \tau_1, \tau_2, \tau_2)\, d\tau_1\, d\tau_2, \tag{27}$$

$$G_5[k_5, x(t)] = \int_{-\infty}^{\infty} \cdots \int_{-\infty}^{\infty} k_5(\tau_1, \cdots, \tau_5) x(t - \tau_1) \cdots x(t - \tau_5)\, d\tau_1 \cdots d\tau_5$$

$$- 10A \int_{-\infty}^{\infty} \cdots \int_{-\infty}^{\infty} k_5(\tau_1, \tau_2, \tau_3, \tau_4, \tau_4) x(t - \tau_1) \cdots x(t - \tau_3)\, d\tau_1 \cdots d\tau_4$$

$$+ 15A^2 \int_{-\infty}^{\infty} \cdots \int_{-\infty}^{\infty} k_5(\tau_1, \tau_2, \tau_2, \tau_3, \tau_3) x(t - \tau_1)\, d\tau_1\, d\tau_2\, d\tau_3. \tag{28}$$

With these results and letting the constant term be

$$G_0[k_0, x(t)] = k_0, \tag{29}$$

Wiener expressed the input-output relation of a nonlinear system whose input is a white Gaussian process by the orthogonal series

$$y(t) = \sum_{n=0}^{\infty} G_n[k_n, x(t)] \tag{30}$$

with the property that

$$\overline{G_m[k_m, x(t)]G_n[k_n, x(t)]} = 0, \qquad m \neq n. \tag{31}$$

The bar over G_mG_n indicates the time average over the interval $(-\infty, \infty)$. The kernels in a G-functional are derived in a systematic manner from the kernel in the leading term. This is clearly seen in the first few terms of the G-functionals given. It can be shown that the coefficients of the terms of a G-functional are the coefficients of the terms of the Hermite polynomials

$$H_n(u) = u^n - C_2{}^nAu^{n-2} + 1 \cdot 3C_4{}^nA^2u^{n-4} - 1 \cdot 3 \cdot 5C_6{}^nA^3u^{n-6} + \cdots \tag{32}$$

in which u is a real variable and $C_m{}^n$ are the binomial coefficients. The problem of convergence has been discussed by Cameron and Martin. Since the Hermite polynomials are a complete orthogonal set, the G-functionals can be shown to be a complete set.

In accordance with (30), the block diagram representation of a nonlinear system with a white Gaussian input is shown in Fig. 1. The output

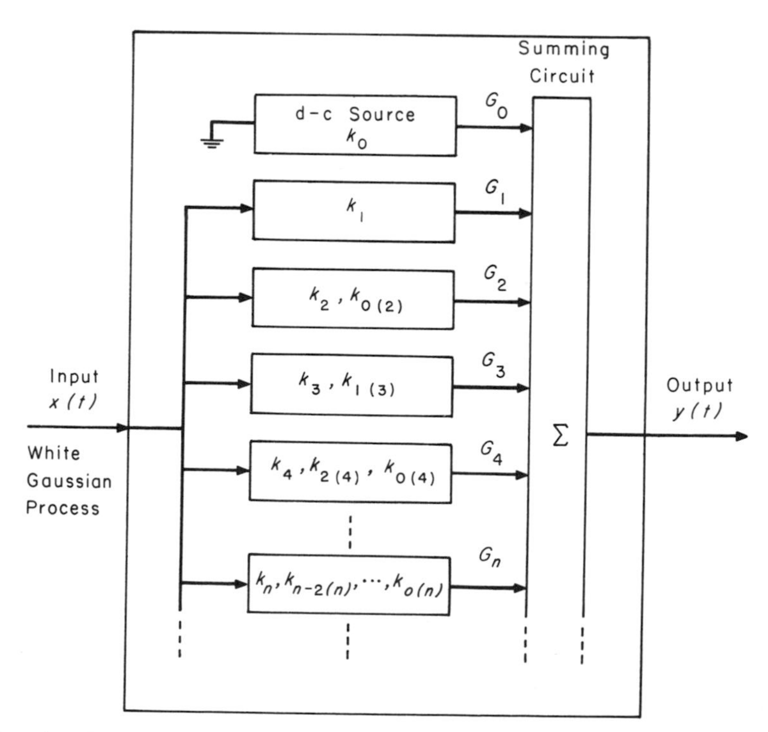

FIG. 1. *Representation of a nonlinear system with a white Gaussian input.*

of the first component is $G_0[k_0, x(t)]$, which is the output of a direct-current source. The second component has the output $G_1[k_1, x(t)]$ as given by (24). This output is that of a linear system whose kernel is k_1. The output $G_2[k_2, x(t)]$ of the third component consists of two parts as expressed by (25), the first being the output of a quadratic system with the kernel k_2 and the second being that of a direct-current source with the value

$$k_{0(2)} = -A \int_{-\infty}^{\infty} k_2(\tau, \tau)\, d\tau.$$

Other components are formed in a similar manner in accordance with (26) to (28) and so on. The output of the system is the sum of the outputs of the components.

One of the applications of the expansion (30) is the analysis and synthesis of a nonlinear system. The nonlinear system is given in the form of a black box with an input terminal to which we may apply a white Gaussian process, and an output terminal from which we may make measurements. We shall call this given system the unknown black box. By synthesis we mean the construction of a system (which we shall call the known black box) with standard parts (such as resistances, capacitances, inductances, d–c sources, multipliers, adders, and amplifiers) in such a manner that for the same white Gaussian input the output of the unknown black box and that of the known black box shall have the minimum mean-square error. When the known black box is constructed in this manner, it is an approximation of the unknown black box, as far as the input-output relation is concerned, for any type of input, not necessarily the Gaussian input.

Since a nonlinear system is characterized by a set of kernels

$$\{k_n(\tau_1, \cdots, \tau_n)\}, \qquad n = 0, 1, 2, \cdots,$$

Wiener's method for its synthesis is to expand the kernels in multidimensional orthonormal functions that can be represented by a series of physical elements. He chose the Laguerre functions because they are well known and they can be represented by a series of phase-shift electrical networks called lattice networks.

The expansion of $k_n(\tau_1, \cdots, \tau_n)$ in a series of Laguerre functions $\ell_m(\tau)$ is

$$(33) \qquad k_n(\tau_1, \cdots, \tau_n) = \sum_{m_1=0}^{\infty} \cdots \sum_{m_n=0}^{\infty} c_{m_1 \cdots m_n} \ell_{m_1}(\tau_1) \cdots \ell_{m_n}(\tau_n)$$

with

$$c_{m_1 \cdots m_n} = \int_{-\infty}^{\infty} \cdots \int_{-\infty}^{\infty} k_n(\tau_1, \cdots, \tau_n) \ell_{m_1}(\tau_1) \cdots \ell_{m_n}(\tau_n) \, d\tau_1 \cdots d\tau_n. \tag{34}$$

In the Wiener input-output relation for a nonlinear system (30), the kernels are symmetrical and

$$\int_{-\infty}^{\infty} \cdots \int_{-\infty}^{\infty} k_n^2(\tau_1, \cdots, \tau_n) \, d\tau_1 \cdots d\tau_n < \infty. \tag{35}$$

Wiener limited his theory to nonlinear systems of "a certain deadbeat character, in which the output is asymptotically independent of the remote-past input."*

The synthesis procedure consists in the determination of the coefficients $c_{m_1 \cdots m_n}$ for $m_1, \cdots, m_n = 0, 1, 2, \cdots$ and the construction of a network in accordance with (30). The coefficients are represented by the gains of a set of amplifiers.

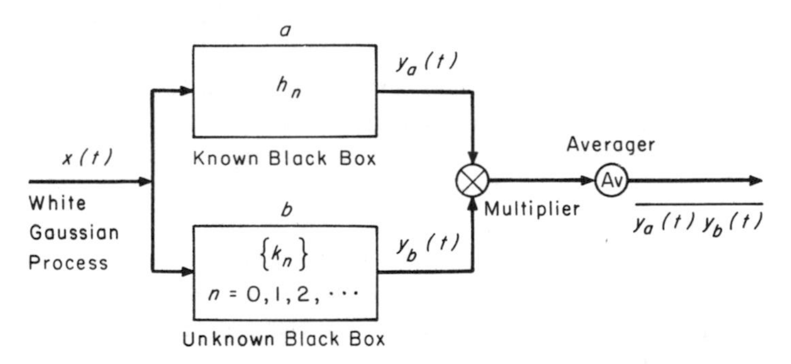

FIG. 2. *Schematic diagram showing a basic property of G-functionals upon which synthesis of a nonlinear system is based.*

An important property of G-functionals upon which the determination of the coefficients of the kernel expansions is based is the following. In Fig. 2, we have a white Gaussian input $x(t)$ to both system a and system b. System b is the unknown black box with the output $y_b(t)$, and system a is a system constructed in such a manner that its output $y_a(t)$ is a G-functional with the leading term containing the kernel h_n. The average value of the product of the outputs of systems a and b can be shown to be

$$\overline{y_a(t) y_b(t)} = n! A^n \int_{-\infty}^{\infty} \cdots \int_{-\infty}^{\infty} h_n(\tau_1, \cdots, \tau_n) k_n(\tau_1, \cdots, \tau_n) \, d\tau_1 \cdots d\tau_n. \tag{36}$$

* Norbert Wiener, *Nonlinear Problems in Random Theory*, p. 89.

To determine a coefficient $c_{m_1 \cdots m_n}$ with a particular set of values for $m_1, \cdots, m_n$, a part of the known black box is constructed in such a manner that its output for the white Gaussian input $x(t)$ is a G-functional with the leading term containing the kernel $\ell_{m_1}(\tau_1) \cdots \ell_{m_n}(\tau_n)$. This kernel is the kernel $h_n(\tau_1, \cdots, \tau_n)$ in (36). Hence (36) becomes

$$\overline{y_a(t) y_b(t)} = n! A^n \int_{-\infty}^{\infty} \cdots \int_{-\infty}^{\infty} k_n(\tau_1, \cdots, \tau_n) \ell_{m_1}(\tau_1) \cdots \ell_{m_n}(\tau_n)\, d\tau_1 \cdots d\tau_n. \tag{37}$$

Since the integral in (37) is in the form of that in (34), which is the expression for the coefficients of the Laguerre expansion of the kernels, we arrive at an expression for the coefficients in terms of the average of the product of the outputs of the partially constructed known black box and of the unknown black box. Thus

$$c_{m_1 \cdots m_n} = \frac{1}{n! A^n} \overline{y_a(t) y_b(t)}. \tag{38}$$

In actual analysis the average value in (38) is obtained by measurement, and in synthesis an amplifier with amplification equal to $c_{m_1 \cdots m_n}$ is connected to the output terminal of the known black box that has been constructed for the determination of the particular coefficient. Theoretically this procedure is repeated for all values of $m_1, \cdots, m_n = 0, 1, 2, \cdots$. However, considerable simplification can be achieved by proper combination of circuit connections. It is not possible to include details of construction and simplification in a short article such as this one.

From studying Wiener's theory, Bose* pointed out that the main features in Wiener's theory of nonlinear systems are given in Fig. 3. These features are clear after the analysis and synthesis procedures just described have been carried out and simplifications made. In the synthesis of a known black box for the representation of an unknown black box, the Wiener theory shows that a nonlinear system consists of three parts. The first part, N_1, is a single-input multiple-output linear system. In the particular case where the Laguerre expansion is chosen for the representation of the kernels, we have a system of lattice networks or their equivalents. Other expansions are possible. This part of the nonlinear system is the memory part. The past of the input at any moment can be reconstructed as a Laguerre expansion whose coefficients are the outputs of the lattice network. Wiener's theory requires that the system

* A. G. Bose, *A Theory of Nonlinear Systems*, Report No. 309, Research Laboratory of Electronics, M.I.T., Cambridge, Mass., May, 1956. The author has made a minor modification of Bose's interpretation.

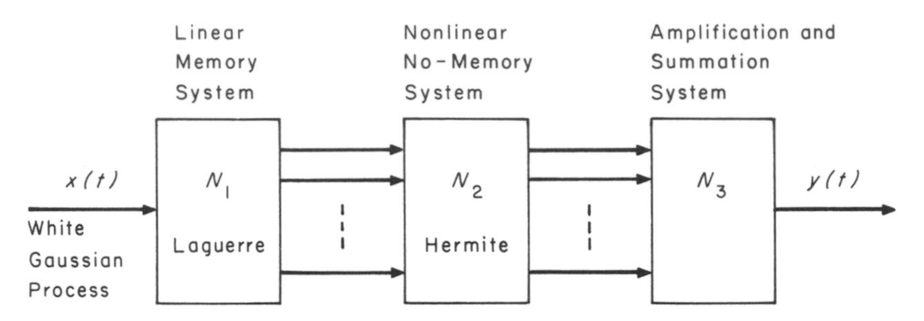

FIG. 3. *Wiener model of a nonlinear system.*

does not depend upon the infinite past. Clearly if the orthogonal linear system is the memory system, it is not appropriate for representing the infinite past of a white Gaussian process. The second part, N_2, is a multiple-input multiple-output nonlinear no-memory system. Its components are multipliers, adders, amplifiers, and direct-current sources. The interconnections within the system are made in accordance with the Hermite expansion. The third part, N_3, of the nonlinear system is a system of amplifiers and adders. All the amplifiers having the values of the coefficients (34) of the Laguerre expansions of the kernels are in this block. We thus see that Wiener's model of a nonlinear system with a white Gaussian input is a multidimensional Hermite expansion of the coefficients of the Laguerre expansion of the past of the input.

In this model the first two parts are the same for all systems representable by the expansion (30) since the first part is a system for the memory of the past of the input, and the second part is a representation of the multidimensional Hermite expansion. Inasmuch as a nonlinear system is synthesized by evaluating the coefficients (34) and properly connecting amplifiers with amplifications equal to these coefficients in system N_3, nonlinear systems represented by this model differ only in the values of amplifications in N_3. Furthermore, just as a linear system is characterized for any type of input once its unit-impulse response (first-order kernel) is determined, irrespective of the type of probe used in its determination, a nonlinear system is characterized for any type of input once its set of kernals $\{k_n\}, n = 0, 1, 2, \cdots$ are determined, irrespective of the type of probe used in their determination. In other words, the fact that a white Gaussian process has been used in the determination of the kernels does not invalidate the input-output relation of the synthesized system when a time function other than the white Gaussian process is the input to the system. Of course, when inputs other than the white Gaussian process are applied to the synthesized system, the outputs of the components in Fig. 1 will no longer be orthogonal.

Although the analysis and synthesis of a nonlinear system in accordance with the Wiener theory is not as yet a common practice since the number of coefficients in (34) becomes extremely large in a practical situation, the theory has been of great value in the field of nonlinear problems. It is conceivable that advances in computer technology will solve the problem of complexity. The model has given researchers a much clearer physical understanding of a nonlinear system. The theory of optimum nonlinear systems has been developed from Wiener's work. There is now much activity in the attempt to extend and apply the nonlinear theory of Wiener to problems in many fields of science and engineering such as mechanical engineering, nuclear engineering, biophysics, geophysics, physics of fluids, control engineering, and communication engineering.

NETS AND THE DIRICHLET PROBLEM

By H. B. Phillips and N. Wiener

1. **Introduction.** In this paper the Dirichlet problem on the plane and in three-space — in general, in n-space, as a matter of fact — is attacked through the analogous problem relating to the potential on a net of equally spaced wires of constant resistance. This method leads at once to the solution of the Dirichlet problem for any continuous boundary conditions on any polygonal or polyhedral boundary of a certain type. Then by means of an important lemma to the effect that if $\{R_n\}$ is any set of polygonal or polyhedral regions of this type, and potential functions U_n are assigned so as to be equicontinuous and uniformly bounded over the boundaries of the R_n's, they shall be equicontinuous over the interiors of the R_n's, we extend our solution of the Dirichlet problem to regions of a much more general character.

Let us start with some definitions. To begin with, a set of functions $\{f_n(P)\}$, each ranging over some corresponding region R_n, is said to be *equicontinuous* if there is a function $\phi(x)$ such that

$$\lim_{x \to 0} \phi(x) = 0,$$

and such that if P and Q lie on R_n,

$$|f_n(P) - f_n(Q)| < \phi(\overline{PQ}). \tag{1}$$

Be it noted that the regions R_n need not be regions having area (or volume) — they may, for example, be arcs of curves. Even in this case, the quantity $\overline{PQ}$ occurring in (1) is to be taken as distance measured *in a straight line.* Curvilinear distances play no part in this paper.

Given a region R in three-space, the *net of order n*[1] over R consists of those portions of the planes $x = a/2^n$, $y = b/2^n$, $z = c/2^n$ which lie within R; a, b, and c assuming all integral values. The *lines* of the net are the intersections of planes of two distinct systems. The *nodes* of the net are the points where planes of all three systems intersect. A function $f(P)$ is defined over the net

[1] Cf. C. Runge, *Göttingische Nachrichten,* Math.-Phys. Klasse 1911. Cf. also R. G. D. Richardson, *Trans. Am. Math. Soc.,* 18 (1917), pp. 489–521.

Reprinted from *J. Math. Phys.,* *2,* **1923**, pp. **105–124**. (Courtesy of the *Journal of Mathematics and Physics,* Massachusetts Institute of Technology and Johnson Reprint Corporation.)

when it is defined at the nodes. A *boundary node* of the net is a node such that some adjacent node does not lie within R—that is, such that the net over R contains less than six nodes adjacent to the one in question. A function $f(P)$ defined over the net is defined within any mesh which contains in its interior no point not on R, as follows:

$$
\begin{aligned}
f(x, y, z) &= f(x_1, y_1, z_1) + 2^n[f(x_2, y_1, z_1) - f(x_1, y_1, z_1)](x - x_1) \\
&+ 2^n [f(x_1, y_2, z_1) - f(x_1, y_1, z_1)](y - y_1) + 2^n[f(x_1, y_1, z_2) - f(x_1, y_1, z_1)](z - z_1) \\
&+ 2^{2n}[f(x_2, y_2, z_1) - f(x_1, y_2, z_1) - f(x_2, y_1, z_1) + f(x_1, y_1, z_1)](x - x_1)(y - y_1) \\
&+ 2^{2n}[f(x_1, y_2, z_2) - f(x_1, y_1, z_2) - f(x_1, y_2, z_1) + f(x_1, y_1, z_1)](y - y_1)(z - z_1) \\
&+ 2^{2n}[f(x_2, y_1, z_2) - f(x_1, y_1, z_2) - f(x_2, y_1, z_1) + f(x_1, y_1, z_1)](x - x_1)(z - z_1) \\
&+ 2^{3n}[f(x_2, y_2, z_2) - f(x_1, y_2, z_2) - f(x_2, y_1, z_2) - f(x_2, y_2, z_1) + f(x_1, y_1, z_2) \\
&\quad + f(x_1, y_2, z_1) + f(x_2, y_1, z_1) - f(x_1, y_1, z_1)](x - x_1)(y - y_1)(z - z_1)
\end{aligned}
\tag{2}
$$

A similar definition holds in the case of two or n dimensions.

2. **The Difference-Equation of Potential on a Net.** Consider a net of order ν in space of three dimensions with nodes at points having coördinates of the form

$$
x = \frac{m}{2^\nu}, \quad y = \frac{n}{2^\nu}, \quad z = \frac{p}{2^\nu}, \tag{3}
$$

m, n, p being integers. We shall now show that there exists a function $f(x, y, z)$ which has given values at the boundary nodes and has at each interior node a value which is the average of its values at the six adjacent nodes with which that one is connected by lines of the net. If the coördinates (3) represent such an interior node, the condition to be satisfied is

$$
\begin{aligned}
6f\left(\frac{m}{2^\nu}, \frac{n}{2^\nu}, \frac{p}{2^\nu}\right) &= f\left(\frac{m-1}{2^\nu}, \frac{n}{2^\nu}, \frac{p}{2^\nu}\right) + f\left(\frac{m+1}{2^\nu}, \frac{n}{2^\nu}, \frac{p}{2^\nu}\right) \\
&+ f\left(\frac{m}{2^\nu}, \frac{n-1}{2^\nu}, \frac{p}{2^\nu}\right) + f\left(\frac{m}{2^\nu}, \frac{n+1}{2^\nu}, \frac{p}{2^\nu}\right) \\
&+ f\left(\frac{m}{2^\nu}, \frac{n}{2^\nu}, \frac{p-1}{2^\nu}\right) + f\left(\frac{m}{2^\nu}, \frac{n}{2^\nu}, \frac{p+1}{2^\nu}\right) .
\end{aligned}
\tag{4}
$$

A function satisfying (4) will be called a potential function on the net. If adjacent nodes are connected by wires of equal resistance and given potentials are maintained at the boundary nodes, Kirchoff's laws show that the electric potential determined at the nodes of the net is such a potential function.

To show the existence of a potential function with given boundary values, consider the function

$$W=\sum\sum\sum\left[g\left(\frac{m+1}{2^\nu},\frac{n}{2^\nu},\frac{p}{2^\nu}\right)-g\left(\frac{m}{2^\nu},\frac{n}{2^\nu},\frac{p}{2^\nu}\right)\right]^2$$

$$+\sum\sum\sum\left[g\left(\frac{m}{2^\nu},\frac{n+1}{2^\nu},\frac{p}{2^\nu}\right)-g\left(\frac{m}{2^\nu},\frac{n}{2^\nu},\frac{p}{2^\nu}\right)\right]^2$$

$$+\sum\sum\sum\left[g\left(\frac{m}{2^\nu},\frac{n}{2^\nu},\frac{p+1}{2^\nu}\right)-g\left(\frac{m}{2^\nu},\frac{n}{2^\nu},\frac{p}{2^\nu}\right)\right]^2,$$

each summation being extended to all values of m, n, p for which the arguments used represent nodes of the net. Let $g(x, y, z)$ have the assigned values at the boundary nodes and values at the interior nodes which make W a minimum. Since W is a definite quadratic form in a finite number of variables (the values of g at the interior nodes) such a minimum exists. Differentiating W with respect to

$$g\left(\frac{m}{2^\nu},\frac{n}{2^\nu},\frac{p}{2^\nu}\right)$$

we find that the function $g(x, y, z)$ satisfies the equation

$$\begin{aligned}&g\left(\frac{m}{2^\nu},\frac{n}{2^\nu},\frac{p}{2^\nu}\right)-g\left(\frac{m+1}{2^\nu},\frac{n}{2^\nu},\frac{p}{2^\nu}\right)\\&+g\left(\frac{m}{2^\nu},\frac{n}{2^\nu},\frac{p}{2^\nu}\right)-g\left(\frac{m-1}{2^\nu},\frac{n}{2^\nu},\frac{p}{2^\nu}\right)\\&+\quad\cdots\cdots\cdots\cdots\cdots\cdots\quad=0\end{aligned}$$

which is equivalent to (4).

There cannot be two distinct solutions of (4) having the same values at the boundary points of the net. For, if there were, their difference would be zero at the boundary points and would satisfy (4) at all interior nodes. Now a function satisfying (4) cannot have a maximum numerical value at an interior node of the net. Hence the difference of the two assumed solutions has its maximum numerical value zero at the boundary, which proves the two solutions indentical.

The equations (4) could be solved in the following way. Take the equation expressing

$$f\left(\frac{m}{2^\nu}, \frac{n}{2^\nu}, \frac{p}{2^\nu}\right)$$

in terms of values at adjacent nodes. For each term on the right of this equation corresponding to an interior node substitute its expression in terms of adjacent values by the proper equation of the type (4). In the result substitute again for all terms corresponding to interior nodes, and continue this process indefinitely. At each stage of this process

$$f\left(\frac{m}{2^\nu}, \frac{n}{2^\nu}, \frac{p}{2^\nu}\right)$$

is expressed as a linear function of values of $f(x, y, z)$ at interior and boundary nodes. From the form of equations (4) it is seen that the coefficients in this linear function are all positive and have a sum equal to 1. If r is the largest number of steps from a node to an adjacent node necessary to pass from any node to the boundary, after r substitutions the coefficients multiplying boundary values in the linear function have a sum not less than

$$\frac{1}{6^r}.$$

The sum of coefficients of interior values is then not greater than

$$1-\frac{1}{6^r}.$$

After rs substitutions, the sum of coefficients of interior values is not greater than

$$\left(1-\frac{1}{6^r}\right)^s$$

which approaches zero as s increases indefinitely. The coefficients of boundary values are less than 1 and never decrease. They therefore approach definite limits. In the limit

$$f\left(\frac{m}{2^\nu}, \frac{n}{2^\nu}, \frac{p}{2^\nu}\right)$$

is thus expressed as a linear function of the boundary values with coefficients all positive and having a sum equal to 1.

3. **The Potential on a Cubical Net.** Let us now consider a net of order ν extending over the interior of a cube which we may without essential restriction of generality consider to be the cube bounded by the coördinate planes and the planes $x=1$, $y=1$, $z=1$. Let us suppose, moreover, that the potential is everywhere zero over the surface of the cube except for the nodes in the xy-plane. On these nodes it is a given continuous function $f(x, y)$. We wish to determine the potential at all the nodes of the cube.

Following the method of Bernoulli let us seek solutions of (4) of the form $X(x)\,Y(y)\,Z(z)$. For such a function (4) becomes

$$\begin{aligned}6X(x)\,Y(y)\,Z(z) = {} & X\left(x-\frac{1}{2^\nu}\right)Y(y)Z(z)+X\left(x+\frac{1}{2^\nu}\right)Y(y)Z(z)\\ & +X(x)\,Y\left(y-\frac{1}{2^\nu}\right)Z(z)+X(x)\,Y\left(y+\frac{1}{2^\nu}\right)Z(z)\\ & +X(x)\,Y(y)Z\left(z-\frac{1}{2^\nu}\right)+X(x)\,Y(y)Z\left(z+\frac{1}{2^\nu}\right). \qquad (5)\end{aligned}$$

By division we get

$$\begin{aligned}3 = {} & \frac{X\left(x-\frac{1}{2^\nu}\right)+X\left(x+\frac{1}{2^\nu}\right)}{2X(x)}+\frac{Y\left(y-\frac{1}{2^\nu}\right)+Y\left(y+\frac{1}{2^\nu}\right)}{2Y(y)}\\ & +\frac{Z\left(z-\frac{1}{2^\nu}\right)+Z\left(z+\frac{1}{2^\nu}\right)}{2Z(z)}. \qquad (6)\end{aligned}$$

Let us note that

$$\left.\begin{aligned} \frac{\sin\lambda\left(x-\frac{1}{2^\nu}\right)+\sin\lambda\left(x+\frac{1}{2^\nu}\right)}{2\sin\lambda x}&=\cos\frac{\lambda}{2^\nu}, \\ \frac{\sinh\rho\left(z-\frac{1}{2^\nu}+\alpha\right)+\sinh\rho\left(z+\frac{1}{2^\nu}+\alpha\right)}{2\sinh\rho(z+\alpha)}&=\cosh\frac{\rho}{2^\nu}. \end{aligned}\right\} \tag{7}$$

Hence

$$\sin(\lambda x)\sin(\mu y)\sinh\rho(z+\alpha) \tag{8}$$

is a solution of (6) provided

$$\cos\frac{\lambda}{2^\nu}+\cos\frac{\mu}{2^\nu}+\cosh\frac{\rho}{2^\nu}=3. \tag{9}$$

If, moreover, $\lambda=r\pi$, $\mu=s\pi$, $\alpha=-1$, where r and s are integers, (8) vanishes when $x=0$, $x=1$, $y=0$, $y=1$, or $z=1$. It thus vanishes over all the surface of the cube except the part in the xy-plane.

Equation (9) may be written in the form

$$1-\cos\frac{\lambda}{2^\nu}+1-\cos\frac{\mu}{2^\nu}=\cosh\frac{\rho}{2^\nu}-1,$$

whence

$$2\sin^2\frac{\lambda}{2^{\nu+1}}+2\sin^2\frac{\mu}{2^{\nu+1}}=2\sinh^2\frac{\rho}{2^{\nu+1}}. \tag{10}$$

From this it is clear that for any given λ and μ

$$\lim_{\nu\to\infty}\rho^2=\lambda^2+\mu^2. \tag{11}$$

Let k, l be positive integers and let m be the positive number such that

$$\lambda=k\pi,\quad \mu=l\pi,\quad \rho=m\pi \tag{12}$$

satisfy (9). Then

$$\sum_{k=1}^{2^\nu-1}\sum_{l=1}^{2^\nu-1}A_{kl}\sin(k\pi x)\sin(l\pi y)\sinh m\pi(z-1) \tag{13}$$

is a solution of (4) which vanishes over all the surface of the cube except the part in the xy-plane where it reduces to

$$-\sum_{k,l=1}^{2^\nu-1}\sum A_{kl}\sinh m\pi \sin(k\pi x)\sin(l\pi y). \tag{14}$$

It is to be noted that m is a function of k and l.

Let us suppose that (14) assumes the same values as the function $f(x, y)$ at the points

$$x=\frac{i}{2^\nu},\quad y=\frac{j}{2^\nu},$$

where

$$0<i<2^\nu,\quad 0<j<2^\nu.$$

From theorems concerning finite Fourier's series, we get

$$-A_{k,l}\sinh m\pi=\frac{1}{2^{2\nu-2}}\sum_{i,j=1}^{2^\nu-1}\sum f\left(\frac{i}{2^\nu},\frac{j}{2^\nu}\right)\sin\frac{ik\pi}{2^\nu}\sin\frac{jl\pi}{2^\nu}. \tag{15}$$

As ν increases indefinitely, this approaches

$$4\int_0^1\int_0^1 f(s,t)\sin(k\pi s)\sin(l\pi t)\,ds\,dt \tag{16}$$

since $f(s, t)$ is continuous. It is never greater in absolute value than 4 max. $|f(x, y)|$.

Using (15), expression (13) can be reduced to the form

$$\frac{1}{2^{2\nu-2}}\sum_{k,l=1}^{2^\nu-1}\sum\left\{\sum_{i,j=1}^{2^\nu-1}\sum f\left(\frac{i}{2^\nu},\frac{j}{2^\nu}\right)\sin\frac{ik\pi}{2^\nu}\sin\frac{jl\pi}{2^\nu}\sin k\pi x\sin l\pi y\right.$$

$$\left.\frac{\sinh m\pi(1-z)}{\sinh(m\pi)}\right\}. \tag{17}$$

If $z\leq 1$,

$$\frac{\sinh m\pi(1-z)}{\sinh m\pi}=e^{-m\pi z}\left(\frac{e^{m\pi}-e^{-m\pi(1-2z)}}{e^{m\pi}-e^{-m\pi}}\right)\leq e^{-m\pi z}. \tag{18}$$

Consider (17) as a double summation with respect to the subscripts k, l. From (18) any term of this double sum is in absolute value not greater than the corresponding term in the series

$$\frac{1}{2^{2\nu-2}} \sum\sum_{k,l=1}^{\infty} \left\{ \sum\sum_{i,j=1}^{2^\nu-1} f\left(\frac{i}{2^\nu}, \frac{j}{2^\nu}\right) \sin\frac{ik\pi}{2^\nu} \sin\frac{jl\pi}{2^\nu} \sin(k\pi x)\sin(l\pi y)e^{-m\pi z} \right\}. \quad (19)$$

Each term in this double series is in absolute value not greater than the corresponding term in

$$\sum\sum_{k,l=1}^{\infty} 4 \max. |f(x, y)| e^{-m\pi z} \quad (20)$$

which in the region $z > \epsilon > 0$ is absolutely and uniformly convergent.

As ν increases indefinitely, from (11) and (12), m approaches the limit

$$m = \sqrt{k^2 + l^2}. \quad (21)$$

The k, l term of (17) approaches the limit

$$4 \int_0^1 \int_0^1 f(r, s) \sin(k\pi r) \sin(l\pi s) dr\, ds \sin(k\pi x) \sin(l\pi y) \frac{\sinh m\pi(1-z)}{\sinh m\pi} \quad (22)$$

uniformly throughout the cube, where m has the value (21). The series

$$\sum\sum_{k,l=1}^{\infty} 4 \int_0^1 \int_0^1 f(r, s) \sin(k\pi r) \sin(l\pi s)\, dr\, ds \sin(k\pi x) \sin(l\pi y) \frac{\sinh m\pi(1-z)}{\sinh m\pi} \quad (23)$$

is term by term in absolute value not greater than (20). It therefore converges absolutely and uniformly in the region

$$1 \geq z > \epsilon > 0. \quad (24)$$

We have then shown that (13) approaches the limit (23) uniformly in the region (24); for the terms in which

$$k,\ l \leq n$$

approach uniformly the corresponding terms in (23) and as n increases indefinitely the remainder in each case approaches zero independently of ν.

Since k, l, m satisfy (21) the function of x, y, z defined by (23) is harmonic within the cube. Hence the net potential function approaches a bounded continuous harmonic function uniformly in the portion of the cube for which $z \geq \epsilon > 0$. If the boundary values are given by a function continuous over the whole surface of the cube, the net potential is the sum of six potentials of the kind just discussed. In that case it therefore has as limit a function harmonic within but not necessarily on the surface of the cube.

Suppose now

$$f(x, y) = 1. \tag{25}$$

Let $f(x, y, z)$ be the harmonic function defined by (23) and let $f_\nu(x, y, z)$ be the net potential on the net of order ν. In a similar way we could start with a function which is 1 on the plane $z=1$ and 0 on the other faces of the cube. The continuous function thus defined is

$$f(x, y, 1-z)$$

and the net potential

$$f_\nu(x, y, 1-z).$$

Similarly we define a pair of functions for each face of the cube. The sum of the six net potentials thus obtained is the net potential with value 1 at all the nodes on the surface of the cube. Therefore

$$\begin{aligned} f_\nu(x, y, z)+f_\nu(x, y, 1-z)+f_\nu(z, y, x)+f_\nu(z, y, 1-x)+f_\nu(x, z, y) \\ +f_\nu(x, z, 1-y)=1. \end{aligned} \tag{26}$$

At any point inside the cube each of the net functions approaches the corresponding continuous function as limit. Hence, at any interior point,

$$\begin{aligned} f(x, y, z)+f(x, y, 1-z)+f(z, y, x)+f(z, y, 1-x)+f(x, z, y) \\ +f(x, z, 1-y)=1. \end{aligned} \tag{27}$$

In the region

$$\left.\begin{array}{r} 0<\epsilon<x<1-\epsilon \\ \epsilon<y<1-\epsilon \\ 0\leq z\leq 1 \end{array}\right\} \quad (28)$$

all the functions of (27) except $f(x, y, z)$ approach 0 uniformly as z approaches 0. Hence $f(x, y, z)$ approaches 1 uniformly. In the same region each term of (26) except $f_\nu(x, y, z)$ approaches uniformly the corresponding term of (27). Hence $f_\nu(x, y, z)$ approaches the continuous harmonic function $f(x, y, z)$ uniformly in the region (28).

If the value 1 is assigned at the nodes of two or more faces of the cube and the value 0 at the others, we show in a similar way that in any closed region belonging to the cube and not containing points where the boundary values are discontinuous the net potential approaches uniformly a continuous potential which has the assigned boundary values.

4. **A Boundary Problem.** Let R be a region whose boundary c is made up entirely of a finite number of finite rectangles lying in the planes.

$$x=a_n, \; y=b_n, \; z=c_n, \quad (29)$$

where a_n, b_n, c_n are terminating binary numbers. With a point P of the surface as center construct a cube Γ_0 with sides parallel to the coördinate planes and edge δ so small that if a cube of this size or smaller is constructed with any point of C as center a part of its surface will lie outside and a part inside R. Let C_0 be the part of C within Γ_0. Let a function F be equal to 1 at the boundary nodes on C_0 and equal to 0 at all other boundary nodes. Let F_ν be the continuous function obtained by constructing the potential determined by F on the net of order ν and continuing it through the cells by trilinear interpolation. We wish to show that a cube can be constructed with P as center within which F_ν differs from 1 by less than an arbitrarily assigned quantity ϵ.

For this purpose construct a series of cubes $\Gamma_1, \Gamma_2, \ldots \Gamma_m$, with centers at P, sides parallel to the coördinate planes, and edges of length $\frac{1}{2}\delta, \frac{1}{4}\delta, \ldots, \frac{1}{2^m}\delta$. Let the part of C within

Γ_m be C_m and let the surface of Γ_m within R be I_m. On the surface I_0 the functions F_ν are zero or positive. We first show that these functions have on I_1 a value everywhere greater than a definite positive number $p>0$. For all finite values of ν these functions are positive on I_1, since they, like the net potentials, have their greatest and least values on the boundary. For sufficiently large values of ν we can construct a cube Γ' within Γ_0 with sides of the form (29) and lying as close to those of Γ_0 as we please. If we construct functions F_ν' which are zero on the surface I' of this cube within R and 1 on the part of C within Γ', we shall have

$$F_\nu > F_\nu'$$

at all the nodes on the surface of Γ' and so at all points of R within Γ'. As ν increases indefinitely, F_ν' approaches uniformly a function F' harmonic within Γ'. Since this function is not zero on I_1, the functions F_ν have a lower limit $p>0$ on I_1. This value p can be taken independently of the position of P on the surface, for the limit of such positions is again a position at which p is not zero. The same value can be used for all cubes of edge $\leq\delta$; for, when the edge is very small, the part of R within the cube is similar to one obtained with a larger cube.

Consider now the functions $F_\nu - p$. These functions are all zero or positive on I_1 and equal to $1-p$ on the part of C within Γ_1. These functions satisfy the same conditions on Γ_1 that F_ν did on Γ_0 except that the value on the surface is $1-p$ instead of 1. Hence on I_2

$$F_\nu - p \geq p(1-p),$$

or

$$F_\nu \geq 1-(1-p)^2.$$

By a continuation of this process we show that on I_m

$$F_\nu \geq 1-(1-p)^m. \tag{30}$$

Since these functions have the value 1 on the remainder of the surface of the region common to Γ_m and R, it follows that (30) is valid for all points of R within Γ_m.

5. **A Special Case of the Dirichlet Problem.** Let R be a region, as in § 4, whose boundary c is finite and made up entirely of rectangles in the planes $x=a_n$, $y=b_n$, $z=c_n$, where a_n, b_n, c_n are terminating binary numbers. Let the function $U(x, y, z)$ be determined on C and let it be continuous in the sense that if (x, y, z) and $(x+\Delta x, y+\Delta y, z+\Delta z)$ are points on C and

$$\text{Lim}\sqrt{(\Delta x)^2+(\Delta y)^2+(\Delta z)^2}=0,$$
$$\text{Lim } U(x+\Delta x, y+\Delta y, z+\Delta z)= U(x, y, z). \tag{31}$$

We shall construct a function $u(x, y, z)$ which has the given values on C, is harmonic over the interior of R, and continuous over $R+C$.

Let us first construct a net of order ν containing every edge and corner of C as line or node. This is clearly possible from the definition of C for all ν's from a certain μ on. Construct the net potential which reduces to $U(x, y, z)$ at the boundary nodes and extend this function through the interior of each cell by trilinear interpolation. Let the resulting function be $U_\nu(x, y, z)$, or briefly, $U_\nu(P)$, where P is the point (x, y, z).

We first show that the functions $U_\nu(x, y, z)$ are equicontinuous over the boundary C. Let

$$f(PQ)=\text{max.}|U(P_1)-U(Q_1)|, \tag{32}$$

where P_1, Q_1 is any pair of points on C such that

$$P_1 Q_1 \leq PQ.$$

Then $f(x)$ is an increasing function of x which by the continuity of $U(P)$ approaches zero with x. If P and Q are two points in the same mesh on C, we can pass from P to Q by two steps P to P_1, and P_1 to Q, each step being parallel to a side of the mesh and not greater than PQ. By the linearity of the interpolation

$$|U_\nu(P)-U_\nu(P_1)|$$

has its maximum value for a given distance PP_1 when P and P_1 are on one of the bounding lines of the mesh. Since U_ν and U have the same values at the ends of this line

$$|U_\nu(P)-U_\nu(P_1)| \leq \text{max.}|U(P)-U(Q)| \leq f(PQ).$$

Similarly

$$|U_\nu(P_1) - U_\nu(Q)| \leq f(PQ).$$

Hence

$$|U_\nu(P) - U_\nu(Q)| \leq 2f(PQ).$$

If P and Q lie in adjacent meshes, we can pass from P to a common edge and from that to Q by two steps of the kind just discussed each of length not greater than PQ. Hence in this case

$$|U_\nu(P) - U_\nu(Q)| \leq 4f(PQ).$$

Finally, if P and Q lie in meshes that are not adjacent, we can pass from P to a corner of its mesh, from that to a corner of the mesh in which Q lies, and from that to Q by three steps, no one of which is greater than PQ. Hence in this case

$$|U_\nu(P) - U_\nu(Q)| \leq 5f(PQ). \tag{33}$$

For any two points P and Q on C this last inequality is valid, which proves the equicontinuity of the functions U_ν on C.

We shall now extend the condition of equicontinuity to the case where P lies on the boundary and Q inside. In § 2 it was shown that the net potential in a linear function of values on the boundary with coefficients whose sum is 1. From the nature of the interpolation, this is evidently true also for the functions U_ν. A cube of side $2PQ$ with center at P contains Q. If $U_\nu(Q)$ is expressed as a linear function of boundary values, the argument of § 4 shows that the sum of coefficients associated with boundary points in a cube of side

$$2^m . 2PQ$$

is not less than

$$1-(1-p)^m.$$

The sum of coefficients associated with points outside this cube is then not greater than $(1-p)^m$. By (33) the maximum variation of U_ν at boundary points within this cube is not greater than

$$5\, f(2^{m+1} . 2PQ).$$

Since $U_\nu(P)$ is one of these boundary points

$$|U_\nu(P) - U_\nu(Q)| \leq 5f(2^{m+2} . PQ) + 2(1-p)^m \max . |U|. \tag{34}$$

Let

$$\frac{1}{2^{2m+1}} \leq PQ \leq \frac{1}{2^{2m}}.$$

As PQ approaches zero, m increases indefinitely, $2^{m+2}\,PQ$ approaches zero and so the right side of (34) is an increasing function $f_1(PQ)$ which approaches zero with PQ.

Finally we wish to show the equicontinuity of the functions U_ν over the whole region $R+C$. If P and Q are nodes, the difference

$$U_\nu(P) - U_\nu(Q)$$

satisfies the net potential equation (4) and its maximum absolute value for a given distance PQ occurs when one of the points P or Q lies on the boundary. This maximum is then not greater than $f_1(PQ)$. If P and Q lie on the same edge of a cell, by the linearity of the interpolation the maximum difference is still not greater than $f_1(PQ)$. If P and Q lie in the same cell the step from P to Q can be broken into three steps parallel to the coördinate axes. In each case the maximum difference occurs when the points are on an edge. Hence in this case the maximum difference is not greater than $3f_1(PQ)$. The case where P and Q lie in adjacent cells can be resolved into two steps of the kind just discussed. Hence the maximum difference is not greater than $6f_1(PQ)$. Finally, if P and Q lie in nonadjacent cells, we can pass from P to a node of its cell, from that to a node of the cell in which Q lies and from that to Q, by three steps not greater than PQ. In any case the difference therefore satisfies the condition

$$|U_\nu(P) - U_\nu(Q)| \leq 13 f_1(PQ).$$

Let

$$13\, f_1(PQ) = \phi(PQ).$$

Then

$$|U_\nu(P) - U_\nu(Q)| \leq \phi(PQ) \tag{35}$$

where P and Q are any points of $R+C$. This function $\phi(x)$ is an increasing function of x and

$$\lim_{x=0} \phi(x) = 0.$$

The functions U_ν are therefore equally continuous on $R+C$. They are also equally bounded, since they lie between the maximum and minimum values of U on the boundary. Hence it is possible[2] to select from an infinite set of the functions U_ν a sequence $\{V_n\}$ converging uniformly over $R+C$ to a continuous function $u(x, y, z)$, which clearly is equal to $U(x, y, z)$ on C.

Let Γ be any cube entirely in R with edges that belong to the net from some ν on. Over the boundary of this cube $\{V_n\}$ converges uniformly to u. The net potential function V_ν' corresponding to values of u on the surface of Γ is the sum of six net potentials of the kind discussed in § 3. In a region within Γ this net potential therefore converges uniformly to a harmonic function. Further when V_n differs from u by less than ϵ on the boundary of the cube, V_n' differs from V_n on the interior by less than ϵ. Hence V_n converges uniformly to a function harmonic in every cube Γ within R. Hence $u(x, y, z)$ is harmonic throughout R.

Any infinite set of the functions U_ν has a sub-set which converges to a function $u(x, y, z)$ harmonic in R and having the given boundary values on C. Since there is only one such harmonic function, U_ν converges uniformly to u. Otherwise there would be an $\epsilon>0$ such that for an infinity of values of ν

$$\max. |U_\nu - u| > \epsilon,$$

implying that the infinite set of the functions U_ν does not have a sub-set converging to u.

6. **The Solution of the Dirichlet Problem.** Let R be an open region in n-space, bounded by a closed set of points C, and within some sphere of n dimensions about the origin. We shall say that R has the property (D) if, whenever a point O belongs to C, there are positive numbers a and b such that if $r<a$ and an n-sphere Γ of radius r is drawn with O as a center, then if Σ is the set of points in Γ but not in R, the projection[3] of Σ on at

[2] *Cf.* M. Fréchet, *Sur quelques points du calcul fonctionnel*, Rend. del Cir. Math. di Palermo, 1906.

[3] The use of the notion of projection in this paper is due to a suggestion of Prof. O. D. Kellogg.

least one of the $n-1$-spaces determined by a set of n perpendicular axes exceeds $b\, r^{n-1}$ in content. Our theorem is that if R has property (D), and $U\,(x, y, \ldots)$ is any continuous function defined for the points of C, then there is a function $u(x, y, \ldots)$, harmonic in R, continuous on $R+C$, and reducing to $U(x, y, \ldots)$ on C. We shall prove this theorem by parallel stages for $n=2$ and $n=3$.

To this end, suppose C overlaid with a net of order ν. The squares [or cubes] of this net which contain points of C will form a set of points K_ν. Then K_ν will, in general, have a part of its boundary within R and a part outside R. Let the part within R be C_ν. For a sufficiently large ν, C_ν will entirely bound a region (not necessarily connected) entirely within R. Let this region be R_ν.

Every node of the net of order ν on C_ν is not further than $d(\nu)=\sqrt{2}/2^\nu$ [or $\sqrt{3}/2^\nu$] from some point of C. Let $U_\nu\,(x, y)$ $[U_\nu, (x, y, z)]$ be defined at a node P of C_ν as the value of $U(x, y)$ $[U(x, y, z)]$ at some point Q of C such that $\overline{PQ}\leq d(\nu)$. On the edges [faces] of C_ν, let U_ν be defined by a process of linear [bilinear] interpolation between the adjacent nodes. Then $U_\nu(x, y)$ $[U_\nu(x, y, z)]$ is continuously defined over C_ν. Now, we have proved that the Dirichlet problem is soluble for regions bounded by contours such as C_ν and for sets of boundary values such as U_ν Hence there is a function $u_\nu(x, y)$ harmonic over the interior of R_ν, continuous over $C_\nu+R_\nu$, and reducing to $U_\nu(x, y)$ $[U_\nu(x, y, z)]$ on C_ν.

In defining the property (D), we have introduced certain positive numbers a and b related to a point O in C. Let $r<a$, and let Γ' be a circle [sphere] about O of radius $2r$. Choose r so small that if P is any point of C within Γ',

$$|U(O)-U(P)|<\epsilon.$$

If ν is so large that $2d(\nu)<r$, then if P lies on C_ν and in a circle Γ [sphere] of radius r about O, we shall have

$$|U(O)-U_\nu(P)|<\epsilon.$$

Let Γ_1 be a circle [sphere] of radius θr about O as center. Let Σ be the part of Γ_1 not in R, and Σ_ν the part of Γ_1 not in

R_ν. Clearly Σ is entirely enclosed in Σ_ν. It then follows from the character of the boundary of Σ_ν and from the fact that R has property (D), that in Σ_ν there can be drawn parallel to one of the coördinate axes [planes] a finite number of linear segments [of regions bounded by a finite number of arcs of circles and straight lines] of length [area] totalling to $b\theta r$ [to $b\theta^2 r^2$], and with projections which do not overlap. Let this set of segments [regions] be σ_ν.

Consider a distribution of matter of density 1 along σ_ν, generating a logarithmic [Newtonian] potential. On the periphery [surface] of Γ, this potential cannot exceed $-b\theta r \log [r(1-\theta)]$ $\left[\text{cannot exceed } \dfrac{b\theta^2 r^2}{r(1-\theta)}\right]$, since $r(1-\theta)$ is less than or equal to the minimum distance of σ_ν from the periphery [surface] of Γ. By a similar argument, the potential in Γ_1 can never be less than $-b\theta r \log 2\theta r$ $\left[\text{or } \dfrac{b\theta r}{2}\right]$. On σ_ν itself, the potential would be increased if the different parts of σ_ν were brought nearer by being brought into the same line [plane], and the maximum potential would be still further increased if they were compressed into a single segment [circle], and the potential taken at the center. This will give us as an upper bound to the potential on σ_ν in the two-dimensional case

$$-2\int_0^{\frac{b\theta r}{2}} \log x \, dx = b\theta r\left(1 - \log \frac{b\theta r}{2}\right),$$

and in the three-dimensional case

$$2\pi \int_0^{\theta r\sqrt{\frac{b}{\pi}}} \frac{1}{\rho}(\rho d\rho) = 2\theta r\sqrt{\pi b}.$$

Let the potential-function just defined be $V(P)$. Let a new harmonic function $W(P)$ be defined as

$$\frac{V(P) + b\theta r \log[r(1-\theta)]}{b\theta r\left\{1 + \log \dfrac{2(1-\theta)}{b\theta}\right\}}$$

in the two-dimensional case, and

$$\frac{V(P)-\dfrac{b\theta^2 r}{1-\theta}}{2\theta r\sqrt{\pi b}-\dfrac{b\theta^2 r}{1-\theta}}$$

in the three-dimensional case. Clearly $W(P)\leq 1$ on σ_ν, and $W(P)\leq 0$ on the periphery of Γ. Over Γ_1 we have in the two-dimensional case

$$W(P)\geq\frac{\log\dfrac{(1-\theta)}{2\theta}}{1+\log\dfrac{2(1-\theta)}{b\theta}},$$

and in the three-dimensional case

$$W(P)\geq\frac{\dfrac{\sqrt{b}}{2}-\dfrac{\theta}{1-\theta}\sqrt{b}}{2\sqrt{\pi}-\dfrac{\theta\sqrt{b}}{1-\theta}}.$$

In both cases, the expression on the right-hand side is positive if $\theta<\frac{1}{3}$. It is, moreover, constant at a given point O if θ is given a definite value, say 1/4. We shall use the symbol α for the expression on the right-hand side of these inequalities.

Now let Γ_2 be a circle [sphere] about O with radius $r/16$, let Γ_3 be a circle [sphere] about O with radius $r/64$, and in general let Γ_n be a circle [sphere] about O with radius $r/4^n$. Let $\xi(P)$ be any function harmonic over R_ν, positive over Γ, and at least 1 over the part of C_ν within Γ. Since $\xi(P)\geq W(P)$, we have $\xi(P)\geq\alpha$ over Γ_1. Form the new harmonic function $\dfrac{\xi(P)-\alpha}{1-\alpha}$

This function is positive over Γ_1, and at least 1 over the part of C_ν within Γ_1. Hence over Γ_2,

$$\frac{\xi(P)-\alpha}{1-\alpha}\geq\alpha,$$

or

$$\xi(P) \geq 1-(1-\alpha)^2.$$

In the same way, over Γ_n,

$$\xi(P) \geq 1-(1-\alpha)^n.$$

Now, over the part of C_ν within Γ, we have

$$|U(O)-U_\nu(P)|<\epsilon.$$

Hence by a change in U_ν of less than ϵ, we can give to U_ν the constant value $U(O)$ over the part of C_ν within Γ. Let the modified function thus obtained from U_ν be V_ν, and let it give rise to a harmonic function v_ν over R_ν. Let the maximum possible absolute value of $U(P)$ be M. Then

$$\frac{v_\nu(P)+M+\epsilon}{U(O)+M+\epsilon} \text{ and } \frac{-v_\nu(P)+M+\epsilon}{-U(O)+M+\epsilon}$$

are functions satisfying the conditions we have laid down for $\xi(P)$. Hence over Γ_n, if we choose n so that $(1-\alpha)^n<\epsilon$

$$\frac{v_\nu(P)+M+\epsilon}{U(O)+M+\epsilon} \geq 1-\epsilon \text{ and } \frac{-v_\nu(P)+M+\epsilon}{-U(O)+M+\epsilon} > 1-\epsilon.$$

Hence over Γ_1

$$|v_\nu(P)-U(O)| \leq 2M\epsilon+\epsilon^2.$$

or

$$|u_\nu(P)-U(O)|<\epsilon^2+(2M+1)\epsilon.$$

By a proper choice of ϵ, we can make $\epsilon^2+(2M+1)\epsilon$ less than any assigned number, say η. Hence about every point O of C we can draw a circle [sphere] Γ_0 such that for a sufficiently large ν if P lies within this circle, and $u_\nu(P)$ exists,

$$|u_\nu(P)-U(O)|<\eta.$$

By the Heine-Borel theorem, every point in C can be made an interior point of one of a finite number of these circles [spheres] Γ_0, for a given value of η. If ν is made sufficiently large, C_ν will lie entirely within this finite set of the Γ_0. It will follow that from some value of μ on, two different u_μ's cannot differ by more than 2η on C_ν, and hence on R_ν. In other words, on any R_ν, the functions $u_\nu(P)$ converge uniformly to a limit, which must thus be a harmonic function $u(P)$ defined over R.

Clearly over Γ_0,

$$|u(P) - U(O)| \leq \eta.$$

Now in Γ_0, η can be made as small as we please. Hence if $\{P_n\}$ is any sequence of points in R approaching a point O on C, we have

$$U(O) = \lim u(P_n).$$

In other words, if u (P) be assigned the same values as $U(P)$ on C, it will constitute a continuous function. Hence we have solved the Dirichlet problem.[4]

[4] This paper was presented at the 1922 Christmas meeting of the American Mathematical Society at Cambridge. At the same meeting a paper involving a somewhat similar method of treatment was presented by Mr. Raynor of Princeton. That paper is to be offered for a doctoral thesis. It was further learned on that occasion that many of the results of this paper were obtained by different methods by Mr. Gleason of Princeton, in a paper likewise to be offered for a doctoral thesis.

DIFFERENTIAL-SPACE

By NORBERT WIENER

§1. **Introduction.** The notion of a function or a curve as an element in a space of an infinitude of dimensions is familiar to all mathematicians, and has been since the early work of Volterra on functions of lines. It is worthy of note, however, that the physicist is equally concerned with systems the dimensionality of which, if not infinite, is so large that it invites the use of limit-processes in which it is treated as infinite. These systems are the systems of statistical mechanics, and the fact that we treat their dimensionality as infinite is witnessed by our continual employment of such asymptotic formulae as that of Stirling or the Gaussian probability-distribution.

The physicist has often occasion to consider quantities which are of the nature of functions with arguments ranging over such a space of infinitely many dimensions. The density of a gas, or one of its velocity-components at a point, considered as depending on the coördinates and velocities of its molecules, are cases in point. He therefore is implicitly, if not explicitly, studying the theory of functionals. Moreover, he generally replaces any of these functionals by some kind of average value, which is essen-

Reprinted from *J. Math. Phys.*, *2*, **1923**, pp. **131–174**. (Courtesy of the *Journal of Mathematics and Physics*, Massachusetts Institute of Technology and Johnson Reprint Corporation.)

tially obtained by an integration in space of infinitely many dimensions.

Now, integration in infinitely many dimensions is a relatively little-studied problem. Apart from certain tentative investigations of Fréchet[1] and E. H. Moore[2], practically all that has been done on it is due to Gâteaux[3], Lévy[4], Daniell[5], and the author of this paper[6]. Of these investigations, perhaps the most complete are those begun by Gâteaux and carried out by Lévy in his *Leçons d'Analyse Fonctionnelle.* In this latter book, the mean value of the functional $U\,|\,[x(t)]\,|$ over the region of function-space

$$\int_0^1 [x(t)]^2 dt \leq 1$$

is considered to be the limit of the mean of the function

$$U\,(x_1, \,.\;.\;.\;.\;, x_n) = U\,|\,[\xi_n(t)]\,|,$$

$$\left(\text{where} \qquad \xi_n(t) = x_k \quad \text{for} \quad \frac{k-1}{n} \leq t < \frac{k}{n}\right)$$

over the sphere

$$x_1{}^2 + x_2{}^2 + \;.\;.\;.\; + x_n{}^2 = n$$

as n increases without limit.

The present paper owes its inception to a conversation which the author had with Professor Lévy in regard to the relation which the two systems of integration in infinitely many dimensions — that of Lévy and that of the author — bear to one another. For this indebtedness the author wishes to give full credit. He also wishes to state that a very considerable part of the substance of the paper has been presented, albeit from a different standpoint and employing different methods, in his previously

[1] *Sur l'intégrale d'une fonctionnelle étendue à un ensemble abstrait,* Bull. Soc. Math. de France, Vol. 43, pp. 249–267.

[2] Cf. American Mathematical Monthly, Vol. 24 (1917), pp. 31, 333.

[3] Two papers, published after his death by Lévy, Bulletin de la Société Mathématique de France, 1919.

[4] P. Lévy, *Leçons d'Analyse Fonctionnelle.* Hereinafter to be referred to as " Lévy."

[5] P. J. Daniell, *A General Form of Integral,* Annals of Mathematics, Series 2 Vol. 19, pp. 279–294. Hereinafter to be referred to as " Daniell." Also paper in Vol. 20.

[6] N. Wiener, *The Average of an Analytic Functional,* Proc. Nat. Acad. of Sci., Vol. 7, pp. 253–26; *The Average of an Analytic Functional and the Brownian Movement, ibid,* 294–298. Also forthcoming paper in Proc. Lond. Math. Soc.

cited publications. It seemed better to repeat a little of what had already been done than to break the continuity of the paper by the constant reference to theorems in other journals and based on treatments so distinct from the present one that it would need a large amount of explanation to show their relevance.

§2. **The Brownian Movement.** When a suspension of small particles in a liquid is viewed under a microscope, the particles seem animated with a peculiar haphazard motion — the Brownian movement. This motion is of such an irregular nature that Perrin[7] says of it: " One realizes from such examples how near the mathematicians are to the truth in refusing, by a logical instinct, to admit the pretended geometrical demonstrations, which are regarded as experimental evidence for the existence of a tangent at each point of a curve." It hence becomes a matter of interest to the mathematician to discover what are the defining conditions and properties of these particle-paths.

The physical explanation of the Brownian movement is that it is due to the haphazard impulses given to the particles by the collisions of the molecules of the fluid in which the particles are suspended. Of course, by the laws of mechanics, to know the motion of a particle, one must know not only the impulses which it receives over a given time, but the initial velocity with which it is imbued. According, however, to the theory of Einstein,[8] this initial velocity over any ordinary interval of time, is of negligible importance in comparison with the impulses received during the time in question. Accordingly, the displacement of a particle during a given time may be regarded as independent of its entire previous history.

Let us then consider the time-equations of the path of a particle subject to the Brownian movement as of the form $x=x(t)$, $y=y(t)$, $z=z(t)$, t being the time and x, y, and z the coördinates of the particle. Let us limit our attention to the function $x(t)$. Since there is no appreciable carrying over of velocity from one instant to another, the difference between $x(t_1)$ and $x(t)$ $[t_1>t]$ may be regarded as the sum of the displacements incurred by the particle

[7] p. 64, *Brownian Movement and Molecular Reality*, tr. by F. Soddy.
[8] Perrin, pp. 51–54.

over a set of intervals constituting the interval from t to t_1. In particular, if the constituent intervals are of equal size, then the probability-distribution of the displacements accrued in the different intervals will be the same. Since positive and negative displacements of the same size will, from physical considerations, be equally likely, it will be seen that for intervals of time large with comparison to the intervals between molecular collisions, by dividing them into many equal parts, which are still large with respect to the intervals between molecular collisions, and breaking up the total incurred displacement into the sum of displacements incurred in these intervals, we get very nearly a Gaussian distribution of our total displacement.[9] That is, the probability that $x(t_1)-x(t)$ lie between a and b is very nearly of the form

$$\frac{1}{\sqrt{\pi\phi(t_1-t)}}\int_a^b e^{-\frac{x^2}{\phi(t_1-t)}}\,dx. \tag{1}$$

Since the error incurred over the interval t to t_1 is the sum of the independent errors incurred over the periods from t to t_2 and from t_2 to t_1, we have

$$\begin{aligned}
&\frac{1}{\sqrt{\pi\phi(t_1-t)}}\,e^{-\frac{x^2}{\phi(t_1-t)}}\\
&=\frac{1}{\pi\sqrt{\phi(t_1-t_2)\phi(t_2-t)}}\int_{-\infty}^{\infty} e^{-\frac{y^2}{\phi(t_2-t)}-\frac{(x-y)^2}{\phi(t_1-t_2)}}\,dy\\
&=\frac{1}{\pi\sqrt{\phi(t_1-t_2)\phi(t_2-t)}}\int_{-\infty}^{\infty} exp\left\{-\left[y\sqrt{\frac{\phi(t_1-t_2)+\phi(t_2-t)}{\phi(t_1-t_2)\phi(t_2-t)}}\right.\right.\\
&\qquad\left.\left.-\frac{x}{\sqrt{\phi(t_1-t_2)}}\sqrt{\frac{\phi(t_1-t_2)\phi(t_2-t)}{\phi(t_1-t_2)+\phi(t_2-t)}}\right]^2-\frac{x^2}{\phi(t_1-t_2)+\phi(t_2-t)}\right\}dy\\
&=\frac{e^{-\frac{x^2}{\phi(t_1-t_2)+\phi(t_2-t)}}}{\pi\sqrt{\phi(t_1-t_2)\phi(t_2-t)}}\int_{-\infty}^{\infty} e^{-y^2\left(\frac{\phi(t_1-t_2)+\phi(t_2-t)}{\phi(t_1-t_2)\phi(t_2-t)}\right)}\,dy\\
&=\frac{e^{-\frac{x^2}{\phi(t_1-t_2)+\phi(t_2-t)}}}{\sqrt{\pi\{\phi(t_1-t_2)+\phi(t_2-t)\}}}.
\end{aligned} \tag{2}$$

[9] Cf. Poincaré, *Le calcul des probabilités*, Ch. XI.

It is readily verified that this cannot be true unless

$$\phi(t_1-t)=\phi(t_1-t_2)+\phi(t_2-t), \tag{3}$$

whence

$$\phi(u)=Au, \tag{4}$$

and the probability that $x(t_1)-x(t)$ lie between a and b is of the form

$$\frac{1}{\sqrt{\pi A(t_1-t)}}\int_a^b e^{-\frac{x^2}{A(t_1-t)}}\,dx. \tag{5}$$

According to Einstein's theory,[10] $A=\frac{RT}{N}\frac{a}{3\pi a\xi}$ where R is the constant of a perfect gas, T is the absolute temperature, N is Avogadro's constant, a the radius of the spherical particles subject to the Brownian movement, and ξ the viscosity of the fluid containing the suspension.

§3. **Differential-Space.** In the Brownian movement, it is not the position of a particle at one time that is independent of the position of a particle at another; it is the displacement of a particle over one interval that is independent of the displacement of the particle over another interval. That is, instead of $f\left(\frac{1}{n}\right), \ldots, f\left(\frac{k}{n}\right), \ldots, f(1)$ representing "dimensions" of $f(t)$, the n quantities

$$x_1=f\left(\frac{1}{n}\right)-f(0)$$

$$x_2=f\left(\frac{2}{n}\right)-f\left(\frac{1}{n}\right)$$

$$\cdots\cdots$$

$$x_k=f\left(\frac{k}{n}\right)-f\left(\frac{k-1}{n}\right)$$

$$\cdots\cdots$$

$$x_n=f(1)-f\left(\frac{n-1}{n}\right)$$

[10] Perrin, p. 53.

are of equal weight, vary independently, and in some degree represent dimensions. It is natural, then, to consider, as Lévy does, the properties of the sphere

$$x_1^2 + \ . \ . \ . + x_n^2 = r_n^2.$$

In particular, let us consider the measure, in terms of the whole sphere, of the region in which $f(a) - f(0)$, assuming it to be representable in terms of the x_s's, lies between the values α and β. Now, we have

$$f(a) - f(0) = \sum_{1}^{na} x_k. \tag{6}$$

Hence, by a change of coördinates, our question becomes: given that

$$\sum_{1}^{n} \xi_k^2 = r_n^2,$$

what is the chance that

$$\alpha \leq \sqrt{na}\, \xi_1 \leq \beta?$$

Letting $\xi_1 = \rho \sin \theta$, $\sqrt{\xi_2^2 + \ . \ . \ . + \xi_n^2} = \rho \cos \theta$, this chance becomes

$$\frac{\displaystyle\int_{\sin^{-1} \frac{\alpha}{r_n \sqrt{na}}}^{\sin^{-1} \frac{\beta}{r_n \sqrt{na}}} \cos^n \theta \, d\, \theta}{\displaystyle\int_{-\frac{\pi}{2}}^{\frac{\pi}{2}} \cos^n \theta \, d\, \theta},$$

as has been indicated by Lévy[11]. This may also be written

$$\frac{\displaystyle\int_{\sqrt{n} \sin^{-1} \frac{\alpha}{r_n \sqrt{na}}}^{\sqrt{n} \sin^{-1} \frac{\beta}{r_n \sqrt{na}}} \cos^n \frac{x}{\sqrt{n}} \, dx}{\displaystyle\int_{-\frac{\sqrt{n}\pi}{2}}^{\frac{\sqrt{n\pi}}{2}} \cos^n \frac{x}{\sqrt{n}} \, dx}.$$

[11] Lévy, p. 266.

Now[12], for n large, $\cos^n \frac{x}{\sqrt{n}}$ converges uniformly to $e^{-\frac{x^2}{2}}$ for $|x| < A$. It may be deduced without difficulty from this fact that the integral in the denominator tends to $\int_{-\infty}^{\infty} e^{-\frac{x^2}{2}} dx$, which is $\sqrt{2\pi}$. As to the numerator, if the limits of integration approach limiting values, it will also converge. If in particular r_n is a constant, then since

$$\lim_{n \to \infty} \sqrt{n} \sin^{-1} \frac{\alpha}{r\sqrt{na}} = \frac{\alpha}{r\sqrt{a}},$$

while a like identity holds for the upper limit, we have for the probability (in the limit) that $f(a)$ lie between α and β.

$$\frac{1}{\sqrt{2\pi}} \int_{\frac{\alpha}{r\sqrt{a}}}^{\frac{\beta}{r\sqrt{a}}} e^{-\frac{x^2}{2}} dx = \frac{1}{r\sqrt{2\pi a}} \int_{\alpha}^{\beta} e^{-\frac{u^2}{2ar^2}} du. \tag{7}$$

Let it be noted that this expression is of exactly the form (5).

We shall call the space of which the constituent points are the functions $f(t)$, and in which the measure of a region is determined as the limit of a measure in u-space in the way in which formula (7) is obtained from (6), by the name *differential-space*. The appropriateness of this name comes from the fact that it is not the values of $f(t)$, but the small differences, that are uniformly distributed, and act as dimensions.

§4. The Non-Differentiability Coefficient of a Function.

It thus appears that if we consider the distribution of $f(a) - f(0) = \sum_1^{na} x_k$ in the sphere $\sum_1^{n} x_k^2 = r^2$, we get in the limit a distribution of the values of $f(a) - f(0)$ essentially like the one indicated in (5). Now, we have

$$\sum_1^n x_k^2 = \sum_1^n \left\{ f\left(\frac{k}{n}\right) - f\left(\frac{k-1}{n}\right) \right\}^2.$$

[12] Lévy, p. 264.

Our result so far suggests, then, that the probability that $\sum_1^n \left\{ f\left(\frac{k}{n}\right) - f\left(\frac{k-1}{n}\right) \right\}^2$ exceed r^2 becomes increasingly negligible as n increases, or at least the probability that it exceed r^2 by a stated amount. On the hypothesis that $f(t_1)-f(t)$ have the distribution indicated in (5), and that the variation of f over one interval be independent of the variation of f over any preceding interval, let us discuss the distribution of

$$\sum_1^n \left\{ f\left(\frac{k}{n}\right) - f\left(\frac{k-1}{n}\right) \right\}^2.$$

Clearly, the chance that $\sum_1^n \left\{ f\left(\frac{k}{n}\right) - f\left(\frac{k-1}{n}\right) \right\}^2$ lie between α^2 and β^2 is the average value of a function of $x_1, \ldots, x_n$ which is 1 when $x_1^2 + \ldots + x_n^2$ lies between α^2 and β^2, and 0 otherwise, given that the weight of the region

$$\xi_k \leq x_k \leq \eta_k \qquad (k=1, 2, \ldots, n)$$

is

$$\prod_1^n \sqrt{\frac{n}{\pi A}} \int_{\xi_k}^{\eta_k} e^{-\frac{nx^2}{A}} dx$$
$$= \left(\sqrt{\frac{n}{\pi A}}\right)^n \int_{\xi_1}^{\eta_1} dx_1 \ldots \int_{\xi_n}^{\eta_n} dx_n \, e^{-\frac{n}{A}(x_1^2 + \ldots + x_n^2)}, \qquad (9)$$

as follows from (5). Now, let us put

$$\left.\begin{aligned} x_1 &= \rho \sin\theta_1 \\ x_2 &= \rho \cos\theta_1 \sin\theta_2 \\ x_3 &= \rho \cos\theta_1 \cos\theta_2 \sin\theta_3 \\ &\cdots\cdots\cdots \\ x_{n-1} &= \rho \cos\theta_1 \cos\theta_2 \ldots \cos\theta_{n-2} \sin\theta_{n-1} \\ x_n &= \rho \cos\theta_1 \cos\theta_2 \ldots \cos\theta_{n-1} \end{aligned}\right\}. \qquad (10)$$

We shall then have

$$\left.\begin{aligned} \rho^2 &= x_1^2 + \ldots + x_n^2 \\ dx_1 \ldots dx_n &= \rho^{n-1}\cos^{n-2}\theta_1 \cos^{n-3}\theta_2 \ldots \cos\theta_{n-2}\, d\rho\, d\theta_1 \ldots d\theta_{n-1} \end{aligned}\right\}, \tag{11}$$

the last of which formulae may be readily demonstrated by an evaluation of the Jacobian

$$\begin{vmatrix} \frac{\partial x_1}{\partial \rho} & \frac{\partial x_1}{\partial \theta_1} & \cdots & \frac{\partial x_1}{\partial \theta_{n-1}} \\ \frac{\partial x_2}{\partial \rho} & \frac{\partial x_2}{\partial \theta_1} & \cdots & \frac{\partial x_2}{\partial \theta_{n-1}} \\ \cdot & \cdot & \cdot & \cdot \\ \frac{\partial x_n}{\partial \rho} & \frac{\partial x_n}{\partial \theta_1} & \cdots & \frac{\partial x_n}{\partial \theta_{n-1}} \end{vmatrix} . \tag{12}$$

Employing a formula of Lévy to the effect that

$$\int_0^{\frac{\pi}{2}} \cos^n \theta d\theta \int_0^{\frac{\pi}{2}} \cos^{n-1} \theta d\theta = \frac{\pi}{2n} \tag{13}$$

[13]

we get for the chance that $\sum_1^n \left\{ f\left(\frac{k}{n}\right) - f\left(\frac{k-1}{n}\right) \right\}^2$ lie between α^2 and β^2

$$\left(\sqrt{\frac{n}{\pi A}}\right)^n \int_\alpha^\beta \rho^{n-1} e^{-\frac{n\rho^2}{A}} d\rho \prod_{k=1}^{n-2} \int_{-\frac{\pi}{2}}^{\frac{\pi}{2}} \cos^k \theta d\theta$$

$$= \left(\sqrt{\frac{n}{\pi A}}\right)^n 2^{n-2} \frac{\pi}{2(n-2)} \frac{\pi}{2(n-4)} \cdots \frac{\pi}{4} \int_\alpha^\beta \rho^{n-1} e^{-\frac{n\rho^2}{A}} d\rho$$

if n is even

$$= \left(\sqrt{\frac{n}{\pi A}}\right)^n 2^{n-2} \frac{\pi}{2(n-2)} \frac{\pi}{2(n-4)} \cdots \frac{\pi}{6} \int_\alpha^\beta \rho^{n-1} e^{-\frac{n\rho^2}{A}} d\rho$$

if n is odd. (14)

[13] Levy, p. 263.

The coefficients of the integral may be reduced in either case to the form

$$\frac{n^{\frac{n}{2}}}{\pi A^{\frac{n}{2}}\,\Gamma\left(\frac{n}{2}\right)},$$

so that the whole expression may be written, by a change of variable,

$$\frac{n^{\frac{n}{2}}}{\pi\Gamma\left(\frac{n}{2}\right)}\int_{\frac{\alpha}{\sqrt{A}}}^{\frac{\beta}{\sqrt{A}}} u^{n-1}\, e^{-nu^2}\, du. \tag{15}$$

The integrand vanishes for u zero and u infinite. Between those points it attains its maximum for $u^2 = \frac{n-1}{2n}$, which for large values of n is in the neighborhood of 1/2. It hence becomes interesting to note how much is contributed to our integral by values of u^2 near 1/2 and how much by values remote from 1/2.

Let us then evaluate

$$\frac{n^{\frac{n}{2}}}{\pi\Gamma\left(\frac{n}{2}\right)}\int_{u^2=\frac{1}{2}+\epsilon}^{\infty} u^{n-1}\, e^{-nu^2} du,$$

remembering that the integrand is a decreasing function. Let us discuss the ratio

$$\frac{(u+1)^{n-1}\, e^{-n(u+1)^2}}{u^{n-1}\, e^{-nu^2}} = \left(1+\frac{1}{u}\right)^{n-1} e^{-n(2u+1)}$$

$$\leq \frac{3^{n-1}}{e^{2n}} < 1/2. \tag{16}$$

We see at once that it follows from this that

$$\frac{n^{\frac{n}{2}}}{\pi\Gamma\left(\frac{n}{2}\right)}\int_{u^2=\frac{1}{2}+\epsilon}^{\infty} u^{n-1}\, e^{-nu^2}\, du < \frac{n^{\frac{n}{2}}}{\pi\Gamma\left(\frac{n}{2}\right)}\sum_{k=0}^{\infty}\frac{1}{2^k}(1/2+\epsilon)^{\frac{n-1}{2}}\, e^{-n(\frac{1}{2}+\epsilon)}$$

$$= \frac{2n^{\frac{n}{2}}}{\pi\Gamma\left(\frac{n}{2}\right)}(1/2+\epsilon)^{\frac{n-1}{2}} e^{-n(\frac{1}{2}+\epsilon)} \,. \tag{17}$$

For n large, by Stirling's theorem, this latter expression is asymptotically represented by

$$\frac{2n^{\frac{n}{2}}}{\pi\left(\frac{n}{2}-1\right)^{\frac{n}{2}-1} e^{1-\frac{n}{2}}\sqrt{2\pi\left(\frac{n}{2}-1\right)}}\left(\frac{1}{2}+\epsilon\right)^{\frac{n-1}{2}} e^{-n\left(\frac{1}{2}+\epsilon\right)}$$

$$= \frac{2}{\pi e}\sqrt{\frac{n-2}{\pi}}\,\frac{(1+2\epsilon)^{\frac{n-1}{2}}\, e^{-n\epsilon}}{\left(1-\frac{n}{2}\right)^{\frac{n}{2}}} . \tag{18}$$

As n grows larger, this in turn may be represented asymptotically by

$$\frac{2}{\pi(1+2\epsilon)^{\frac{1}{2}}}\sqrt{\frac{n-2}{\pi}}\left(\frac{1+2\epsilon}{e^{2\epsilon}}\right)^{\frac{n}{2}} . \tag{19}$$

Now, $e^{2\epsilon} > 1+2\epsilon$, as may be seen directly from the Maclaurin series for e^x. Hence (19) may be written

$$K\sqrt{n-2}\, C^n,$$

where K is a constant and C is a positive constant less than 1. This, of course, tends to vanish for n large.

We may prove in a similar way that

$$\frac{n^{\frac{n}{2}}}{\pi\Gamma\left(\frac{n}{2}\right)}\int_0^{u^2=\frac{1}{2}-\epsilon} u^{n-1}\, e^{-nu^2}\, du$$

tends to vanish for n large. Clearly, for n sufficiently large, $\frac{n-1}{2n}$ will be greater than $1/2-\epsilon$. Under these circumstances,

$$\frac{n^{\frac{n}{2}}}{\pi\Gamma\left(\frac{n}{2}\right)}\int_0^{u^2=\frac{1}{2}-\epsilon} u^{n-1}\, e^{-nu^2}\, du \leq \frac{n^{\frac{n}{2}}}{2\pi\Gamma\left(\frac{n}{2}\right)}\left(\frac{1}{2}-\epsilon\right)^{\frac{n-1}{2}} e^{-n(\frac{1}{2}-\epsilon)}. \quad (20)$$

This latter quantity is asymptotically represented by

$$\frac{n^{\frac{n}{2}}}{2\pi\left(\frac{n}{2}-1\right)^{\frac{n}{2}-1} e^{1-\frac{n}{2}}\sqrt{2\pi\left(\frac{n}{2}-1\right)}}\left(\frac{1}{2}-\epsilon\right)^{\frac{n-1}{2}} e^{-n(\frac{1}{2}-\epsilon)}$$

$$= \frac{1}{2\pi\epsilon}\sqrt{\frac{n-2}{\pi}}\,\frac{(1-2\epsilon)^{\frac{n}{2}-1}\, e^{n\epsilon}}{\left(1-\frac{2}{n}\right)^{\frac{n}{2}}}. \quad (21)$$

This is in turn asymptotically represented by

$$\frac{1}{2\pi(1-2\epsilon)^{\frac{1}{2}}}\sqrt{\frac{n-2}{\pi}}\left[(1-2\epsilon)\, e^{2\epsilon}\right]^{\frac{n}{2}}. \quad (22)$$

Now, since by Taylor's theorem with remainder, $e^{-2\epsilon}=1-2\epsilon+\frac{\theta^2}{2}$, θ being a positive number less than 2ϵ, we get $(1-2\epsilon)e^{2\epsilon}<1$. Hence expression (22) is again of the form $K\sqrt{n-2}\;C^n$, C being a positive constant less than 1, and approaches zero as n increases.

It will be seen, then, that for n sufficiently large, the chance that $\sum_1^n\left\{f\left(\frac{k}{n}\right)-f\left(\frac{k-1}{n}\right)\right\}^2$ diverge from $A/2$ by more than ϵ is less than an expression of the form $K\sqrt{n-2}\;C^n$, in which K

and C are positive constants independent of n, and $C<1$. It follows that for n sufficiently large, the chance that *any* $\sum_{1}^{N}\left\{f\left(\frac{k}{N}\right)-f\left(\frac{k-1}{N}\right)\right\}^{2}$ differ from $A/2$ by more than ϵ, for $N \geq n$, is less than

$$K\sum_{n}^{\infty}\sqrt{N-2}\,C^{N},$$

which is the remainder of a convergent series, and hence vanishes as n becomes infinite. In other words, the chance that

$$\left|\lim_{n\to\infty}\sum_{1}^{n}f\left\{\left(\frac{k}{n}\right)-f\left(\frac{k-1}{n}\right)\right\}^{2}-A/2\right|>\epsilon$$

is less than any assignable positive number, if ϵ is any positive quantity.

Let $f(t)$ be a continuous function of limited total variation T between 0 and 1. Then clearly

$$\sum_{1}^{n}\left\{f\left(\frac{k}{n}\right)-f\left(\frac{k-1}{n}\right)\right\}^{2}\leq \max\left|f\left(\frac{k}{n}\right)-f\left(\frac{k-1}{n}\right)\right|\sum_{1}^{n}\left|f\left(\frac{k}{n}\right)\right.$$

$$\left.-f\left(\frac{k-1}{n}\right)\right|\leq T\max\left|f\left(\frac{k}{n}\right)-f\left(\frac{k-1}{n}\right)\right|. \tag{23}$$

Hence

$$\lim_{n\to\infty}\sum_{1}^{n}\left\{f\left(\frac{k}{n}\right)-f\left(\frac{k-1}{n}\right)^{2}\right\}=0. \tag{24}$$

This will in particular be the case when $f(t)$ possesses a derivative bounded over the closed interval $(0, 1)$. Hence it is infinitely improbable, under our distribution of functions $f(t)$, that $f(t)$ be a continuous function of limited total variation, and in particular that it have a bounded derivative. We may regard $\lim_{n\to\infty}\sum_{1}^{n}\left\{f\left(\frac{k}{n}\right)-f\left(\frac{k-1}{n}\right)^{2}\right\}$ as in some sort a nondifferentiability coefficient of f.

§5. **The Maximum Gain in Coin-Tossing.** We have now investigated the differentiability of the functions $f(t)$; it behoves us to inquire as to their continuity. To this end we shall discuss the maximum value of $f(t)-f(t_0)$ incurred for $t_0 \leq t \leq t_1$, and shall determine its distribution. As a simple model of this rather complex situation, however, we shall consider a problem in coin-tossing.

A gambler stakes one dollar on each throw of a coin, which is tossed n times, losing the dollar if the throw is heads, and gaining it if the throw is tails. At the beginning he has lost nothing and won nothing; after m throws he will be k_m dollars ahead, k_m being positive or negative. The question here asked is, what is the distribution of the maximum value of k_m for $m \leq n$?

Let n throws be made in all the 2^n possible manners. Let $A_n(p)$ be the number of throw-sequences in which $\max(k_m)=p$. Clearly $A_n(-\mu)=0$, if μ is positive, since $k_0=0$. Furthermore, any throw-sequence in which the maximum gain is zero consists of a throw-sequence in which the maximum gain is either 1 or 0 preceded by a throw of heads. That is,

$$A_n(0)=A_{n-1}(0)+A_{n-1}(1). \tag{25}$$

A throw-sequence in which the maximum gain is $p>0$ consists either of a throw-sequence in which the maximum gain is $p-1$ preceded by a throw of tails, or a throw-sequence in which the maximum gain is $p+1$, preceded by a throw of heads. That is,

$$A_n(p)=A_{n-1}(p-1)+A_{n-1}(p+1). \tag{26}$$

Let us tabulate the first few values of $A_n(p)$. We get

$A_n(p)$	$n=1$	2	3	4	5	6
$p=0$	1	2	3	6	10	20
1	1	1	3	4	10	15
2	0	1	1	4	5	15
3	0	0	1	1	5	6
4	0	0	0	1	1	6
5	0	0	0	0	1	1
6	0	0	0	0	0	1

It will be seen that the various numbers in the table are the binomial coefficients, beginning in each column with the middle coefficient for n even and the next coefficient beyond the middle for n odd, and with every coefficient in the expansion $(1+1)^n$ repeated twice, with the exception of the middle coefficient for n even. That is, as far as the table is carried, we have

$$\left.\begin{aligned} A_{2n}(2p) &= A_{2n}(2p-1) = \frac{(2n)!}{(n+p)!\,(n-p)!} \\ A_{2n+1}(2p) &= A_{2n+1}(2p+1) = \frac{(2n+1)!}{(n+p+1)!\,(n-p)!} \end{aligned}\right\}. \tag{27}$$

We may then verify (25) and (26) by direct substitution, thus proving that the values given for $A_m(q)$ in (27) are valid for all values of m and q.

Let us now consider a series of n successive runs of m throws each, for points of $1/\sqrt{m}$ dollars, m being even. The chance that the amount gained in a single run lie between α and β dollars is then

$$1/2^m \sum_{\alpha'\sqrt{m}}^{\beta'\sqrt{m}} B_m(k), \tag{28}$$

where $\alpha'\sqrt{m}$ is the even integer next greater than or equal to $\alpha\sqrt{m}$, $\beta'\sqrt{m}$ is the even integer next less than $\beta\sqrt{m}$ and $B_m(k)/2^m$ is the chance that of m throws of a coin just k more should be tails than heads — that is,

$$B_m(k) = \frac{m!}{\frac{m-k}{2}!\;\frac{m+k}{2}!}. \tag{29}$$

If we begin by assuming α and β positive and less than γ, expression (28) will lie between the values

$$\frac{(\beta'-\alpha')\sqrt{m}}{2^{m+1}} \; \frac{m!}{\left(\frac{m-\beta'\sqrt{m}}{2}\right)!\;\left(\frac{m+\beta'\sqrt{m}}{2}\right)!} \quad \text{and}$$

$$\frac{\beta('-\alpha')\sqrt{m}}{2^{m+1}} \; \frac{m!}{\left(\frac{m-\alpha'\sqrt{m}}{2}\right)!\;\left(\frac{m+\alpha'\sqrt{m}}{2}\right)!}.$$

Now, by Stirling's theorem, we have uniformly for $\alpha<\gamma$, $\beta<\gamma$

$$\lim_{m\to\infty} \frac{(\beta'-\alpha')\sqrt{m}}{2^{m+1}} \quad \frac{m!}{\left(\frac{m-\beta'\sqrt{m}}{2}\right)!\left(\frac{m+\beta'\sqrt{m}}{2}\right)!}$$

$$= \lim_{m\to\infty} \frac{(\beta-\alpha)\sqrt{m} \quad m^m e^{-m}\sqrt{2\pi m}}{2^{m+1}\left(\frac{m-\beta\sqrt{m}}{2}\right)^{\frac{m-\beta\sqrt{m}}{2}}\left(\frac{m+\beta\sqrt{m}}{2}\right)^{\frac{m+\beta\sqrt{m}}{2}} e^{-m}\pi\sqrt{m^2-\beta^2 m}}$$

$$= \lim_{m\to\infty} \frac{\beta-\alpha}{\sqrt{2\pi}}\left(\frac{m^2}{m^2-\beta^2 m}\right)^{\frac{m}{2}}\left(\frac{m-\beta\sqrt{m}}{m+\beta\sqrt{m}}\right)^{\frac{\beta\sqrt{m}}{2}}$$

$$= \lim_{m\to\infty} \frac{\beta-\alpha}{\sqrt{2\pi}}(1-\beta^2/m)^{\frac{m}{2}}(1-\beta/\sqrt{m})^{\frac{\beta\sqrt{m}}{2}}(1+\beta/\sqrt{m})^{-\frac{\beta\sqrt{m}}{2}}$$

$$= \frac{\beta-\alpha}{\sqrt{2\pi}} e^{-\frac{\beta^2}{2}}. \tag{30}$$

We have, then, for m sufficiently large, by combining (30) with an analogous theorem,

$$\frac{\beta-\alpha}{\sqrt{2\pi}} e^{-\frac{\beta^2}{2}} < 1/2^m \sum_{\alpha'\sqrt{m}}^{\beta'\sqrt{m}} B_m(k) < \frac{\beta-\alpha}{\sqrt{2\pi}} e^{-\frac{\alpha^2}{2}}. \tag{31}$$

Moreover, since (30) holds uniformly, the value of m necessary to make (31) valid depends only on $\beta-\alpha$.

Now, let us divide the interval (α, β) into the p equal parts (α, α_1), (α_1, α_2), ... , (α_{n-1}, β), and let us choose m sufficiently large for us to have over each interval

$$\frac{\alpha_{k+1}-\alpha_k}{\sqrt{2\pi}} e^{-\frac{\alpha^2_{k+1}}{2}} < 1/2^m \sum_{\alpha'_k\sqrt{m}}^{\alpha''_{k+1}\sqrt{m}} B_m(k) < \frac{\alpha_{k+1}-\alpha_k}{\sqrt{2\pi}} e^{-\frac{\alpha^2_k}{2}},$$

where $\alpha_k'\sqrt{m}$ is $\alpha_k\sqrt{m}$ if this is an even integer, or otherwise the next larger integer, and $\alpha_k''\sqrt{m}=\alpha_k\sqrt{m}-2$. Then we have

$$\frac{1}{\sqrt{2\pi}}\sum_1^n(\alpha_{k+1}-\alpha_k)\, e^{-\frac{\alpha^2_{k+1}}{2}} < 1/2^m \sum_{\alpha'\sqrt{m}}^{\beta'\sqrt{m}} B_m(k)$$

$$< \frac{1}{\sqrt{2\pi}}\sum_1^n(\alpha_{k+1}-\alpha_k)\, e^{-\frac{\alpha^2_k}{2}},$$

where the value of m needed depends only on $\alpha_{k+1}-\alpha_k$. Since $e^{-\frac{x^2}{2}}$ is a decreasing function, we hence have

$$\left|\frac{1}{\sqrt{2\pi}}\int_\alpha^\beta e^{-\frac{x^2}{2}}\,dx-1/2^m\sum_{\alpha'\sqrt{m}}^{\beta'\sqrt{m}}B_m(k)\right|<\frac{\gamma}{\sqrt{2\pi}}\max\left(e^{-\frac{\alpha^2_k}{2}}-e^{-\frac{\alpha^2_{k+1}}{2}}\right)$$

$$<\frac{\gamma}{\sqrt{2\pi e}}(\alpha_{k+1}-\alpha_k), \qquad (32)$$

where a minimum value of m may be determined in terms of $\alpha_{h+1}-\alpha_h$ alone. In other words, $1/2^m \sum_{\alpha'\sqrt{m}}^{\beta'\sqrt{m}} B_m(k)$ converges *uniformly* in α and β to

$$\frac{1}{\sqrt{2\pi}}\int_\alpha^\beta e^{-\frac{x^2}{2}}dx,$$

provided only α and β are positive and smaller than γ. It is obvious that the requirement of positiveness is superfluous, if only $|\alpha|<\gamma$, $|\beta|<\gamma$.

This last restriction can be removed. In the first place

$$B_m(2k)/B_m(k)=\frac{(m-k)!\,(m+k)!}{(m-2k)!\,(m+2k)!}$$

$$=\frac{(m-2k+1)(m-2k+2)\ \ .\ \ .\ \ .\ \ (m-k)}{(m+k+1)\ (m+k+2)\ \ .\ \ .\ \ .\ \ (m+2k)}$$

$$<\left(\frac{m-k}{m+2k}\right)^k. \qquad (33)$$

If now $k>l\sqrt{m}$, we have

$$B_m(2k)/B_m(k)<\left(\frac{\sqrt{m}-l}{\sqrt{m}+2l}\right)^{\sqrt{m}}<\left(1-\frac{l}{\sqrt{m}}\right)^{\sqrt{m}}<l^{-l}, \qquad (34)$$

for $k\leq m$. If $k>m$, we have clearly

$$B_m(2k)=B_m(k)=0.$$

Hence in general, for all $k > l\sqrt{m}$,

$$B_m(2^n k) \leq e^{-nl} B_m(k). \tag{34}$$

It follows that by making γ sufficiently large, we can make

$$\frac{1}{2^m}\sum_{\gamma'\sqrt{m}}^{\infty} B_m(k) \leq \frac{1}{2^m}\sum_{0}^{\gamma'\sqrt{m}} B_m(k)\ (1+e^{-l}+e^{-2l}+\ .\ .\ .)$$

$$\leq \frac{1}{2^m}\sum_{0}^{\gamma'\sqrt{m}} B_m(k)\ \frac{1}{1-e^{-l}}, \tag{35}$$

where $\gamma'\sqrt{m}$ is the integer next larger than $\gamma\sqrt{m}$ and l is independent of m. The result is that

$$\lim_{\gamma\to\infty} \frac{1}{2^m}\sum_{\gamma'\sqrt{m}}^{\infty} B_m(k) = 0 \tag{36}$$

uniformly in m. Hence if we consider

$$\left| \frac{1}{\sqrt{2\pi}} \int_{\alpha} e^{-\frac{x^2}{2}} dx - 1/2^{m+1} \sum_{\alpha'\sqrt{m}}^{\beta'\sqrt{m}} B_m(k) \right|, \tag{37}$$

we may first choose γ so large that the difference in this expression made by replacing α or β by $\pm\gamma$, in case they lie outside $(-\gamma, \gamma)$, is less than $\epsilon/2$, and then choose m so large that within the region $(-\gamma, \gamma)$, the expression is less than $\epsilon/2$. That is, by a choice of m alone, independently of α and β, we can make (37) less than ϵ.

Now let us write

$$\left.\begin{aligned} \frac{1}{2^{m+1}} \sum_{-\infty}^{\beta'\sqrt{m}} B_m(k) &= \phi_m(\beta) \\ \frac{1}{\sqrt{2\pi}} \int_{-\infty}^{\beta} e^{-\frac{x^2}{2}}\, dx &= \phi(\beta) \end{aligned}\right\}. \tag{38}$$

Clearly, the probability that the maximum amount won for any value of k at the end of the first km throws out of mn for $\sqrt{m}$

dollars each lie between 0 and u will be the Stieltjes integral expression

$$\int_{0-0}^{u} d\phi_m(x_1) \int_{-\infty}^{-x_1} d\phi_m(x_2) \int_{-\infty}^{-x_1-x_2} d\phi_m(x_3) \ldots \int_{-\infty}^{-x_1-x_2-\ldots-x_{n-1}} d\phi_m(x_n)$$

$$+\int_{0+0}^{u} d\phi_m(x_1) \int_{0+0}^{u-x_1} d\phi_m(x_2) \int_{-\infty}^{-x_1-x_2} d\phi_m(x_3) \ldots \int_{-\infty}^{-x_1-x_2-\ldots-x_{n-1}} d\phi_m(x_n)$$

$$+\int_{0+0}^{u} d\phi_m(x_1) \int_{-\infty}^{0} d\phi_m(x_2) \int_{x_1+0}^{u} d\phi_m(x_3) \ldots \int_{-\infty}^{-x_1-x_2-\ldots-x_{n-1}} d\phi_m(x_n)$$

$$+\int_{0+0}^{u} d\phi_m(x_1) \int_{0+0}^{u-x_1} d\phi_m(x_2) \int_{x_1+x_2+0}^{u} d\phi_m(x_3) \ldots \int_{-\infty}^{-x_1-x_2-\ldots-x_{n-1}} d\phi_m(x_n)$$

$$+ \quad \ldots\ldots\ldots\ldots\ldots\ldots\ldots\ldots\ldots$$

$$+\int_{0+0}^{u} d\phi_m(x_1) \int_{0+0}^{u-x_1} d\phi_m(x_2) \quad \ldots \quad \int_{0+0}^{u-x_1-x_2-\ldots-x_{n-1}} d\phi_m(x_n), \tag{39}$$

consisting of $1+1+2!+3!+\ldots+n!$ terms. If in this expression we substitute ϕ for ϕ_m, it results from what we have just said of the uniformity of the convergence of ϕ_m to ϕ that the expression we obtain will be uniformly the limit of (39). This expression will represent the chance that after n independent games, in each of which the chance of winning a sum between α dollars and β dollars is $\frac{1}{\sqrt{2\pi}}\int_{\alpha}^{\beta} e^{-\frac{x^2}{2}}\,dx$, the maximum gain incurred lies between 0 and u. We shall denominate this quantity $G(u, n)$, and expression (39) $G_m(u, n)$.

It is obvious from the definition that $G(u, n)$ is more than

$$1/2^{mn}\sum_{0}^{u'\sqrt{m}} A_{mn}(k) - 1/2^{mn-1}\sum_{k=0}^{k=u'\sqrt{m}} \frac{mn!}{\left(\frac{mn-k}{2}\right)!\left(\frac{mn+k}{2}\right)!}, \tag{40}$$

where $u'\sqrt{m}$ is the integer next larger than $u\sqrt{m}$. By going through an argument precisely analogous to that by which (32) was obtained, we see that

$$\lim_{m \to \infty} 1/2^{mn} \sum_{0}^{u'\sqrt{mn}} A_{mn}(k) = \sqrt{\frac{2}{\pi}} \int_0^{u\sqrt{n}} e^{-\frac{x^2}{2}} dx$$

$$= \sqrt{\frac{2}{n\pi}} \int_0^{u} e^{-\frac{x^2}{2n}} dx. \tag{41}$$

Hence

$$G\ (u,n) > \sqrt{\frac{2}{n\pi}} \int_0^{u} e^{-\frac{x^2}{2n}} dx. \tag{42}$$

§6. **Measure in Differential-Space.** To revert to differential-space, let us consider a functional $F|f|$, f being defined for arguments between zero and one, and let us see if we can frame a definition of its average. To begin with, let us suppose that F only depends on the values of f for the arguments $t_1 < t_2, \ldots < t_n$. F will then be unchanged if we alter f to any other function, assuming the same values for these arguments — in particular, to a step-function $f_\nu(t)$ with steps all of length $1/\nu$, ν being sufficiently large.

Let us now call the difference between the height of the kth step and that of the $(k-1)$st by the name x_k. We shall then have

$$\begin{aligned} F|f_\nu| &= F(f_\nu(t_1), f_\nu(t_2), \ldots, f_\nu(t_n)) \\ &= F(x_1+x_2+ \ldots +x_{T_1}, x_1+x_2+ \ldots +x_{T_2}, \ldots, \\ &\qquad x_1+x_2+ \ldots +x_{T_n}), \end{aligned} \tag{43}$$

where T_k is νt_k if this is an integer, and otherwise the next smaller integer. Now, the region of the space $(x_1, x_2, \ldots, x_n)$ which corresponds to differential-space is the interior of the sphere

$$x_1^2+x_2^2+ \ldots +x_n^2=r^2.$$

Over the interior of this sphere we may take the average of expression (43), let it approach a limit as ν increases indefinitely, and call this limit the average of F in differential-space.

By the change in coördinates

$$\left.\begin{aligned}
&x_1+x_2 \quad + \ . \ . \ . \ +x_{T_1} \ =\sqrt{T_1}\,\xi_1 \\
&x_{T_1+1} \quad + \ . \ . \ . \ +x_{T_2} \ =\sqrt{T_2-T_1}\,\xi_2 \\
&\cdot \quad \cdot \quad \cdot \quad \cdot \quad \cdot \quad \cdot \quad \cdot \quad \cdot \quad \cdot \quad \cdot \quad \cdot \\
&x_{T_{n-1}+1} + \ . \ . \ . \ . +x_{T_n} =\sqrt{T_n-T_{n-1}}\,\xi_n, \\
&\xi_{n+1}, \ . \ . \ ., \xi_n \text{ represent coördinates orthogonal to } \xi_1, \ . \ . \ ., \xi_n
\end{aligned}\right\} \quad (44)$$

we change our sphere into

$$\xi_1{}^2+ \qquad +\xi_\nu{}^2-r^2,$$

and have also

$$F|f_\nu|=F(\sqrt{T_1}\xi_1, \sqrt{T_2-T_1}\ \xi_2, \ . \ . \ . \ , \ \sqrt{T_n-T_{n-1}}\ \xi_n).$$

By a transformation like (10), this will become

$$F|f_\nu|=F(\sqrt{T_1}\,\rho \sin \theta_1, \sqrt{T_2-T_1}\,\rho \cos \theta_1 \sin \theta_2, \ . \ . \ . \ \sqrt{T_n-T_{n-1}}\,\rho \cos \theta_1, \ . \ . \ . \ \cos \theta_{n-1}\sin \theta_n).$$

The average value of this over our sphere is

$$\frac{\left\{\begin{aligned}&\int_0^r d\rho \int_{-\frac{\pi}{2}}^{\frac{\pi}{2}} d\theta_1 \ . \ . \ . \int_{-\frac{\pi}{2}}^{\frac{\pi}{2}} d\theta_{\nu-1}\, \rho^{\nu-1} \cos^{\nu-2} \theta_1 \cos^{\nu-3} \theta_2 \ . \ . \ . \ \cos \theta_{\nu-2} \\ &F(\sqrt{T_1}\,\rho \sin \theta_1, \ . \ . \ ., \sqrt{T_n-T_{n-1}}\,\rho \cos \theta_1 . \ . \ . \ \sin \theta_n)\end{aligned}\right\}}{\int_0^r d\rho \int_{-\frac{\pi}{2}}^{\frac{\pi}{2}} d\theta_1, \ . \ . \ . \int_{-\frac{\pi}{2}}^{\frac{\pi}{2}} d\theta_{\nu-1}\, \rho^{\nu-1} \cos^{\nu-2} \theta_1 \ . \ . \ . \ \cos \theta_{\nu-2}}$$

$$=\frac{\left\{\begin{aligned}&\int_0^r d\rho \int_{-\frac{\pi}{2}}^{\frac{\pi}{2}} d\theta_1 \ . \ . \ . \int_{-\frac{\pi}{2}}^{\frac{\pi}{2}} d\theta_n \rho^{\nu-1} \cos^{\nu-2} \theta_1 \ . \ . \ . \ \cos^{\nu-n-1} \theta_n \\ &F(\sqrt{T_1}\,\rho \ \sin \theta_1, \ . \ . \ . \ , \sqrt{T-T_{n-1}}\,\rho \cos \theta_1 \ . \ . \ . \ \sin \theta_n)\end{aligned}\right\}}{\int_0^r d\rho \int_{-\frac{\pi}{2}}^{\frac{\pi}{2}} d\theta_1 \ . \ . \ . \int_{-\frac{\pi}{2}}^{\frac{\pi}{2}} d\theta_n\, \rho^{\nu-1} \cos^{\nu-2} \theta_1 \ . \ . \ . \ \cos^{\nu-n-1} \theta_n}$$

$$
= \frac{\left\{ \begin{array}{l} \int_0^r d\rho \int_{-\frac{\pi\sqrt{\nu}}{2}}^{\frac{\pi\sqrt{\nu}}{2}} du_1 \ldots \int_{-\frac{\pi\sqrt{\nu}}{2}}^{\frac{\pi\sqrt{\nu}}{2}} du_n \rho^{\nu-1} \cos^{\nu-2}\frac{u_1}{\sqrt{\nu}} \ldots \cos^{\nu-n-1}\frac{u_n}{\sqrt{\nu}} \\ F\left(\sqrt{T}\rho \sin \frac{u_1}{\sqrt{\nu}}, \ldots, \sqrt{T_n - T_{n-1}}\,\rho \cos \frac{u_1}{\sqrt{\nu}}, \ldots \sin \frac{u_n}{\sqrt{\nu}}\right) \end{array} \right\}}{\int_0^r d\rho \int_{-\frac{\pi\sqrt{\nu}}{2}}^{\frac{\pi\sqrt{\nu}}{2}} du_1 \ldots \int_{-\frac{\pi\sqrt{\nu}}{2}}^{\frac{\pi\sqrt{\nu}}{2}} du_n \rho^{\nu-1} \cos^{\nu-2}\frac{u_1}{\sqrt{\nu}} \ldots \cos^{\nu-n-1}\frac{u_n}{\sqrt{\nu}}} \tag{45}
$$

If we now consider the integrands in the numerator and the denominator for $|u_1| < U, \ldots, |u_n| < U$, and let ν increase indefinitely, we see[14] that

$$
\begin{aligned}
\lim_{\nu \to \infty} \cos^{\nu-2}\frac{u_1}{\sqrt{\nu}} \ldots \cos^{\nu-n-1}\frac{u_n}{\sqrt{\nu}} F(\sqrt{T_1}\,\rho \sin \frac{u_1}{\sqrt{\nu}}, \ldots, \\
\sqrt{T_n - T_{n-1}}\,\rho \cos \frac{u_1}{\sqrt{\nu}} \ldots \sin \frac{u_n}{\sqrt{\nu}} \\
= e^{-\frac{u_1^2}{2} - \cdots - \frac{u_n^2}{2}} F(\rho u_1 \sqrt{t_1}, \ldots, \rho u_n \sqrt{t_n - t_{n-1}}),
\end{aligned} \tag{46}
$$

and

$$
\lim_{\nu \to \infty} \cos^{\nu-2}\frac{u_1}{\sqrt{\nu}} \ldots \cos^{\nu-n-1}\frac{u_n}{\sqrt{\nu}} = e^{-\frac{u_1^2}{2} - \cdots - \frac{u_n^2}{2}} \tag{47}
$$

uniformly, provided only F is continuous. If in addition F vanishes for $|u_1| > U, \ldots, |U_n| > U$, we have for the limit of the expression in (45)

$$
\lim_{\nu \to \infty} \frac{\left\{ \begin{array}{l} \int_0^r \rho^{\nu-1} d\rho \int_{-\infty}^{\infty} du_1 \ldots \int_{-\infty}^{\infty} du_n\, e^{-\frac{u_1^2}{2} - \cdots - \frac{u_n^2}{2}} \\ F(\rho u_1 \sqrt{t_1}, \ldots, \rho u_n \sqrt{t_n - t_{n-1}}) \end{array} \right\}}{\int_0^r \rho^{\nu-1} d\rho \int_{-\infty}^{\infty} du_1 \ldots \int_{-\infty}^{\infty} du_n\, e^{-\frac{u_1^2}{2} - \cdots - \frac{u_n^2}{2}}}
$$

14 Cf. note 12.

This will be the case even if F has a set of discontinuities of zero measure, provided F vanishes for $|u_1| > U, \ldots, |U_n| > U$, or even, as may be shown without difficulty by a limit argument, provided merely that $|F|$ is bounded. It follows at once that, under these conditions, the expression in (45) becomes

$$\frac{\int_{-\infty}^{\infty} du_1 \ldots \int_{-\infty}^{\infty} du_n \, e^{-\frac{u_1^2}{2} - \cdots - \frac{u_n^2}{2}} F(ru_1\sqrt{t_1}, \ldots, ru_n\sqrt{t_n - t_{n-1}})}{\int_{-\infty}^{\infty} du_1 \ldots \int_{-\infty}^{\infty} du_n \, e^{-\frac{u_1^2}{2} - \cdots - \frac{u_n^2}{2}}}$$

$$= (2\pi)^{\frac{n}{2}} \int_{-\infty}^{\infty} du_1 \ldots \int_{-\infty}^{\infty} du_n \, e^{-\frac{u_1^2}{2} - \cdots - \frac{u_n^2}{2}} F(ru_1\sqrt{t_1}, \ldots, ru_n\sqrt{t_n - t_{n-1}})$$

$$= (2\pi)^{-\frac{n}{2}} r^{-n} [t_1(t_2 - t_1) \ldots (t_n - t_{n-1})]^{-\frac{1}{2}} \int_{\infty}^{\infty} dy_1 \ldots \int_{-\infty}^{\infty} dy_n \exp \sum_{t}^{n} \frac{y_k^2}{2r^2(t_k - t_{k-1})} F(y_1, \ldots y_n), \tag{48}$$

to being zero. In particular, if $F(y_1, \ldots, y_n)$ is 1 when

$$\left.\begin{array}{l} y_{11} < y_1 < y_{12} \\ y_{21} < y_2 < y_{22} \\ \cdot \quad \cdot \quad \cdot \quad \cdot \quad \cdot \\ y_{n1} < y_n < y_{n2} \end{array}\right\}, \tag{49}$$

we get for the average value of F

$$(2\pi)^{-\frac{n}{2}} r^{-n} [t_1 (t_2 - t_1) \ldots (t_n - t_{n-1})]^{-\frac{1}{2}} \int_{y_{11}}^{y_{12}} dy_1 \ldots \int_{y_{n1}}^{y_{n2}} dy_n \exp \sum_{1}^{n} \frac{y_k^2}{2r^2(t_k - t_{k-1})} . \tag{50}$$

This we shall term the *measure* or *probability* of region (49).

§7. **Measure and Equal Continuity.** Let us now consider all those functions $f(t)$ $(0 \leq t \leq 1)$ for which $f(0)=0$ and which for some pair of rational arguments, $0 \leq t_1 < t_2 \leq 1$, satisfy the inequality

$$|f(t_2)-f(t_1)| > ar(t_2-t_1)^{\frac{1}{2}-\epsilon} \tag{51}$$

If this inequality is satisfied, we can certainly find an n such that

$$\frac{k-1}{n} \leq t_1 \leq \frac{k}{n} \leq t_2 \leq \frac{k+1}{n}, \quad t_2-t_1 \geq 1/n, \tag{52}$$

while the variation of f over the interval $\left(\frac{k-1}{n}, \frac{k+1}{n}\right)$ will exceed $ar(t_2-t_1)^{\frac{1}{2}-\epsilon}$. That is, the class of functions satisfying the inequality (51) is a sub-class of the class of functions for which the variation over some interval $\left(\frac{k-1}{n}, \frac{k}{n}\right)$ exceeds $\frac{ar}{2} n^{\epsilon-\frac{1}{2}}$. Now, every such function must have either the difference between some *positive* value in the interval and its initial value exceed $\frac{ar}{4} n^{\epsilon-\frac{1}{2}}$, or the difference between its initial value and some *negative* value in the interval must exceed $\frac{ar}{4} n^{\epsilon-\frac{1}{2}}$.

To discuss what we may interpret as the measure of the set of functions (51), we may hence investigate the measure of the set of functions which over an interval of length $1/n$ depart from their initial value (for some rational argument, be it understood) by more than $\frac{ar}{4} n^{\epsilon-\frac{1}{2}}$. Now, if we subdivide the interval of length $1/n$ into m parts, each of length $1/mn$, the probability that in a given one of these sub-intervals the total difference between the initial and the final value of the function lie between α and β is, by (50)

$$\frac{\sqrt{mn}}{r\sqrt{2\pi}} \int_{\alpha}^{\beta} e^{-\frac{mnx^2}{2r^2}} dx. \tag{53}$$

Hence the probability that the maximum difference between the value of f at the end of one of these sub-intervals and its initial value be greater than $\frac{ar}{4} n^{\epsilon-\frac{1}{2}}$ is by (42) less than

$$\frac{1}{r}\sqrt{\frac{2n}{\pi}}\int_{\frac{ar}{4}n^{\epsilon-\frac{1}{2}}}^{\infty} e^{-\frac{nx^2}{2r^2}}\,dx$$

$$< \frac{1}{r}\sqrt{\frac{2n}{\pi}}\int_{\frac{ar}{4}n^{\epsilon-\frac{1}{2}}}^{\infty} e^{-\sqrt{\frac{nx^2}{2r^2}}}\,dx \left(\text{for } \frac{n^{2\epsilon}a^2r^2}{16} \geq 1\right)$$

$$= \frac{1}{r}\sqrt{\frac{2n}{\pi}}\sqrt{\frac{n}{2r^2}}\, e^{-\frac{n^{\epsilon}ar}{4}}$$

$$= \frac{n}{r^2\sqrt{\pi}}\, e^{-\frac{n^{\epsilon}ar}{4}}. \tag{54}$$

This will be true no matter how large m is. Hence those functions f for which the maximum difference between $f\left(\frac{k}{n}+\frac{l}{mn}\right)$ and $f\left(\frac{k}{n}\right)$ is greater than $\frac{ar}{4}n^{\epsilon-\frac{1}{2}}$, l being an integer not greater than m, which is any integer whatever, can be included in a denumerable set of regions such as those contemplated in the last section, of total measure less than $\frac{n}{r^2\sqrt{\pi}}e^{-\frac{n^{\epsilon}ar}{4}}$. If now we let k vary from 1 to n, we shall find that those functions whose positive excursion in one of the intervals $\left(\frac{k}{n}, \frac{k+1}{n}\right)$ for some rational argument exceeds $\frac{ar}{4}n^{\epsilon-\frac{1}{2}}$ will have a total measure less than $\frac{n^2}{r^2\sqrt{\pi}}e^{-\frac{n^{\epsilon}ar}{4}}$, in the sense that they can be included in a denumerable assemblage of sets measurable in accordance with the provisions of the last section, and of total measure less than this amount. If in this last sentence we replace the words " positive excursion " by the words " positive or negative excursion," we get for the total measure of our set of functions a quantity less than $\frac{2n^2}{r^2\sqrt{\pi}} e^{-\frac{n^\epsilon ar}{4}}$. It follows that all the functions satisfying (51) can be included in a denumerable set of regions of total measure less than

$$\frac{2}{r^2\sqrt{\pi}} \sum_1^\infty n^2 e^{-\frac{n^\epsilon ar}{4}}. \tag{55}$$

This series converges if a has a sufficiently large value independent of n, as may be seen by comparing it with the series.

$$\frac{2}{r^2\sqrt{\pi}} \sum_1^\infty n^2 (n^\epsilon)^{-e^{\frac{ar}{4}}},$$

which converges for a sufficiently large. Moreover, if (55) converges for $a=a_1$, we may write (55) in the form

$$\frac{2}{r^2\sqrt{\pi}} \sum_1^\infty n^2 e^{-\frac{n^\epsilon ar}{4}} = \frac{2}{r^2\sqrt{\pi}} \sum_1^\infty n^2 e^{-\frac{n^\epsilon a_1 r}{4}} e^{\frac{n^\epsilon (a_1-a)r}{4}}$$

$$< \left[\frac{2}{r^2\sqrt{\pi}} \sum_1^\infty n^2 e^{-\frac{n^\epsilon a_1 r}{4}}\right] e^{\frac{(a_1-a)r}{4}} \tag{56}$$

for $a>a_1$. Hence we have

$$\lim_{a\to\infty} \frac{2}{r^2\sqrt{\pi}} \sum_1^\infty n^2 e^{-\frac{n^\epsilon ar}{4}} = 0. \tag{57}$$

That is, by making a in (51) sufficiently large, the functions satisfying (51) may be included in a denumerable set of such regions as those discussed in §6, of total measure as small as may be desired.

Let us now consider the sphere

$$x_1^2 + \ldots + x_n^2 = r^2,$$

and on this sphere, a sector subtended at the center by a given area of the surface. Let us cut off a portion of this sector by a concentric sphere of radius $r_1 < r$, and let us measure the volume of this new sector in terms of the sphere of radius r. Let us compare the quantity thus obtained with

$$(2\pi)^{-\frac{n}{2}} r^{-n} n^{\frac{n}{2}} \int dx_1 \ldots \int dx_n\, e^{-\frac{(x_1^2+\ldots+x_n^2)n}{2r^2}}, \tag{58}$$

the integral being taken over the same sector. The ratio between these two quantities may readily be shown to be

$$\frac{\dfrac{r^{-n}\left(\frac{n}{2}\right)!}{\pi^{\frac{n}{2}}}\displaystyle\int_0^{r_1}\rho^{n-1}\,d\rho}{(2\pi)^{-\frac{n}{2}}r^{-n}n^{\frac{n}{2}}\displaystyle\int_0^{r_1}\rho^{n-1}e^{-\frac{\rho^2 n}{2r^2}}\,d\rho}$$

$$=\frac{\left(\frac{n}{2}\right)!\displaystyle\int_0^{\frac{r_1\sqrt{n}}{r}}u^{n-1}du}{\left(\frac{n}{2}\right)^{\frac{n}{2}}\displaystyle\int_0^{\frac{r_1\sqrt{n}}{r}}u^{n-1}\,e^{-\frac{u^2}{2}}\,du} \tag{59}$$

Since $e^{-\frac{u^2}{2}}$ is a decreasing function, we have

$$\frac{\left(\frac{n}{2}\right)!\displaystyle\int_0^{\frac{r_1\sqrt{n}}{r}}u^{n-1}\,du}{\left(\frac{n}{2}\right)^{\frac{n}{2}}\displaystyle\int_0^{\frac{r_1\sqrt{n}}{r}}u^{n-1}\,e^{-\frac{u^2}{2}}\,du}$$

$$>\frac{\left(\frac{n}{2}\right)!\displaystyle\int_0^{\sqrt{n}}u^{n-1}du}{\left(\frac{n}{2}\right)^{\frac{n}{2}}\displaystyle\int_0^{\sqrt{n}}u^{n-1}\,e^{-\frac{u^2}{2}}\,du}$$

$$>\frac{\left(\frac{n}{2}\right)!\,n^{\frac{n}{2}-1}}{\left(\frac{n}{2}\right)^{\frac{n}{2}}\displaystyle\int_0^{\infty}u^{n-1}\,e^{-\frac{u^2}{2}}\,du}$$

$$=\frac{\left(\frac{n}{2}\right)!\,n^{\frac{n}{2}-1}}{\left(\frac{n}{2}\right)^{\frac{n}{2}}\left(\frac{n}{2}-1\right)!\,2^{\frac{n}{2}-1}}=1. \tag{60}$$

Hence if we consider a region of the sphere $\Sigma x_k^2=r^2$ bounded by radii and the sphere $\Sigma x_k^2=r_1^2$, its measure in terms of the volume of the sphere $\Sigma x_k^2=r^2$ is greater than its measure as given by an expression such as (58). In this statement we may replace the sphere $\Sigma x_k^2=r_1^2$ by any region such that if it contains a point $(x_1, \ldots, x_n)$ it contains $(\theta x_1, \ldots, \theta x_n)$ $(0\leq\theta\leq 1)$, since such a region may be approximated to within any desired degree of accuracy by a sum of sectors. It hence results that the measure of all the points in $\Sigma x_k^2=r^2$ but *without* such a region is less than the measure after the fashion of (58) of all the points without such a region. We may conclude from this fact and (57) that in the space $\Sigma x_k^2=r^2$, the measure of all of the points $(x_1, \ldots, x_n)$ such that for some k and $l>k$,

$$\left|\sum_{k+1}^{l} x_i\right| > a\, r\left(\frac{l-k}{n}\right)^{\frac{1}{2}-\epsilon}. \tag{61}$$

vanishes uniformly in n as a increases.

§8. **The average of a Bounded, Uniformly Continuous Functional.** Let $F\,|f|$ be a functional, defined in the first instance for all step-functions constant over each of the intervals $\left(\frac{k}{n}, \frac{k+1}{n}\right)$ for some n, and having the following properties:

(1) $F|f|<A$ for all f,

(2) There is a function $\phi(x)$ such that

$$\lim_{x\to 0} \phi(x)=0,$$

and

$$\left|\, F\,|f|-F\,|f+g|\,\right| < \phi(\max|g|).$$

We shall speak of the function $F(x_1, \ldots, x_n)$, which is $F\,|f_n|$, where

$$f_n(t) = \sum_{1}^{k} x_i \quad \text{for} \quad \frac{k-1}{n} < t \leq \frac{k}{n},$$

as the *nth section* of $F\,|f|$. I say that for any functional $F\,|f|$

satisfying (1) and (2), the average of the nth section over the sphere $\sum_{1}^{n} x_k^2 = r^2$ approaches a limit as n increases indefinitely, which limit we shall term the *average of* $F|f|$.

To begin with, let us have given a positive number η; we shall construct a value of n such that the difference between the averages of the nth and the $(n+p)$th sections of F is always less than 2η. First choose a so great that the measure of the points in $\Sigma x_k^2 \leq r^2$ satisfying (61) is for all n less than $\eta/2A$. Next find a number ξ so small that

$$\phi(ar\xi^{1/2-\epsilon}) < \eta/4. \tag{62}$$

Divide the interval (0, 1) into portions of width no greater than ξ, and let $t_1, \ldots, t_m$ be the boundaries of these intervals. Let

$$\psi(t) = f(t_n) \qquad (t_n \leq t \leq t_{n+1}),$$

and let

$$G|f| = F|\psi|.$$

Form the average of G as in §6, and let n be so large that this average differs from the average of

$$G(x_1+x_2+\ldots+x_{T_1}, x_1+x_2+\ldots+x_{T_2}, \ldots, x_1+x_2+\ldots+x_{T_m})^{15}$$

by less than $\eta/4$. Then it is immediately obvious that the difference between the averages of the nth and the $(n+p)$th sections of F is less than 2ϵ. In other words, the average of a functional satisfying (1) and (2) necessarily exists.

An example of such a functional is

$$\frac{1}{1+\int_0^1 [f(x)]^2 dx}.$$

Another method of defining the average of a bounded, uniformly continuous functional F is as

$$\lim_{n\to\infty} (2\pi)^{-\frac{n}{2}} r^{-n} n^{\frac{n}{2}} \int_{-\infty}^{\infty} dx_1 \ldots \int_{-\infty}^{\infty} dx_n\, e^{-\frac{(x_1^2+\ldots+x_n^2)n}{2r^2}} F(x_1, \ldots, x_n). \tag{63}$$

[15] Cf. (43).

There is no particular difficulty in showing the equivalence of the two definitions with the aid of (57). A fuller discussion of this definition of the average of a functional is to be found in the papers of the author which have been already cited. It is easy to demonstrate, as the author has done, that the division into parts that are exactly equal plays no essential rôle in definition (63), and can be replaced by a much more general type of division.

For the average of $F\,|f|$ as here defined, we shall write

$$A_r\{F\}. \tag{64}$$

§9. **The Average of an Analytic Functional.** Let $F|f|$ be a functional such that:

(1) Given any positive number B, there is an increasing function $A(B)$, such that if

$$\max\,|f(t)|\leq B,$$

then

$$|F|f||<A(B).$$

(2) Given any positive number B, there is a function $\phi(x)$ such that

$$\lim_{x\to 0}\phi(x)=0,$$

while if

$$\max\,|f(t)|\leq B,\quad \max\,|g(t)|\leq B$$

then

$$|F|f|-F|g||<\phi(\max\,|f-g|).$$

Let us write

$$\left.\begin{aligned} F_H|f| &= F|f| && \text{for } \max\,|f(t)|\leq H \\ F_H|f| &= F\left|\frac{f(H)}{\max\,|f(t)|}\right| && \text{for } \max\,|f(t)|> H \end{aligned}\right\}. \tag{65}$$

$F_H|f|$ is clearly bounded and uniformly continuous, and as such comes under the class of those functionals which have averages in the sense of the last paragraph.

Let us define for any functional $G|f|$ the function $G(x_1, \ldots, x_n)$ as in the last paragraph, and let us write $A_r^n\{F\}$ for the average of $F(x_1, \ldots, x_k)$ over $\Sigma x_k^2=r^2$. Let H and $K>H$ be any two positive numbers. Clearly by (42) and the argument which follows (60), the set of functions for which $\max |f(t)| \geq H$ can be enclosed over $\Sigma x_k^2 = r^2$ in a region of measure

$$\frac{2}{r}\sqrt{\frac{2}{\pi}}\int_H^\infty e^{-\frac{x^2}{2r^2}}\,dx.$$

Hence

$$|A_r^n\{F_H\} - A_r^n\{F_K\}| < \frac{4}{r}\sqrt{\frac{2}{\pi}}\int_H^\infty e^{-\frac{x^2}{2r^2}}\,dx\; A(K). \tag{66}$$

In particular, if $\theta \leq 1$

$$|A_r^n\{F_H\} - A_r^n\{F_{H+\theta}\}| < \frac{4}{r}\sqrt{\frac{2}{\pi}}\int_H^\infty e^{-\frac{x^2}{2r^2}}\,dx\; A(H+1).$$

Hence if $K - H < m$, m being an integer, we have

$$|A_r^n\{F_H\} - A_r^n\{F_K\}| < \frac{4}{r}\sqrt{\frac{2}{\pi}}\sum_1^m \int_{H+i}^\infty e^{-\frac{x^2}{2r^2}}\,dx\; A(H+i+1). \tag{67}$$

It follows that if

$$\lim_{p\to\infty} \frac{A(p+1)\displaystyle\int_p^\infty e^{-\frac{x^2}{2r^2}}\,dx}{A(p)\displaystyle\int_{p-1}^\infty e^{-\frac{x^2}{2r^2}}\,dx} < 1, \tag{68}$$

then

$$\lim_{H\to\infty} |A_r^n\{F_H\} - A_r^n\{F_K\}| = 0 \tag{69}$$

uniformly in n and K. This will be the case, for example, if $A(p)$ is a polynomial in p, or is of the form a^p. Under these circumstances

$$\lim_{n\to\infty} A_r^n\{F\} = \lim_{n\to\infty}\lim_{H\to\infty} A_r^n\{F_H\}$$

$$= \lim_{H\to\infty}\lim_{n\to\infty} A_r^n\{F_H\} = \lim_{H\to\infty} A_r\{F_H\} \tag{70}$$

exists. We shall call this quantity $A_r\{F\}$. There is no difficulty in showing that formula (63) holds in this case also.

Now let us have an $F|f|$ of the form

$$F|f| = a_0 + \int_0^1 K_1(t)f(t)dt + \dots$$
$$+ \int_0^1 dt_1 \dots \int_0^1 dt_n K_n(t_1, \dots, t_n)f(t_1) \dots f(t_n) + \dots, \tag{71}$$

given that

$$A(u) = a_0 + u \int_0^1 |K_1(t)|dt + \dots$$
$$+ u^n \int_0^1 dt_1 \dots \int_0^1 dt_n | K_n(t_1, \dots, t_n)| + \dots \tag{72}$$

exists for all u and satisfies (68). Then if

$$\max f|(t)| \leq B,$$

we have as an obvious result that (1) at the beginning of this section is satisfied. Moreover

$$|F|f| - F|g|| \leq \Big| \sum_1^\infty \int_0^1 dt_1 \dots$$
$$\int_0^1 dt_n K_n(t_1, \dots, t_n)[f(t_1) \dots f(t_n) - g(t_1) \dots g(t_n)]$$
$$\leq \sum_1^\infty \int_0^1 dt_1 \dots \int_0^1 dt_n | K_n(t_1, \dots, t_n)|$$
$$\times \{ [\max|G(x)| + \max|f(x) - g(x)|]^n - [\max|g(x)|]^n \}. \tag{73}$$

If $\max|f(x)| < B$, and $\max|f(x)-g(x)| < \epsilon$,

$$|F|f|-F|g|| \leq \sum_1^\infty [(B+\epsilon)^n - B^n] \int_0^1 dt_1, \ldots \int_0^1 dt_n |K_n(t_1, \ldots, t_n)|$$

$$= \sum_1^\infty \int_B^{B+\epsilon} ny^{n-1} \int_0^1 dt_1 \ldots \int_0^1 dt_n |K_n(t_1, \ldots, t_n)|$$

$$\leq \sum_1^\infty \epsilon n (B+\epsilon)^{n-1} \int_0^1 dt_1 \ldots \int_0^1 dt_n |K_n(t_1, \ldots, t_n)|. \tag{74}$$

Series (74) converges for all ϵ, since series (72) has a convergent derivative series. Hence condition (2) may be proved to be satisfied. In other words, $F|f|$ has an average in the sense of this paper.

This result may be much generalized without any great difficulty. It may be extended to functionals containing Stieltjes integrals such as

$$a_0 + \int_0^1 f(t) dQ_1(t) + \quad \ldots \ldots$$

$$+ \int_0^1 \ldots\ldots \int_0^1 f(t_1) \quad \ldots \quad f(t_n)\, d^n Q_n(t_1, \quad \ldots \quad, t_n) + \quad \ldots \quad, \tag{75}$$

provided

$$A(u) = a_0 + \sum_1^\infty u^n \int_0^1 \ldots \int_0^1 |d^n Q(t_1, \quad \ldots \quad t_n)| \text{ [16]} \tag{76}$$

exists for all u and satisfies (68). This class of functionals includes such expressions as

$$\int_0^1 \int_0^1 \int_0^1 [f(t_1)]^m [f(t_2)]^n [f(t_3)]^p K(t_1, t_2, t_3)\, dt_1\, dt_2\, dt_3.$$

[16] Cf. Daniell, *Functions of Limited Variation in an Infinite Number of Dimension*, Annals of Mathematics, Series 2, Vol. 21, pp. 30–38.

On the other hand, $A(u)$ in (72) may be replaced by

$$a_0+\sqrt{u}\sqrt{\int_0^1 [K_1(t)]^2\,dt} + \ldots + u^{\frac{n}{2}}\sqrt{\int_0^1 \ldots \int_0^1 dt_1 \ldots dt_n\,[K_n(t_1, \ldots, t_n)]^2}, \tag{77}$$

the K's being all summable and of summable square, and by the use of the Schwarz inequality, (1) and (2) may be deduced. Again, we may under the proper conditions concerning ϕ show that

$$\phi(F_1|f|, \ldots, F_n|f|) \tag{78}$$

has an average in the sense of this paper, if $F_1, \ldots, F_n$ are such functionals as we have already described.

I here wish to discuss only such a functional as

$$F|f| = \int_0^1 \ldots \int_0^1 [f(t_1)]^{\mu_1} \ldots [f(t_\nu)]^{\mu_\nu} K(t_1, \ldots t_\nu)\,dt_1 \ldots dt_\nu.$$

The average of this functional will be the limit as n increases indefinitely of

$$\sum_{k_1=1}^{n} \cdots \sum_{k_\nu=1}^{n} (2\pi)^{-\frac{n}{2}} r^{-n} n^{-\frac{n}{2}} \int_{-\infty}^{\infty} dx_1 \ldots \int_{-\infty}^{\infty} dx_n\, e^{-\frac{(x_1^2+\cdots+x_n^2)n}{2r^2}}$$
$$(x_1+\ldots+x_{k_1})^{\mu_1} \ldots (x_1+\ldots+x_{k_\nu})^{\mu_\nu}$$
$$\int_{\frac{k_1-1}{n}}^{\frac{k_1}{n}} dt_1 \ldots \int_{\frac{k_\nu-1}{n}}^{\frac{k_\nu}{n}} dt_\nu\, K(t_1, \ldots, t_\nu)$$

$$= \sum_{k_1=1}^{n} \cdots \sum_{k_\nu=1}^{n} \frac{(2\pi)^{-\frac{\nu}{2}} r^{-\nu} n^{\frac{\nu}{2}}}{[k_1(k_2-k_1)\ldots(k_\nu-k_{\nu-1})]^{\frac{1}{2}}}$$
$$\int_{-\infty}^{\infty} d\xi_1 \ldots \int_{-\infty}^{\infty} d\xi_\nu\, e^{-\frac{\xi_1^2 n}{2r^2k_1} - \cdots \frac{\xi_\nu^2 n}{2r^2(k_\nu-k_{\nu-1})^{\mu_1}}}$$
$$\xi_1(\xi_1+\xi_2)^{\mu_2} \ldots (\xi_1+\ldots+\xi_\nu)^{\mu_2}$$
$$\int_{\frac{k_1-1}{n}}^{\frac{k_1}{n}} dt_1 \ldots \int_{\frac{k_\nu-1}{n}}^{\frac{k_\nu}{n}} dt_n\, K(t_1, \ldots, t_\nu), \tag{80}$$

because of (63). Expression (80) clearly approaches the limit

$$\int_0^1 dt_1 \ldots \int_0^1 dt_n K(t_1, \ldots, t_n) \frac{(2\pi)^{-\frac{\nu}{2}} r^{-\nu}}{[t_1(t_2-t_1) \ldots (t_\nu - t_{\nu-1})]^{\frac{1}{2}}}$$

$$\int_{-\infty}^{\infty} d\xi_1 \ldots \int_{-\infty}^{\infty} d\xi_\nu \, exp\left(\frac{-\xi_1^2}{2r^2 t_1} - \ldots - \frac{\xi_\nu^2}{2r^2(t_\nu - t_{\nu-1})}\right)$$

$$\xi_1^{\mu_1}(\xi_1+\xi_2)^{\mu_2} \ldots (\xi_1 + \ldots + \xi_\nu)^{\mu_\nu}, \tag{81}$$

provided that K is of limited total variation. This agrees with a definition already obtained by the author.[17]

§10. **The Average of a Functional as a Daniell Integral.** Daniell[18] has discussed a generalized definition of an integral in the following manner: he starts with a set of functions $f(p)$ of general elements p. He assumes a class T_0 of such functions which is closed with respect to the operations, multiplication by a constant, addition, and taking the modulus. He also assumes that to each f of class T_0 there corresponds a number K, independent of p, such that

$$|f(p)| \leq K,$$

and that to each f there corresponds a finite "integral" $U(f)$ having the properties

(C) $\quad Uc(f) = c\,U(f),$

(A) $\quad U(f_1+f_2) = U(f_1) + U(f_2),$

(L) If $\quad f_1 \geq f_2 \geq \ldots \geq 0 = \lim f_n,$ then $\lim \; U(f_n) = 0,$

(P) $\quad U(f) \geq 0$ if $f \geq 0.$

There is no difficulty in showing that if T_0 be taken as the set of all functionals F such as those defined by (1) and (2) of §8, and the operator A_r is taken as U, all these conditions are ful-

[17] *The Average of an Analytic Functional*, p. 256.
[18] Daniell, p. 280.

filled save possibly (L). I here wish to discuss the question as to whether (L) is fulfilled.

Let it be noted that (L) involves the knowledge of those entities which form the arguments to the members of T_0. Up to this point we have regarded the arguments of our functionals F as step-functions. However, if $F|f|$ is a bounded, uniformly continuous functional of all step-functions constant over every interval $\left(\frac{k-1}{n}, \frac{k}{n}\right)$, it may readily be shown that if f is the uniform limit of a sequence of such step-functions f_n, then $\lim F|f_n|$ exists, is independent of the particular sequence f_n chosen, and is bounded and uniformly continuous. We shall call this limit $F|f|$. This extension of the arguments of F does not vitiate the validity of any of the Daniell conditions. In fact, no condition save possibly (L) is vitiated if we then restrict the arguments of our functionals of T_0 to continuous functions $f(t)$ such that $f(0)=0$.

We wish, then, to show that if $F_1 \geq F_2 \geq \ldots \geq 0 = \lim F_n$, for every f that is continuous, then

$$\lim A_r\{F_n\} = 0.$$

Let us notice that it follows from (50), (55) and (63) that

$$A_r\{F_n\} \leq [\max F_1] \frac{2}{r^2 \pi} \sum_1^{\infty} n^2 . e^{-\frac{n^{\epsilon} a r}{4}} + \max_{S_a} F_n, \tag{82}$$

where S_a is the set of all functions f for which for every t_1 and t_2 between 0 and 1,

$$|f(t_2) - f(t_1)| \leq ar\,(t_2 - t_1)^{1/2-\epsilon}.$$

By (57), the first term in (82) can be made as small as we please by taking a large enough. Accordingly, we can prove (L) if we can show that for every a, the set of functions Σ_n consisting in all the functions f in S_a for which $F_n|f| \geq \eta$ can be made to become null by making n large enough, whatever η may be. We shall prove this by a reductio ad absurdum.

Σ_n is in the language of Fréchet an extremal set, in that it is closed, equicontinuous, and uniformly bounded.[19] Hence either all the Σ_n's from a certain stage are null, or there is an element common to every Σ_n.[20] Let this element be $f(t)$. Then for all n, we have

$$F_n|f| \geq \eta \qquad (83)$$

This however contradicts our hypothesis.

Our operation of taking the average is hence a Daniell integration, and as such capable of the extensions which Daniell develops. Daniell first proves that if $f_1 \leq f_2 \leq \ldots$ is a sequence from T_0, then the sequence $U(f_n)$ is an increasing sequence, and hence either becomes positively infinite or has a limit. If then $f = \lim f_n$ exists, Daniell defines $U(f)$ as $\lim U(f_n)$, and says that f belongs to T_1. Our T_1 will contain all the functions discussed in §9, and it admits of an easy proof that the definition of $A_r\{F\}$ in §9 accords with the definition arising from the Daniel extension of A_r whenever the former definition is applicable. Daniell then defines for any function f the upper semi-integral $\dot{U}(f)$ as the lower bound of $U(g)$ for g in T_1 and $g > f$. He defines $\underset{.}{U}(f)$ as $-\dot{U}(-f)$, and $U(f)$ as $\dot{U}(f)$ if $\dot{U}(f) = \underset{.}{U}(f) =$ finite, f being then called summable. All these extensions are applicable to our average operator A_r, as is also Daniell's theorem to the effect that if $f_1, \ldots, f_n, \ldots$ is a sequence of summable functions with limit f, and if a summable function ϕ exists such that $|f_n| \leq \phi$ for all n, f is summable, $\lim U(f_n)$ exists and $= U(f)$.

It may be shown from theorems of Daniell that if the measure of a set of functions be held to be the average of a functional 1 over the set and zero elsewhere, and the outer measure the upper semi-average of a functional 1 over the set and zero elsewhere, then the definition in §6 will coincide with this whenever it is applicable. §4 may be interpreted as saying that the measure of the set of functions for which the non-differentiability coefficient differs from r by more than ϵ is zero, and §7 as saying that the outer measure of the functions satisfying (51) is less than (55).

[19] *Sur quelques points du calcul fonctionnel*, Rend. Cir. Math. di Palermo Vol. 22, pp. 7, 37.
[20] Ibid, p. 7.

§11. **Independent Linear Functionals.** Let us consider a functional

$$F|f| = \phi\left(\int_0^1 K(x)f(x)dx, \int_0^1 Q(x)f(x)dx\right),$$

K and Q being summable of summable square, and ϕ being bounded and uniformly continuous for all arguments from $-\infty$ to ∞. We then have

$$F(x_1, \ldots, x_n) = \phi\left(\sum_1^n x_k \int_{\frac{k-1}{n}}^{1} K(x)dx, \sum_1^n x_k \int_{\frac{k-1}{n}}^{1} Q(x)dx\right).$$

$$= \phi\left\{\sum_1^n x_k \int_{\frac{k-1}{n}}^{1} K(x)dx, \frac{\sum_1^n \int_{\frac{k-1}{n}}^{1} K(x)dx \int_{\frac{k-1}{n}}^{1} Q(x)dx}{\sum_1^n \left[\int_{\frac{k-1}{n}}^{1} K(x)dx\right]^2} \times \sum_1^n x_k \int_{\frac{k-1}{n}}^{1} K(x)dx \right.$$

$$\left. + \sum_1^n x_k \left[\int_{\frac{k-1}{n}}^{1} Q(x)dx - \frac{\sum_1^n \int_{\frac{j-1}{n}}^{1} K(x)dx \int_{\frac{j-1}{n}}^{1} Q(x)dx}{\sum_1^n \left[\int_{\frac{j-1}{n}}^{1} K(x)dx\right]^2} \times \int_{\frac{k-1}{n}}^{1} K(x)dx\right]\right\}$$

$$= \phi\left\{\xi \sqrt{\sum_1^n \left[\int_{\frac{k-1}{n}}^{1} K(x)dx\right]^2}, \xi \frac{\sum_1^n \int_{\frac{k-1}{n}}^{1} K(x)dx \int_{\frac{k-1}{n}}^{1} Q(x)dx}{\sqrt{\sum_1^n \left[\int_{\frac{k-1}{n}}^{1} K(x)dx\right]^2}}\right.$$

$$+ \eta \sqrt{\sum_1^n \left[\int_{\frac{k-1}{n}}^1 Q(x)dx\right]^2 - \frac{\left[\sum_1^n \int_{\frac{k-1}{n}}^1 K(x)dx \int_{\frac{k-1}{n}}^1 Q(x)dx\right]^2}{\sum_1^n \left[\int_{\frac{k-1}{n}}^1 K(x)dx\right]^2}}\Bigg\},$$

ξ and η being two orthogonal unit functions. As η increases, the arguments of ϕ approach uniformly

$$\xi \sqrt{\int_0^1 \left[\int_x^1 K(x)dx\right]^2 dx}$$

and

$$\xi \frac{\int_0^1 \left[\int_x^1 K(x)dx \int_x^1 Q(x)dx\right]dx}{\sqrt{\int_0^1 \left[\int_x^1 K(x)dx\right]^2 dx}}$$

$$+ \eta \sqrt{\int_0^1 \left[\int_x^1 Q(x)dx\right]^2 dx - \frac{\left[\int_0^1 \left[\int_x^1 K(x)dx \int_x^1 Q(x)dx\right]dx\right]^2}{\int_0^1 \left[\int_x^1 K(x)dx\right]^2 dx}}.$$

If in particular $\int_x^1 K(x)dx$ and $\int_x^1 Q(x)dx$ are normal and orthogonal,

$$A_r\{F\} = \lim_{n\to\infty} \frac{\int_{-\frac{\pi}{2}}^{\frac{\pi}{2}} \int_{-\frac{\pi}{2}}^{\frac{\pi}{2}} \cos^{n-1}\theta_1 \cos^{n-2}\theta_2\, \phi(\sin\theta_1, \sin\theta_2)\, d\theta_1\, d\theta_2}{\int_{-\frac{\pi}{2}}^{\frac{\pi}{2}} \int_{-\frac{\pi}{2}}^{\frac{\pi}{2}} \cos^{n-1}\theta_1 \cos^{n-2}\theta_2\, d\theta_1\, d\theta_2}$$

$$= \frac{1}{2\pi r^2} \int_{-\infty}^{\infty} dx_1 \int_{-\infty}^{\infty} dx_2\, e^{-\left(\frac{x_1^2 + x_2^2}{2r^2}\right)} \phi(x_1, x_2). \tag{84}$$

That is, the average of F with respect to f can be obtained by first forming the average with respect to f of

$$\phi\left(\int_0^1 K(x)f(x)dx, \int_0^1 Q(x)g(x)dx\right),$$

then forming the average of this average with respect to g, and finally putting $f=g$. A similar theorem can be shown by the same means to hold of

$$\phi\left(\int_0^1 K_1(x)f(x)dx, \ . \ . \ . \ , \int_0^1 K_n(x)f(x)dx\right),$$

in case the set of functions $\left[\int_x^1 K_k(x)dx\right]$ is normal and orthogonal.

If now ϕ is not a bounded function, but is merely uniformly continuous over any finite ranges of its arguments, and

$$\frac{1}{(2\pi r^2)^{\frac{n}{2}}}\int_{\infty}^{\infty} dx_1 \ . \ . \ . \int_{-\infty}^{\infty} dx_n e^{-\frac{\sum_1^n x_k^2}{2r^2}} \phi(x_1, \ . \ . \ . \ , x_n) \qquad (85)$$

exists, it may be proved by a simple limit argument that (85) represents the average of

$$\phi\left(\int_0^1 K_1(x)f(x)dx, \ . \ . \ . \ , \int_0^1 K_n(x)f(x)dx\right).$$

§12. **Fourier Coefficients and the Average of a Functional.** The functions

$$-\sqrt{2} \sin n\pi x = \int_x^1 n\pi\sqrt{2} \cos n\pi x$$

form a normal and orthogonal set. Accordingly if we have a functional

$$\phi(a_1, \ . \ . \ . \ , a_n)$$

of the function

$$f(x) = a_0 + a_1\sqrt{2} \cos \sqrt{\pi x} + \ . \ . \ . \ + a_n\sqrt{2} \cos n\pi x +, \ . \ . \ .$$

its average will be

$$\frac{1}{(2\pi r^2)^{\frac{n}{2}}}\int_{-\infty}^{\infty} dx_1 \;.\;.\;.\; \int_{-\infty}^{\infty} dx_n\, e^{-\frac{\sum_1^n x_k^2}{2r^2}}\,\phi\left(\frac{x_1}{\pi},\;.\;.\;.\;.\;,\frac{x_n}{\pi n}\right) \tag{86}$$

provided this exists and ϕ is continuous, or even provided ϕ is a sum of step-functions. In particular if $F|f(x)|=1$ if $a_m^2+\;.\;.\;.\;+a_n^2+\;.\;.\;.\;\geq a^2$ and zero otherwise, its average will be less than the average of a functional which is 1 if for some k between m and n included,

$$a_k^2 > \frac{a^2 k^{-\frac{3}{2}}}{\sum_1^{\infty} j^{-\frac{3}{2}}}$$

and zero otherwise. The upper average of this latter functional is by (86) not greater than

$$\sum_m^{\infty} \frac{2}{(2\pi r^2)^{\frac{1}{2}}} \int_{\frac{a k^{\frac{1}{4}}\pi}{\left(\sum_1^{\infty} j^{-\frac{3}{2}}\right)^{\frac{1}{2}}}}^{\infty} e^{-\frac{x^2}{2}}\,dx$$

$$< \frac{2}{(2\pi r^2)^{\frac{1}{2}}} \sum_m^{\infty} L^{-k^{\frac{1}{2}}}, \tag{87}$$

for m sufficiently large, where

$$L = e^{\frac{a\pi}{\left(\sum_1^{\infty} j^{-\frac{3}{2}}\right)^{\frac{1}{2}}}} > 1.$$

Series (87) converges. Hence as m increases, the measure of the set of functions for which $a_m^2+\;.\;.\;.\;+a_n^2+\;.\;.\;.>a^2$ approaches zero.

In this demonstration, we have made use of several theorems of Daniell which it did not seem worth while to enumerate in detail. They may all be found in his discussion of measure and integration.[21]

[21] Daniell, loc. cit.

Let us now consider a bounded functional $F|f|$ that is uniformly continuous in the more restrictive sense that for any positive η, there is a positive θ such that

$$|F|f|-F|g||<\eta$$

whenever

$$\int_0^1 [f(t)-g(t)]^2 dt<\theta.$$

Let $F|f|$ be invariant, moreover, when a constant is added to f. It will then be possible to write $F|f|$ in the form

$$F|f|=\phi(a_1, \ldots, a_n, \ldots),$$

where

$$|\phi(a_1, \ldots, a_n, 0, 0 \ldots)-\phi(a_1, \ldots, a_n, a_{n+1}, \ldots)|<\eta$$

whenever

$$\sum_{n+1}^{\infty} a_k^2<\theta.$$

Hence

$$\left| A_r\{F\} - \frac{1}{(2\pi r^2)^{\frac{n}{2}}} \int_{-\infty}^{\infty} dx_1, \ldots \int_{-\infty}^{\infty} dx_n e^{-\frac{\sum_1^n x_k^2}{2r^2}} \right.$$

$$\left. \times\phi\left(\frac{x_1}{\pi}, \ldots, \frac{x_n}{\pi_n}, 0, 0, \ldots\right)\right| < \eta+2 \max |F| M(\theta). \quad (87)$$

where $M(\theta)$ is the outer measure of all the functions for which $\sum_{n+1}^{\infty} a_k^2 \geq \theta$. N can, as we have seen, be made arbitrarily small by making n sufficient y large. It can hence be shown that

$$A_r\{F\} = \lim_{n\to\infty} \frac{1}{(2\pi r^2)^{\frac{n}{2}}} \int_{-\infty}^{\infty} dx_1 \ldots \int_{-\infty}^{\infty} dx_n e^{-\frac{\sum_1^n x_k^2}{2r^2}}$$

$$\times\phi\left(\frac{x_1}{\pi}, \ldots, \frac{x_n}{\pi n}, 0, 0, \ldots\right) \quad (88)$$

This theorem may be established for the more general case of *f* functional F which is uniformly continuous for all functions a for which

$$\int_0^1 [f(x)]^2 \, dx < \theta,$$

whatever θ may be, and for which there is an increasing $\psi(u)$ such that

$$F|f| \leq \psi\left[\int_0^1 [f(x)]^2 dx\right],$$

while

$$\sum_1^\infty \psi(n+1)[M(n+1)-M(n)]$$

converges.

Now let us consider a functional of the form

$$F|f| = H_{i_1}\left(\frac{x_1\pi}{r\sqrt{2}}\right) \quad . \quad . \quad . \quad H_{i_n}\left(\frac{nx_n\pi}{r\sqrt{2}}\right), \tag{89}$$

where the H's are any set of Hermite polynomials corresponding to normalized Hermite functions. If $G|f|$ is another such function, it is easy to show that

$$A_r\{F|f|G|f|\} = 0, \tag{90}$$

and to compute

$$A_r\{F|f|\}^2. \tag{91}$$

Now let us arrange all these functionals in a progression $\{F_n\}$ It follows from theorems analogous to those familiar in the ordinary theory of orthogonal functions that if $G|f|$ is a continuous function ϕ of $a_1, \ . \ . \ . \ , a_n$ for which expression (86) exists, then

$$\sum_1^\infty F_k|f| \frac{A_r\{F_k|f|G|f|\}}{\sqrt{\{A_r F_k|f|\}^2}} \tag{92}$$

converges in the mean to $G|f|$ in the sense that if $G_n|f|$ is its nth partial sum,

$$\lim_{n\to\infty} A_r\{G|f| - G_n|f|\}^2 = 0. \tag{93}$$

With the aid of generalizations of familiar theorems concerning orthogonal functions, it may be shown that this result remains valid if $\{G|f|\}^2$ fulfils the conditions laid down for F in (88), and G does likewise. The functionals (89) are then in a certain sense a complete set of normal and orthogonal functions.

One final remark. By methods which exactly duplicate §4, it can be shown that the set of functions f for which it is not true that

$$\left| \frac{r^2}{\pi^2} - \lim_{n\to\infty} \sum_1^n \frac{k^2 a_k^2}{n} \right| < \epsilon \tag{94}$$

is of zero measure, whatever ϵ. This suggests interesting questions relating to the connection between the coefficient of non-differentiability of a function and $\lim_{n\to\infty} \sum_1^n \frac{k^2 a_k^2}{n}$.

THE DIRICHLET PROBLEM

By Norbert Wiener

In a recent paper in the *Comptes rendus*,[1] Lebesgue points out that the Dirichlet problem divides itself into two parts, the first of which is the determination of a harmonic function corresponding to certain boundary conditions, while the second is the investigation of the behaviour of this function in the neighborhood of the boundary. In a paper appearing the same week[2] the author of the present paper made the same remark independently, and went on to give a precise definition of the sense in which the harmonic function depends on the boundary conditions. He proved, moreover, that his method assigns a definite harmonic function to any continuous boundary condition on any bounded set of points in any number of dimensions. In particular, he proved the theorem that the potential corresponding to boundary values 1 on a given bounded set of points and (in three or more dimensions) 0 at infinity determines in his sense a harmonic function generated by a distribution of charge (represented by a Stieltjes integral) over the bounded set. The charge on no region will be negative. The total amount of the charge will be known as the capacity of the boundary set of points. Thus every bounded set of points, whether a surface or not, and indeed whether measurable or not, will have a definite, finite capacity. Moreover, if C_2 is a set of points containing as a part C_1, the capacity of C_2 will at least equal that of C_1.

To return to Lebesque: like the author, he points out that the solubility of the Dirichlet problem in the classical sense for a given region depends only on the *im Kleinen* properties of the boundary, and may be reduced to the investigation of what he terms the regularity of the individual points of the boundary. A point O of the boundary B of a set of points D is termed regular

[1] January 21, 1924.
[2] *Certain Notions in Potential Theory, Jour. Math. Phys. M. I. T.*, January, 1924.

Reprinted from *J. Math. Phys.*, *3*, 1924, pp. 127–147. (Courtesy of the *Journal of Mathematics and Physics,* Massachusetts Institute of Technology and Johnson Reprint Corporation.)

if whenever $F(P)$ is a function continuous on B, and $f(P)$ is the corresponding function harmonic over D,

$$\lim_{P \to O} f(P) = F(O).$$

This must hold for *every* function f continuous on B.

The oldest fairly general condition for the regularity of a point on a boundary in three-space is that of Poincaré and Zaremba.[3] This is stated by Lebesgue as follows: *A point O of the boundary of a domain D is regular if it is the vertex of a closed conical surface the interior of which is exterior to D.* A result obtained by G. E. Raynor[4] may be stated as follows: *A point O of the boundary of a domain D is regular if there is a number λ greater than zero such that there exist an infinity of spheres S with O as center and such that the measure of the set of points on S and exterior to D exceeds λ times the measure of the surface of S.* This anticipates and contains condition A of Lebesgue, which reads: *The point O is regular if on each sphere with center O and radius r there is a point exterior to D and distant from D by a quantity at least equal to Kr, K being a given positive quantity.* Lebesgue also quotes a condition due to Bouligand,[5] which reads: *The point O is regular if it is the vertex of a conical surface, open or closed, exterior to D.*

All the conditions so far stated, with the exception of that of Raynor, are special cases of one published by H. B. Phillips and the author of the present paper[6] prior to all these conditions except that of Poincaré and Zaremba. In its three-dimensional form, this reads as follows: *The point O is regular whenever there are positive numbers a and b such that if $r<a$ and a sphere S of radius r is constructed with O as center, the content of the projection of the points in S and not in D on some plane exceeds br^2.* In the proof of this theorem, only a denumerable set of spheres S are actually used. If the theorem be stated in the somewhat generalized

[3] Poincaré, *Sur les équations aux derivées partielles de la physique mathématique*, *Am. J. of M.*, V. 12, 1890; Zaremba, *Sur le principe du minimum*, *Bull. de l'Ac. des Sc. de Cracovie*, 1909.

[4] *Dirichlet's Problem*, *Annals of Mathematics*, dated March, 1922, but not actually published until the end of March, 1923.

[5] *Sur le problème de Dirichlet harmonique*, *Comptes rendus*, January 2, 1924

[6] *Nets and the Dirichlet Problem*, *Jour. Math. Phys. M. I. T.*, March, 1923.

form involving only a denumerable set of spheres, it will include that of Raynor.

In his paper on *Certain Notions in Potential Theory*, the author gives the following sufficient condition for the regularity of a point: *The point O of the boundary of a domain D is regular if there is a sequence of values of r tending to 0 such that either:*

(*a*) *The inner Lebesgue measure of the points on a sphere of radius r exterior to D in terms of the surface of the sphere itself; exceeds a positive quantity independent of r; or*

(*b*) *The capacity of the part of the boundary of D interior to a sphere of radius r exceeds M times the capacity of the sphere, M being independent of r.*

This condition includes as special cases all those previously mentioned. It is to be noted that it introduces the concept of capacity. This is rendered possible by an antecedent proof that every bounded set of points has a capacity.

Up to this point the theory has dealt with conical points and their immediate generalizations. We next turn to the discussion of cuspidal points and their generalizations. The published work on this subject, so far as is known by the author, is entirely due to Lebesgue,[7] who gave the first example[8] of a simply connected boundary in three-space with an irregular point. Lebesgue's results in this direction have been collected in his recent paper, and read as follows: *If there is a segment of a straight line (or analytic curve) terminating at the point O of the boundary of the domain D, if the segment is entirely exterior to D, and if there exist positive numbers A and B such that if P is a point of the segment, its distance from D exceeds $A\,\overline{OP}^B$, O is regular. If there exist positive numbers A and B such that when P lies on the segment, $\overline{OP} = \overline{OQ}$, and*

$$\overline{PQ} > Ae^{\frac{-B}{\overline{OP}}},$$

Q lies in D, O is irregular. Lebesgue states that he has been unable to generalize these results as he has generalized that of Zaremba, by constructing an infinite set of spheres about the

[7] Mr. Gleason of Princeton, however, verbally communicated to me last year several very interesting results in this direction which he obtained several years ago, before Lebesgue had published similar results.

[8] *Comptes rendus des séances de la société mathématique de France*, 1913.

point O, marking on each the region excluded by D, and rotating each independently about O. He also states no results on cuspidal points reducing to flat tongues.

We now come to conditions of a much more general character. It was stated by Poincaré and repeated by Lebesgue that if a point is regular for a given boundary, it is regular for a boundary entirely interior to the first in the neighborhood of this point. Lebesgue gives the following necessary and sufficient condition for the regularity of the point O: *For a point O of the boundary of a domain D to be regular, it is necessary and sufficient that there exist a function $F(x, y, z)$ continuous at O, attaining its lower bound at O and O only, and such that everywhere in D*

$$\frac{\partial^2 F}{\partial x^2}+\frac{\partial^2 F}{\partial y^2}+\frac{\partial^2 F}{\partial z^2}\leq 0.$$

Other forms of this condition are stated by Lebesgue and Kellogg.[9] They all suffer from the defect of involving the geometrical character of the boundary only in a very indirect and devious manner. From the geometrical point of view, indeed, they are scarcely more than restatements of the regularity of O.

It is the purpose of this paper to develop a complete necessary and sufficient characterization of regular points which shall be at least quasi-geometrical. Its statement reads as follows: *Let O be a point of C, the boundary of an open set of points D. Let λ be any positive quantity less than 1. Let γ_n be the capacity of the set of all points Q not belonging to D such that*

$$\lambda^n \leq \overline{OQ} \leq \lambda^{n-1}.$$

Then O is regular or irregular according as

$$\frac{\gamma_1}{\lambda}+\frac{\gamma_2}{\lambda^2}+\ \cdots\ +\frac{\gamma_n}{\lambda^n}+\ \cdots$$

diverges or converges.

In order to prove this theorem, it will be necessary to repeat from the author's previous paper a few theorems relating to the determination of a harmonic function by continuous boundary conditions on an arbitrary boundary. He there proved that if

[9] *An Example in Potential Theory, Proc. Am. Ac. Arts and Sciences*, 1923.

C is any closed bounded set of points, and $F(P)$ any function continuous on this set of points, there is a function $f(P)$ harmonic except on C, and determined by F in the following manner. We first construct[10] a continuous function $g(P)$ defined throughout space and reducing to $F(P)$ on C. We then construct a sequence of boundaries for which the Dirichlet problem is soluble, each within its predecessor, and together sharing no point not on C. This is always possible. We finally form the sequence of harmonic functions assuming the boundary values $g(P)$ on these boundaries. Outside C this sequence will converge to a harmonic function independent of $g(P)$ and the sequence of boundaries, and entirely determined by C and $F(P)$. In the particular case where $F(P)$ is identically 1, $g(P)$ may also be taken as identically 1.

It may be remarked at once that if $f(P)$ and $g(P)$ are any two harmonic functions corresponding to boundary values $F(P)$ and $G(P)$, respectively, on the same boundary C, and if throughout C we have $F(P) \geq G(P)$, then we shall always have $f(P) \geq g(P)$. This may be considerably generalized: let Q be a point interior to the boundary C, and let C' be a boundary containing Q in its interior. Let $F(P) \geq G(P)$ at every point of C in or on C', and let $f(P) \geq g(P)$ at every point of C' interior to C. Then $f(Q) \geq g(Q)$.

If $f(P)$ is the harmonic function corresponding to 1 on the boundary C, if C' is another boundary such that every point of C is on or exterior to C', and if $g(P)$ is the harmonic function corresponding to 1 on C', while both f and g correspond to a potential 0 at infinity, in case the region over which they are defined is unbounded, then $g(P) \geq f(P)$ at any point where both are defined. From this it may be concluded at once that if a given set of points completely surrounds another set, the former set has a capacity at least as large as the latter.

The lemmas of the last two paragraphs are readily proved from the definition given by the author of the sense in which the generalized Dirichlet problem is soluble. They will be presupposed without explicit reference in all that follows.

10 For the possibility of this construction, see Lebesgue, *Sur le problème de Dirichlet*, *Rendiconti di Palermo*, v. 24 (1907), pp. 371–402.

We now proceed to the proof of our main theorem. We first wish to show that if

$$\frac{\gamma_1}{\lambda}+\frac{\gamma_2}{\lambda^2}+\ \cdot\ \cdot\ \cdot\ +\frac{\gamma_n}{\lambda^n}+\ \cdot\ \cdot\ \cdot$$

diverges, O is regular. It is to be noted that under this condition either

$$\frac{\gamma_1}{\lambda}+\frac{\gamma_3}{\lambda^3}+\ \cdot\ \cdot\ \cdot\ +\frac{\gamma_{2n+1}}{\lambda^{2n+1}}+\ \cdot\ \cdot\ \cdot$$

or

$$\frac{\gamma_2}{\lambda^2}+\frac{\gamma_4}{\lambda^4}+\ \cdot\ \cdot\ \cdot\ +\frac{\gamma_{2n}}{\lambda^{2n}}+\ \cdot\ \cdot\ \cdot$$

diverges.

Let C_n be the set of all points exterior to D in the zone between a sphere of radius λ^n about O as center and a sphere of radius λ^{n-1}. Its capacity, as we have seen, will be γ_n. Now consider the set of points consisting of all points contained in a sphere of radius $\lambda^{n-3/2}$ about O as center and exterior to C_n. We wish to consider the harmonic function corresponding to boundary values O over the sphere and 1 over C_n. This function we shall term $F_n(P)$.

To begin with let C_n consist entirely of regular points. Let the harmonic function $V_n(P)=1$ on C_n and 0 at infinity be represented by the Stieltjes integral

$$\int\int\int \overline{PQ}^{-1}dM(Q).$$

$dM(Q)$ will here be everywhere positive, and will represent the charge over the rectangular parallelepiped $dx\ dy\ dz$. It will vanish when this parallelepiped contains no point of C_n.

The Green's function of the sphere of radius C about O is

$$G_C(P,Q)=\frac{1}{\overline{PQ}}-\sqrt{\frac{1}{C^2+\dfrac{\overline{OP}^2\,\overline{OQ}^2}{C^2}-2\overline{OP}\ \overline{OQ}\cos(OP,OQ)}}.$$

Let it be noted that

$$G_C(P,Q)\leq 1/\overline{PQ},$$

and that if P and Q both lie at a distance from O not greater than μC $(\mu<1)$,

$$G_C(P,Q)\geq(1-\mu^2)^2/8\overline{PQ}.$$

Hence over a sphere of radius λ^{n-1} about O,

$$\iiint \overline{PQ}^{-1} dM(Q) \geq \iiint G_{\lambda^{n-3/2}}(P, Q) dM(Q)$$
$$\geq \frac{(1-\lambda)^2}{8} \iiint \overline{PQ}^{-1} dM(Q).$$

The function

$$\iiint G_{\lambda^{n-3/2}}(P, Q) dM(Q)$$

is harmonic within our sphere of radius $\lambda^{n-3/2}$ and exterior to C_n It may be verified at once that it vanishes on the sphere. It certainly does not exceed 1 on C_n. Hence

$$F_n(P) \geq \iiint G_{\lambda^{n-3/2}}(P, Q) dM(Q).$$

This can be generalized at once to the case where C_n contains irregular points, by taking a set of regular points C_n' surrounding C_n, assigning boundary values 1 on C_n' and 0 on the sphere, and letting C_n' shrink to C_n.

On a sphere of radius $\lambda^{n+½}$ with O as center we shall have

$$F_n(P) \geq \frac{(1-\lambda)^2}{8} \iiint \overline{PQ}^{-1} dM(Q)$$
$$\geq \frac{(1-\lambda)^2}{8} \iiint \tfrac{1}{2} \lambda^{1-n} dM(Q).$$
$$= \frac{(1-\lambda)^2}{16\lambda} \frac{\gamma_n}{\lambda^n}.$$

Now let us consider the potential on a sphere of radius $\lambda^{n+5/2}$ corresponding to a potential of 0 on a sphere of radius $\lambda^{n-3/2}$ about O and a potential 1 on C_n and C_{n+2}. This will be a potential corresponding to a potential at least

$$\frac{(1-\lambda)^2}{16\lambda} \frac{\gamma_n}{\lambda^n}$$

on a sphere of radius $\lambda^{n+½}$ about O and 1 on C_{n+2}. Hence it will be at least the sum of

$$\frac{(1-\lambda)^2}{16\lambda} \frac{\gamma_n}{\lambda^n}$$

and the potential due to a boundary potential of 0 on a sphere of radius $\lambda^{n+\frac{1}{2}}$ about O and

$$1-\frac{(1-\lambda)^2}{16\lambda}\frac{\gamma_n}{\lambda^n}$$

on C_{n+2}. In other words, it will at least equal

$$\frac{(1-\lambda)^2}{16\lambda}\frac{\gamma_n}{\lambda^n}+\frac{(1-\lambda)^2}{16\lambda}\frac{\gamma_{n+2}}{\lambda^{n+2}}\left[1-\frac{(1-\lambda)^2}{16\lambda}\frac{\gamma_n}{\lambda^n}\right]$$

$$=1-\left[1-\frac{(1-\lambda)^2}{16\lambda}\frac{\gamma_n}{\lambda^n}\right]\left[1-\frac{(1-\lambda)^2}{16\lambda}\frac{\gamma_{n+2}}{\lambda^{n+2}}\right]$$

on C_{n+4}. In a similar manner, the potential within some sphere about O corresponding to a potential 1 on C_n, C_{n+2}, ... , C_{n+2m} is at least

$$1-\left[1-\frac{(1-\lambda)^2}{16\lambda}\frac{\gamma_n}{\lambda^n}\right]\left[1-\frac{(1-\lambda)^2}{16\lambda}\frac{\gamma_{n+2}}{\lambda^{n+2}}\right]\cdots\left[1-\frac{(1-\lambda)^2}{16\lambda}\frac{\gamma_{n+2m}}{\lambda^{n+2m}}\right]$$

Thus the potential corresponding to a potential 1 on C_n, C_{n+2}, ... , C_{n+2m}, ... assumes continuously the boundary value 1 at O if

$$1-\prod_{m=0}^{\infty}\left[1-\frac{(1-\lambda)^2}{16\lambda}\frac{\gamma_{n+2m}}{\lambda^{n+2m}}\right]=1$$

By a familiar theorem in analysis, this will be the case when and only when

$$\sum_{m=0}^{\infty}\frac{\gamma_{n+2m}}{\lambda^{n+2m}}$$

diverges.

We are now in a position to proceed to a demonstration of the regularity of O. It may be shown immediately by the alternating process, as in my previous paper, that the harmonic function corresponding to boundary values 1 on C_n, C_{n+2}, . . . and 0 on the part of C at a distance at least λ^{n-2} from O is continuous at O. It is easy to show, however, that this latter function is dominated by the harmonic function corresponding to any continuous boundary conditions on C that are non-negative and are 1 over that part of C at a distance from O not exceeding λ^{n-2}.

It follows that any harmonic function corresponding to such boundary conditions on C is continuous at O. However, any positive boundary conditions continuous at O and there alone attaining their maximum value 1 can be reduced to boundary conditions of this type by a modification not exceeding the arbitrarily small quantity ϵ. Thus the harmonic function corresponding to these new boundary conditions cannot have an oscillation of more than ϵ at O, and so must assume its boundary values continuously. Since any continuous boundary condition may be represented linearly in terms of two boundary conditions of this sort, O is regular. Hence O is regular if either

$$\frac{\gamma_1}{\lambda}+\frac{\gamma_3}{\lambda^3}+\quad\cdots\quad+\frac{\gamma_{2n+1}}{\lambda^{2n+1}}+\quad\cdots$$

or

$$\frac{\gamma_2}{\lambda^2}+\frac{\gamma_4}{\lambda^4}+\quad\cdots\quad+\frac{\gamma_{2n}}{\lambda^{2n}}+\quad\cdots$$

diverges, and consequently if

$$\frac{\gamma_1}{\lambda}+\frac{\gamma_2}{\lambda^2}+\quad\cdots\quad+\frac{\gamma_n}{\lambda^n}+\quad\cdots$$

diverges.

We now proceed to the converse of this theorem. For this we need a lemma, to the effect that if O is a regular point of the boundary of the domain D, then given any positive numbers η and R, there is a positive number r less than R such that the potential at O corresponding to a boundary potential 0 at infinity and 1 on the set of points exterior to D and lying at a distance from O between r and R inclusive exceeds $1-\eta$. In the first place, the harmonic function $F(P)$ corresponding to boundary values 0 at infinity and 1 on the part of C, the boundary of D, at a distance from O not exceeding R assumes its boundary value tinuously at O, as may be seen by a comparison of this harmonic function with one corresponding to boundary values continuous on C, never exceeding 1, 1 at O, and 0 on the part of C at a distance at least R from O. Hence there is a sphere about O such that in the part of this sphere within D, $F(P)$ exceeds $1-\eta/2$. Furthermore, there is a smaller sphere about O such that the potential on the larger sphere corresponding to a potential

0 at infinity and 1 on the smaller sphere nowhere exceeds $\eta/2$. Let the radius of the smaller sphere be r. Then the potential on the larger sphere corresponding to a boundary potential of 0 at infinity and 1 on the set of points exterior to D at a distance from O between r and R, inclusive, exceeds $1-\eta$. From this fact, and the fact that it corresponds to boundary values 1 over the set of points within the larger sphere, outside the sphere of radius r, and exterior to D, it may readily be concluded that it exceeds $1-\eta$ at O.

Now suppose that

$$\frac{\gamma_1}{\lambda}+\frac{\gamma_2}{\lambda^2}+\ \ldots\ +\frac{\gamma_n}{\lambda^n}+\ \ldots$$

converges. Then given any number ϵ, it is possible to find an n such that

$$\frac{\gamma_n}{\lambda^n}+\frac{\gamma_{n+1}}{\lambda^{n+1}}+\ \ldots\ +\frac{\gamma_{n+p}}{\lambda^{n+p}}<\epsilon$$

whatever p may be. The potential at O due to a potential 1 on C_m cannot exceed γ_m/λ^m, since no part of the charge on C_m corresponding to this potential is nearer to O than λ^m. Hence the potential at O corresponding to a potential 1 on $C_n, \ldots$ C_{n+p} cannot exceed

$$\frac{\gamma_n}{\lambda^n}+\ \ldots\ +\frac{\gamma_{n+p}}{\lambda^{n+p}}<\epsilon$$

Since however ϵ is arbitrarily small, the lemma of the last paragraph shows that O is irregular.

We thus have obtained a necessary and sufficient condition for the regularity of O that explicitly involves the boundary. From this all the more special conditions for regularity may be deduced at once. I shall not, however, discuss them all in detail, but shall proceed at once to the theory of monotone cuspidal points of revolution. We shall say that O is a monotone cuspidal point of revolution of the boundary of D if this boundary in the neighborhood of O is a reëntrant surface with equation

$$\phi=f(\rho)$$

in some scheme of spherical coördinates, where $f(\rho)$ is a monotone increasing function of ρ never exceeding π and vanishing for $\rho=0$. Let us write ϕ_m for the value of ϕ corresponding to $\rho=\lambda^m$, and let us define C_m and γ_m as before. Then C_m will be entirely included in a right circular cylinder with base of radius $\lambda^{m-1}\sin\phi_{m-1}$ and altitude λ^{m-1}, and hence in a prolate spheroid with semi-axes $2\lambda^{m-1}\sin\phi_{m-1}$, $2\lambda^{m-1}\sin\phi_{m-1}$, and $2\lambda^{m-1}$. On the other hand, C will ultimately contain a right circular cylinder with base of radius $\lambda^m \sin\phi_m$ and altitude $\frac{1-\lambda}{2}\lambda^{m-1}$, and hence a prolate spheroid of semi-axes $\frac{\lambda(1-\lambda)}{4}\lambda^{m-1}\sin\phi_m$, $\frac{\lambda(1-\lambda)}{4}\lambda^{m-1}\sin\phi_m$, and $\frac{\lambda(1-\lambda)}{4}\lambda^{m-1}$. The capacities of these spheroids are respectively[11]

$$\frac{4\lambda^{m-1}\cos\phi_{m-1}}{\log\cot\phi_{m-1}/2}$$

and

$$\frac{\lambda^m(1-\lambda)\cos\phi_m}{2\log\cot\phi_m/2}.$$

Hence

$$\frac{4\cos\phi_{m-1}}{\lambda\log\cot\phi_{m-1}/2}\geq\frac{\gamma_m}{\lambda^m}\geq\frac{(1-\lambda)\cos\phi_m}{2\log\cot\phi_m/2}.$$

It then follows from our general theorem that O *is regular or irregular according as*

$$\sum_1^\infty\frac{\cos\phi_m}{\log\cot\phi_m/2}$$

diverges or converges. This series, however, converges or diverges with

$$\sum_1^\infty\frac{-1}{\log\phi_m}$$

since the ratio between corresponding terms tends to the definite limit 1. In particular, if $f(\rho)>A\rho^n (A>0)$, this series dominates a harmonic progression and diverges. On the other hand, if

[11] Jeans, *Electricity and Magnetism*, p. 248.

$f(\rho) < Ae^{-\frac{n}{\rho}}$, this series is dominated by a geometric progression with ratio between 0 and 1, and converges. From these facts, Lebesgue's results as to cuspidal points with finite and exponential order of contact follow at once.

We can generalize this result considerably with the aid of a few lemmas. To begin with, if we replace the boundary in the neighborhood of the cuspidal point which we have just discussed by a certain sequence of zones of one base, the regularity of O will remain unchanged. These zones are chosen in the following manner: the first consists in those points exterior to D at a distance ρ_1 from O. The second consists in those points exterior to D at a distance $\rho_1 - f(\rho_1)$ from O. The third consists in those points at a distance $\rho_1 - f(\rho_1) - f(\rho_1 - f(\rho_1))$ from O, and so on. Let the set of all points exterior to D and at a distance between λ^m and λ^{m-1}, inclusive, from O be C_m as before. Let the set of all points on the zones just mentioned between a distance of λ^m and λ^{m-1} from O, inclusive be C_m'. It will be noted that to each point of C_m, there can be assigned a zone of C_m' or C_{m-1}' bounded by a circle of radius say α, such that the point in question is remote from the furthest point of the zone by less than 3α. The capacity of the zone will exceed or equal $2\alpha/\pi$. Hence to a potential of 1 on C_m' and C_{m-1}' there will correspond a potential at least $2/3\pi$ on C_m. It may be concluded from this at once by a comparison of the potentials at distant points corresponding respectively to a unit potential on C_m and to a unit potential on C_m' and C_{m-1}' that the capacity of C_m is not greater than $3\pi/2$ times the sum of the capacities of C_m' and C_{m-1}'. From this the regularity of O with respect to the boundary consisting of the zones just defined follows at once.

The next lemma reads as follows: Let $S_1, S_2, \ldots$ be a set of sets of points with capacities $\sigma_1, \sigma_2, \ldots$. Let $T_1, T_2, \ldots$ be a set of sets of points with capacities $\tau_1, \tau_2, \ldots$. Let there be positive numbers a and b such that for every n, $\sigma_n \geq a\,\tau_n$, while for every m and n, the lower bound of the distance between a point of S_m and a point of S_n exceeds b times the upper bound of the distance of a point of T_m from a point of T_n. Then the capacity of the set of points

$$S_1 + S_2 + \ldots$$

is at least as large as the capacity of the set of points

$$T_1+T_2+ \ . \ . \ .$$

multiplied by

$$\frac{ab}{a+b}.$$

The proof of this is simple. Suppose $T_1+T_2+ \ . \ . \ .$ brought to potential 1 by a distribution of charge, and let the charge then found on T_m be ω_m. Transfer this charge to S_m, and it will raise the latter set of points to a potential not exceeding ω_m/a. The remainder of the charge is similarly transferred, and cannot raise any point of S_m to a potential of more than ω_m/b. We thus have a positive distribution of charge over $S_1+S_2+ \ . \ . \ .$ which nowhere on the boundary of this set gives rise to a potential greater than $\Sigma\omega_m(1/a+1/b)$, while the total amount of charge is $\Sigma\omega_m$. A comparison with the charge necessary to bring the whole of $S_1+S_2+ \ . \ . \ .$ to a potential $1/a+1/b$ completes the proof of the theorem.

Combining this theorem with the theorem we have just proved concerning a boundary consisting of spherical caps, and remembering that of all sets of points on the surface of a sphere which have a given area, the spherical cap has the least capacity, we obtain the following theorem: *Let O be a point of the boundary of a region D. Let O' be a monotone cuspidal point of revolution on the boundary of a region D'. Let O' be regular. Let us construct spheres about O and O' respectively as centers with radius r. Let the areas of the parts of D and D' respectively on the surface of these spheres be A and A'. Then if for all sufficiently small values of r we have $A<A'$, O is regular.*

This is the theorem generalizing Lebesgue's result concerning cuspidal points as he generalized Zaremba's result concerning conical points. There is another generalization of the Lebesgue result which resembles rather that of Bouligand concerning flat conical points. Let us return to our sequence of zones of one base terminating in a regular point O. It will be noted that the solid angle subtended by one of these zones will approach 0 as we tend to O, and that consequently these zones will tend in shape to flat discs. Their capacity will hence tend to the capacity of a flat disc bounded by the same circle, and the regularity of O will not be affected if we replace each zone by the corresponding flat disc. If we replace each disc by a disc in the same plane and with the same center, but with a third the radius, we shall have reduced the capacity of each disc to just a third its original value. If we now rotate each disc about its center in such a manner as to bring them all into a single plane passing through O, the least distance between two points of two adjacent discs will be at least a ninth of the greatest distance between two points of the corresponding zones in the original figure. Hence O will still be regular.

It is manifestly true, as Lebesgue has pointed out, that if two boundaries have a point in common, and one boundary is in the neighborhood of this point entirely exterior to another, then if the point is regular for the exterior boundary, it is also regular for the interior boundary. This indeed follows directly from our fundamental theorem. Thus by a comparison on the one hand with the set of discs we have just discussed, and on the other with a monotone cuspidal point of revolution, we get the following criterion for the regularity of a cuspidal point on a lamina: *Let the boundary of the region D in the neighborhood of a point O consist in a plane lamina containing all the points such that*

$$\theta = f(\rho)$$

and no others, in a scheme of polar coördinates in the plain of the lamina with O as origin. Let $f(\rho)$ be a monotone increasing function of ρ never exceeding π and vanishing for $\rho = 0$. Then O is regular or irregular according as

$$\sum_1^\infty \frac{1}{\log f(\lambda^m)}$$

diverges or converges.

We may indeed obtain a still more general theorem concerning flat cuspidal points. To begin with, a unit charge uniformly distributed over any region with a plane projection of area A with respect to its projection can never produce a greater potential at any point than if the region is a circle of area A, and the point where the potential is taken is at the center of the circle. The potential at the center of the circle will then be $2\sqrt{\frac{\pi}{A}}$. This means that the capacity of the region with projection of area A is at least $\frac{1}{2}\sqrt{\frac{A}{\pi}}$, since a charge of this amount will correspond to a potential nowhere exceeding 1. Now the capacity of a circular disc of area A is $\frac{2}{\pi}\sqrt{\frac{A}{\pi}}$. Thus the capacity of any region of projection A is at least $\pi/4$ times as great as the capacity of a circular disc of the same area. Combining this fact with the lemma we have just proved concerning a sequence of discs, we get the following theorem: *Let O be a point of the boundary of a region D. Let O' be a monotone cuspidal point of revolution on the boundary of a region D'. Let O' be regular. Let C be the part of the boundary of D lying at a distance of from r to r', inclusive, from O, and let C' be the part of the boundary of D' similarly situated with reference to O'. Then if for all sufficiently small values of r and r', some projection of C exceeds C' in area, O is regular.*

So much for the three-dimensional case: the theory for the n-dimensional case is closely similar. For $n>3$, the fundamental theorem reads as follows: *Let O be a point of C, the boundary of an open set of points D, in a space of n dimensions. Let λ be any positive quantity less than 1. Let γ_m be the capacity of the set of all points Q not belonging to D, such that*

$$\lambda^m \leq \overline{OQ} \leq \lambda^{m-1}.$$

Then O is regular or irregular according as

$$\frac{\gamma_1}{\lambda^{n-2}} + \frac{\gamma_2}{\lambda^{2(n-2)}} + \cdots + \frac{\gamma_m}{\lambda^{m(n-2)}} + \cdots$$

diverges or converges.

Here the proof differs in no essential point from that in the three-dimensional case. In the two-dimensional case, a slight complication is introduced by the fact that the potential due to an isolated charge has a logarithmic singularity at infinity instead of vanishing there. This necessitates a recasting of the definition of capacity. The fundamental existence theorem in my previous paper is stated in an incorrect fashion, although the subsequent work of the paper is perfectly correct, and follows immediately from the correct theorem. To begin with, let R be a bounded set of points in the plane, and let it contain a point O in its interior. Form the harmonic function $f(P)$ corresponding to the boundary values $\log \overline{OP}$ for points P on R and finite at infinity. Let this function assume the value $-a$ at infinity. Then we shall say that the function

$$\frac{f(P)-\log \overline{OP}+a}{a}$$

corresponds to boundary values 1 on R and behaves like $\log \overline{OP}$ at infinity. This function will always exist if R contains points properly in its interior and is of sufficiently small linear dimensions. The condition that R should contain points properly in its interior is inessential, if we follow out the spirit of my last paper by regarding the harmonic function corresponding to boundary values 1 on R and behaving like $\log \overline{OP}$ at infinity as the lower bound (which will always exist if R is small enough) of the harmonic functions corresponding to boundary values 1 on contours near to R and containing R and behaving like $\log \overline{OP}$ at infinity. The harmonic function thus obtained may readily be proved unique, and independent of O.

The proper form of the existence theorem for capacity in the two-dimensional case is the legitimate conclusion of the argument used in deriving the erroneous theorem of my previous paper, and reads as follows: *Let R be any bounded set of points in the plane, entirely within a rectangle with opposite corners* (a_1, a_2) *and* (b_1, b_2), *and let*

$$|a_1-b_1|<1, \quad |a_2-b_2|<1.$$

Let $u(P)$ behave like $\log \overline{OP}$ *at infinity, and correspond to boundary values* 1 *over R. Then there will be a function $M(P)$ representing the charge on that portion of R consisting of points with both coördinates less than those of P. $M(P)$ will be an increasing function of the coördinates of P. Over every point Q exterior to R we shall have*

$$u(Q)=\int_{a_1}^{b_1}\int_{a_2}^{b_2}(-\log \overline{PQ})dM(P)$$

Moreover, if the rectangle with opposite vertices (x_1, x_2) *and* (y_1, y_2) *is entirely exterior to* R, *then*

$$\int_{x_1}^{y_1}\int_{x_2}^{y_2} dM(P) = 0,$$

We shall term

$$\int_{a_1}^{b_1}\int_{a_2}^{b_2} dM(P)$$

the capacity of R.

With this definition, our cardinal theorem is the following: *Let* O *be a point of* C, *the boundary of an open set of points* D *in the plane. Let* λ *be any positive quantity less than* 1. *Let* γ_m *be the capacity of the set of all points* Q *not belonging to* D, *such that*

$$\lambda^{2^m} \leq \overline{OQ} \leq \lambda^{2^{m-1}}.$$

Then O *is regular or irregular according as*

$$\gamma_1 + 2\gamma_2 + \ . \ . \ . \ + 2^m\gamma_m + \ . \ . \ .$$

diverges or converges.

The proof follows that in the three-dimensional case step by step. It is of course necessary to take account of the fact that the logarithm assumes positive as well as negative values, but any difficulty from this source may be avoided by confining our attention to the interior of a circle of radius less than 1/2 about O. The fact that the bounds of $\overline{OQ}$ for points Q on γ_m are λ^{2^m} and $\lambda^{2^{m-1}}$ is accounted for by the fact that the potential at O due to a unit charge at Q will then lie between the numbers $2^m \log \frac{1}{\lambda}$ and $2^{m-1} \log \frac{1}{\lambda}$, which bear to one another a fixed ratio. There is one stage of the proof which perhaps needs a little comment: that in which we determine the potential on a given circle about the origin corresponding to a potential 0 on a circle of larger radius and 1 on γ_m. We shall take the radius of the outer circle as $\lambda^{2^{m-3/2}}$, and that of the smaller circle as $\lambda^{2^{m+1/2}}$. We desire a theorem correlating the potential due to a potential 1 on γ_m and 0 at infinity with the potential in question. As before, we do this by establishing certain inequalities connecting the Green's function of the outer circle with $-\log \overline{PQ}$, the Green's function of the entire plane.

If we write c for the radius of the outer circle, its Green's function will be

$$G(P,Q)=\tfrac{1}{2}\log\left\{\frac{c^2+\dfrac{\overline{OP}^2\,\overline{OQ}^2}{c^2}-2\overline{OP}\,\overline{OQ}\cos(\overline{OP},\overline{OQ})}{\overline{OP}^2+\overline{OQ}^2-2\overline{OP}\,\overline{OQ}\cos(\overline{OP},\overline{OQ})}\right\}$$

$$=\tfrac{1}{2}\log\left\{\frac{1+\left(c-\dfrac{\overline{OP}^2}{c}\right)\left(c-\dfrac{\overline{OQ}^2}{c}\right)}{\overline{PQ}^2}\right\}.$$

We shall always have $G(P,Q)\leq-\log\overline{PQ}$. If $c=\lambda^{2^{m-3/2}}$, while P and Q lie within a circle about O of radius $\lambda^{2^{m-1}}$, we shall have

$$G(P,Q)\geq\tfrac{1}{2}\log\left\{[1-\lambda^{(2-\sqrt{2})2^{m-1}}]^2\,\frac{\lambda^{2^{m-1/2}}}{\overline{PQ}^2}\right\}$$

$$\geq\tfrac{1}{2}\log\frac{\lambda^{2^{m-1/2}}}{4\overline{PQ}^2}$$

$$\geq(\log\lambda)2^{m-3/2}-\log 4-\log\overline{PQ}$$

for all sufficiently large values of m. Furthermore,

$$-\log\overline{PQ}\geq-\log 2-\log\lambda^{2^{m-1}}$$
$$=-\log 2-\log\lambda^{[2^{m-3/2}]\sqrt{2}}$$
$$=-\log 2-\sqrt{2}\,\log\lambda^{2^{m-3/2}}$$

Hence for all sufficiently large values of m,

$$(-\log\lambda)2^{m-3/2}+\log 4\leq\frac{-1}{1.4}\log\overline{PQ}$$

It follows that for all sufficiently large values of m, we shall have

$$G(P,Q)\geq-.2\log\overline{PQ}.$$

With this we are in a position to proceed with our two-dimensional theorem in a fashion exactly paralleling that employed in the demonstration of our theorem in three dimensions.

The classical sufficient condition for the solubility of the plane Dirichlet problem is that of Lebesgue[12]. It may be stated as follows: *A point O of the boundary of a region D is regular unless it is possible to construct a circle of arbitrarily small radius about O*

[12] *Rendiconti di Palermo*, loc. cit.

which will lie entirely interior to D. This is completely contained in two conditions, given by Phillips and the author, and the author alone, respectively, which are entirely parallel to the three-dimensional conditions given earlier in this paper. They may readily be deduced from the general condition here given, the series whose divergence is to be investigated reducing to the form

$$A+A+ \ldots +A+ \ldots$$

A plane problem of particular interest has been discussed by Kellogg in the memoir already cited. He discusses the boundary consisting of all points on the segment (0, 1) which have a representation in ternary fractions not containing the digit 2, and shows that every point of it is regular. If we wish to bring this under our general theorem, we must first show that this set has a non-zero capacity, or in other words that it is possible to distribute over it a positive charge without producing anywhere an infinite potential. This we do as follows: like Kellogg, we form the function $f(x)$ which has the value 0 for $x=0$, 1 for $x=1$, 1/2 for $1/3 \leq x \leq 2/3$, 1/4 for $1/9 \leq x \leq 2/9$, 3/4 for $7/9 \leq x \leq 8/9$, and so on. Then the function

$$-\int_0^1 \log x df(x)$$

is bounded, for it never exceeds

$$\log 6+1/2 \log 18+1/4 \log 54+ \ldots$$
$$=\log 2(1+1/2+1/4+\ldots)+\log 3\left(1+1+\frac{3}{4}+\frac{4}{8}+\frac{5}{16}+\ldots\right)$$
$$\leq 2 \log 2 + 5 \log 3.$$

Hence *a fortiori* the function

$$-\int_0^1 \log \overline{OP}\, df(O)$$

is bounded.

If the capacity of Kellogg's set of points is at least C, the capacity of each non-null third is at least $\dfrac{C}{1+C \log 3}$, of each non-null ninth $\dfrac{C}{1+C \log 9}$, and so on indefinitely. Moreover if O is a point of Kellogg's set, γ_m will contain at least one non-null

$1/3^{2+k}$th for all sufficiently large values of m, where $1/3^{k-1} \geq \lambda^{2^{m-1}} \geq 1/3^k$. Hence the test-series dominates

$$+\ldots+\frac{2^m}{1-2^{m-1}C\log\lambda+\log 27}+\frac{2^{m+1}}{1-2^m C\log\lambda+\log 27}+\ldots$$

which diverges. Thus Kellogg's case comes under our general theory, even though it can be treated somewhat more elegantly by Kellogg's direct methods[13].

[13] Since this paper has gone to the printer, the author has become more closely acquainted with the very important researches of M. Georges Bouligand. These go back to his Paris thesis of 1914 on the Green and Neumann functions of the cylinder, and culminate in a paper in the *Comptes rendus* for March 24, 1924, yielding conditions of regularity comparable in generality to those here given. In the latter paper are contained very many references to his previous work, which has appeared for the most part in the *Comptes rendus*. His treatment depends upon the discussion of infinite regions, and the use of inversions and Kelvin transformations.

GENERALIZED HARMONIC ANALYSIS.[1]

By

NORBERT WIENER

of Cambridge, Mass., U. S. A.

Reprinted from *Acta Math.*, *55*, **1930**, pp. **117–258**. (Courtesy of the Institut Mittag-Leffler.)

Introduction.

Generalized harmonic analysis represents the culmination and combination of a number of very diverse mathematical movements. The theory of almost periodic functions finds its precursors in the theory of Dirichlet series, and in the quasiperiodic functions of Bohl and Esclangon. These latter, in turn, are an answer to the demands of the theory of orbits in celestial mechanics; the former take their origin in the analytic theory of numbers. Quite independent of the regions of thought just enumerated, we have the order of ideas associated with the names of Lord Rayleigh, of Gouy, and above all, of Sir Arthur Schuster; these writers concerned themselves with the problems of white light, of noise, of coherent and incoherent sources. More particularly, Schuster was able to point out the close analogy between the problems of the harmonic analysis of light and the statistical analysis of hidden periods in such scientific data as are common in meteorology and astronomy, and developed the extremely valuable theory of the periodogram. The work of G. I. Taylor on diffusion represents another valuable anticipation of theories here developed, from the standpoint of an applied mathematician of the British school, with preoccupations much the same as those of Schuster.

The work of Hahn seems to have a much more definitely pure-mathematics motivation. To the pure mathematician in general, however, and the worker in real function theory in particular, we owe, not so much the setting of our problem, as the chief tool in its attack: the famous theorem of Plancherel, the proof of which Titchmarsh has extended and improved.

It may seem a little strange to the reader that the present paper should contain yet another proof of this much proved theorem. In view, however, of the centralness of the Plancherel theorem in all that is to follow, and more expecially of the fact that the proof here given furnishes an excellent introduction to the meaning and motivation of our proofs in more complicated cases, it has seemed worth while to prove the Plancherel theorem in full.

The germs of the generalized harmonic analysis of this paper are already in the work of Schuster, but only the germs. To make the Schuster theory assume a form suitable for extension and generalization, a radical recasting is necessary. This recasting brings out the fact that the expression

$$\varphi(x) = \lim_{T \to \infty} \frac{1}{2T} \int_{-T}^{T} f(x+t)\bar{f}(t)\,dt \tag{0.1}$$

plays a fundamental part in Schuster's theory, as does also

$$S(u) = \frac{1}{2\pi} \int_{-\infty}^{\infty} \varphi(x) \frac{e^{iux} - 1}{ix}\,dx. \tag{0.2}$$

Accordingly, section 3 is devoted to the independent study of these two expressions, and to the definition of $S(u)$ under appropriate assumption as the *spectrum* of $f(x)$.

There are some interesting relations between the total spectral intensity of $f(x)$ as represented by $S(u)$ and the other expressions of the theory. Some of these demand for their proper appreciation a mode of connecting various weighted means of a positive quantity. The appropriate tool for this purpose is the general theory of Tauberian theorems developed by the author and applied to these problems by Mr. S. B. Littauer.

These latter Tauberian theorems enable us to correlate the mean square of the modulus of a function and the »quadratic variation» of a related function which determines its harmonic analysis. The theory of harmonic analysis here indicated has been extended by Bochner to cover the case of very general functions behaving algebraically at infinity. A somewhat similar theory is due to Hahn, who is, however, more interested in questions of ordinary convergence than in those clustering about the Parseval theorem.

The theory of generalized harmonic analysis is itself capable of extension in very varied directions. Mr. A. C. Berry has recently developed a vectorial extension of the theory to n dimensions, while on the other hand, the author himself has extended the theory to cover the simultaneous harmonic analysis of a set of functions and the notions of coherent and incoherent sources of light. A third extension depends on the replacement of the translation group, fundamental in all harmonic analysis, by another group.

To prove that the theory is not vacuous and trivial, it is of importance to give examples of different types of spectra. We do this, both by direct methods, and by methods involving an infinite series of choices between alternatives of equal probability. The latter method, of course, involves the assumption of the

Zermelo axiom: on the other hand, it yields a most interesting probability theory of spectra. This theory may be developed to cover the case where the infinite sequence of choices is replaced by a haphazard motion of the type known as Brownian.

The spectrum theory of the present paper has as one very special application the theory of almost periodic functions. It is not difficult to prove that the spectrum of such a function contains a discrete set of lines and no continuous part, and to deduce from this, Bohr's form of the Parseval theorem. The transition from the Parseval theorem to the Weierstrassian theorem that it is possible to approximate uniformly to any almost periodic function by a sequence of trigonometrical polynomials follows essentially lines laid down by Weyl, though it differs somewhat in detail.

Besides the well-known generalizations of almost periodic functions due to Stepanoff, Besicovitch, Weyl, and the author, there is the little explored field of extensions of almost periodic functions containing a parameter. These have been used by Mr. C. F. Muckenhoupt to prove the closure of the set of the Eigenfunktionen of certain linear vibrating systems. This is one of the few applications of almost periodic functions of a fairly general type to definite mathematicophysical problems. Our last section is devoted to this, and to related matters.

CHAPTER I.

1. Plancherel's theorem.

Plancherel's theorem reads as follows: *Let $f(x)$ be quadratically summable over $(-\infty, \infty)$ in the sense of Lebesgue — that is, let it be measurable, and let*

$$\int_{-\infty}^{\infty} |f(x)|^2 \, dx \tag{1.01}$$

exist and be finite. (i.e. $f \subset L_2$). *Then*

$$g(u) = \underset{A \to \infty}{\text{l.i.m.}} \frac{1}{\sqrt{2\pi}} \int_{-A}^{A} f(x) e^{iux} \, dx \tag{1.02}$$

(*where* l.i.m. *stands for* »*limit in the mean*») *will exist, and*

$$f(x)=\underset{A\to\infty}{\mathrm{l.i.m.}}\frac{1}{\sqrt{2\pi}}\int_{-A}^{A} g(u)e^{iux}\,du. \tag{1.03}$$

$g(u)$ is known as the »Fourier transform» of $f(x)$. To prove this, let us put

$$f_A(x)=\begin{cases} f(x) & \text{if } |x|<A,\\ 0 & \text{if } |x|\geq A.\end{cases} \tag{1.04}$$

Let us represent $f_A(x)$ over $(-2A, 2A)$ by the Fourier series

$$f_A(x)\sim\sum_{-\infty}^{\infty} a_n e^{\frac{in\pi x}{2A}}. \tag{1.05}$$

Then

$$\int_{-\infty}^{\infty} f_A(x+\xi)\bar{f}_A(\xi)\,d\xi=\int_{-2A}^{2A} f_A(x+\xi)\bar{f}_A(\xi)\,d\xi$$

$$=\sum_{-\infty}^{\infty} a_n\bar{a}_n\int_{-2A}^{2A} e^{\frac{in\pi(x+\xi)}{2A}}e^{-\frac{in\pi\xi}{2A}}\,d\xi$$

$$=\sum_{-\infty}^{\infty} 4A\,|a_n|^2 e^{\frac{in\pi x}{2A}}. \tag{1.06}$$

This series of equations merits several comments. First, the infinite integrals which appear are infinite in appearance only, as the integrand vanishes beyond a certain point. Secondly, the period chosen for the Fourier representation of $f_A(x)$ is twice the length of the interval over which $f_A(x)$ may differ from 0, so that one period of $f_A(x+\xi)$ may overlap not more than one corresponding period of $\bar{f}_A(\xi)$. Third, the function $\int_{-\infty}^{\infty} f_A(x+\xi)\bar{f}_A(\xi)\,d\xi$ has a Fourier development which possesses only positive coefficients, and is absolutely and uniformly convergent, as follows at once from the Hurwitz theorem. The positiveness of the Fourier coefficients of this function forms the point of departure for the greater part of the present paper.

16—29764. *Acta mathematica*. 55. Imprimé le 7 avril 1930.

It follows at once that

$$\int_{-\infty}^{\infty} f_A(x+\xi) f_A(\xi)\, d\xi = \lim_{N\to\infty} \frac{1}{4A} \int_{-2A}^{2A} \frac{\sin \frac{2N+1}{4A}\pi(x-y)}{\sin \frac{\pi(x-y)}{4A}}\, dy \int_{-\infty}^{\infty} f_A(y+\xi) \bar{f}_A(\xi)\, d\xi. \quad (1.07)$$

However, Lebesgue's fundamental theorem on the Fourier coefficients, to the effect that they always tend to zero, yields us

$$\lim_{N\to\infty} \frac{1}{4A} \int_{-2A}^{2A} \sin \frac{2N+1}{4A}\pi(x-y) \left[\frac{1}{\frac{\pi(x-y)}{4A}} - \frac{1}{\sin \frac{\pi(x-y)}{4A}} \right] \cdot$$

$$\cdot\, dy \int_{-\infty}^{\infty} f_A(y+\xi) \bar{f}_A(\xi)\, d\xi = 0. \quad (1.08)$$

Combining these two relations, we see that

$$\int_{-A}^{A} |f(x)|^2\, dx = \int_{-\infty}^{\infty} |f_A(x)|^2\, dx$$

$$= \lim_{N\to\infty} \frac{1}{\pi} \int_{-2A}^{2A} \frac{\sin \frac{2N+1}{4A}\pi y}{y}\, dy \int_{-\infty}^{\infty} f_A(y+\xi) \bar{f}_A(\xi)\, d\xi$$

$$= \lim_{N\to\infty} \frac{1}{\pi} \int_{-\infty}^{\infty} dy \int_{0}^{\frac{2N+1}{4A}\pi} \cos uy\, du \int_{-\infty}^{\infty} f_A(y+\xi) \bar{f}_A(\xi)\, d\xi$$

$$= \frac{1}{2\pi} \int_{-\infty}^{\infty} du \int_{-\infty}^{\infty} f_A(\eta)\, d\eta \int_{-\infty}^{\infty} \bar{f}_A(\xi)\, e^{iu(\eta-\xi)}\, d\xi$$

$$= \frac{1}{2\pi} \int_{-\infty}^{\infty} du \left| \int_{-A}^{A} f(\eta)\, e^{iu\eta}\, d\eta \right|^2. \quad (1.09)$$

The inversions of the order of integration are here justified by the fact that all the infinite limits are merely apparent, and are introduced to simplify the formal work of inversion. If we replace $f_A(x)$ by the function $f_B(x) - f_A(x)$ which has essentially the same properties, we see that

$$\int_{-B}^{B} |f(x)|^2 \, dx - \int_{-A}^{A} |f(x)|^2 \, dx = \frac{1}{2\pi} \int_{-\infty}^{\infty} du \left| \int_{-B}^{B} f(\eta) e^{iu\eta} \, d\eta - \int_{-A}^{A} f(\eta) e^{iu\eta} \, d\eta \right|^2. \quad (1.10)$$

In case $\int_{-\infty}^{\infty} |f(x)|^2 \, dx$ exists,

$$\lim_{B, A \to \infty} \int_{-\infty}^{\infty} du \left| \int_{-B}^{B} f(\eta) e^{iu\eta} \, d\eta - \int_{-A}^{A} f(\eta) e^{iu\eta} \, d\eta \right|^2 = 0 \quad (1.11)$$

and we may use Weyl's lemma to the Riesz-Fischer theorem to prove that

$$g(u) = \underset{A \to \infty}{\text{l.i.m.}} \frac{1}{\sqrt{2\pi}} \int_{-A}^{A} f(\eta) e^{iu\eta} \, d\eta \quad (1.12)$$

exists, and is »quadratically summable». Combining this definition of $g(u)$ with (1.09), we see that

$$\int_{-\infty}^{\infty} |g(u)|^2 \, du = \int_{-\infty}^{\infty} |f(x)|^2 \, dx. \quad (1.13)$$

That is, the integral of the square of the modulus of a function is invariant under a Fourier transformation.

To complete the proof of Plancherel's theorem, it is merely necessary to show that for functions $f(x)$ of some closed set,

$$f(x) = \underset{A \to \infty}{\text{l.i.m.}} \frac{1}{\sqrt{2\pi}} \int_{-A}^{A} g(u) e^{-iux} \, du. \quad (1.14)$$

A particular choice of $f(x)$ is the following:

$$f(x) = \begin{cases} 0; & [x < \alpha] \\ 1; & [\alpha < x < \beta] \\ 0; & [\beta < x]. \end{cases} \quad (1.15)$$

Here

$$g(u)=\frac{1}{\sqrt{2\pi}}\int_{\alpha}^{\beta} e^{iux}\,dx=\frac{e^{iu\beta}-e^{iu\alpha}}{iu\sqrt{2\pi}}. \tag{1.16}$$

Hence

$$\begin{aligned}\underset{A\to\infty}{\text{l.i.m.}}\frac{1}{\sqrt{2\pi}}\int_{-A}^{A} g(u)e^{-iux}\,du&=\underset{A\to\infty}{\text{l.i.m.}}\frac{1}{2\pi}\int_{-A}^{A}\frac{e^{iu(\beta-x)}-e^{iu(\alpha-x)}}{iu}\,du\\&=\underset{A\to\infty}{\text{l.i.m.}}\frac{1}{2\pi}\int_{-A}^{A}\frac{\sin u(\beta-x)-\sin u(\alpha-x)}{u}\,du\\&=\frac{1}{2}[\operatorname{sgn}(\beta-x)-\operatorname{sgn}(\alpha-x)]\\&=f(x)\end{aligned} \tag{1.17}$$

except possibly at the two points α and β, a set of zero measure. This completes the proof of Plancherel's theorem.

Plancherel states this theorem somewhat differently. He essentially defines $g(u)$ as

$$g(u)=\frac{d}{du}\frac{1}{\sqrt{2\pi}}\int_{-\infty}^{\infty} d\eta\int_{0}^{u} f(\eta)\,e^{iu\eta}\,dv. \tag{1.18}$$

If we retain our definition, it follows from an elementary use of the Schwarz inequality that

$$\int_{0}^{u} g(v)\,dv=\frac{1}{\sqrt{2\pi}}\int_{-\infty}^{\infty} d\eta\int_{0}^{u} f(\eta)\,e^{iv\eta}\,dv. \tag{1.19}$$

To see this, les us reflect that

$$\begin{aligned}\left|\int_{0}^{u} g(v)\,dv-\frac{1}{\sqrt{2\pi}}\int_{-A}^{A} d\eta\int_{0}^{u} f(\eta)e^{iv\eta}\,dv\right|&=\lim_{B\to\infty}\left|\frac{1}{\sqrt{2\pi}}\int_{0}^{u} dv\left[\int_{A}^{B}+\int_{-B}^{-A}\right]f(\eta)\,e^{iv\eta}\,d\eta\right|\\&\le\lim_{B\to\infty}\left\{\frac{1}{2\pi}\int_{0}^{u} dv\int_{0}^{u}\left|\left[\int_{A}^{B}+\int_{-B}^{-A}\right]f(\eta)\,e^{iv\eta}\,d\eta\right|^{2}dv\right\}^{1/2}\end{aligned}$$

$$\leq u \lim_{B\to\infty} \left\{ \frac{1}{2\pi} \int_{-\infty}^{\infty} \left| \left[\int_{A}^{B} + \int_{-B}^{-A} \right] f(\eta)\, e^{iu\eta}\, d\eta \right|^2 dv \right\}^{1/2}$$

$$= u \left\{ \left[\int_{A}^{\infty} + \int_{-\infty}^{-A} \right] |f(\eta)|^2\, d\eta \right\}^{1/2},$$

and since $\int_{-\infty}^{\infty} |f(\eta)|^2\, d\eta$ is finite, it follows that

$$\lim_{A\to\infty} \left| \int_0^u g(v)\, dv - \frac{1}{\sqrt{2\pi}} \int_{-A}^{A} d\eta \int_0^u f(\eta)\, e^{iv\eta}\, dv \right| = 0.$$

From this (1. 19) follows at once. Since a summable function is almost everywhere the derivative of its integral, the transition to Plancherel's form of the definition is immediate.

It follows at once from Plancherel's theorem that if $f_1(x)$ and $f_2(x)$ are quadratically summable,

$$F_1(u) = \underset{A\to\infty}{\text{l.i.m.}}\, \frac{1}{\sqrt{2\pi}} \int_{-A}^{A} f_1(x)\, e^{iux}\, dx \tag{1. 20}$$

and

$$F_2(u) = \underset{A\to\infty}{\text{l.i.m.}}\, \frac{1}{\sqrt{2\pi}} \int_{-A}^{A} f_2(x)\, e^{iux}\, dx \tag{1. 21}$$

exist, and that

$$\int_{-\infty}^{\infty} |F_1(u) \pm F_2(u)|^2\, du = \int_{-\infty}^{\infty} |f_1(x) \pm f_2(x)|^2\, dx \tag{1. 22}$$

and

$$\int_{-\infty}^{\infty} |F_1(u) \pm iF_2(u)|^2\, du = \int_{-\infty}^{\infty} |f_1(x) \pm if_2(x)|^2\, dx. \tag{1. 23}$$

Combining the last four formulae with one another, we have

$$\int_{-\infty}^{\infty} F_1(u)\,\overline{F}_2(u)\,du = \int_{-\infty}^{\infty} f_1(x)\bar{f}_2(x)\,dx. \qquad (1.\ 24)$$

This we may know as the Parseval theorem for the Fourier integral. Since

$$\overline{F}_2(-u) = \underset{A\to\infty}{\text{l.i.m.}}\ \frac{1}{\sqrt{2\pi}} \int_{-A}^{A} \bar{f}_2(x)\,e^{iux}\,dx \qquad (1.\ 25)$$

we may deduce at once that

$$\int_{-\infty}^{\infty} F_1(u)\,F_2(-u)\,du = \int_{-\infty}^{\infty} f_1(x) f_2(x)\,dx. \qquad (1.\ 26)$$

Since furthermore

$$\overline{F}_2(v-u) = \underset{A\to\infty}{\text{l.i.m.}}\ \frac{1}{\sqrt{2\pi}} \int_{-A}^{B} \bar{f}_2(x)\,e^{ivx}\,e^{iux}\,dx \qquad (1.\ 27)$$

it follows that

$$\int_{-\infty}^{\infty} F_1(u)\,F_2(v-u)\,du = \int_{-\infty}^{\infty} f_1(x) f_2(x)\,e^{ivx}\,dx. \qquad (1.\ 28)$$

As a consequence, if $f_1(x) f_2(x)$ is quadratically summable, its Fourier transform is

$$\sqrt{2\pi}\int_{-\infty}^{\infty} F_1(u)\,F_2(v-u)\,du. \qquad (1.\ 29)$$

This theorem lies at the basis of the whole operational calculus.

2. Schuster's periodogram analysis.

The two theories of harmonic analysis embodied in the classical Fourier series development and the theory of Plancherel do not exhaust the possibilities of harmonic analysis. The Fourier series is restricted to the very special class of periodic functions, while the Plancherel theory is restricted to functions which are quadratically summable, and hence tend on the average to zero as their argument tends to infinity. Neither is adequate for the treatment of a ray of white light which is supposed to endure for an indefinite time. Nevertheless, the physicists who first were faced with the problem of analyzing white light into its components had to employ one or the other of these tools. Gouy accordingly represented white light by a Fourier series, the period of which he allowed to grow without limit, and by focussing his attention on the average values of the energies concerned, he was able to arrive at results in agreement with the experiments. Lord Rayleigh on the other hand, achieved much the same purpose by using the Fourier integral, and what we now should call Plancherel's theorem. In both cases one is astonished by the skill with which the authors use clumsy and unsuitable tools to obtain the right results, and one is led to admire the unfailing heuristic insight of the true physicist.

The net outcome of the work of these writers was to dispel the idea that white light consist in some physical, supermathematical way of homogeneous monochromatic vibrations. Schuster in particular, was led to the conclusion that when white light is analyzed by a grating, the monochromatic components are created by the grating rather than selected by it. Thus a great stimulus was given to the investigation of the sense in which any phenomenon may be said to contain hidden periodic components. The successful completion of this investigation is also due to Schuster.

Schuster sums up his conclusions as follows[2]: »Let y be a function of t, such that its values are regulated by some law of probability, not necessarily the exponential one, but acting in such a manner that if a large number of t be chosen at random, there will always be a definite fraction of that number depending on t_1 only, which lie between t_1 and t_1+T, where T is any given time interval.

»Writing

$$A=\int_{t_1}^{t_1+T} y \cos \varkappa t \, dt \quad \text{and} \quad B=\int_{t_1}^{t_1+T} y \sin \varkappa t \, dt,$$

and forming

$$R=\sqrt{A^2+B^2},$$

the quantity R will, with increasing values of T, fluctuate about some mean value, which increases proportionally to $\sqrt{T}$ provided T is taken sufficiently large.

»If this theorem is taken in conjunction with the two following well-known propositions,

(1) If $y=\cos \varkappa t$, R will, apart from periodical terms, increase proportionally to T;

(2) If $y=\cos \lambda t$, λ being different from $\varkappa$, the quantity R will fluctuate about a constant value;

it is seen that we have means at our disposal to separate any true periodicity of a variable from among its irregular changes, provided we can extend the time limits sufficiently. . . . The application of the theory of probability to the investigation of what may be called »hidden» periodicities . . . may be further extended . . .»

While Schuster's statement is perhaps not in all respects clear, it contains the germs of all subsequent generalizations of harmonic analysis. First among these is the emphasis on the notion of the *mean*. The operator which yields

$$A_1 = \lim_{T \to \infty} \frac{1}{T} \int_{t_1}^{t_1+T} y \cos \varkappa t \, dt \text{ or } B_1 = \lim_{T \to \infty} \frac{1}{T} \int_{t_1}^{t_1+T} y \sin \varkappa t \, dt \tag{2.01}$$

annihilates all functions $y(t)$ made up in a purely fortuitous or haphazard manner, as well as all trigonometrical functions other than $\cos \varkappa t$ or $\sin \varkappa t$, respectively. Hence we may take A_1 and B_1 to indicate the amounts of $\cos \varkappa t$ or $\sin \varkappa t$ contained in y. As a simultaneous indication of these two quantities, neglecting phase, Schuster takes $\sqrt{A_1^2+B_1^2}$ which he supersedes in his later papers by the somewhat simpler expression $A_1^2+B_1^2$.

It is possible to lend a certain plausibility to this later choice of Schuster as contrasted with his earlier, by considering the expression

$$\varphi(x) = \lim_{T \to \infty} \frac{1}{2T} \int_{-T}^{T} f(x+t) \bar{f}(t) \, dt. \tag{2.02}$$

If

$$f(t) = \sum_1^N a_n e^{i\lambda_n t}, \tag{2.03}$$

we have

$$\varphi(x) = \sum_1^N |a_n|^2 e^{i\lambda_n x}. \tag{2.04}$$

Accordingly,

$$\varrho(\lambda_n) = |a_n|^2 = \lim_{T\to\infty} \frac{1}{2T} \int_{-T}^{T} \varphi(x)\, e^{-i\lambda_n x}\, dx. \tag{2.05}$$

This function $\varphi(t)$ differs from $f(t)$ in that every emplitude of a trigonometric term in $f(t)$ is replaced by the square of its modulus.

The expression $|a_n|^2$ is necessarily positive. It is, however, unobservable in any actual case, as we only have a finite interval of time at our disposal. Let it be noted that if we put

$$\varphi'_T(x) = \frac{1}{2T} \int_{-T}^{T} f(x+t)\bar{f}(t)\, dt \tag{2.06}$$

and

$$\varrho'_T(\lambda_n) = \frac{1}{2T} \int_{-T}^{T} \varphi'_T(x)\, e^{-i\lambda_n x}\, dx, \tag{2.07}$$

it is *not* necessarily true that ϱ'_T is non-negative. On the other hand, if we put

$$f_A(t) = f(t)\ [|t| < A];\ f_A(t) = 0 \text{ otherwise} \tag{2.08}$$

and

$$\varphi_A(x) = \frac{1}{2A} \int_{-\infty}^{\infty} f_A(x+t)\bar{f}_A(t)\, dt \tag{2.09}$$

then

$$\varrho_A(\lambda_n) = \frac{1}{2A} \int_{-\infty}^{\infty} \varphi_A(x)\, e^{-i\lambda_n x}\, dx = \frac{1}{4A^2} \left| \int_{-\infty}^{\infty} f_A(t)\, e^{-i\lambda_n t}\, dt \right|^2 \geq 0. \tag{2.10}$$

This suggests an improved method of treating the approximate periodogram of a function under observation for a finite time.

The periodogram of a function — that is, the graph of the discontinuous function $\varrho(\lambda_n)$ or its approximate continuous analyses $\varrho_A(\lambda_n)$ — contains but a small amount of the information which the complete graph of the original function is able to yield. Not only do we deliberately discard all phase relations, but a large part of the original function — often the most interesting and important part — is thrown away as the aperiodic residue. The chief reason for this that any measure for a continuous spectral density becomes infinite at a

17—29764. *Acta mathematica.* 55. Imprimé le 7 avril 1930.

spectral line, while any measure for the intensity of a spectral line becomes zero over the continuous spectrum.

This is a difficulty, however, which has had to be faced in many other branches of mathematics and physics. Impulses and forces are treated side by side in mechanics, although they have no common unit. We are familiar in potential theory with distributions of charge containing point, line, and surface distributions, as well as continuous volume distributions. The basic theory of all these problems is that of the Stieltjes integral.

Let us put

$$S(u) = \frac{1}{2\pi} \int_{-\infty}^{\infty} \varphi(x) \frac{e^{ixu} - 1}{ix} \, dx. \tag{2. 11}$$

Here the term 1 is introduced to cancel the singularity which we should otherwise find for $x=0$. We have formally and heuristically

$$\begin{aligned}
S(u+0) - S(u-0) &= \lim_{\varepsilon \to 0} \int_{-\infty}^{\infty} \varphi(x) \frac{e^{ix(u+\varepsilon)} - e^{ix(u-\varepsilon)}}{2\pi i x} \, dx \\
&= \lim_{\varepsilon \to 0} \frac{1}{\pi} \int_{-\infty}^{\infty} \varphi(x) e^{iux} \frac{\sin \varepsilon x}{x} \, dx \\
&= \lim_{\eta \to 0} \frac{1}{\pi \eta} \int_{0}^{\eta} d\varepsilon \int_{-\infty}^{\infty} \varphi(x) e^{iux} \frac{\sin \varepsilon x}{x} \, dx \\
&= \lim_{\eta \to 0} \frac{\eta}{\pi} \int_{-\infty}^{\infty} \varphi(x) e^{iux} \frac{1 - \cos \eta x}{\eta^2 x^2} \, dx \\
&= \lim_{T \to \infty} \frac{1}{\pi T} \int_{-\infty}^{\infty} \varphi(x) e^{iux} \frac{T^2 \left(1 - \cos \frac{x}{T}\right)}{x^2} \, dx.
\end{aligned} \tag{2. 12}$$

Now,

$$\frac{T^2 \left(1 - \cos \frac{x}{T}\right)}{x^2} \tag{2. 13}$$

is a positive function assuming the value $1/2$ for $x=0$, with a graph with a scale in the x direction proportional to T, and with a finite integral. Hence, it does not seem amiss to consider $S(u+0)-S(u-0)$ except for a constant factor, as the same expression as $\varrho(u)$. We shall later verify this fact in more detail and with more rigor. On the other hand, again formally,

$$S'(u)=\frac{1}{2\pi}\int_{-\infty}^{\infty}\varphi(x)\,e^{ixu}\,dx. \tag{2. 14}$$

Thus in case $\varphi(x)$ is of too small an order of magnitude to possess a line spectrum, $S(u)$ still has a significance. We shall interpret its derivative as meaning the density of the continuous portion of the spectrum of $f(t)$.

The graph of $S(u)$ shall be called the *integrated periodogram* of $f(t)$. We shall show later that under very general conditions, it may be so chosen as to be a monotone non-decreasing curve. The amount of rise of this curve between the arguments indicates the total intensity of the part of the spectrum lying between the frequencies. This shift of our attention from the periodogram itself to the integrated periodogram, which is monotone but not necessarily everywhere differentiable, is as we have said of the same nature as the shift from $g(x)$ in

$$\int f(x)\,g(x)\,dx \tag{2. 15}$$

to $\alpha(x)$ in

$$\int f(x)\,d\alpha(x). \tag{2. 16}$$

I wish to remark in passing that the formulae for the integrated periodogram are at least as convenient for computational purposes as the formulae of the Schuster analysis, that the monotony of the intergrated periodogram avoids the possibility of overlooking important periods by an insufficient search, while it gives an immediate indication of empty parts of the spectrum which need no further exploration; and that the computation of $\varphi(x)$ and $S(u)$ may be performed by such instruments as the product integraph of V. Bush. I also wish to call attention to a practical study of these modified periodogram methods by Mr. G. W. Kenrick of the Massachusetts Institute of Technology.

CHAPTER II.

3. The spectrum of an arbitrary function of a single variable.

The present section is devoted to the rigorous delimitation and demonstration of the theorems heuristically indicated in section 2. Let $f(t)$ be a measurable function such that

$$\varphi(x) = \lim_{T\to\infty} \frac{1}{2T} \int_{-T}^{T} f(x+t)\bar{f}(t)\,dt \tag{3.01}$$

exists for every x. This is the sole assumption necessary in the present section. By the Schwarz inequality

$$|\varphi(x)| \leq \lim_{T\to\infty} \frac{1}{2T} \left\{ \int_{-T}^{T} |f(x+t)|^2\,dt \int_{-T}^{T} |f(t)|^2\,dt \right\}^{1/2}. \tag{3.02}$$

It follows from this that $\varphi(x)$ is bounded. To show this, it is only necessary to prove that

$$\lim_{T\to\infty} \frac{1}{2T} \int_{-T}^{T} |f(x+t)|^2\,dt = \lim_{T\to\infty} \frac{1}{2T} \int_{-T}^{T} |f(t)|^2\,dt = \varphi(0). \tag{3.03}$$

We have

$$\left| \frac{1}{2T} \int_{-T}^{T} |f(x+t)|^2\,dt - \frac{1}{2T} \int_{-T}^{T} |f(t)|^2\,dt \right|$$

$$= \left| \frac{1}{2T} \int_{T}^{T+x} |f(t)|^2\,dt - \frac{1}{2T} \int_{-T}^{-T+x} |f(t)|^2\,dt \right|$$

$$\leq \left| \frac{1}{2T} \int_{T}^{T+x} |f(t)|^2\,dt + \frac{1}{2T} \int_{-T}^{-T+x} |f(t)|^2\,dt \right|$$

$$\leq \left| \frac{1}{2T} \int_{-T-x}^{T+x} |f(t)|^2\,dt - \frac{1}{2T} \int_{-T+x}^{T-x} |f(t)|^2\,dt \right|$$

$$= \left| \left(1 + \frac{x}{T}\right) \frac{1}{2(T+x)} \int_{-T-x}^{T+x} |f(t)|^2\,dt - \left(1 - \frac{x}{T}\right) \frac{1}{2(T-x)} \int_{-T+x}^{T-x} |f(t)|^2\,dt \right|. \tag{3.04}$$

Hence

$$\overline{\lim_{T\to\infty}} \left| \frac{1}{2T} \int_{-T}^{T} |f(x+t)|^2 \, dt - \frac{1}{2T} \int_{-T}^{T} |f(t)|^2 \, dt \right|$$

$$\leq \lim_{T\to\infty} \left| \left(1 + \frac{x}{T}\right) \frac{1}{2(T+x)} \int_{-T-x}^{T+x} |f(t)|^2 \, dt - \left(1 - \frac{x}{T}\right) \frac{1}{2(T-x)} \int_{-T+x}^{T-x} |f(t)|^2 \, dt \right| = 0. \qquad (3.05)$$

Therefore

$$|\varphi(x)| \leq \varphi(0), \qquad (3.06)$$

and $\varphi(x)$ is bounded.

As before, we put

$$\varphi_A(x) = \frac{1}{2A} \int_{-\infty}^{\infty} f_A(x+t) \bar{f}_A(t) \, dt. \qquad (3.07)$$

By the Schwarz inequality

$$|\varphi_A(x)| \leq \frac{1}{2A} \sqrt{\int_{-\infty}^{\infty} |f_A(x+t)|^2 \, dt \int_{-\infty}^{\infty} |f_A(t)|^2 \, dt}$$

$$= \frac{1}{2A} \int_{-\infty}^{\infty} |f_A(t)|^2 \, dt, \qquad (3.08)$$

and $\varphi_A(x)$ is uniformly bounded in x and A for all values of A larger than some given value. Furthermore, if $x > 0$,

$$\varphi_A(x) = \frac{1}{2A} \int_{-A}^{A-x} f(x+t) \bar{f}(t) \, dt$$

$$= \frac{1}{2A} \int_{-A}^{A} f(x+t) \bar{f}(t) \, dt - \frac{1}{2A} \int_{A-x}^{A} f(x+t) \bar{f}(t) \, dt. \qquad (3.09)$$

We shall have a similar formula in case x is negative. We have furthermore

$$\left|\frac{1}{2A}\int_{A-x}^{A} f(x+t)\bar{f}(t)\,dt\right| \leq \frac{1}{2A}\sqrt{\int_{A}^{A+x}|f(t)|^2\,dt\int_{A-x}^{A}|f(t)|^2\,dt}$$

$$\leq \frac{1}{2A}\left|\int_{A-x}^{A+x}|f(t)|^2\,dt\right|. \qquad (3.10)$$

Since

$$\lim_{A\to\infty}\frac{1}{2A}\left|\int_{A-x}^{A+x}|f(t)|^2\,dt\right| = 0, \qquad (3.11)$$

it follows at once that

$$\lim_{A\to\infty}\varphi_A(x) = \lim_{A\to\infty}\frac{1}{2A}\int_{-A}^{A} f(x+t)\bar{f}(t)\,dt = \varphi(x). \qquad (3.12)$$

Thus $\varphi(x)$ is the limit of a uniformly bounded sequence of measurable functions, and is measurable. Since it is also bounded, it is quadratically summable over any finite range, while $\varphi(x)/x$ is quadratically summable over any range excluding the origin. It is, moreover, easy to prove that

$$\varphi(x) = \underset{A\to\infty}{\text{l.i.m.}}\ \varphi_A(x) \qquad (3.13)$$

over any finite range, and that

$$\frac{\varphi(x)}{x} = \underset{A\to\infty}{\text{l.i.m.}}\ \frac{\varphi_A(x)}{x} \qquad (3.14)$$

over any range excluding the origin. Hence,

$$\varphi(x)\frac{\sin\mu x}{x} = \underset{A\to\infty}{\text{l.i.m.}}\ \varphi_A(x)\frac{\sin\mu x}{x}. \qquad (3.15)$$

In as much as the Fourier transformation leaves invariant the integral of the square of the modulus of a function, and hence leaves invariant all properties of convergence in the mean,

$$\underset{N\to\infty}{\text{l.i.m.}}\int_{-N}^{N}\varphi(x)\frac{\sin\mu x}{x}e^{iux}\,dx = \underset{A\to\infty}{\text{l.i.m.}}\int_{-\infty}^{\infty}\varphi_A(x)\frac{\sin\mu x}{x}e^{iux}\,dx$$

$$= \underset{A\to\infty}{\text{l.i.m.}} \int_{-\infty}^{\infty} \varphi_A(x)\, e^{iux}\, dx \int_0^{\mu} \cos \xi x\, d\xi$$

$$= \frac{1}{2} \underset{A\to\infty}{\text{l.i.m.}} \int_0^{\mu} d\xi \int_{-\infty}^{\infty} \varphi_A(x) [e^{i(u+\xi)x} + e^{i(u-\xi)x}]\, dx. \qquad (3.16)$$

The inversion of the order of integration is justified as usual by the fact that the infinite integral is only apparently infinite. This, let me remark parenthetically, is the case also in the next set of formulae.

The last expression is the limit in the mean of a real non-negative quantity, for

$$\int_{-\infty}^{\infty} \varphi_A(x) e^{ivx}\, dx = \frac{1}{2A} \int_{-\infty}^{\infty} e^{ivx}\, dx \int_{-\infty}^{\infty} f_A(x+t) \bar{f}_A(t)\, dt$$

$$= \frac{1}{2A} \int_{-\infty}^{\infty} \bar{f}_A(t)\, dt \int_{-\infty}^{\infty} f_A(x+t)\, e^{ivx}\, dx$$

$$= \frac{1}{2A} \int_{-\infty}^{\infty} \bar{f}_A(t)\, dt \int_{-\infty}^{\infty} f_A(w)\, e^{iv(w-t)}\, dw$$

$$= \frac{1}{2A} \left| \int_{-\infty}^{\infty} f_A(w)\, e^{ivw}\, dw \right|^2 \geq 0. \qquad (3.17)$$

The limit in the mean of a function is determined with the exception of a set of points of zero measure, but the limit in the mean of a non-negative function may always be so chosen as to be non-negative. If we make this choice,

$$\underset{N\to\infty}{\text{l.i.m.}} \int_{-N}^{N} \varphi(x) \frac{\sin \mu x}{x} e^{iux}\, dx \geq 0. \qquad (3.18)$$

The expression

$$\sigma_1(u) = \underset{A\to\infty}{\text{l.i.m.}} \frac{1}{2\pi} \left[\int_1^A + \int_{-A}^{-1} \right] \varphi(x) \frac{e^{ixu}}{ix}\, dx \qquad (3.19)$$

exists, as the Fourier transform of a quadratically summable function. Moreover,

$$\sigma_2(u) = \frac{1}{2\pi}\int_{-1}^{1} \varphi(x)\frac{e^{ixu}-1}{ix}\,dx \tag{3. 20}$$

exists. If we put

$$\sigma(u)=\sigma_1(u)+\sigma_2(u) \tag{3. 21}$$

we have

$$\sigma(u+\mu)-\sigma(u-\mu)=\underset{A\to\infty}{\mathrm{l.i.m.}}\ \frac{1}{\pi}\int_{-A}^{A}\varphi(x)\frac{\sin\mu x}{x}e^{iux}\,dx \geq 0. \tag{3. 22}$$

Of course, when we say that a limit in the mean is non-negative, we merely mean that it can be so chosen. Thus the expression $\sigma(u)$ is monotone, or at least can be so chosen, for example, by putting

$$\sigma(u)=\frac{d}{du}\int_{0}^{u}\sigma(u)\,du \tag{3. 23}$$

at every point where the latter expression is defined. Here we introduce (3. 23), because $\sigma(u)$ is now almost everywhere the limit of the difference quotient of $\int_0^u \sigma(u)\,du$, namely, $\frac{1}{2\varepsilon}\int_{u-\varepsilon}^{u+\varepsilon}\sigma(u)\,du$, which is monotone as a consequence of (3. 22). Thus, except at a set of zero measure, $\sigma(u)$ is the limit, not merely the limit in the mean, of a set of monotone functions, and is monotone. Elsewhere, at a set of zero measure, we put

$$\sigma(u)=\frac{1}{2}\left[\sigma(u+0)+\sigma(u-0)\right]. \tag{3. 24}$$

It follows that $\sigma(u+\mu)-\sigma(u-\mu)$ is of limited total variation over any finite interval. We shall show in the next paragraph that

$$\lim_{\mu\to\infty}\left[\sigma(u+\mu)-\sigma(u-\mu)\right]$$

is finite, and that hence $\sigma(u+\mu)-\sigma(u-\mu)$ is of limited total variation over $(-\infty, \infty)$. It is moreover, quadratically summable, as the Fourier transform of a quadratically summable function. It tends to 0 as $u \to \pm \infty$ and hence, by a theorem of Hobson[3], we have

$$\frac{1}{2}[\sigma(u+0+\mu)-\sigma(u+0-\mu)+\sigma(u-0+\mu)-\sigma(u-0-\mu)]$$

$$= \frac{1}{\pi}\int_{-\infty}^{\infty} \varphi(x) \frac{\sin \mu x}{x} e^{iux}\, dx. \qquad (3.25)$$

In particular, if $u = \mu = v/2$,

$$\frac{1}{2}[\sigma(v+0)+\sigma(v-0)] - \frac{1}{2}[\sigma(+0)+\sigma(-0)] = \frac{1}{2\pi}\int_{-\infty}^{\infty} \varphi(x) \frac{e^{ivx}-1}{ix}\, dx. \qquad (3.26)$$

If therefore we define

$$S(u) = \frac{1}{2\pi}\int_{-\infty}^{\infty} \varphi(x) \frac{e^{iux}-1}{ix}\, dx, \qquad (3.27)$$

$S(u)$ will exist, and

$$S(u) - \sigma(u) = \text{constant}. \qquad (3.28)$$

4. The total spectral intensity.

It is manifest that $\lim_{\mu \to \infty} [S(u+\mu) - S(u-\mu)]$, or as we shall write it, $S(\infty) - S(-\infty)$, if it exists, is a measure of the total spectral intensity of $f(x)$. We shall prove that this quantity exists and is finite.

We have

$$\frac{1}{A}\int_0^A [\sigma(u+\mu)-\sigma(u-\mu)]\, d\mu = \frac{1}{\pi A}\int_0^A d\mu \underset{B\to\infty}{\text{l.i.m.}} \int_{-B}^{B} \varphi(x) \frac{\sin \mu x}{x} e^{iux}\, dx. \qquad (4.01)$$

The limit in the mean is here taken with u as the fundamental variable, and with μ as parameter. It is not difficult to deduce from the boundedness of

$$\int_0^A d\mu \int_{-\infty}^{\infty} |\varphi(x)|^2 \frac{\sin^2 \mu x}{x^2}\, dx$$

that we may invert the order of integration, and get

18—29764. *Acta mathematica*. 55. Imprimé le 7 avril 1930.

$$\frac{1}{A}\int_0^A [\sigma(u+\mu)-\sigma(u-\mu)]\,d\mu = \frac{1}{\pi A}\,\underset{B\to\infty}{\text{l.i.m.}}\int_{-B}^{B}\varphi(x)\frac{1-\cos Ax}{x^2}e^{iux}\,dx. \tag{4.02}$$

To show this, let us remark that

$$\int_{-\infty}^{\infty}\left|\frac{1}{\pi A}\int_0^A d\mu\,\underset{B\to\infty}{\text{l.i.m.}}\int_{-B}^{B}\varphi(x)\frac{\sin\mu x}{x}e^{iux}\,dx-\frac{1}{\pi A}\int_0^A d\mu\int_{-C}^{C}\varphi(x)\frac{\sin\mu x}{x}e^{iux}\,dx\right|^2 du$$

$$=\frac{1}{\pi^2 A^2}\int_{-\infty}^{\infty}\left|\int_0^A d\mu\,\underset{B\to\infty}{\text{l.i.m.}}\left[\int_C^B+\int_{-B}^{-C}\right]\varphi(x)\frac{\sin\mu x}{x}e^{iux}\,dx\right|^2 du$$

$$\leq\frac{1}{\pi^2 A^2}\int_{-\infty}^{\infty}du\int_0^A d\mu_1\int_0^A\left|\underset{B\to\infty}{\text{l.i.m.}}\left[\int_C^B+\int_{-B}^{-C}\right]\varphi(x)\frac{\sin\mu x}{x}e^{iux}\,dx\right|^2 d\mu$$

$$=\frac{1}{\pi^2 A}\int_0^A d\mu\int_{-\infty}^{\infty}du\left|\underset{B\to\infty}{\text{l.i.m.}}\left[\int_C^B+\int_{-B}^{-C}\right]\varphi(x)\frac{\sin\mu x}{x}e^{iux}\,dx\right|^2$$

$$=\frac{2}{\pi A}\int_0^A d\mu\left[\int_C^{\infty}+\int_{-\infty}^{-C}\right]|\varphi(x)|^2\frac{\sin^2\mu x}{x^2}\,dx. \tag{4.03}$$

Inasmuch as this latter expression tends to 0 with increasing C,

$$\frac{1}{\pi A}\int_0^A d\mu\,\underset{B\to\infty}{\text{l.i.m.}}\int_{-B}^{B}\varphi(x)\frac{\sin\mu x}{x}e^{iux}\,dx=\underset{C\to\infty}{\text{l.i.m.}}\frac{1}{\pi A}\int_0^A d\mu\int_{-C}^{C}\varphi(x)\frac{\sin\mu x}{x}e^{iux}\,dx$$

$$=\underset{B\to\infty}{\text{l.i.m.}}\frac{1}{\pi A}\int_{-B}^{B}\varphi(x)\frac{e^{iux}}{x}\,dx\int_0^A\sin\mu x\,d\mu$$

$$=\frac{1}{\pi A}\underset{B\to\infty}{\text{l.i.m.}}\int_{-B}^{B}\varphi(x)\frac{1-\cos Ax}{x^2}e^{iux}\,dx, \tag{4.04}$$

thus proving our statement.

Our limit in the mean may be replaced by an ordinary limit, as this limit exists, owing to the boundedness of $\varphi(x)$. Therefore

$$\lim_{A\to\infty}\frac{1}{A}\int_0^A[\sigma(u+\mu)-\sigma(u-\mu)]\,d\mu=\lim_{A\to\infty}\frac{1}{\pi A}\int_{-\infty}^{\infty}\varphi(x)\frac{1-\cos Ax}{x^2}e^{iux}\,dx. \tag{4.05}$$

It follows from the monotony of $\sigma(u+\mu)-\sigma(u-\mu)$ in μ that we may write

$$\begin{aligned}\lim_{\mu\to\infty}[\sigma(u+\mu)-\sigma(u-\mu)]&=\lim_{A\to\infty}\frac{1}{\pi}\int_{-\infty}^{\infty}\varphi\left(\frac{x}{A}\right)e^{iu\frac{x}{A}}\frac{1-\cos x}{x^2}\,dx\\&\leq\max\left|\varphi\left(\frac{x}{A}\right)\right|\frac{1}{\pi}\int_{-\infty}^{\infty}\frac{1-\cos x}{x^2}\,dx\\&=\varphi(0).\end{aligned} \tag{4.06}$$

This yields us the existence of $\sigma(\infty)-\sigma(-\infty)$ and hence, according to the last paragraph, of $S(\infty)-S(-\infty)$. We have

$$\begin{aligned}S(\infty)-S(-\infty)&-\lim_{A\to\infty}\frac{1}{\pi}\int_{-\infty}^{\infty}\varphi\left(\frac{x}{A}\right)\frac{1-\cos x}{x^2}\,dx\\&=\varphi(0)+\lim_{A\to\infty}\frac{1}{\pi}\int_{-\infty}^{\infty}\left[\varphi\left(\frac{x}{A}\right)-\varphi(0)\right]\frac{1-\cos x}{x^2}\,dx.\end{aligned} \tag{4.07}$$

Hence for sufficiently large A

$$\begin{aligned}&|S(\infty)-S(-\infty)-\varphi(0)|\\&\leq\frac{2}{\pi}\max|\varphi(\xi)|\left[\int_{A^{1/2}}^{\infty}+\int_{-\infty}^{-A^{1/2}}\right]\frac{1-\cos x}{x^2}\,dx+\max_{|\xi|<A^{-1/2}}|\varphi(\xi)-\varphi(0)|+\varepsilon.\end{aligned} \tag{4.08}$$

Since A is arbitrary,

$$|S(\infty)-S(-\infty)-\varphi(0)|\leq\overline{\lim_{|\xi|\to 0}}|\varphi(\xi)-\varphi(0)|. \tag{4.09}$$

In case $\varphi(x)$ is continuous at the origin,

$$\varphi(0)=S(\infty)-S(-\infty). \tag{4.10}$$

However, $\varphi(x)$ need not be continuous at the origin, even if $f(t)$ is everywhere continuous. Thus let $f(t)=\sin t^2$.

Then

$$\varphi(0)=\lim_{T\to\infty}\frac{1}{2T}\int_{-T}^{T}\sin^2 t^2\,dt$$

$$=\lim_{T\to\infty}\frac{1}{T}\int_{0}^{T}\frac{1-\cos 2t^2}{2}\,dt$$

$$=\frac{1}{2}-\lim_{T\to\infty}\frac{\sqrt{2}}{T}\int_{0}^{T\sqrt{2}}\cos u^2\,du$$

$$=\frac{1}{2}. \tag{4.11}$$

since $\int_0^\infty \cos u^2\,du$ is a Fresnel integral, and equals $\frac{1}{2}\sqrt{\frac{\pi}{2}}$.

On the other hand, if $x\neq 0$, we have

$$\varphi(x)=\lim_{T\to\infty}\frac{1}{2T}\int_{-T}^{T}\sin(t+x)^2\sin t^2\,dt$$

$$=\lim_{T\to\infty}\frac{1}{4T}\int_{-T}^{T}[-\cos(2t^2+2tx+x^2)+\cos(2tx+x^2)]\,dt.$$

The second part of this mean obviously vanishes. Hence

$$\varphi(x)=-\lim_{T\to\infty}\frac{1}{4T}\int_{-T}^{T}\cos(2t^2+2tx+x^2)\,dt$$

$$=-\lim_{U\to\infty}\frac{1}{2U}\int_{0}^{U}\cos\left(u^2+\frac{x^2}{2}\right)du$$

$$= \text{o}\left[\cos\frac{x^2}{2}\int_0^\infty \cos u^2\, du - \sin\frac{x^2}{2}\int_0^\infty \sin u^2\, du\right]$$

$$= \text{o}. \tag{4. 12}$$

since $\int_0^\infty \cos u^2\, du = \int_0^\infty \sin u^2\, du = \frac{1}{2}\sqrt{\frac{\pi}{2}}$.

Thus $\varphi(x)$ vanishes almost everywhere, $S(u)$ vanishes identically, and

$$\varphi(0) \neq S(\infty) - S(-\infty). \tag{4. 13}$$

5. Tauberian theorems and spectral intensity.

In a recent paper, the author has proved the following general Tauberian theorem: *Let $M_1(x)$ and $M_2(x)$ be two functions bounded over every range* $(\varepsilon, 1/\varepsilon)$, *which are $O\left(\frac{1}{x(\log x)^2}\right)$ at* 0 *and* ∞. *Let $M_1(x)$ be measurable and non-negative, and let*

$$\int_0^\infty M_1(x)\, x^{iu}\, dx \neq 0. \qquad [-\infty < u < \infty] \tag{5. 01}$$

Let $M_2(x)$ be continuous, except for a finite number of finite jumps. Let $f(x)$ be a measurable function bounded below and such that

(a)
$$\lim_{\lambda \to 0\,[\infty]} \int_0^\infty f(\lambda x)\, M_1(x)\, dx = A \int_0^\infty M_1(x)\, dx;$$

(b)
$$\int_0^\infty f(\lambda x)\, M_1(x)\, dx \text{ is bounded.} \qquad [0 < \lambda < \infty]$$

Then

$$\lim_{\lambda \to 0\,[\infty]} \int_0^\infty f(\lambda x)\, M_2(x)\, dx = A \int_0^\infty M_2(x)\, dx. \tag{5. 02}$$

Here ∞ is put into brackets to indicate that at these points it may be consistently substituted for 0. There is manifestly no restriction in assuming

$f(x)$ non-negative, as the theorem, if true for a given $f(x)$, is unchanged as to its validity by the addition to $f(x)$ of a constant. The theorem assumes a more understandable form under the transformations

$$\left.\begin{aligned} &x = e^{\xi}; \quad \lambda = e^{-\eta}; \\ &e^{\xi} M_1(e^{\xi}) = N_1(\xi); \quad e^{\xi} M_2(e^{\xi}) = N_2(\xi). \end{aligned}\right\} \tag{5.03}$$

It then becomes: *Let* $N_1(\xi)$ *and* $N_2(\xi)$ *be two bounded functions which are* $O(\xi^{-2})$ *at* $\pm\infty$. *Let* $N_1(\xi)$ *be measurable and non-negative, and let*

$$\int_{-\infty}^{\infty} N_1(\xi)\, e^{iu\xi}\, d\xi \neq 0. \qquad [-\infty < u < \infty] \tag{5.04}$$

Let $N_2(\xi)$ *be continuous, except for a finite number of finite jumps. Let* $g(\xi)$ *be a non-negative measurable function such that*

(a) $$\lim_{\eta\to\infty} \int_{-\infty}^{\infty} g(\eta-\xi)\, N_1(\xi)\, d\xi = A \int_{-\infty}^{\infty} N_1(\xi)\, d\xi;$$

(b) $$\int_{-\infty}^{\infty} g(\eta-\xi)\, N_1(\xi)\, d\xi \text{ is bounded.} \qquad [-\infty < \eta < \infty]$$

Then

(c) $$\lim_{\eta\to\infty} \int_{-\infty}^{\infty} g(\eta-\xi)\, N_2(\xi)\, d\xi = A \int_{-\infty}^{\infty} N_2(\xi)\, d\xi. \tag{5.05}$$

The proof proceeds as follows: We shall symbolize by C the class of all functions $N_2(\xi)$, bounded and $O(\xi^{-2})$ at $\pm\infty$, and continuous except for a finite number of finite jumps, for which (c) is a consequence of (a) and (b) for all non-negative measurable functions $g(\xi)$. Among the functions in C are all functions $N_2(\xi)$ of the form

$$N_2(\xi) = \int_{-\infty}^{\infty} N_1(\eta)\, R(\eta-\xi)\, d\eta \tag{5.06}$$

for which $\int_{-\infty}^{\infty} |R(\eta)|\, d\eta$ converges, inasmuch as the double integral

$$\int_{-\infty}^{\infty} R(\zeta)\, d\zeta \int_{-\infty}^{\infty} g(\eta-\xi-\zeta)\, N_1(\xi)\, d\xi$$

is absolutely convergent, so that

$$\int_{-\infty}^{\infty} N_2(\xi)\, g(\eta-\xi)\, d\xi = \int_{-\infty}^{\infty} g(\eta-\xi)\, d\xi \int_{-\infty}^{\infty} N_1(\zeta+\xi)\, R(\zeta)\, d\zeta$$

$$= \int_{-\infty}^{\infty} R(\zeta)\, d\zeta \int_{-\infty}^{\infty} g(\eta-\xi-\zeta)\, N_1(\xi)\, d\xi, \tag{5.07}$$

and

$$\lim_{\eta\to\infty} \int_{-\infty}^{\infty} N_2(\xi)\, g(\eta-\xi)\, d\xi = \lim_{\eta\to\infty} \int_{-\infty}^{\infty} R(\zeta)\, d\zeta \int_{-\infty}^{\infty} g(\eta-\xi-\zeta)\, N_1(\xi)\, d\xi$$

$$= A \int_{-\infty}^{\infty} R(\zeta)\, d\zeta \int_{-\infty}^{\infty} N_1(\xi)\, d\xi = A \int_{-\infty}^{\infty} N_2(\xi)\, d\xi. \tag{5.08}$$

A particular example of such a function is furnished by

$$N_2(\xi) = \int_{-B}^{B} \nu_2(u)\, e^{-iu\xi}\, du \tag{5.09}$$

where $\nu_2(u)$ is continuous over $(-B, B)$, while its first derivative is continuous except for a finite number of discontinuities of the first kind, and

$$\nu_2(B) = \nu_2(-B) = 0. \tag{5.10}$$

To prove this, let us reflect that $N_1(\xi)$ and $N_2(\xi)$ are quadratically summable by assumption, and that

$$\nu_1(u) = \int_{-\infty}^{\infty} N_1(\xi)\, e^{iu\xi}\, d\xi \tag{5.11}$$

exists, as well as

$$\nu'_1(u) = \underset{E\to\infty}{\text{l.i.m.}} \int_{-E}^{E} i\xi\, N_1(\xi)\, e^{iu\xi}\, d\xi. \tag{5.12}$$

By our hypothesis (5.04)

$$\nu_1(u) \neq 0. \qquad [-\infty < u < \infty] \tag{5.13}$$

Let us put

$$\mu(u) = \nu_2(u)/\nu_1(u). \tag{5.14}$$

Inasmuch as $\mu(u)$ is absolutely continuous, its derivative may be computed by the rules, and

$$\mu'(u) = \frac{\nu_1(u)\,\nu'_2(u) - \nu_2(u)\,\nu'_1(u)}{[\nu_1(u)]^2}. \tag{5.15}$$

I now say that we shall have

$$R(\zeta) = \int_{-B}^{B} \mu(u)\, e^{iu\zeta}\, du$$

$$= \frac{1}{i\zeta} \int_{-B}^{B} \mu'(u)\, e^{iu\zeta}\, du. \tag{5.16}$$

Inasmuch as

$$\int_{-B}^{B} \mu'(u)\, e^{iu\zeta}\, du \tag{5.17}$$

is quadratically summable,

$$\int_{-\infty}^{\infty} |R(\zeta)|\, d\zeta \tag{5.18}$$

exists. Since the integrals involved converge absolutely,

$$\int_{-\infty}^{\infty} N_1(\eta)\, R(\eta - \xi)\, d\eta = \int_{-\infty}^{\infty} N_1(\eta)\, d\eta \int_{-B}^{B} \mu(u)\, e^{iu(\eta-\xi)}\, du$$

$$= \int_{-B}^{B} \mu(u)\, e^{-iu\xi}\, du \int_{-\infty}^{\infty} N_1(\eta) e^{iu\eta}\, d\eta$$

$$= \int_{-B}^{B} \mu(u)\, \nu_1(u)\, e^{-iu\xi}\, du$$

$$= \int_{-B}^{B} \nu_2(u)\, e^{-iu\xi}\, du = N_2(\xi). \tag{5.19}$$

This justifies our evaluation of $R(\zeta)$, and proves that $N_2(\xi)$ belongs to C. The following are particular cases which may be probed to belong to C in this manner:

$$T_B(\xi) = \int_{-B}^{B} [e^{-|u|} - e^{-B}]\, e^{-i\xi u}\, du$$

$$= \int_{0}^{B} [e^{-u} - e^{-B}] \cos \xi u\, du$$

$$- \frac{1}{\xi} \int_{0}^{B} e^{-u} \sin \xi u\, du$$

$$= \frac{1 - e^{-B} \cos B\xi}{1 + \xi^2} - \frac{e^{-B} \sin B\xi}{\xi(1 + \xi^2)}. \tag{5.20}$$

Again,

$$Q_B(\xi) = \frac{1}{\pi} \int_{-B}^{B} \left(1 - \frac{|u|}{B}\right) \frac{\sin u}{u}\, e^{-iu\xi}\, du$$

$$= \int_{B(\xi-1)}^{B(\xi+1)} \frac{1 - \cos z}{\pi z^2}\, dz$$

$$= \frac{1}{\pi B(\xi - 1)} - \frac{1}{\pi B(\xi + 1)} + O\left(\frac{1}{B^2 \xi^2}\right). \qquad [B\xi \to \pm \infty] \tag{5.21}$$

If we already know certain members of the class C, we may obtain new members of the class in the following manner: Let $V(\xi)$ be a function continuous, except for a finite number of finite jumps, such that, when any positive ξ is given, we can find two members of C, $V_1(\xi)$ and $V_2(\xi)$, such that

$$V_1(\xi) \leq V(\xi) \leq V_2(\xi), \tag{5.22}$$

while

$$\int_{-\infty}^{\infty} [V_2(\xi) - V_1(\xi)]\, d\xi < \varepsilon. \tag{5.23}$$

Then $V(\xi)$ itself belongs to C. For

19—29764. *Acta mathematica.* 55. Imprimé le 7 avril 1930.

$$\overline{\lim_{\eta\to\infty}}\left|\int_{-\infty}^{\infty} g(\eta-\xi)\, V(\xi)\, d\xi - A\int_{-\infty}^{\infty} V(\xi)\, d\xi\right|$$

$$\leq \overline{\lim_{\eta\to\infty}}\left|\int_{-\infty}^{\infty} g(\eta-\xi)\, V_1(\xi)\, d\xi - A\int_{-\infty}^{\infty} V_2(\xi)\, d\xi\right|$$

$$+ \overline{\lim_{\eta\to\infty}}\left|\int_{-\infty}^{\infty} g(\eta-\xi)\, V_2(\xi)\, d\xi - A\int_{-\infty}^{\infty} V_1(\xi)\, d\xi\right|$$

$$< 2A\varepsilon. \tag{5.24}$$

and since ε is arbitrarily small, this limit is 0. Furthermore, any linear combination of a finite number of members of C belongs to C.

As a particular case, we have

$$[1+(1+B)e^{-B}]^{-1}\, T_B(\xi) < \frac{1}{1+\xi^2} < [1-(1-B)\, e^{-B}]^{-1}\, T_B(\xi). \tag{5.25}$$

Inasmuch as

$$\lim_{B\to\infty} (1+B)\, e^{-B} = 0, \tag{5.26}$$

it follows at once that $\frac{1}{1+\xi^2}$ belongs to C. An exactly similar proof will show that the same thing holds of

$$\frac{p}{(\xi-q)^2-r^2}.$$

Again,

$$\lim_{A\to\infty} Q_A(\xi) = \operatorname{sgn}(\xi+1) - \operatorname{sgn}(\xi-1) = V(\xi), \tag{5.27}$$

and this convergence is uniform except in the neighborhood of ± 1 while we always have for $B > 0$

$$Q_A(\xi) < \frac{1}{B(\xi^2+1)} \text{ over } (1+\eta,\ \infty) \text{ and } (-\infty,\ -1-\eta) \quad [A \text{ large enough}] \tag{5.28}$$

Furthermore,

$$\int_{-\infty}^{\infty} Q_A(\xi)\, d\xi = 4. \tag{5.29}$$

Let us put

$$\left.\begin{aligned} (1+\eta)\, Q_A(\xi(1-\eta)) &= V_2(\xi); \\ (1-\eta)\, Q_A(\xi(1+\eta)) - \frac{1}{B(\xi^2+1)} &= V_1(\xi). \end{aligned}\right\} \tag{5.30}$$

We can so determine a large A when η and B are given, that for that A and all larger ones,

$$V_2(\xi) \geq V(\xi) \geq V_1(\xi). \tag{5.22}$$

We have

$$\int_{-\infty}^{\infty} [V_2(\xi) - V_1(\xi)]\, d\xi = 2\left[\frac{1+\eta}{1-\eta} - \frac{1-\eta}{1+\eta}\right] + \frac{\pi}{B} \tag{5.31}$$

which we may make arbitrarily small. Hence

$$\operatorname{sgn}(\xi+1) - \operatorname{sgn}(\xi-1)$$

belongs to C. As an immediate consequence, since

$$\operatorname{sgn}(\xi+\alpha) - \operatorname{sgn}(\xi+\beta)$$

may be shown by the same means to belong to C, any step function vanishing for large positive and negative arguments belongs to C, and hence any function continuous except for a finite set of discontinuities of the first kind, and vanishing outside of a finite interval, since the latter function may be penned in between two step functions enclosing an arbitrarily small area.

Now let $N_2(\xi)$ be a bounded function which is $O(\xi^{-2})$ at $\pm\infty$, and which is continuous except for a finite number of finite jumps. Let

$$|N_2(\xi)| < P/(\xi^2+1), \tag{5.32}$$

for all ξ. We put

$$V_1(\xi) = \begin{cases} N_2(\xi); & [|\xi| < M] \\ -P/(\xi^2+1); & [|\xi| \geq M] \end{cases} \tag{5.33}$$

$$V_2(\xi) = \begin{cases} N_2(\xi); & [|\xi| < M] \\ P/(\xi^2+1). & [|\xi| \geq M] \end{cases} \tag{5.34}$$

The functions $V_1(\xi)$ and $V_2(\xi)$ are sums of functions of C and functions of the form $\pm P/(\xi^2+1)$ which also belong to C. Hence, they themselves belong to C. We have

$$V_2(\xi) > N_2(\xi) > V_1(\xi) \tag{5.35}$$

and

$$\int_{-\infty}^{\infty} [V_2(\xi) - V_1(\xi)]\, d\xi < 2P[\pi - \tan^{-1} M] \tag{5.36}$$

which we may make as small as we like. Hence $N_2(\xi)$ belongs to C. This concludes the proof of our generalized Tauberian theorem.

As a corollary of our Tauberian theorem, Mr. S. B. Littauer has given a proof of the following theorem of Jacob: *If $f(t)$ is a measurable function, integrable in every finite interval of* $(0, \infty)$ *and if for some given* α $(0 \leq \alpha < 1)$

(a) $$\frac{1}{T^{1-\alpha}} \int_0^T |f(t)|\, dt < B \text{ for every } T;$$

(b) $$\lim_{T \to \infty} \frac{1}{T^{1-\alpha}} \int_0^T f(t)\, dt = A;$$

then

(c) $$\lim_{\varepsilon \to 0} \frac{\varepsilon^{1-\alpha}}{\gamma_\alpha \pi} \int_0^\infty f(t) \left(\frac{\sin \varepsilon t}{\varepsilon t}\right)^2 dt = A,$$

where

$$\gamma_\alpha = \frac{2^{\alpha-1}(1-\alpha)}{\Gamma(2+\alpha) \cos \dfrac{\pi\alpha}{2}}.$$

Furthermore, if $f(t)$ is measurable and non-negative (*or bounded below*), (c) *implies* (b).

The particular case of this theorem where $\alpha=0$ had already been treated by Bochner, Hardy, and the present author.

In all theorems of this type, there is a close relation between the theorem which one obtains by letting λ become infinite and that which one obtains by letting λ become 0. This is to be explained by the fact that the general Tauberian theorem assumes a perfectly symmetric form when we make the substitutions

$$x = e^\xi;\ \lambda = e^\eta;\ M_1(x) = N_1(\xi)\, e^\xi;\ M_2(x) = N_2(\xi)\, e^\xi. \tag{5.03}$$

If we take

$$\left.\begin{aligned} &M_1(x)\, [\text{or } M_2(x)] = 1 \text{ if } 0 < x < 1;\ = 0 \text{ otherwise;} \\ &M_2(x)\, [\text{or } M_1(x)] = \frac{4 \sin^2 x/2}{x^2} \end{aligned}\right\} \tag{5.37}$$

in our general Tauberian theorem, since $\int_0^\infty M_1(x)\,e^{iux}\,dx \neq 0$ and $\int_0^\infty M_2(x)\,e^{iux}\,dx \neq 0$, we may deduce the conclusions of the theorem. We thus get the following result: *Let*

$$|f(x)| < B. \qquad [0 < x < \infty]$$

Then the two propositions,

(a)
$$\lim_{T\to 0}\frac{1}{T}\int_0^T f(x)\,dx = A$$

and

(b)
$$\lim_{T\to 0}\frac{2}{\pi}\int_0^\infty f(Tx)\frac{1-\cos x}{x^2}\,dx = A$$

are equivalent.

In the particular case where $f(x)$ is replaced by $\frac{1}{2}[\varphi(x)+\varphi(-x)]$, we see that

$$A = \lim_{r\to\infty}\frac{1}{\pi}\int_{-\infty}^{\infty}\varphi\left(\frac{x}{r}\right)\frac{1-\cos x}{x^2}\,dx \tag{5.38}$$

implies, and is implied by

$$A = \lim_{\varepsilon\to 0}\frac{1}{2\varepsilon}\int_{-\varepsilon}^{\varepsilon}\varphi(x)\,dx. \tag{5.39}$$

We have (see (3.27))

$$\int_{-\infty}^{\infty} e^{-i\lambda u}\,dS(u) = \lim_{\mu\to 0}\frac{1}{2\pi\mu}\int_{-\infty}^{\infty} e^{-i\lambda u}\,du \operatorname*{l.i.m.}_{A\to\infty}\int_{-A}^{A}\varphi(x)\frac{\sin\mu x}{x}e^{iux}\,dx$$

$$= \lim_{\mu\to 0}\frac{1}{\mu\lambda}\varphi(\lambda)\sin\mu\lambda = \varphi(\lambda), \tag{5.40}$$

except possibly at a set of points of zero measure. To see this, it is only necessary to reflect that it follows from the definition of the Stieltjes integral that if $\alpha(x)$ is of limited total variation over $(-\infty, \infty)$,

$$\int_{-\infty}^{\infty} f(x)\,d\alpha(x) = \lim_{\varepsilon\to 0}\sum_{-\infty}^{\infty} f(u+2n\varepsilon)[\alpha(u+(2n+1)\varepsilon)-\alpha(u+(2n-1)\varepsilon)]$$

$$= \lim_{\varepsilon \to 0} \frac{1}{2\varepsilon} \int_{-\varepsilon}^{\varepsilon} du \sum_{-\infty}^{\infty} f(u+2n\varepsilon)\left[\alpha(u+(2n+1)\varepsilon) - \alpha(u+(2n-1)\varepsilon)\right]$$

$$= \lim_{\varepsilon \to 0} \frac{1}{2\varepsilon} \int_{-\varepsilon}^{\varepsilon} f(u)\left[\alpha(u+\varepsilon) - \alpha(u-\varepsilon)\right] du. \tag{5.41}$$

Let us put

$$\Phi(\lambda) = \int_{-\infty}^{\infty} e^{-i\lambda u}\, d\, S(u). \tag{5.42}$$

This function will be defined for all real arguments, and we shall have

$$\Phi(\lambda+\varepsilon) - \Phi(\lambda) = -2i \int_{-\infty}^{\infty} e^{-iu\left(\lambda+\frac{\varepsilon}{2}\right)} \sin\frac{u\varepsilon}{2}\, d\, S(u). \tag{5.43}$$

Since the function $\sin u\varepsilon/2$ is uniformly bounded, and tends to 0 over every finite range of u as $\varepsilon \to 0$, while $e^{-iu\left(\lambda+\frac{\varepsilon}{2}\right)}$ has modulus 1, it follows that $\Phi(\lambda+\varepsilon) - \Phi(\lambda)$ is less than the sum of two terms, one of which is the total variation of $S(u)$ over a region receding towards infinity, while the other is less than the total variation of $S(u)$ multiplied by a factor tending to 0. Hence

$$\lim_{\varepsilon \to 0} \left[\Phi(\lambda+\varepsilon) - \Phi(\lambda)\right] = 0. \tag{5.44}$$

Thus the function $\Phi(\lambda)$ is continuous, and indeed, this proof shows it to be uniformly continuous. Hence

$$\lim_{\varepsilon \to 0} \frac{1}{2\varepsilon} \int_{x-\varepsilon}^{x+\varepsilon} \varphi(\xi)\, d\xi = \lim_{\varepsilon \to 0} \frac{1}{2\varepsilon} \int_{x-\varepsilon}^{x+\varepsilon} \Phi(\xi)\, d\xi = \Phi(x). \tag{5.45}$$

This gives another proof that

$$\lim_{\varepsilon \to 0} \frac{1}{2\varepsilon} \int_{-\varepsilon}^{\varepsilon} \varphi(\xi)\, d\xi = S(\infty) - S(-\infty), \tag{5.46}$$

and indeed proves considerably more.

It is thus possible to dispense with Tauberian theorems for this part of the theory. There is another point, however, where they play a more essential rôle. That is in the study of the generalized Fourier transform of a function. Let $\lim_{T\to\infty} \frac{1}{2T}\int_{-T}^{T} |f(x)|^2\, dx$ exist. Then

$$\left[\int_1^T + \int_{-T}^{-1}\right] \frac{|f(x)|^2}{x^2}\, dx = \int_1^T \frac{|f(x)|^2 + |f(-x)|^2}{x^2}\, dx$$

$$- \int_1^T \frac{1}{x^2}\, d \int_0^x [|f(\xi)|^2 + |f(-\xi)|^2]\, d\xi$$

$$= \frac{1}{T^2}\int_{-T}^{T} |f(x)|^2\, dx - \int_{-1}^{1} |f(x)|^2\, dx + \int_1^T \frac{2\, dx}{x^3}\int_{-x}^{x} |f(\xi)|^2\, d\xi$$

$$= O(1) + \int_1^T 2\, O(1)\frac{dx}{x^2}$$

$$= O(1). \tag{5.47}$$

Consequently

$$\left[\int_1^\infty + \int_{-\infty}^{-1}\right] \frac{|f(x)|^2}{x^2}\, dx \tag{5.48}$$

exists. It follows from this that

$$\psi_\mu(u) = \frac{1}{\pi} \mathop{\text{l.i.m.}}_{A\to\infty} \int_{-A}^{A} f(x)\frac{\sin \mu x}{x}\, e^{iux}\, dx \tag{5.49}$$

exists, and that

$$\psi_\mu(u) = s(u+\mu) - s(u-\mu), \tag{5.50}$$

where

$$s(u) = \frac{1}{2\pi}\int_{-1}^{1} f(x)\frac{e^{iux} - 1}{ix}\, dx + \frac{1}{2\pi} \mathop{\text{l.i.m.}}_{A\to\infty} \left[\int_1^A + \int_{-A}^{-1}\right] \frac{f(x)\, e^{iux}}{ix}\, dx. \tag{5.51}$$

$s(u)$ has a somewhat artificial appearance, due to the fact that it is necessary to avoid the consequences of the vanishing of the denominator at the origin.

We shall see later, however, that we always actually work with $\psi_\mu(u)$ rather than with $s(u)$.

As a result of the Plancherel theory,

$$\frac{1}{2\mu}\int_{-\infty}^{\infty} |s(u+\mu)-s(u-\mu)|^2\,du = \frac{1}{\pi\mu}\int_{-\infty}^{\infty} |f(x)|^2 \frac{\sin^2 \mu x}{x^2}\,dx. \tag{5.52}$$

It follows from this by an immediate application of the Tauberian theorem associated with the names of Bochner, Hardy, Jacob, Littauer, and the author, and already proved in this section, that

$$\lim_{\mu\to 0}\frac{1}{2\mu}\int_{-\infty}^{\infty} |s(u+\mu)-s(u-\mu)|^2\,du = \lim_{T\to\infty}\frac{1}{2T}\int_{-T}^{T} |f(x)|^2\,dx. \tag{5.53}$$

The meaning of (5.53) is that *if $f(x)$ is quadratically summable over every finite range, and $f(x)/x$ is quadratically summable over any infinite range excluding the origin, then if either side of* (5.53) *exists, the other side exists and assumes the same value.*

This formula is worthy of some detailed attention. If $s(u)$ is of limited total variation, we shall always have

$$\frac{1}{2\mu}\int_{-\infty}^{\infty} |s(u+\mu)-s(u-\mu)|\,du \leq V(s). \tag{5.54}$$

Accordingly, if in addition

$$\lim_{T\to\infty}\frac{1}{2T}\int_{-T}^{T} |f(x)|^2\,dx \neq 0, \tag{5.55}$$

the function $s(u)$ cannot be uniformly continuous. Again, if

$$s(u) = A_n, \quad [\lambda_n < u < \lambda_{n+1}] \tag{5.56}$$

we shall have

$$\lim_{\mu\to 0}\frac{1}{2\mu}\int_{-\infty}^{\infty} |s(u+\mu)-s(u-\mu)|^2\,du = \Sigma |A_{n+1}-A_n|^2, \tag{5.57}$$

so that $\lim_{T\to\infty} \frac{1}{2T} \int_{-T}^{T} |f(x)|^2 dx$ represents the sum of the squares of the moduli of the jumps of $s(u)$. Let it be noted that if $f(x)$ is a periodic function with the period 2π,

$$\lim_{T\to\infty} \frac{1}{2T} \int_{-T}^{T} |f(x)|^2 dx = \frac{1}{2\pi} \int_{-\pi}^{\pi} |f(x)|^2 dx, \tag{5.58}$$

while if

$$a_n = \frac{1}{2\pi} \int_{-\pi}^{\pi} f(x) e^{inx} dx, \tag{5.59}$$

then

$$\psi_\mu(u) = s(u+\mu) - s(u-\mu) = \sum_{[u-\mu]+1}^{[u+\mu]} a_n. \tag{5.60}$$

Thus our formula (5.53) is a generalization of the Parseval formula for the Fourier series, though it is not a direct generalization of the Parseval formula for the Fourier integral. For the Fourier integral,

$$\lim_{\mu\to 0} \frac{1}{2\mu} \int_{-\infty}^{\infty} |s(u+\mu) - s(u-\mu)|^2 du = 0, \tag{5.61}$$

although $s(u)$ exists, and indeed becomes the integral of the Fourier transform of $f(x)$. In this case, $s(u)$ is of limited total variation over every finite interval.

6. Bochner's generalization of harmonic analysis.

The study of the function $s(u)$ and its generalizations was first undertaken by Hahn, although, as we shall see later, on a basis insufficiently general to cover the needs of physics. The present author developed the theory for functions $f(x)$ with a finite mean square modulus, but the complete generalization of the theory is due to Bochner.

We have so far been interested in the problem of proceeding from $f(x)$ to $s(x)$. The question now arises, can we go backward, and determine $f(x)$ from $s(x)$? We should formally expect

20—29764. *Acta mathematica* 55. Imprimé le 8 avril 1930.

$$f(x) = \int_{-\infty}^{\infty} e^{-ixu}\, ds(u), \tag{6.01}$$

though the integral in question cannot be an ordinary Stieltjes integral, as $s(u)$ is not in general of limited total variation.

We may, however, develop this integral by a formal integration by parts, and we get

$$\int_{-A}^{A} e^{-ixu}\, ds(u) = e^{-iAx} s(A) - e^{iAx} s(-A) + ix \int_{-A}^{A} s(u)\, e^{-ixu}\, du$$

$$= e^{-iAx}\left[\frac{1}{2\pi}\int_{-1}^{1} f(\xi)\frac{e^{iA\xi}-1}{i\xi}\, d\xi + \underset{B\to\infty}{\text{l.i.m.}}\, \frac{1}{2\pi}\left[\int_{-B}^{-1} + \int_{1}^{B}\right] f(\xi)\frac{e^{iA\xi}}{i\xi}\, d\xi\right]$$

$$- e^{iAx}\left[\frac{1}{2\pi}\int_{-1}^{1} f(\xi)\frac{e^{-iA\xi}-1}{i\xi}\, d\xi + \underset{B\to\infty}{\text{l.i.m.}}\, \frac{1}{2\pi}\left[\int_{-B}^{-1} + \int_{1}^{B}\right] f(\xi)\frac{e^{-iA\xi}}{i\xi}\, d\xi\right]$$

$$+ ix\int_{-A}^{A} e^{-ixu}\, du\left[\frac{1}{2\pi}\int_{-1}^{1} f(\xi)\frac{e^{iu\xi}-1}{i\xi}\, d\xi + \underset{B\to\infty}{\text{l.i.m.}}\, \frac{1}{2\pi}\left[\int_{-B}^{-1} + \int_{1}^{B}\right]\frac{f(\xi)e^{iux}}{i\xi}\, d\xi\right]$$

$$= \underset{B\to\infty}{\text{l.i.m.}}\, \frac{1}{2\pi}\int_{-B}^{B}\frac{f(\xi)}{i\xi}\left[e^{iA(\xi-x)} - e^{-iA(\xi-x)} + ix\int_{-A}^{A} e^{iu(\xi-x)}\, du\right] d\xi$$

$$= \underset{B\to\infty}{\text{l.i.m.}}\, \frac{1}{\pi}\int_{-B}^{B} f(\xi)\frac{\sin A(\xi-x)}{\xi-x}\, d\xi. \tag{6.02}$$

Even this expression fails in general to converge in the mean as $A\to\infty$. A natural device to choose to compel the desired convergence in the mean is to replace this integral by its Cesaro sum, and to investigate the behavior of

$$\frac{1}{D}\int_{0}^{D} dA \int_{-A}^{A} e^{-ixu}\, ds(u) = \frac{1}{\pi D}\int_{-\infty}^{\infty} f(\xi)\frac{1-\cos D(\xi-x)}{(\xi-x)^2}\, d\xi. \tag{6.03}$$

This is the familiar Fejér expression for the partial Cesaro sum of a Fourier series, at least in form. The classical Fejér argument will prove that at any point of continuity of $f(x)$ we shall have

$$f(x) = \lim_{D\to\infty} \frac{1}{\pi D} \int_{-N}^{N} f(\xi) \frac{1-\cos D(\xi-x)}{(\xi-x)^2} d\xi, \tag{6.04}$$

and indeed, that this will in any case be true almost everywhere. To proceed to

$$\lim_{D\to\infty} \frac{1}{D} \int_{0}^{D} dA \int_{-A}^{A} e^{-ixu}\, ds(u) = f(x) \tag{6.05}$$

requires only the reflection that $\left[\int_{-\infty}^{-1} + \int_{1}^{\infty}\right] \frac{f(t)}{t^2} dt$ converges.

In a manner similar to that in which we have proved

$$\int_{-A}^{A} e^{-ixu}\, ds(u) = \underset{B\to\infty}{\text{l.i.m.}} \frac{1}{\pi} \int_{-B}^{B} f(\xi) \frac{\sin A(\xi-x)}{\xi-x} d\xi \tag{6.02}$$

we may show that

$$\int_{P}^{A} e^{-ixu}\, ds(u) = \underset{B\to\infty}{\text{l.i.m.}} \frac{1}{2\pi i} \int_{-B}^{B} f(\xi+x) \frac{e^{iA\xi} - e^{iP\xi}}{\xi} d\xi \tag{6.06}$$

as a function of A and P. Thus, except for an additive constant, $\int_{P}^{A} e^{-ixu}\, ds(u)$ bears to $f(x+\xi)$ the relation which $s(A)$ bears to $f(\xi)$. Similarly, to

$$\left.\begin{array}{l} f(x) \pm f(t+x) \quad \text{there corresponds} \int_{P}^{A} (1 \pm e^{-itu})\, ds(u) \\ \text{and to} \\ f(x) \pm i f(t+x) \quad \text{there corresponds} \int_{P}^{A} (1 \pm i e^{-itu})\, ds(u). \end{array}\right\} \tag{6.07}$$

By an obvious linear combination of the formulae relating to these four functions separately (cf. (1. 24)), we obtain

$$\lim_{\varepsilon \to 0} \frac{1}{2\varepsilon} \int_{-\infty}^{\infty} \left[\int_{u-\varepsilon}^{u+\varepsilon} e^{-itv}\, ds(v) \right] [\bar{s}(u+\varepsilon) - s(u-\varepsilon)]\, du$$

$$= \lim_{T \to \infty} \frac{1}{2T} \int_{-T}^{T} f(x+t) \bar{f}(x)\, dx = \varphi(t). \tag{6. 08}$$

Here, by definition,

$$\int_{u-\varepsilon}^{u+\varepsilon} e^{-itv}\, ds(v) = e^{-it(u+\varepsilon)} s(u+\varepsilon) - e^{-it(u-\varepsilon)} s(u-\varepsilon) + it \int_{u-\varepsilon}^{u+\varepsilon} s(v) e^{-itv}\, dv$$

$$= e^{-iut} (s(u+\varepsilon) - s(u-\varepsilon))$$

$$+ it \left\{ \int_{u}^{u+\varepsilon} [s(v) - s(u+\varepsilon)] e^{-itv}\, dv + \int_{u-\varepsilon}^{u} [s(v) - s(u-\varepsilon)] e^{itv}\, dv \right\}. \tag{6. 09}$$

However, we have

$$\frac{1}{\varepsilon} \left[\int_{u-\varepsilon}^{u+\varepsilon} e^{-itv}\, ds(v) - e^{-iut}(s(u+\varepsilon) - s(u-\varepsilon)) \right]$$

$$= \operatorname*{l.i.m.}_{B \to \infty} \frac{1}{2\pi i \varepsilon} \int_{-B}^{B} f(\xi+t) \left[\frac{e^{i(u+\varepsilon)\xi} - e^{i(u-\varepsilon)\xi}}{\xi} - e^{-iut} \frac{e^{i(u+\varepsilon)(\xi+t)} - e^{i(u-\varepsilon)(\xi+t)}}{\xi+t} \right] d\xi$$

$$= \operatorname*{l.i.m.}_{B \to \infty} \frac{1}{2\pi i \varepsilon} \int_{-B}^{B} f(\xi+t) \left\{ e^{i(u+\varepsilon)\xi} \left(\frac{1}{\xi} - \frac{e^{i\varepsilon t}}{\xi+t} \right) - e^{i(u-\varepsilon)\xi} \left(\frac{1}{\xi} - \frac{e^{-i\varepsilon t}}{\xi+t} \right) \right\} d\xi$$

$$= \operatorname*{l.i.m.}_{B \to \infty} \frac{1}{\pi} \int_{-B}^{B} f(\xi+t) e^{iu\xi} \left[\frac{\sin \varepsilon\xi}{\varepsilon\xi} - \frac{\sin \varepsilon(\xi+t)}{\varepsilon(\xi+t)} \right] d\xi. \tag{6. 10}$$

Now,

$$\left| \frac{\sin \varepsilon\xi}{\varepsilon\xi} - \frac{\sin \varepsilon(\xi+t)}{\varepsilon(\xi+t)} \right|$$

$$= \left| \frac{-2\xi \sin \frac{\varepsilon t}{2} \cos \varepsilon \left(\xi + \frac{t}{2} \right) + t \sin \varepsilon\xi}{\varepsilon\xi(\xi+t)} \right| \leq \left| \frac{2t}{\xi+t} \right| \text{and also} \leq 2. \tag{6. 11}$$

Thus, by the use of a Fourier transformation, it follows at once that

$$\int_{-\infty}^{\infty} \frac{1}{\varepsilon^2} \left| \int_{u-\varepsilon}^{u+\varepsilon} e^{-itv}\, ds(v) - e^{-iut}(s(u+\varepsilon)-s(u-\varepsilon)) \right|^2 du = O(1). \tag{6.12}$$

Inasmuch as we may readily show that

$$\lim_{\varepsilon\to 0} \int_{-\infty}^{\infty} |s(u+\varepsilon)-s(u-\varepsilon)|^2\, du = 0, \tag{6.13}$$

and since

$$\varphi(t) = \lim_{\varepsilon\to 0} \frac{1}{2\,\varepsilon} \int_{-\infty}^{\infty} e^{-itu} |s(u+\varepsilon)-s(u-\varepsilon)|^2\, du$$

$$+ \lim_{\varepsilon\to 0} \frac{1}{2\,\varepsilon} \int_{-\infty}^{\infty} \left[\int_{u-\varepsilon}^{u+\varepsilon} e^{-itv}\, ds(v) - e^{-iut}(s(u+\varepsilon)-s(u-\varepsilon)) \right] [\bar{s}(u+\varepsilon)-\bar{s}(u-\varepsilon)]\, du, \tag{6.14}$$

it follows by an elementary use of the Schwarz inequality that

$$\varphi(t) = \lim_{\varepsilon\to 0} \frac{1}{2\,\varepsilon} \int_{-\infty}^{\infty} e^{-itu} |s(u+\varepsilon)-s(u-\varepsilon)|^2\, du. \tag{6.15}$$

This formula holds in the same sense as (5.53) *for each t independently.*

We may deduce from (6.06) the analogous formula

$$\int_{-\lambda}^{\lambda} dy \int_{P}^{A} e^{-iyu}\, ds(u) = \int_{P}^{A} \frac{2\sin\lambda u}{u}\, ds(u) \tag{6.16}$$

by an easily justified inversion of the order of integration. Furthermore,

$$\int_{-\lambda}^{\lambda} dy \underset{B\to\infty}{\text{l.i.m.}} \frac{1}{2\pi i} \int_{-B}^{B} f(\xi+y) \frac{e^{iA\xi}-e^{iP\xi}}{\xi}\, d\xi$$

$$= \underset{B\to\infty}{\text{l.i.m.}} \frac{1}{2\pi i} \int_{-B}^{B} \frac{e^{iA\xi}-e^{iP\xi}}{\xi}\, d\xi \int_{-\lambda}^{\lambda} f(\xi+y)\, dy, \tag{6.17}$$

as may be deduced from the fact that

$$\int_{N}^{\infty} \frac{|f(\xi+y)|^2}{\xi^2}\,d\xi \quad \text{and} \quad \int_{-\infty}^{-N} \frac{|f(\xi+y)|^2}{\xi^2}\,d\xi$$

tend uniformly to 0 with increasing N for all y in $(-\lambda, \lambda)$. Thus to

$$\int_{-\lambda}^{\lambda} f(x+y)\,dy \quad \text{there corresponds} \quad \int_{P}^{A} \frac{2\sin\lambda u}{u}\,ds(u)$$

in the same sense in which to $f(x)$ there corresponds $s(A)$. It follows from (5. 53) that

$$\lim_{\mu\to 0} \frac{1}{2\mu}\int_{-\infty}^{\infty}\left|\int_{A-\mu}^{A+\mu} \frac{2\sin\lambda u}{u}\,ds(u)\right|^2 dA = \lim_{T\to\infty}\frac{1}{2T}\int_{-T}^{T}\left|\int_{-\lambda}^{\lambda} f(x+y)\,dy\right|^2 dx. \tag{6. 18}$$

As in (6. 09),

$$\int_{A-\mu}^{A+\mu} \frac{2\sin\lambda u}{u}\,ds(u) = \frac{2\sin\lambda(A+\mu)}{A+\mu}\,s(A+\mu) - \frac{2\sin\lambda(A-\mu)}{A-\mu}\,s(A-\mu)$$

$$-\int_{A-\mu}^{A+\mu} s(u)\,\frac{d}{du}\left(\frac{2\sin\lambda u}{u}\right)du, \tag{6. 19}$$

and as in (6. 10),

$$\frac{1}{\mu}\int_{A-\mu}^{A+\mu} \frac{2\sin\lambda u}{u}\,ds(u) - \frac{2\sin\lambda A}{A\mu}\,[s(A+\mu)-s(A-\mu)]$$

$$= \frac{1}{\mu}\int_{A-\mu}^{A+\mu} ds(u)\int_{-\lambda}^{\lambda} e^{-i\sigma u}\,d\sigma - 2\int_{-\lambda}^{\lambda} e^{-i\sigma A}\,d\sigma\,[s(A+\mu)-s(A-\mu)]$$

$$= \int_{-\lambda}^{\lambda} d\sigma \operatorname*{l.i.m.}_{B\to\infty} \frac{1}{\pi}\int_{-B}^{B} f(\xi+\sigma)\,e^{iA\sigma}\left[\frac{\sin\mu\xi}{\mu\xi} - \frac{\sin\mu(\xi+\sigma)}{\mu(\xi+\sigma)}\right] d\xi \tag{6. 20}$$

because (6. 10) holds uniformly over a finite range of σ, because of the fact that

$$\left|\frac{\sin \mu\xi}{\mu\xi} - \frac{\sin \mu(\xi+\sigma)}{\mu(\xi-\sigma)}\right| = O(\xi^{-1}) \tag{6. 21}$$

uniformly over such a range. Hence, as in (6. 12), we may show that

$$\int_{-\infty}^{\infty} \frac{1}{\mu^2} \left| \int_{A-\mu}^{A+\mu} \frac{2 \sin \lambda u}{u} \, ds(u) - \frac{2 \sin \lambda A}{A} [s(A+\mu) - s(A-\mu)] \right|^2 dA = O(1); \tag{6. 22}$$

and by an argument exactly similar to that leading to (6. 15), we obtain

$$\lim_{T\to\infty} \frac{1}{2T} \int_{-T}^{T} \left| \int_{-\lambda}^{\lambda} f(x+y)\, dy \right|^2 dx = \lim_{\mu\to 0} \frac{1}{2\mu} \int_{-\infty}^{\infty} \frac{4 \sin^2 \lambda u}{u^2} \, | s(u+\mu) - s(u-\mu) |^2 \, du \tag{6. 23}$$

for each λ *independently, in the same sense as that in which we have proved* (5. 53) *and* (6. 15).

We now proceed to the part of the theory that is specifically Bochner's. We wish to discuss the harmonic analysis of functions which are no longer bounded on the average (for so we may interpret the finiteness of $\lim_{T\to\infty} \frac{1}{2T} \int_{-T}^{T} |f(x)|^2 dx$), but instead have on the average an algebraic rate of growth as the argument proceeds to $\pm\infty$. That is, we assume the existence of

$$\lim_{T\to\infty} \frac{1}{2T^n} \int_{-T}^{T} |f(x)|^2 \, dx \tag{6. 24}$$

and we shall take n to be a positive odd integer. By arguments following identically the lines laid down for $n=1$, we show that

$$\left[\int_{1}^{T} + \int_{-T}^{-1}\right] \frac{|f(x)|^2}{x^{n+1}} \, dx = O(1), \tag{6. 25}$$

and hence that

$$\left[\int_{1}^{\infty} + \int_{-\infty}^{-1}\right] \frac{|f(x)|^2}{x^{n+1}} \, dx \tag{6. 26}$$

exists. We can then show the existence of

$$s(u) = s_1(u) + s_2(u)$$

$$= \frac{1}{2\pi} \int_{-1}^{1} f(x) \left[\left(\int_0^u \right)^{\frac{n+1}{2}} e^{ivx}\, dv \right] dx + \frac{1}{2\pi} \underset{A\to\infty}{\text{l.i.m.}} \left[\int_1^A + \int_{-A}^{-1} \right] f(x) \frac{e^{iux}}{(ix)^{\frac{n+1}{2}}}\, dx. \qquad (6.27)$$

If we now put

$$\Delta_\mu^{\frac{n+1}{2}}(s) = s\left(u + \mu\left(\frac{n+1}{2}\right)\right) - ns\left(u + \mu\left(\frac{n-3}{2}\right)\right) + \frac{n(n-1)}{2!} s\left(u + \mu\left(\frac{n-7}{2}\right)\right)$$

$$- \cdots \pm s\left(u - \mu\left(\frac{n+1}{2}\right)\right)$$

$$= \frac{2^{\frac{n+1}{2}}}{2\pi} \underset{A\to\infty}{\text{l.i.m.}} \int_{-A}^{A} f(x)\, e^{iux} \frac{\sin^{\frac{n+1}{2}} \mu x}{x^{\frac{n+1}{2}}}\, dx, \qquad (6.28)$$

we can show by a Plancherel argument that

$$\frac{1}{\mu} \int_{-\infty}^{\infty} \left| \Delta_\mu^{\frac{n+1}{2}}(s(u)) \right|^2 du = \frac{2^n}{\pi\mu} \int_{-\infty}^{\infty} |f(x)|^2 \frac{\sin^{n+1} \mu x}{x^{n+1}}\, dx. \qquad (6.29)$$

It is easy to show that for any function $F(x)$,

$$\frac{1}{T^n} \int_0^T F(x)\, dx = \frac{1}{T^n} \int_0^T \frac{F(x)}{x^{n-1}} x^{n-1}\, dx = \int_0^1 \frac{F(Ty)}{(Ty)^{n-1}} y^{n-1}\, dy. \qquad (6.30)$$

Inasmuch as

$$\int_0^1 y^{n-1}\, dy = 1/n, \qquad (6.31)$$

if we put

$$F(x)/nx^{n-1} = G(x), \qquad (6.32)$$

the above integral is a mean of G, in the sense in which means enter into our general Tauberian theorem. Furthermore,

$$\frac{2^{n+1}}{\pi\mu} \int_0^\infty F(x) \frac{\sin^{n+1} \mu x}{x^{n+1}}\, dx = \frac{2^{n+1} n}{\pi\mu} \int_0^\infty G(x) \frac{\sin^{n+1} \mu x}{x^2}\, dx$$

$$= \frac{2^{n+1} n}{\pi} \int_0^\infty G(y/\mu) \frac{\sin^{n+1} y}{y^2}\, dy \qquad (6.33)$$

and if we put

$$\frac{2^{n+1}n}{\pi}\int_0^\infty \frac{\sin^{n+1}y}{y^2}\,dy = P_n, \tag{6.34}$$

then

$$\frac{2^{n+1}}{\pi\mu P_n}\int_0^\infty F(x)\frac{\sin^{n+1}\mu x}{x^{n+1}}\,dx \tag{6.35}$$

is another mean of the quantity G. If we set

$$M_1(x)\;[\text{or } M_2(x)] = \begin{cases} n x^{n-1}; & [0 < x < 1] \\ 0 \quad ; & [x \geq 1] \end{cases} \qquad M_2(x)\;[\text{or } M_1(x)] = \frac{2^{n+1}n}{\pi P_n}\frac{\sin^{n+1}x}{x^2}, \tag{6.36}$$

then

$$\int_0^\infty M_1(x)\,x^{iu}\,dx = \frac{n}{n+iu} \neq 0. \tag{6.37}$$

As to

$$\int_0^\infty M_2(x)\,x^{iu}\,dx,$$

we have

$$\int_0^\infty \sin^{n+1}x\, x^{iu-2}\,dx$$

$$= \int_0^\infty \frac{(n+1)!}{2^{n-1}}\left[\frac{\sin^2 x}{\frac{n-1}{2}!\,\frac{n+3}{2}!} - \frac{\sin^2 2x}{\frac{n-3}{2}!\,\frac{n+5}{2}!} + \frac{\sin^2 3x}{\frac{n-5}{2}!\,\frac{n+7}{2}!} - \cdots\right] x^{iu-2}\,dx.$$

Hence

$$\int_0^\infty M_2(x)\,x^{iu}\,dx = \frac{4n(n+1)!}{\pi P_n}\cdot$$

$$\left[\frac{1}{\frac{n-1}{2}!\,\frac{n+3}{2}!} - \frac{2^{iu-1}}{\frac{n-3}{2}!\,\frac{n+5}{2}!} + \frac{3^{iu-1}}{\frac{n-5}{2}!\,\frac{n+7}{2}!} - \cdots\right]\int_0^\infty \sin^2 x\, x^{iu-2}\,dx. \tag{6.38}$$

We have already seen that

$$\int_0^\infty \sin^2 x \, x^{iu-2} \, dx \neq 0. \tag{6.39}$$

Thus the question of the possibility that

$$\int_0^\infty M_2(x) \, x^{iu} \, dx$$

should vanish depends on the possibility of the vanishing of

$$\frac{1}{\frac{n-1}{2}! \, \frac{n+3}{2}!} - \frac{2^{iu-1}}{\frac{n-3}{2}! \, \frac{n+5}{2}!} + \frac{3^{iu-1}}{\frac{n-5}{2}! \, \frac{n+7}{2}!} - \cdots. \tag{6.40}$$

It is easy enough to prove that this cannot vanish for $n=1, 3, 5, 7, 9, 11$ but the author has not yet been able to produce a proof in the general case.

In case this expression does not vanish, we may apply our Tauberian theorem, replacing $f(x)$ by $|f(x)|^2+|f(-x)|^2$. We shall assume to begin with that $f(x)$ vanishes in the neighbourhood of the origin. Then

$$\lim_{T\to\infty} \frac{1}{2\,T^n} \int_{-T}^{T} |f(x)|^2 \, dx = A \tag{6.41}$$

and

$$\lim_{\mu\to 0} \frac{1}{P_n \mu} \int_{-\infty}^{\infty} \left| \Delta_\mu^{\frac{n+1}{2}} s(u) \right|^2 du = \lim_{\mu\to 0} \frac{2^n}{P_n \pi \mu} \int_{-\infty}^{\infty} |f(x)|^2 \frac{\sin^{n+1} \mu x}{x^{n+1}} \, dx \tag{6.42}$$

are equivalent.

We have put $f(x)=0$ in the neighbourhood of the origin to be sure of the boundedness of

$$\frac{1}{2\,T^n} \int_{-T}^{T} |f(x)|^2 \, dx \quad \text{and} \quad \frac{2^n}{P_n \pi \mu} \int_{-\infty}^{\infty} |f(x)|^2 \frac{\sin^{n+1} \mu x}{x^{n+1}} \, dx. \tag{6.43}$$

In any case,

$$\lim_{T\to\infty} \frac{1}{2\,T^n} \int_{-T}^{T} |f(x)|^2 \, dx$$

will not be changed if $f(x)$ is made to vanish in this neighbourhood. Moreover,

$$\frac{1}{\mu}\int_{-B}^{B}|f(x)|^2\frac{\sin^{n+1}\mu x}{x^{n+1}}dx \leq \mu^n\int_{-B}^{B}|f(x)|^2\,dx. \tag{6. 44}$$

Thus

$$\lim_{\mu\to 0}\frac{1}{P_n\mu}\int_{-\infty}^{\infty}\left|\Delta_{\mu}^{\frac{n+1}{2}}s(u)\right|du$$

will not be changed either, and we may always write:

$$\lim_{\mu\to 0}\frac{1}{P_n\mu}\int_{-\infty}^{\infty}\left|\Delta_{\mu}^{\frac{n+1}{2}}s(u)\right|^2du = \lim_{T\to\infty}\frac{1}{2T^n}\int_{-T}^{T}|f(x)|^2\,dx. \tag{6. 45}$$

We now come to the problem of returning from $s(u)$ to $f(x)$. We should expect formally

$$f(x)=\int_{-\infty}^{\infty}e^{-iux}\,d^{\frac{n+1}{2}}s(u)/du^{\frac{n-1}{2}} \tag{6. 46}$$

Again, we need to interpret

$$\int_{-\infty}^{\infty}e^{-iux}\,d^{\frac{n+1}{2}}s(u)/du^{\frac{n-1}{2}}$$

by an integration by parts, and again some form of summation is necessary to get a more manageable expression. One method is the following: Let us replace

$$\int_{-\infty}^{\infty}e^{-iux}\,d^{\frac{n+1}{2}}s(u)/du^{\frac{n-1}{2}}$$

by

$$\int_{-\infty}^{\infty}e^{-\lambda u^2-iux}\,d^{\frac{n+1}{2}}s(u)/du^{\frac{n-1}{2}} \tag{6. 47}$$

and let us investigate its behavior as $\lambda\to 0$. This expression is to be interpreted by $\frac{n+1}{2}$ formal integrations by parts, which convert it to

$$(-1)^{\frac{n+1}{2}} \int_{-\infty}^{\infty} s_2(u) d^{\frac{n+1}{2}} (e^{-\lambda u^2 - iux})/du^{\frac{n+1}{2}} du. \qquad (6.48)$$

It then becomes obvious that this expression will have the same limit in the mean as

$$\int_{-B}^{B} s_2(u)(ix)^{\frac{n+1}{2}} e^{-iux} du \qquad [B \to \infty] \qquad (6.49)$$

It here will, however, be $f(x)$ if $|x| > A$, and will be 0 otherwise. To see this, we need only compare $\frac{d^{\frac{n+1}{2}}}{dx^{\frac{n+1}{2}}}(e^{-\lambda u^2 - iux})$ with $\frac{d^{\frac{n+1}{2}}}{dx^{\frac{n+1}{2}}} e^{-iux}$. A similar result holds for $s_1(u)$, and the final result is that over any finite interval not including the origin,

$$f(x) = \underset{\lambda \to 0}{\text{l.i.m.}} \int_{-\infty}^{\infty} e^{-\lambda u^2 - iux} d^{\frac{n+1}{2}} s(u)/du^{\frac{n-1}{2}} \qquad (6.50)$$

7. The Hahn generalization of harmonic analysis.

Up to the present we have concerned ourselves rather with questions of convergence in the mean than of ordinary convergence. Retaining our previous notation, in the case where $n=1$, we may raise the further questions: when does

$$\int_{-A}^{A} e^{-iux} ds(u)$$

exist for all u as an ordinary Stieltjes integral, rather than a generalized Stieltjes integral such as we have treated in the last section? When does

$$\lim_{A \to \infty} \int_{-A}^{A} e^{-iux} ds(u)$$

converge in the ordinary sense? These questions furnish the vital link between the generalized harmonic analysis of Hahn, and that developed here. In the

$$= \underset{A\to\infty}{\text{l.i.m.}} \frac{1}{8\pi\eta^2}\int_{-A}^{A}\varphi(x)\,dx\int_{u-\mu}^{u+\mu}dw\int_{-2\eta}^{2\eta}(2\eta-|y|)\,e^{iw(x+y)}\,dy$$

$$= \underset{A\to\infty}{\text{l.i.m.}} \frac{1}{4\pi\eta^2}\int_{-A}^{A}\varphi(x)\,dx\int_{u-\mu}^{u+\mu}\frac{e^{iwx}(1-\cos 2w\eta)}{w^2}\,dw$$

$$= \underset{A\to\infty}{\text{l.i.m.}} \frac{1}{4\pi\eta^2}\int_{u-\mu}^{u+\mu}\frac{1-\cos 2w\eta}{w^2}\,d_w\int_{-A}^{A}\varphi(x)\frac{e^{iux}-1}{ix}\,dx$$

$$= \frac{1}{4\pi\eta^2}\int_{u-\mu}^{u+\mu}\frac{1-\cos 2w\eta}{w^2}\,d_w\int_{-\infty}^{\infty}\varphi(x)\frac{e^{iwx}-1}{ix}\,dx$$

$$= \frac{1}{2\eta^2}\int_{u-\mu}^{u+\mu}\frac{1-\cos 2w\eta}{w^2}\,dS(w) = \int_{u-\mu}^{u+\mu}\frac{\sin^2\eta w}{\eta^2 w^2}\,dS(w); \qquad (7.13)$$

$$s^{(\eta)}(u+\varepsilon)-s^{(\eta)}(u-\varepsilon) = \underset{A\to\infty}{\text{l.i.m.}} \frac{1}{\pi}\int_{-A}^{A} f^{(\eta)}(x)\frac{\sin\varepsilon x}{x}e^{iux}\,dx$$

$$= \underset{A\to\infty}{\text{l.i.m.}} \frac{1}{2\pi\eta}\int_{-\eta}^{\eta}d\xi\int_{-A}^{A}f(x+\xi)\frac{\sin\varepsilon x}{x}e^{iux}\,dx$$

$$= \frac{1}{2\eta}\int_{-\eta}^{\eta}d\xi\int_{u-\varepsilon}^{u+\varepsilon}e^{-i\xi u}\,ds(u) \text{ [by (6.06)].} \qquad (7.14)$$

Although this is not in general an ordinary Stieltjes integral, we may integrate by parts and then invert the order of integration, and thus obtain

$$s^{(\eta)}(u+\varepsilon)-s^{(\eta)}(u-\varepsilon) = \frac{1}{2\eta}\int_{u-\varepsilon}^{u+\varepsilon}ds(u)\int_{-\eta}^{\eta}e^{-i\xi u}\,d\xi = -\int_{u-\varepsilon}^{u+\varepsilon}\frac{\sin\eta u}{\eta u}\,ds(u). \qquad (7.15)$$

Hence, by (5.53) and (6.23),

$$\varphi^{(\eta)}(0) = \lim_{T\to\infty}\frac{1}{2T}\int_{-T}^{T}|f^{(\eta)}(x)|^2\,dx = \lim_{\varepsilon\to 0}\frac{1}{2\varepsilon}\int_{-\infty}^{\infty}\frac{\sin^2\eta u}{\eta^2 u^2}|s(u+\varepsilon)-s(u-\varepsilon)|^2\,du. \qquad (7.16)$$

22—29764. *Acta mathematica.* 55. Imprimé le 14 avril 1930.

may now be carried out, and all our quasi-Stieltjes integrals become ordinary Stieltjes integrals. If, on the other hand, $s(u)$ is not of limited total variation, Hahn adopts a generalized definition of the Stieltjes integral identical in content with that here given. In either case, we have already seen that

$$\int_{-A}^{A} e^{-ixu}\, ds(u) = \underset{B\to\infty}{\text{l.i.m.}}\, \frac{1}{\pi} \int_{-B}^{B} f(\xi) \frac{\sin A(\xi-x)}{\xi-x}\, d\xi, \tag{7.04}$$

and that

$$f(x) = \lim_{A\to\infty} \frac{1}{2\pi A} \int_{-\infty}^{\infty} f(\xi) \frac{1-\cos 2A(\xi-x)}{(\xi-x)^2}\, d\xi. \tag{7.05}$$

Moreover,

$$\underset{B\to\infty}{\text{l.i.m.}}\, \frac{1}{2\pi A} \int_{-B}^{B} f(\xi) \frac{1-\cos 2A(\xi-x) - 2A(\xi-x)\sin A(\xi-x)}{(\xi-x)^2}\, d\xi$$

$$= \frac{1}{2\pi} \int_{-\infty}^{\infty} f\left(\xi + \frac{w}{A}\right) \frac{1-\cos 2w - 2w \sin w}{w^2}\, dw$$

$$= \frac{1}{2\pi} \int_{-\infty}^{\infty} \left[f\left(\xi + \frac{w}{A}\right) - f(\xi)\right] \frac{1-\cos 2w - 2w\sin w}{w^2}\, dw$$

$$+ \frac{f(\xi)}{2\pi} \int_{-\infty}^{\infty} \frac{1-\cos 2w - 2w\sin w}{w^2}\, dw$$

$$= \frac{1}{2\pi} \int_{-D}^{D} \left[f\left(\xi + \frac{w}{A}\right) - f(\xi)\right] \frac{1-\cos 2w - 2w\sin w}{w^2}\, dw$$

$$+ \frac{1}{2\pi} \left[\int_{-\infty}^{-D} + \int_{D}^{\infty}\right] \left[f\left(\xi + \frac{w}{A}\right) - f(\xi)\right] \frac{1-\cos 2w - 2w\sin w}{w^2}\, dw. \tag{7.06}$$

Thus a sufficient condition for

$$\int_{-\infty}^{\infty} e^{-ixu}\, ds(u) = f(x) \tag{7.07}$$

is that $f(x)$ should satisfy locally one of the sufficient conditions for the convergence of a Fourier series, and that

$$\left[\int_{-\infty}^{-1} + \int_{1}^{\infty}\right]\left|\frac{f(w)}{w}\right| dw \text{ converge}$$

or what is the same, that

$$\int_{-\infty}^{\infty}\left|\frac{f(w)}{\sqrt{1+w^2}}\right| dw \tag{7.08}$$

exist. This condition thus constitutes a sufficient solution of the Hahn problem. To such a function we may add any function with a convergent Fourier series, without destroying the fact that it solves Hahn's problem.

Another condition under which (7.07) holds is that $f(x)$ should be of the form

$$g(x) \cdot h(x)$$

where $g(x)$ is periodic, and $h(x)$ bounded and monotone at $\pm\infty$, and that $f(x)$ should satisfy one of the sufficient conditions for the convergence of a Fourier series to its function. As in the previous case, the proof of his assertion made by Hahn depends simply on the fact that the second term of the last line of (7.06) will then vanish with increasing D, while the first term is asymptotically equal to the difference between the Cesaro and ordinary partial sums of the Fourier integral of the function $f(x)$ mutilated by being made to vanish outside the interval $(-D, D)$. It then follows from the Fejér theorem and the fact that the conditions for the convergence of such a Fourier integral and of a Fourier series are the same, that Hahn's theorem holds.

The class of functions for which $s(u)$ exists as an ordinary Stieltjes integral is too narrow to cover the physically interesting cases of continuous spectra. To see this, let $f(x)$ be a function for which $\varphi(x)$ and $S(u)$ exist, and let

$$f^{(\eta)}(x) = \frac{1}{2\eta} \cdot \int_{-\eta}^{\eta} f(x+\xi)\, d\xi. \tag{7.09}$$

Let us suppose that

$$\varphi^{(\eta)}(x) = \lim_{T\to\infty} \frac{1}{2T} \int_{-T}^{T} f^{(\eta)}(x+t) \bar{f}^{(\eta)}(t)\, dt \tag{7.10}$$

exists for every x, and hence that

$$S^{(\eta)}(u) = \frac{1}{2\pi} \int_{-\infty}^{\infty} \varphi^{(\eta)}(x) \frac{e^{iux}-1}{ix}\, dx \tag{7.11}$$

exists. Then

$$\varphi^{(\eta)}(x) = \lim_{T\to\infty} \frac{1}{2T} \frac{1}{4\eta^2} \int_{-\eta}^{\eta} d\xi \int_{-\eta}^{\eta} d\zeta \int_{-T}^{T} f(x+t+\xi) \bar{f}(t+\zeta)\, dt$$

$$= \frac{1}{4\eta^2} \int_{-\eta}^{\eta} d\xi \int_{-\eta}^{\eta} d\zeta\, \varphi(x+\xi-\zeta)$$

$$= \frac{1}{4\eta^2} \int_{-\eta}^{\eta} d\xi \int_{-\eta-\xi}^{\eta-\xi} \varphi(x-y)\, dy$$

$$= \frac{1}{4\eta^2} \left\{ \eta \int_{-2\eta}^{0} \varphi(x-y)\, dy + \eta \int_{0}^{2\eta} \varphi(x-y)\, dy - \int_{-\eta}^{\eta} \xi [\varphi(x+\eta+\xi) - \varphi(x-\eta+\xi)]\, d\xi \right\}$$

$$= \frac{1}{4\eta^2} \int_{-2\eta}^{2\eta} \varphi(x-y)(2\eta - |y|)\, dy; \tag{7.12}$$

$$S^{(\eta)}(u+\mu) - S^{(\eta)}(u-\mu) = \frac{1}{\pi} \underset{A\to\infty}{\text{l.i.m.}} \int_{-A}^{A} \varphi^{(\eta)}(x) \frac{\sin \mu x}{x} e^{iux}\, dx$$

$$= \underset{A\to\infty}{\text{l.i.m.}} \frac{1}{4\pi\eta^2} \int_{-2\eta}^{2\eta} (2\eta - |y|)\, dy \int_{-A-y}^{A-y} \varphi(x-y) \frac{\sin \mu x}{x} e^{iux}\, dx$$

$$= \underset{A\to\infty}{\text{l.i.m.}} \frac{1}{4\pi\eta^2} \int_{-A}^{A} \varphi(x) dx \int_{-2\eta}^{2\eta} (2\eta - |y|) \frac{\sin \mu(x+y)}{x+y} e^{iu(x+y)}\, dy$$

$$= \underset{A\to\infty}{\text{l.i.m.}} \frac{1}{8\pi\eta^2}\int_{-A}^{A}\varphi(x)\,dx\int_{u-\mu}^{u+\mu}dw\int_{-2\eta}^{2\eta}(2\eta-|y|)\,e^{iw(x+y)}\,dy$$

$$= \underset{A\to\infty}{\text{l.i.m.}} \frac{1}{4\pi\eta^2}\int_{-A}^{A}\varphi(x)\,dx\int_{u-\mu}^{u+\mu}\frac{e^{iwx}(1-\cos 2w\eta)}{w^2}\,dw$$

$$= \underset{A\to\infty}{\text{l.i.m.}} \frac{1}{4\pi\eta^2}\int_{u-\mu}^{u+\mu}\frac{1-\cos 2w\eta}{w^2}\,dw\int_{-A}^{A}\varphi(x)\frac{e^{iwx}-1}{ix}\,dx$$

$$= \frac{1}{4\pi\eta^2}\int_{u-\mu}^{u+\mu}\frac{1-\cos 2w\eta}{w^2}\,dw\int_{-\infty}^{\infty}\varphi(x)\frac{e^{iwx}-1}{ix}\,dx$$

$$= \frac{1}{2\eta^2}\int_{u-\mu}^{u+\mu}\frac{1-\cos 2w\eta}{w^2}\,d\,S(w) = \int_{u-\mu}^{u+\mu}\frac{\sin^2\eta w}{\eta^2 w^2}\,d\,S(w); \tag{7.13}$$

$$s^{(\eta)}(u+\varepsilon)-s^{(\eta)}(u-\varepsilon) = \underset{A\to\infty}{\text{l.i.m.}} \frac{1}{\pi}\int_{-A}^{A}f^{(\eta)}(x)\,\frac{\sin\varepsilon x}{x}\,e^{iux}\,dx$$

$$= \underset{A\to\infty}{\text{l.i.m.}} \frac{1}{2\pi\eta}\int_{-\eta}^{\eta}d\xi\int_{-A}^{A}f(x+\xi)\,\frac{\sin\varepsilon x}{x}\,e^{iux}\,dx$$

$$= \frac{1}{2\eta}\int_{-\eta}^{\eta}d\xi\int_{u-\varepsilon}^{u+\varepsilon}e^{-i\xi u}\,ds(u)\ \ [\text{by (6.06)}]. \tag{7.14}$$

Although this is not in general an ordinary Stieltjes integral, we may integrate by parts and then invert the order of integration, and thus obtain

$$s^{(\eta)}(u+\varepsilon)-s^{(\eta)}(u-\varepsilon) = \frac{1}{2\eta}\int_{u-\varepsilon}^{u+\varepsilon}ds(u)\int_{-\eta}^{\eta}e^{-i\xi u}\,d\xi = -\int_{u-\varepsilon}^{u+\varepsilon}\frac{\sin\eta u}{\eta u}\,ds(u). \tag{7.15}$$

Hence, by (5.53) and (6.23),

$$\varphi^{(\eta)}(0) = \lim_{T\to\infty}\frac{1}{2T}\int_{-T}^{T}|f^{(\eta)}(x)|^2\,dx = \lim_{\varepsilon\to 0}\frac{1}{2\varepsilon}\int_{-\infty}^{\infty}\frac{\sin^2\eta u}{\eta^2u^2}\,|s(u+\varepsilon)-s(u-\varepsilon)|^2\,du. \tag{7.16}$$

By (7. 14),

$$\varphi^{(\eta)}(0) = \frac{1}{4\eta^2}\int_{-2\eta}^{2\eta} \varphi(\xi)(2\eta - |\xi|)\, d\xi$$

$$= \frac{1}{4\eta^2}\int_{0}^{2\eta} 2\lambda d\lambda \left[\frac{1}{2\lambda}\int_{-\lambda}^{\lambda}\varphi(\xi)\, d\xi\right], \tag{7. 17}$$

which in combination with (5. 47) leads us to

$$\lim_{\eta\to 0} \varphi^{(\eta)}(0) = S(\infty) - S(-\infty). \tag{7. 18}$$

In other words,

$$\lim_{\eta\to 0}\lim_{\varepsilon\to 0}\frac{1}{2\varepsilon}\int_{-\infty}^{\infty}\frac{\sin^2 \eta u}{\eta^2 u^2}|s(u+\varepsilon) - s(u-\varepsilon)|^2\, du = S(\infty) - S(-\infty). \tag{7. 19}$$

If we now assume that

$$\lim_{\eta\to 0}\varphi^{(\eta)}(0) = \varphi(0), \tag{7. 20}$$

it follows from (7. 21), (7. 22), and (5. 54) that

$$\lim_{\eta\to 0}\lim_{\varepsilon\to 0}\frac{1}{2\varepsilon}\int_{-\infty}^{\infty}\left[1 - \frac{\sin^2\eta u}{\eta^2 u^2}\right]|s(u+\varepsilon) - s(u-\varepsilon)|^2 du = 0, \tag{7. 21}$$

or since $\sin \eta u/\eta u$ tends to 1 as $\eta \to 0$

$$\lim_{A\to\infty}\lim_{\varepsilon\to 0}\frac{1}{2\varepsilon}\left[\int_{A}^{\infty} + \int_{-\infty}^{-A}\right]|s(u+\varepsilon) - s(u-\varepsilon)|^2 du = 0. \tag{7. 22}$$

If $\varphi(x)$ is a continuous function, not only shall we have (7. 20) as a consequence of (7. 17), but all the functions $\varphi^{(\eta)}(x)$ will be continuous, as follows from (7. 12). Then, by (6. 15), we shall have

$$\lim_{\varepsilon\to 0}\frac{1}{2\varepsilon}\int_{-\infty}^{\infty} e^{-iux}|s(u+\varepsilon) - s(u-\varepsilon)|^2\, du = \int_{-\infty}^{\infty} e^{-iux}\, d\, S(u). \tag{7. 23}$$

It follows at once that if $P(u)$ is a trigonometrical polynomial, or the uniform limit of such a polynomial,

$$\lim_{\varepsilon\to 0}\frac{1}{2\varepsilon}\int_{-\infty}^{\infty} P(u)\,|\,s(u+\varepsilon)-s(u-\varepsilon)\,|^2\,du = \int_{-\infty}^{\infty} P(u)\,dS(u). \tag{7.24}$$

For the transition to the case where $P(u)$ is a uniform limit of a polynomial we need only make appeal to the boundedness of

$$\lim_{\varepsilon\to 0}\frac{1}{2\varepsilon}\int_{-\infty}^{\infty} |\,s(u+\varepsilon)-s(u-\varepsilon)\,|^2\,du. \tag{7.25}$$

Hence, by Fejér's theorem, $P(u)$ may be any continuous periodic function, and because of (7.22), any continuous function differing from 0 only over a finite range, as the change in the left side of this expression due to making $P(u)$ artificially periodic tends to 0 as the period increases.

It follows at once that under the hypotheses:

(a) $\varphi^{(\eta)}(x)$ exists for every x and η;

(b) $\varphi(x)$ is continuous;

if

$$\lim_{\varepsilon\to 0}\frac{1}{2\varepsilon}\int_{\alpha}^{\beta} |\,s(u+\varepsilon)-s(u-\varepsilon)\,|^2\,du=0, \tag{7.26}$$

over any interval, $S(u)$ is a constant over any interior interval. Thus if $s(u)$ is of limited total variation over any finite interval, and is continuous, $S(u)$ reduces to a constant, and the spectrum of $f(x)$ vanishes. In other words, the very natural hypotheses (a) and (b) are inconsistent with the existence of a continuous spectrum, provided $s(u)$ is of limited total variation. To see this, we need only notice that almost everywhere

$$S(u_2)-S(u_1) = \lim_{\varepsilon\to 0}\frac{1}{2\varepsilon}\int_{u_1}^{u_2} |\,s(u+\varepsilon)-s(u-\varepsilon)\,|^2\,du$$

$$= \text{sum of squares of jumps of } S(u) \text{ between } u_1 \text{ and } u_2. \tag{7.27}$$

Thus the expansion of $f(x)$ in an ordinary Stieltjes integral is not adequate to the discussion of such continuous spectra as occur in physics, inasmuch, as in these cases, as we shall see in section 13, conditions analogous to (a) and (b) are fulfilled.

CHAPTER III.

8. **Harmonic analysis in more than one dimension.**

The elementary function of harmonic analysis in one dimension is e^{iux}. In n dimensions, this is replaced by

$$e^{i(u_1x_1+\cdots+u_nx_n)}$$

which we may write vectorially

$$e^{i(U\cdot X)},$$

where the vector X represents the argument of the function to be analyzed, and the vector U the vectorial frequency. If we keep the term $(U.X)$ invariant, and X varies cogrediently, U varies contragrediently. Thus the familiar duality relation between Fourier transforms is intimately connected with the point-plane duality of geometry. This is why the relation between position-coordinates and momentum-coordinates in modern quantum physics appears as a Fourier duality, while the same relation appears in the theory of relativity as the relation between a certain cogredient tensor and a certain contragredient tensor.

Practically the whole generalized theory of harmonic analysis so far developed is susceptible to a generalization to n dimensions. This generalization has been carried out by Mr. A. C. Berry, who has been kind enough to furnish me with the following summary of his Harvard doctoral thesis.

It is necessary to introduce certain notations at once and to make a few preliminary remarks. Let there be given a real n-dimensional space and, in it, some fixed reference point, or origin, O. If X is an arbitrary point of this space, the symbol X shall be used to denote not only this point, but also any real n-dimensional vector equivalent to the directed line segment OX. Let $f(X)$ be any complex, measurable function defined for all such real arguments X. Hereafter it will be assumed that all functions with which we start satisfy these requirements. If R is any measurable point-set, then

$$\int_{R} f(X)\, d\, V_X \tag{8.01}$$

shall mean the n-dimensional volume integral, in the sense of Lebesgue, of $f(X)$ taken over the region R. Since, in general, n-dimensional »spheres» will be employed as regions of integration, it will be convenient to use the notation $(r;\, X)$ to signify a sphere of radius r and having its center at X. The vector interpretation of X enables us to write

$$\int_{(r;\, Y)} f(X)\, d\, V_X - \int_{(r;\, 0)} f(X+Y)\, d\, V_X. \tag{8.02}$$

The »volume» or measure, of an n-dimensional sphere of radius r is known to be

$$\frac{\pi^{\frac{n}{2}}\, r^n}{\Gamma\left(\frac{n}{2}+1\right)} \tag{8.03}$$

which quantity, hereafter, shall be denoted by the symbol $v(r)$. Thus the average of a function $f(x)$ over a sphere of radius r about the point Y is

$$\frac{1}{v(r)} \int_{(r;\, Y)} f(X)\, d\, V_X. \tag{8.04}$$

Corresponding to the theorem, in one dimension, that a function is almost everywhere the derivative of its integral, there is here the fact that, for almost every Y,

$$\lim_{r\to 0} \frac{1}{v(r)} \int_{(r;\, Y)} f(X)\, d\, V_X = \lim_{r\to 0} \frac{1}{v(r)} \int_{(r;\, O)} f(X+Y)\, d\, V_X = f(Y). \tag{8.05}$$

For any positive integer m it readily follows that, almost everywhere,

$$\lim_{r\to 0} \frac{1}{[v(r)]^m} \int_{(r;\, Y)} d\, V_{X_m} \cdots \int_{(r;\, X_3)} d\, V_{X_2} \int_{(r;\, X_2)} f(X_1)\, d\, V_{X_1}$$

$$= \lim_{r\to 0} \frac{1}{[v(r)]^m} \int_{(r;\, O)} d\, V_{X_m} \cdots \int_{(r;\, O)} d\, V_{X_2} \int_{(r;\, O)} f(X_1+X_2+\cdots+X_m+Y)\, d\, V_{X_1}$$

$$= f(Y). \tag{8.06}$$

The classical Stieltjes integral, for functions of a single variable,

$$\int_{-\infty}^{\infty} f(x)\, d\varphi(x),$$

may be defined under suitable conditions as the following limit:

$$\lim_{r \to 0} \frac{1}{2r} \int_{-\infty}^{\infty} f(x)\{\varphi(x+r) - \varphi(x-r)\}\, dx. \tag{8.07}$$

If one denotes by $(r; x)$ that interval of length $2r$ which has its center at the point x and if one constructs the following function of an interval:

$$M(r; x) = \varphi(x+r) - \varphi(x-r), \tag{8.08}$$

then this limit may be written in the form

$$\lim_{r \to 0} \frac{1}{2r} \int_{-\infty}^{\infty} f(x)\, M(r; x)\, dx, \tag{8.09}$$

which may be called the Stieltjes integral of the point-function f with respect to the point-set-function M. Proceding by analogy we shall call the expression

$$\lim_{r \to 0} \frac{1}{v(r)} \int_{\infty} f(X)\, M(r; X)\, d V_X, \tag{8.10}$$

when it exists, the Stieltjes integral of the n-dimensional point-function f with respect to the region-function M, and shall denote it by the symbol

$$\int_{\infty} f(X)\, d_X M. \tag{8.11}$$

It will be necessary to introduce a generalization of this integral in order to handle certain region-functions M to appear later. If the expression

$$\lim_{r \to 0} \frac{1}{[v(r)]^m} \int_{\infty} f(X)\, M(r; X)\, d V_X \tag{8.12}$$

exists it shall be called the mth Stieltjes integral of f with respect to M and will be denoted by

$$\int_{\infty} f(X)\, d_X^m M. \tag{8.13}$$

A function $f(X)$ shall be said to be quadratically summable over all space, or q. s. over ∞, provided it satisfies the requirements laid down above and provided that

$$\int_{\infty} |f(X)|^2\, d\, V_X \tag{8.14}$$

exists. A one-parameter family of functions $f_n(X)$, each of which is q. s. over ∞, is said to converge in mean to a function $f(X)$, also q. s. over ∞, as $n \to \infty$, if

$$\lim_{\lambda \to \infty} \int_{\infty} |f(X) - f_n(X)|^2\, d\, V_X = 0. \tag{8.15}$$

We are now in a position to discuss the harmonic analysis of a given function $f(X)$. The fundamental theorem is, of course, Plancherel's theorem on the Fourier transform. In n dimensions this reads as follows:

If $f(X)$ is q. s. over ∞, there exists its Fourier transform

$$g(U) = \underset{R \to \infty}{\text{l.i.m.}} \frac{1}{(2\pi)^{n/2}} \int_{R} f(X)\, e^{i(X \cdot U)}\, d\, V_X, \tag{8.16}$$

where the expanding region R is selected from an arbitrary one-parameter family of regions which are such that ultimately R covers and continues to cover almost any given point of space. All such transforms are equivalent; i.e. any pair of such functions can differ at most on a set of zero measure. Furthermore, $g(U)$ is q. s. over ∞ and satisfies the equality

$$\int_{\infty} |g(U)|^2\, d\, V_U = \int_{\infty} |f(X)|^2\, d\, V_X. \tag{8.17}$$

For any integer $m \geq 1$, there exists as an absolutely convergent integral

$$\int_{(r;\,0)} d\,V_{T_m} \cdots \int_{(r;\,0)} d\,V_{T_2} \int_{(r;\,0)} g(T_1+T_2+\cdots+T_m+U)\,d\,V_{T_1}$$

$$= \frac{1}{(2\,\pi)^{n/2}} \int_{\infty} f(X) \left\{ \int_{(r;\,0)} e^{i(X\cdot T)}\,d\,V_T \right\}^m e^{i(X\cdot U)}\,d\,V_X, \tag{8. 18}$$

which yields the following explicit formula for $g(U)$:

$$g(U) = \lim_{r\to 0} \frac{1}{(2\,\pi)^{n/2}} \cdot \frac{1}{[v(r)]^m} \int_{\infty} f(X) \left\{ \int_{(r;\,0)} e^{i(X\cdot T)}\,d\,V_T \right\}^m e^{i(X\cdot U)}\,d\,V_X;\ m=1, 2, \ldots; \tag{8. 19}$$

this limit existing for almost every U. Conversely, $g(U)$ possesses a Fourier transform and this is $f(-X)$:

$$f(X) = \underset{R\to\infty}{\mathrm{l.i.m.}} \frac{1}{(2\,\pi)^{n/2}} \int_{R} g(U)\,e^{-i(X\cdot U)}\,d\,V_U. \tag{8. 20}$$

An equivalent statement is that $\bar{f}(X)$ is the Fourier transform of $\bar{g}(U)$. As above, we may write explicitly, for almost every X,

$$f(X) = \lim_{r\to 0} \frac{1}{(2\,\pi)^{n/2}} \cdot \frac{1}{[v(r)]^m} \int_{\infty} g(U) \left\{ \int_{(r;\,0)} e^{-i(U\cdot T)}\,d\,V_T \right\}^m e^{-i(X\cdot U)}\,d\,V_U;\ m=1, 2, \ldots. \tag{8. 21}$$

Finally, if $f_1(X)$ and $f_2(X)$ are each q. s. over ∞, and if $g_1(U)$ and $g_2(U)$ are their respective Fourier transforms, then

$$\int_{\infty} f_1(X)\,g_2(X)\,d\,V_X = \int_{\infty} f_2(U)\,g_1(U)\,d\,V_U. \tag{8. 22}$$

We are thus possessed of a harmonic analysis of any quadratically summable function. For a given such function $f(X)$ this consists in associating with almost every vector frequency U a complex amplitude $g(U)$. It is suggestive to imagine this $g(U)$ as the density of a complex mass distribution in the U space. The converse procedure by which we rebuild the given $f(X)$ from this mass distribution may be described formally as follows: We begin by multiplying the density of this distribution by the factor,

$$e^{-i(X\cdot U)}$$

thus altering the complex phase of the density in an simply periodic fashion. This done, we calculate the total mass of the resulting distribution and find it to be $f(X)$.

Now, there exists a very important class of functions to which exactly this harmonic analysis cannot be given, namely the n-tuply periodic functions. If such a function $f(X)$ be also q. s. over an arbitrary finite region, it is known to possess an n-tuple Fourier series representation:

$$f(X) \sim \Sigma a_k e^{i(X \cdot U_k)}; \tag{8.23}$$

where the summation is effected for all points U_k which are vertices of a certain rectangular network. This analysis can also be interpreted as a mass distribution. Here, however, the spread is not continuous but consists of masses a_k concentrated at the corresponding frequency points U_k. Yet the process by which $f(X)$ is reobtained is again that of calculating the total mass of a distribution.

If one seeks a uniform method of treating these two types of mass spreads, one naturally is led to construct a region-function: the total mass in an arbitrary region. Knowledge of this function is equivalent to knowledge of the particular distribution in question. The advantage derived in employing it is that it is of the same order of magnitude for the various types of spreads whereas the densities are not. We shall see that this region-function can readily be calculated. To effect a return to the given function $f(X)$ we shall employ the Stieltjes integral which will simultaneously correctly modify the complex phases of the spread and determine the total resulting mass. Let us note how easily all this is carried out.

However, we can handle at once a more general problem. Let $f(X)$ be a function such that, for some integer m, the product

$$f(X) \left\{ \int\limits_{(r;\, O)} e^{i(X \cdot T)} \, d\, V_T \right\}^m \tag{8.24}$$

is q. s. over ∞. Let the Fourier transform of this product be denoted by

$$M^{(m)}(r;\, U) = \underset{R \to \infty}{\text{l.i.m.}} \frac{1}{(2\pi)^{n/2}} \int\limits_{r} f(X) \left\{ \int\limits_{(r;\, O)} e^{i(X \cdot T)} \, d\, V_T \right\}^m e^{i(X \cdot U)} \, d\, V_X. \tag{8.25}$$

Hence

23 — 29764. *Acta mathematica.* 55. Imprimé le 14 avril 1930.

$$f(X)\left\{\int\limits_{(r;\,O)} e^{i(X\cdot T)}\,d\,V_T\right\}^m = \underset{R\to\infty}{\text{l.i.m.}}\ \frac{1}{(2\,\pi)^{n/2}}\int\limits_R e^{-i(X\cdot U)}\,M^{(m)}(r;\,U)\,d\,V_U. \qquad (8.\,26)$$

Integrating the above expression we obtain the result:

$$\int\limits_{(s;\,Y)} f(X)\left\{\int\limits_{(r;\,O)} e^{i(X\cdot T)}\,d\,V_T\right\}^m d\,V_X$$

$$= \frac{1}{(2\,\pi)^{n/2}}\int\limits_{\infty}\left\{\int\limits_{(s;\,O)} e^{-i(X\cdot U)}\,d\,V_X\right\} e^{-i(Y\cdot U)}\,M^{(m)}(r;\,U)\,d\,V_U. \qquad (8.\,27)$$

Now since

$$\lim_{r\to 0}\frac{1}{[v(r)]^m}\left\{\int\limits_{(r;\,O)} e^{i(X\cdot T)}\,d\,V_T\right\}^m = 1, \qquad (8.\,28)$$

and since $f(X)$ is summable, it follows that

$$\int\limits_{(s;\,Y)} f(X)\,d\,V_X = \lim_{r\to 0}\frac{1}{(2\,\pi)^{n/2}}\frac{1}{[v(r)]^m}\int\limits_{\infty}\left\{\int\limits_{(s;\,O)} e^{-i(X\cdot U)}\,d\,V_X\right\} e^{-i(Y\cdot U)}\,M^{(m)}(r;\,U)\,d\,V_U. \quad (8.\,29)$$

The right hand side is precisely an mth order Stieltjes integral. We have, then,

$$\int\limits_{(s;\,Y)} f(X)\,d\,V_X = \frac{1}{(2\,\pi)^{n/2}}\int\limits_{\infty}\left\{\int\limits_{(s;\,O)} e^{-i(X\cdot U)}\,d\,V_X\right\} e^{-i(Y\cdot U)}\,d_U^m\,M^{(m)}, \qquad (8.\,30)$$

and, therefore, for almost every X,

$$f(X) = \lim_{s\to 0}\frac{1}{(2\,\pi)^{n/2}}\frac{1}{v(r)}\int\limits_{\infty}\left\{\int\limits_{(s;\,O)} e^{-i(T\cdot U)}\,d\,V_T\right\} e^{-i(X\cdot U)}\,d_U^m\,M^{(m)}. \qquad (8.\,31)$$

That the function $M^{(m)}(r;\,U)$ constitutes a harmonic analysis of the given $f(X)$ is established by the following considerations. It is a matter of simple calculation to show that if $f(X)$ is itself q. s. over ∞, $M^{(1)}(r;\,U)$ exists and has for its value the total mass in the sphere $(r;\,U)$ of a distribution of density $g(U)$. It is similarly easy to show that if $f(X)$ is possessed of a Fourier series development, $M^{(1)}(r;\,U)$ exists as a limit in the mean and is equal to the sum of those Fourier coefficients a_k which correspond to frequency points U lying in $(r;\,U)$. As we have seen above, $M^{(1)}(r;\,U)$ determines the mass distributions which constitute the harmonic analysis of $f(X)$. In the same fashion $M^{(2)}(r; U)$ determines $M^{(1)}(r; U)$; etc.

While, then, we are justified in considering that $M^{(m)}(r; U)$ is a harmonic analysis of $f(X)$, we must yet determine for what class of functions $f(X)$ the region-function M will exist at least for some integer m. It can be shown that the function

$$\int_{(r;\,O)} e^{i(X\cdot T)}\, d\,V_T \tag{8. 32}$$

is bounded for all X and is of the order of $|X|^{-\frac{n+1}{2}}$ for large values of $|X|$. This at once establishes the fact that if, for some positive p, the quotient

$$\frac{f(X)}{1+|X|^r} \tag{8. 33}$$

is q. s. over ∞, then a value of m can be determined for which $M^{(m)}$ will exist. Furthermore, a generalization to n dimensions of a Tauberian theorem such as given by Jacob readily shows that if $f(X)$ is q. s. over all finite regions and is such that for some positive p the expression

$$\frac{1}{[v(r)]^p}\int_{(r;\,O)} |f(X)|^2\, d\,V_X \tag{8. 34}$$

is bounded for sufficiently large values of r, then it too belongs to the class in question. An n-tuply periodic function is included in this last type of function $f(X)$. Essentially, then, the functions which we can harmonically analyze are those which are »algebraic on the average» as $|X|\to\infty$.

When we come to the study of the energy spectrum of a given function $f(X)$ we again subject the function to a harmonic analysis but not in such great detail as before. In the corresponding mass spread we are no longer interested in the complex phase of the various masses and densities present, but solely in their absolute values, or, more precisely, in the squares of these absolute values. Obviously, again, there will be different orders of density, or, as we may say, different orders of energy in the component oscillations. For the functions of the class described above there will exist for some value of $k\geq 0$ the limit

$$\varphi(X)=\lim_{r\to\infty}\frac{1}{[v(r)]^k}\int_{(r;\,O)} f(T)\bar{f}(T-X)\, d\,V_T. \tag{8. 35}$$

The harmonic analysis of this latter function will show that its mass distribution consists of the squares of the absolute values of the highest order densities that appear in the distribution corresponding to $f(X)$. Terms of lower energy level do not appear. Thus, the total mass of the distribution corresponding to $\varphi(X)$ does not necessarily coincide with although it will never exceed the total energy which could be calculated from the spread associated with $f(X)$.

Precisely as in the one-dimensional case it can be shown that

$$|\varphi(X)| \leq \varphi(0) = \lim_{r \to \infty} \frac{1}{[v(r)]^k} \int_{(r;\,O)} |f(T)|^2 \, d\,V_T. \tag{8. 36}$$

From this boundedness it follows that the product

$$\varphi(X) \left\{ \int_{(r;\,O)} e^{i(X \cdot T)} \, d\,V_T \right\} \tag{8. 37}$$

is q. s. over ∞, and hence that there exists $M^{(1)}(r;\,U)$ formed with respect to $\varphi(X)$. This function is at present only defined as a limit in the mean. It is desired to show that it can be so defined that for any given U it will be a monotonic non-decreasing function of r. This is carried out as when $n=1$ by a series of limiting-processes which do not alter any monotonic properties. To begin with one shows that if

$$\left.\begin{aligned} \varphi_s(X) &= \frac{1}{[v(s)]^k} \int_{\infty} f_s(T) \bar{f}_s(T - X) \, d\,V_T; \\ f_s(T) &= \begin{cases} f(T); & [T \text{ in } (s;\,0)] \\ 0\,; & [T \text{ elsewhere}] \end{cases} \end{aligned}\right\} \tag{8. 38}$$

then

$$\varphi(X) = \lim_{s \to \infty} \varphi_s(X). \tag{8. 39}$$

Furthermore one notes that

$$\int_{\infty} \varphi_s(X) \, e^{i(X \cdot Y)} \, d\,V_X = \frac{1}{[v(s)]^k} \left| \int_{\infty} f_s(T) \, e^{i(T \cdot Y)} \, d\,V_T \right|^2 \geq 0, \tag{8. 40}$$

precisely as in the one-dimensional case. From this it follows that the function

$$M_s(r;\ U) = \frac{1}{(2\pi)^n} \int_{\infty} \varphi_s(X) \left\{ \int_{(r;\ O)} e^{i\,(X \cdot T)}\, d\,V_T \right\} e^{i\,(X \cdot U)}\, d\,V_X$$

$$= \frac{1}{(2\pi)^n} \int_{(r;\ U)} d\,V_T \int_{\infty} \varphi_s(X)\, e^{i(X \cdot T)}\, d\,V_X \tag{8.41}$$

for given U is monotone non-decreasing in r. It is readily seen that

$$\int_{(r';\ U')} M(r;\ U)\, d\,V_U = \lim_{S \to \infty} \int_{(r';\ U')} M_s(r;\ U)\, d\,V_U, \tag{8.42}$$

and hence that for given r' and U' the integral on the left has the same monotone property with respect to r. Finally since, for almost every U',

$$M(r;\ U') = \lim_{r' \to 0} \frac{1}{v(r')} \int_{(r';\ U')} M(r;\ U)\, d\,V_U, \tag{8.43}$$

we see that, in so far as it is thus defined, $M(r;\ U)$ has the desired property. Since the point-set on which M is not defined constitutes at most a set of zero measure, it is a simple matter to define M for all r and U so that for each U will be a monotonic non-decreasing function of r.

The total spectral intensity of $f(X)$, or more accurately, the total spectral intensity of those components which are associated with the maximum energy level, is given by the limit

$$\lim_{r \to \infty} M(r;\ U), \tag{8.44}$$

if this exists. One fairly readily shows that

$$M(r;\ U) \leq \varphi(0), \tag{8.45}$$

and, because of the monotony, hence that the limit in question exists and lies between 0 and $\varphi(0)$. The details of the argument whereby one next shows that

$$\lim_{r \to \infty} M(r;\ U) = \lim_{r' \to 0} \frac{1}{v(r')} \int_{(r';\ O)} \varphi(X)\, d\,V_X \leq \varphi(0) \tag{8.46}$$

will be omitted here to avoid too much complication.

9. Coherency matrices.

The spectrum theory of our earlier sections is a theory of the spectrum of an individual function. There are, however, many phenomena intimately connected with harmonic analysis which refer to several functions considered simultaneously. Chief among these are the phenomena of coherency and incoherency, of interference, and of polarization.

It is known to every beginner in physics that two rays of light from the same source may interfere: that is, they may be superimposed to form a darkness, or else a light more intense than is ordinarily formed by two rays of light of their respective intensities. On the other hand, two rays of light from independent sources or from different parts of the same source never exhibit this phenomenon. The former rays are said to be *coherent*, and the latter to be *incoherent*. Although it is mathematically impossible for two truly sinusoidal oscillations to be incoherent, even the most purely monochromatic light which we can sensibly produce never coheres with similar light from another source.

The physicist's explanation of incoherency is the following: the interference pattern produced by two sources of light depends on their relative displacement in phase. Now, the relative phase of two sensibly monochromatic sources of light is able to assume all possible values, and since light probably consists in a series of approximately sinusoidal trains of oscillations each lasting but a small portion of a millionth of a second, this relative phase assumes in any sensible interval all possible values with a uniform distribution which averages out light and dark bands into a sensibly uniform illumination.

This explanation of incoherency is unquestionable adequate to account for the phenomenon which it was invented to explain. Nevertheless, it is desirable to have a theory of coherency and incoherency which does not postulate a hypothetical set of constituent harmonic trains of oscillations, which at any rate must become merged in the general electromagnetic oscillation constituting the light. The present section is devoted to the development of a theory of coherency which is as direct as the theory of this paper concerning the harmonic analysis of a single function, and indeed forms a natural extension of the latter.

In interference, the components of the electromagnetic field of the constituent light rays combine additively. Accordingly, the theory of coherency and interference must be a theory of the harmonic analysis of all functions which can be obtained from a given set by linear combination. Let us see what the outlines of this theory are.

We start from a class of functions, $f_k(t)$, in general complex, and defined for all real arguments between $-\infty$ and ∞. For the present we shall assume this class of functions to be finite, although this restriction is not essential. Let

$$f(t)=a_1 f_1(t)+a_2 f_2(t)+\cdots+a_n f_n(t) \tag{9.01}$$

be the general linear combination of functions of the set. We shall have

$$\varphi(\tau)=\lim_{T\to\infty}\frac{1}{2T}\int_{-T}^{T} f(t+\tau)\bar{f}(t)\,dt=\sum_{j,k=1}^{n}\sum a_j\bar{a}_k\lim_{T\to\infty}\frac{1}{2T}\int_{-T}^{T} f_j(t+\tau)\bar{f}_k(t)\,dt \tag{9.02}$$

in case the latter limits exist. The necessary and sufficient condition for this to exist for all linear combinations $f(t)$ of functions of the set is that

$$\varphi_{jk}(\tau)=\lim_{T\to\infty}\frac{1}{2T}\int_{-T}^{T} f_j(t+\tau)\bar{f}_k(t)\,dt \tag{9.03}$$

should exist for every j, k, and τ. Then

$$\varphi(\tau)=\sum_j\sum_k a_j\bar{a}_k\varphi_{jk}(\tau). \tag{9.04}$$

Again, we shall have formally

$$S(u)=\frac{1}{2\pi}\int_{-\infty}^{\infty}\varphi(\tau)\frac{e^{iu\tau}-1}{i\tau}\,d\tau=\sum_j\sum_k a_j\bar{a}_k\frac{1}{2\pi}\int_{-\infty}^{\infty}\varphi_{jk}(\tau)\frac{e^{iu\tau}-1}{i\tau}\,d\tau, \tag{9.05}$$

where we may write

$$S_{jk}(u)=\frac{1}{2\pi}\int_{-\infty}^{\infty}\varphi_{jk}(\tau)\frac{e^{iu\tau}-1}{i\tau}\,d\tau. \tag{9.06}$$

If $\varphi_{jk}(\tau)$ exists for every j, k, and τ, we may readily show that each $S_{jk}(u)$ exists. Clearly

$$\varphi_{kj}(\tau)=\lim_{T\to\infty}\frac{1}{2T}\int_{-T}^{T} f_k(t+\tau)\bar{f}_j(t)\,dt=\lim_{T\to\infty}\frac{1}{2T}\int_{-T}^{T}\bar{f}_j(t-\tau)f_k(t)\,dt=\bar{\varphi}_{jk}(-\tau) \tag{9.07}$$

and

$$S_{kj}(u) = \frac{1}{2\pi}\int_{-\infty}^{\infty} \bar{\varphi}_{jk}(-\tau)\frac{e^{iu\tau}-1}{i\tau}\,d\tau = \frac{1}{2\pi}\int_{-\infty}^{\infty} \bar{\varphi}_{jk}(\tau)\frac{e^{-iu\tau}-1}{-i\tau}\,d\tau = \bar{S}_{jk}(u), \qquad (9.08)$$

so that the matrix

$$\| S_{jk}(u) \|$$

is Hermitian. This matrix determines the spectra of all possible linear combinations of $f_1(t), \ldots, f_n(t)$. Since it determines the precise coherency relations of the functions in question, we shall call it the coherency matrix.

Let us subject the functions $f_k(t)$ to the linear transformation

$$g_j(t) = \sum_k a_{jk} f_k(t). \qquad (9.09)$$

Then

$$\psi_{jk}(\tau) = \lim_{T\to\infty} \frac{1}{2T}\int_{-T}^{T} g_j(t+\tau)\,\bar{g}_k(t)\,dt = \sum_l \sum_m a_{jl}\bar{a}_{km}\,\varphi_{lm}(\tau) \qquad (9.10)$$

and

$$T_{jk}(u) = \frac{1}{2\pi}\int_{-\infty}^{\infty} \psi_{jk}(\tau)\frac{e^{iu\tau}-1}{i\tau}\,d\tau = \sum_l \sum_m a_{jl}\bar{a}_{km}\,S_{lm}(u). \qquad (9.11)$$

Thus the new coherency matrix $\| T_{jk}(u) \|$ may be written

$$M_1 \cdot \| S_{jk}(u) \| \cdot M_2, \qquad (9.12)$$

where

$$M_1 = \| a_{jk} \|, \qquad (9.13)$$

and

$$M_2 = \| \bar{a}_{kj} \|. \qquad (9.14)$$

In case the transformation with matrix M_1 has the property

$$M_1 \cdot M_2 = M_2 \cdot M_1 = 1, \qquad (9.15)$$

it is said to be *unitary*. For such transformations,

$$\| T_{jk} \| = M_1 \cdot \| S_{jk} \| \cdot M_1^{-1}. \qquad (9.16)$$

A matrix $\| a_{jk} \|$ is said to be in diagonal form if all the terms a_{jk} for which $j \neq k$ are identically zero. By a theorem of Weyl[5], every Hermitian matrix may be transformed into a diagonal matrix by a unitary transformation. Since we may regard a diagonal matrix as representing a set of completely incoherent phenomena, this transformation is of fundamental importance in the characterization of the state of coherency of the functions determining the matrix. Together with the numerical values of the diagonal elements of the diagonal matrix, it indeed consitutes a complete characterization of the state of coherency of the original function. In the case the values of the diagonal elements are distinct, this characterization is indeed to be carried through in but a single way.

The production of incoherent functions is a simple matter, when we have once settled the existence theory of functions with given types of spectra. Let $f_1(t)$ be any bounded function such that $\varphi(\tau)$ and consequently $S(u)$ exists. Then if

$$f_2(t) = f_1(t)\, e^{i\sqrt{|t|}}, \tag{9.17}$$

we have

$$\begin{aligned}\lim_{T\to\infty} \frac{1}{2T}\int_{-T}^{T} f_2(t+\tau)\bar{f}_2(t)\, dt &= \lim_{T\to\infty} \frac{1}{2T}\int_{-T}^{T} f_1(t+\tau)\bar{f}_1(t)\, e^{i(\sqrt{|t+\tau|}-\sqrt{|t|})}\, dt \\ &= \lim_{T\to\infty} \frac{1}{2T}\int_{-T}^{T} f_1(t+\tau)\bar{f}_1(t) \exp\left(i\frac{|t+\tau|-|t|}{\sqrt{|t+\tau|}+\sqrt{|t|}}\right) dt,\end{aligned} \tag{9.18}$$

and hence, since $\lim_{t\to\pm\infty} \exp\left(i\dfrac{|t+\tau|-|t|}{\sqrt{|t+\tau|}+\sqrt{|t|}}\right) = 1$,

$$\varphi_{22}(\tau) = \varphi_{11}(\tau) \tag{9.19}$$

and

$$S_{22}(u) = S_{11}(u). \tag{9.20}$$

On the other hand, if for example $f(t) = e^{i\lambda t}$,

$$\left.\begin{aligned}\varphi_{12}(\tau) &= \lim_{T\to\infty} \frac{1}{2T}\int_{-T}^{T} f_1(t+\tau)\bar{f}_1(t)\, e^{-i\sqrt{|t|}}\, dt \equiv 0; \\ S_{12}(u) &= S_{21}(u) = 0.\end{aligned}\right\} \tag{9.21}$$

Thus the coherency matrix of $f_1(t)$ and $f_2(t)$ is

$$S_{11}(u)\begin{Vmatrix} 1 & 0 \\ 0 & 1 \end{Vmatrix}. \tag{9.22}$$

The coherency matrix of $\sqrt{2}\,f_1(t)$ and 0 is

$$S_{11}(u)\begin{Vmatrix} 2 & 0 \\ 0 & 0 \end{Vmatrix}; \tag{9.23}$$

that of $f_1(t)$ and $f_1(t)$ is

$$S_{11}(u)\begin{Vmatrix} 1 & 1 \\ 1 & 1 \end{Vmatrix}; \tag{9.24}$$

that of $f_1(t)$ and $if_1(t)$ is

$$S_{11}(u)\begin{Vmatrix} 1 & -i \\ i & 1 \end{Vmatrix}. \tag{9.25}$$

Let it be noted that the coherency matrices of real functions are in general complex. Thus if

$$f_2(t) = f_1(t+\lambda), \tag{9.26}$$

we have

$$\left.\begin{aligned} \varphi_{22}(\tau) &= \varphi_{11}(\tau); \\ \varphi_{12}(\tau) &= \lim_{T\to\infty} \frac{1}{2T}\int_{-T}^{T} f_1(t+\lambda+\tau) f_1(t)\,dt \\ &= \varphi_{11}(\tau+\lambda); \end{aligned}\right\} \tag{9.27}$$

and hence

$$\left.\begin{aligned} S_{22}(u) &= S_{11}(u); \\ S_{12}(u) &= \int_{-\infty}^{\infty} \varphi_{11}(\tau+\lambda)\frac{e^{iu\tau}-1}{i\tau}\,d\tau = \int_{-\infty}^{u} e^{-iv\lambda}\,dS_{11}(v) + S_{12}(-\infty) \end{aligned}\right\} \tag{9.28}$$

giving the coherency matrix with derivative

$$S'_{11}(u)\begin{Vmatrix} 1 & e^{-iv\lambda} \\ e^{iv\lambda} & 1 \end{Vmatrix}. \tag{9.29}$$

In optics, coherency is generally considered for light of one particular frequency. From that standpoint, the coherency of a set of functions $f_1(t), \ldots, f_n(t)$

with continuous differentiable spectra may be regarded as determined for frequency u by the matrix

$$\begin{Vmatrix} S^1_{11}(u), & \ldots, & S^1_{1n}(u) \\ \cdot & \cdot & \cdot \\ \cdot & \cdot & \cdot \\ S^1_{n1}(u), & \ldots, & S^1_{nn}(u) \end{Vmatrix}; \tag{9.30}$$

or in the case of functions with line spectra, by

$$\begin{Vmatrix} S_{11}(u+0)-S_{11}(u-0), & \ldots, & S_{1n}(u+0)-S_{1n}(u-0) \\ \cdot & \cdot & \cdot \\ \cdot & \cdot & \cdot \\ S_{n1}(u+0)-S_{n1}(u-0), & \ldots, & S_{nn}(u+0)-S_{nn}(u-0) \end{Vmatrix}. \tag{9.31}$$

We may regard these matrices in a secondary sense as coherency matrices.

Coherency matrices of two functions are of particular interest in connection with the characterization of the state of polarization of light. As everyone knows, this characterization is identically the characterization of the state of coherency between two components of the electric vector at right angles to one another. With this interpretation, matrix (9. 22) represents unpolarized light, matrix (9. 23) light polarized completely in one plane, while

$$\begin{Vmatrix} 0 & 0 \\ 0 & 2 \end{Vmatrix} \tag{9.32}$$

represents light completely polarized in a plane perpendicular to the first. Matrix (9. 24) and matrix

$$\begin{Vmatrix} 1 & -1 \\ -1 & 1 \end{Vmatrix} \tag{9.33}$$

represent light polarized respectively at 45° and at 135° to the first direction. Matrix (9. 25) and matrix

$$\begin{Vmatrix} 1 & i \\ -i & 1 \end{Vmatrix} \tag{9.34}$$

represent respectively light polarized circularly in a counter-clockwise and a clockwise direction.

When the matrix of completely polarized light, whether linearly, elliptically, or circularly polarized, is brought into diagonal form by a linear unitary transformation, the resulting diagonal matrix will have only one element distinct from 0. On the other hand, completely unpolarized light has the diagonal terms equal. This suggests as a measure of the amount of polarization of the diagonal matrix

$$\left\| \begin{matrix} a & 0 \\ 0 & b \end{matrix} \right\|, \tag{9.35}$$

or of any other matrix equivalent to it under a unitary transformation, the quantity

$$a - b. \tag{9.36}$$

If we subtract from our original diagonal matrix the completely incoherent matrix

$$\left\| \begin{matrix} \frac{a+b}{2} & 0 \\ 0 & \frac{a+b}{2} \end{matrix} \right\|, \tag{9.37}$$

which is invariant under every unitary transformation, we get the matrix

$$\left\| \begin{matrix} \frac{a-b}{2} & 0 \\ 0 & \frac{b-a}{2} \end{matrix} \right\|, \tag{9.38}$$

which may be regarded as a representative of the quantity $a-b$. This suggests that given any coherency matrix

$$\left\| \begin{matrix} A & B+Ci \\ B+Ci & D \end{matrix} \right\|, \tag{9.39}$$

we may take $A+D$ to represent the intensity of the corresponding light, and the matrix

$$\left\| \begin{matrix} \frac{A-D}{2} & B-Ci \\ B+Ci & \frac{D-A}{2} \end{matrix} \right\| \tag{9.40}$$

as its polarization. Thus horizontal polarization is represented by the matrix

$$\left\| \begin{matrix} 1 & 0 \\ 0 & -1 \end{matrix} \right\|; \tag{9.41}$$

polarization at 45° by the matrix

$$\left\| \begin{matrix} 0 & 1 \\ 1. & 0 \end{matrix} \right\|; \tag{9.42}$$

and circular polarization by the matrix

$$\left\| \begin{matrix} 0 & i \\ -i & 0 \end{matrix} \right\|. \tag{9.43}$$

These are the same matrices which Jordan, Dirac and Weyl have employed to such advantage in the theory of quanta.

Since the most general Hermitian matrix of the second order may be written

$$\left\| \begin{matrix} \alpha+\beta & \gamma+\delta i \\ \gamma-\delta i & \alpha-\beta \end{matrix} \right\| = \alpha \left\| \begin{matrix} 1 & 0 \\ 0 & 1 \end{matrix} \right\| + \beta \left\| \begin{matrix} 1 & 0 \\ 0 & -1 \end{matrix} \right\| + \gamma \left\| \begin{matrix} 0 & 1 \\ 1 & 0 \end{matrix} \right\| + \delta \left\| \begin{matrix} 0 & i \\ -i & 0 \end{matrix} \right\|, \tag{9.44}$$

it appears that all light may characterized as to its state of polarization by given the total amount of light it contains, the excess of the amount polarized at 0° over that polarized at 90°, the excess of the amount polarized at 45° over that polarized at 135° and the excess of that polarized circularly to the right over that polarized circularly to the left. This characterization is complete and univocal. The total intensity of the light may be read off any sort of a photometer. The excess of light polarized horizontally over that polarized vertically may be determined by a doubly refracting crystal in one orientation, and the excess of light polarized at 45° over that polarized at 135° by the same crystal in another orientation. It is possible furthermore to devise an instrument which will read off the amount of circular polarization in the light in question. The three latter instruments possess some very remarkable group properties with respect to one another. Either portion of the light emerging from any one of the instruments will behave towards the other two exactly like unpolarized light. Rotation of the plane of polarization of the light through 45° will change the reading of the first of the last three instruments into that of the second, and the reading of the second into minus that of the first, leaving the reading of the third unchanged. There are precisely analogous unitary transformations interchanging any other pair of the three instruments, leaving the reading of the remaining instrument untouched. These transformations together with their powers and the identical transformation form a group.

A fact concerning polarized light which is so apparently obvious that it is generally regarded as not needing any proof is that all light is a case or limiting case of partially elliptically polarized light. It is nevertheless desirable to prove this statement. Completely elliptically polarized light with the coordinate axes as principal axes has a coherency matrix of the form

$$\begin{Vmatrix} A^2 & -ABi \\ ABi & B^2 \end{Vmatrix}; \tag{9.45}$$

and hence partially polarized light with the same principal axes has a coherency matrix of the form

$$P = \begin{Vmatrix} A^2+D^2 & -ABi \\ ABi & B^2+D^2 \end{Vmatrix}. \tag{9.46}$$

We wish to show that the general coherency matrix

$$M = \begin{Vmatrix} \alpha & \gamma-\delta i \\ \gamma+\delta i & \beta \end{Vmatrix} \tag{9.47}$$

may be transformed into this form by a real unitary transformation in such a way that

$$T \,.\, M \,.\, T^{-1} = P.$$

To do this, we need only put

$$T = \begin{Vmatrix} \cos\varphi & \sin\varphi \\ \sin\varphi & -\cos\varphi \end{Vmatrix}; \text{ where } \tan 2\varphi = \frac{2\gamma}{\alpha-\beta}. \tag{9.48}$$

Thus the axes of polarization of M are 1 and 2 directions when we replace $f_1(t)$ and $f_2(t)$ by

$$\left.\begin{aligned} g_1(t) &= f_1(t)\cos\varphi + f_2(t)\sin\varphi; \\ g_2(t) &= -f_2(t)\cos\varphi + f_1(t)\sin\varphi; \end{aligned}\right\} \tag{9.49}$$

the »lengths» of these axes are respectively

$$\left.\begin{aligned} A &= \left\{\frac{1}{2}\left[(\alpha-\beta)^2+4(\gamma^2+\delta^2)\right]^{1/2}+\frac{1}{2}\left[(\alpha-\beta)^2+4\gamma^2\right]^{1/2}\right\}^{1/2}; \\ B &= \left\{\frac{1}{2}\left[(\alpha-\beta)^2+4(\gamma^2+\delta^2)\right]^{1/2}-\frac{1}{2}\left[(\alpha-\beta)^2+4\gamma^2\right]^{1/2}\right\}^{1/2}; \end{aligned}\right\} \tag{9.50}$$

and the percentage of polarization

$$100\left(1-\frac{4(\gamma^2+\delta^2)}{(\alpha+\beta)^2}\right)^{1/2}. \tag{9.51}$$

The connection between coherency matrices and optical instruments, which we have already mentioned in the case of polarized light, is of far more general applicability. An optical instrument is a method, linear in electric and magnetic field vectors, of transforming a light input into a light output. In general, this transformation, in the language of Volterra[5], belongs to the group of the closed cycle with respect to the time, in the sense that it is independent of the position of our initial instant in time. Such a transformation leaves a simple harmonic input still simply harmonic in the time, although in general with a shift in phase.

An example of an optical instrument is a microscope. This may be regarded as a means of making an electromagnetic disturbance in the image plane depend linearly on a given electromagnetic disturbance in the object plane. Telescopes, spectroscopes, Nicol prisms, etc., serve as further examples of optical instruments in the sense in question. Among these, a particularly interesting ideal type is the conservative optical instrument, in which the power of the input and the power of the output are identical. This power depends quadratically on the electric and magnetic vectors, so that a conservative optical instrument has a quadratic invariant for the corresponding transformations. When only terms of the one frequency of e^{iut} are considered, this quadratic positive invariant becomes Hermitian, and has essentially the same properties as the expression

$$x_1\bar{x}_1+x_2\bar{x}_2+\cdots+x_n\bar{x}_n \tag{9.52}$$

which is invariant under all unitary transformations of $x_1, \ldots, x_n$. Thus the theory of the group of unitary transformations is physically applicable, not only in quantum mechanics, where Weyl has already employed it so successfully, but even in classical optics. It is the conviction of the author that this analogy is not merely an accident, but is due to a deep-lying connection between the two theories.

In quantum mechanics, while all the terms of a matrix enter in an essential way into its transformation theory, only diagonal terms are given an immediate physical significance. This is also in precise accord with the optical situation. Every optical observation ends with the measurement of an energy or power, either by direct bolometric or thermometric means, or by the observations of a visual intensity or the blackening of a photographic plate. Every such observation means the more or less complete determination of some diagonal term. The non-diagonal terms of a coherency matrix of light only have signifiance in so far as they enable us to predict the energies or intensities which the light will show after having been subjected to some linear transformation or optical instrument. This fact that new diagonal terms after a transformation cannot be read off from the old diagonal terms before a transformation, without the intervention of non-diagonal terms, is the optical analogue for the principle of indetermination in quantum theory, according to which observations on the momentum of an electron alone cannot yield a single value if its position is known, and vice versa. The statement that every observation of an electron affects its properties has the following analogy: if two optical instruments are arranged in series, the taking of a reading from the first will involve the interposition of a ground-glass screen or photographic plate between the two, and such a plate will destroy the phase relations of the coherency matrix of the emitted light, replacing it by the diagonal matrix with the same diagonal terms. Thus the observation of the output of the first instrument alters the output of the second. In this case, the possibility of taking part of the output of the first instrument for reading by a thinly silvered mirror warns us not to try to push the analogy with quantum theory too far.

Coherency matrices form a close analogue to the correlation matrices long familiar in statistical theory. If we have a set of n observations $x^{(1)}, x^{(2)}, \ldots, x^{(n)}$ all made together, and this set of observations is repeated on the occasions 1, 2, ..., m thus yielding sets $x_1^{(1)}, \ldots, x_1^{(n)}$; $x_2^{(1)}, \ldots, x_2^{(n)}$; ...; ...; $x_m^{(1)}, \ldots, x_m^{(n)}$, the correlation matrix of these observations is

$$\left\| \begin{array}{cccc} \sum_1^m (x_k^{(1)})^2, & \sum_1^m x_k^{(1)} x_k^{(2)}, & \cdots\cdots, & \sum_1^m x_k^{(1)} x_k^{(n)} \\ \sum_1^m x_k^{(2)} x_k^{(1)}, & \sum_1^m (x_k^{(2)})^2, & \cdots\cdots, & \sum_1^m x_k^{(2)} x_k^{(n)} \\ \cdot & \cdot & \cdot & \cdot \\ \cdot & \cdot & \cdot & \cdot \\ \sum_1^m x_k^{(n)} x_k^{(1)}, & \sum_1^m x_k^{(n)} x_k^{(2)}, & \cdots\cdots, & \sum_1^m (x_k^{(n)})^2 \end{array} \right\| . \tag{9.53}$$

This symmetrical matrix represents the entire amount and kind of linear relationship to be observed between the different observations in question. The further analysis of the information yielded by a correlation matrix depends on the nature of the data to be analysed. Thus if the two observations of each set are the x and y coordinates of the position of a shot on a target, the rotations of the x and y axes have a concrete geometrical meaning, and the question of reducing the matrix to diagonal form by a rotation of axes is the significant question of determining the ellipse which best represents the distribution of holes in the target. On the other hand, if the quantities whose correlation we are investigating are the price of wheat x in dollars per bushel and the marriage rate y, rotations have no significance, as there is no common scale, while on the other hand, the significant information yielded by the matrix must be invariant under the transformations

$$\left. \begin{array}{l} x_1 = kx; \\ y_1 = ly. \end{array} \right\} \tag{9.54}$$

The so-called coefficients of correlation and of partial correlation and the lines of regression of x on y and of y on x have this type of invariance.

Correlation matrices and their derived quantities are the tool for the statistical analysis of what is known as frequency series, series of data where no such variable as the time enters as a parameter. In the study of meteorology, of business cycles, and of many other phenomena of interest to the statistician, on the other hand, we must discuss time series, where the relations of the data in time are essential. The proper analysis of these has long been a moot point among statisticians and economists. As far as it is linear relationships which we are seeking for, it is only reasonable to suppose that coherency matrices

25—29764. *Acta mathematica.* 55. Imprimé le 14 avril 1930.

must play the same rôle for time series which correlation matrices play for frequency series. In statistical work, the group of transformations which will most frequently be permissible is as before

$$g_1(t) = A f_1(t); \quad g_2(t) = B f_2(t). \qquad [A \text{ and } B \text{ real}] \tag{9.55}$$

Under this group, the significant invariants of the Hermitian matrix

$$\begin{Vmatrix} S^1_{11}(u) & S^1_{12}(u) \\ S^1_{21}(u) & S^1_{22}(u) \end{Vmatrix} \tag{9.56}$$

are

$$r(u) = S^1_{12}(u)\,[S^1_{11}(u)\,S^1_{22}(u)]^{-1/2}; \tag{9.57}$$

which we may call the coefficient of coherency of f_1 and f_2 for frequency u, and

$$\sigma_1(u) = \frac{S^1_{12}(u)\sqrt{S^1_{11}(u)}}{S^1_{22}(u)} \text{ and } \sigma_2(u) = \frac{S^1_{21}(u)\sqrt{S^1_{22}(u)}}{S^1_{11}(u)}, \tag{9.58}$$

the coefficients of regression respectively of f_1 on f_2 and of f_2 on f_1. The modulus of the coefficient of coherency represents the amount of linear coherency between $f_1(t)$ and $f_2(t)$ and the argument the phase-lag of this coherency. The coefficients of regression determine in addition the relative scale for equivalent changes of f_1 and f_2.

The computation of coefficients of coherency and of regression is to be done in the steps indicated in their definition. In the case where only a finite set of real data are at our disposal, distributed at equal unit intervals from 0 to n, say $x_0, x_1, \ldots, x_n$ and $y_0, y_1, \ldots, y_n$, the steps of our computation are:

$$\left.\begin{aligned} (\varphi_k)_{11} &= \frac{1}{n}\sum_0^{n-k} x_j x_{j+k}; \\ (\varphi_k)_{12} &= \frac{1}{n}\sum_0^{n-k} x_j y_{j+k}; \\ (\varphi_k)_{21} &= \frac{1}{n}\sum_0^{n-k} y_j x_{j+k}; \\ (\varphi_k)_{22} &= \frac{1}{n}\sum_0^{n-k} y_i y_{j+k}; \end{aligned}\right\} \qquad [0 \leq k \leq n] \tag{9.59}$$

$$r(u) = \frac{\sum_0^n \{[(\varphi_k)_{21} + (\varphi_k)_{12}] \cos ku + i\,[(\varphi_k)_{21} - (\varphi_k)_{12}] \sin ku\} - (\varphi_0)_{12}/2}{2\left[\sum_0^n (\varphi_k)_{11} \cos ku - (\varphi_0)_{11}/2\right]^{1/2}\left[\sum_0^n (\varphi_k)_{22} \cos ku - (\varphi_0)_{22}/2\right]^{1/2}}; \tag{9.60}$$

$$\left.\begin{aligned} \sigma_1(u) &= r(u)\frac{\left[\sum_0^n (\varphi_k)_{11} \cos ku - (\varphi_0)_{11}/2\right]^{1/2}}{\left[\sum_0^n (\varphi_k)_{22} \cos ku - (\varphi_0)_{22}/2\right]^{1/2}}; \\ \sigma_2(u) &= \bar{r}(u)\frac{\left[\sum_0^n (\varphi_k)_{22} \cos ku - (\varphi_0)_{22}/2\right]^{1/2}}{\left[\sum_0^n (\varphi_k)_{11} \cos ku - (\varphi_0)_{11}/2\right]^{1/2}}. \end{aligned}\right\} \tag{9.61}$$

In case we have at our disposal methods for performing the Fourier transformation, we may compute these coefficients directly from graphs. Several devices for this purpose are now being developed in the laboratory of Professor V. Bush in the Department of Electrical Engineering of the Massachusetts Institut of Technology.

10. **Harmonic analysis and transformation groups.**

Inasmuch, as the theory of Fourier series forms a special chapter in the theory of expansions in general sets of normal and orthogonal functions, it is reasonable to expect that the theory developed in the present paper is but a special chapter in the theory of general orthogonal developments. An attempt to translate the present theory into more general terms, however, incurs at once somewhat serious difficulties. This is due to the fact that the theory of the Fourier series involves only one fundamental Hermitian form,

$$\frac{1}{2\pi}\int_{-\pi}^{\pi} f(x)\bar{f}(x)\,dx; \tag{10.01}$$

the closely related theory of the Fourier integral involves only the analogous form

$$\int_{-\infty}^{\infty} f(x)\bar{f}(x)\,dx; \tag{10.02}$$

while the present paper involves besides this latter form the singular quadratic form

$$M_x(f(x)) = \lim_{T\to\infty} \frac{1}{2T}\int_{-T}^{T} f(x)\bar{f}(x)\,dx. \tag{10.03}$$

The forms (10.02) and (10.03) are quite independent of one another in their formal properties, but the complex exponentials e^{iux} stand in close relation to both of them: to (10.02) because if $a < b$, $c < d$,

$$\int_{-\infty}^{\infty} dx \int_a^b e^{iux}\,du \int_c^d e^{-ivx}\,dv = 2\pi \text{ [length common to } (a, b) \text{ and } (c, d)]; \tag{10.04}$$

and to (10.03) because

$$\lim_{T\to\infty} \frac{1}{2T}\int_{-T}^{T} e^{iux}\,e^{-ivx}\,dx = \begin{cases} 0; & [u \neq v] \\ 1 & [u = v]. \end{cases} \tag{10.05}$$

In the classical Plancherel theory, only the first form is in evidence; in the Bohr theory of almost periodic functions, only the second; in the theory of the present paper, the two play an equal rôle.

Weyl has developed in some detail the relations between the theory of unitary groups and the theory of periodic and almost periodic functions. The groups which he introduces are one parameter groups of linear functional transformations leaving (10.03) invariant. The Weyl theory is manifestly not susceptible to an extension to more general forms of harmonic analysis, unless a way is found to take cognizance of the invariance of (10.02) as well. This is the purpose of the present section.

Let us restrict the functions $f(x)$ which we discuss in the present section to those for which

$$\int_{-\infty}^{\infty} \frac{|f(x)|^2}{1 + x^2}\,dx \tag{10.06}$$

is finite. Let

$$f_A(x) = \begin{cases} 0; & [|x| > A] \\ f(x); & [|x| \le A] \end{cases} \tag{10.07}$$

and let s_A be the transformation leading from $f(x)$ to $f_A(x)$. Let T be a transformation which is linear, and with an inverse, and is defined for all functions $f(x)$ subject to the finiteness of (10.06). Let T preserve (10.02) invariant, and in case (10.03) is finite, let

$$\int_{-\infty}^{\infty} |(Ts_A - s_A T)f(x)|^2 \, dx = o(A). \tag{10.08}$$

Then since

$$\lim_{A\to\infty} \frac{1}{2} \int_{-\infty}^{\infty} \left| \frac{Ts_A f(x)}{\sqrt{A}} \right|^2 dx = M_x(|f(x)|^2), \tag{10.09}$$

and

$$\lim_{A\to\infty} \int_{-\infty}^{\infty} \left| \frac{Ts_A f(x)}{\sqrt{A}} - \frac{s_A T f(x)}{\sqrt{A}} \right|^2 dx = 0, \tag{10.10}$$

it follows that

$$M_x(|Tf(x)|^2) = \lim_{A\to\infty} \frac{1}{2} \int_{-\infty}^{\infty} \left| \frac{s_A T f(x)}{\sqrt{A}} \right|^2 dx = M_x(|f(x)|^2). \tag{10.11}$$

The transformations T form a group. If T_1 and T_2 are of this form,

$$\int_{-\infty}^{\infty} |(T_1 T_2 s_A - s_A T_1 T_2) f(x)|^2 \, dx$$

$$= \int_{-\infty}^{\infty} |[(T_1 T_2 s_A - T_1 s_A T_2) + (T_1 s_A T_2 - s_A T_1 T_2)] f(x)|^2 \, dx$$

$$\le 2 \int_{-\infty}^{\infty} |T_1 T_2 s_A - T_1 s_A T_2) f(x)|^2 \, dx + 2 \int_{-\infty}^{\infty} |(T_1 s_A T_2 - s_A T_1 T_2) f(x)|^2 dx$$

$$= 2 \int_{-\infty}^{\infty} |(T_2 s_A - s_A T_2) f(x)|^2 \, dx + 2 \int_{-\infty}^{\infty} |(T_1 s_A - s_A T_1) T_2 f(x)|^2 \, dx. \tag{10.12}$$

Furthermore,

$$\int_{-\infty}^{\infty} |(T^{-1}s_A - s_A T^{-1}) f(x)|^2 \, dx = \int_{-\infty}^{\infty} |(s_A - T s_A T^{-1}) f(x)|^2 \, dx$$

$$= \int_{-\infty}^{\infty} |(T s_A - s_A T) T^{-1} f(x)|^2 \, dx. \qquad (10.13)$$

Thus the product of two transformations satisfying (10.08) and the inverse of a transformation satisfying (10.08) likewise satisfy (10.08).

An example of a transformation satisfying (10.08) is

$$T^\lambda f(x) = f(x+\lambda), \qquad (10.14)$$

for

$$\int_{-\infty}^{\infty} |(T^\lambda s_A - s_A T^\lambda) f(x)|^2 \, dx$$

$$\leq \int_0^\lambda [|f(x-A-\lambda)|^2 + |f(x+A-\lambda)|^2] \, dx = o(A). \qquad (10.15)$$

If T satisfies (10.08) and (10.02) is invariant under it, we shall call it *properly unitary*. Let us consider a one parameter group consisting of all properly unitary transformations U^λ, where

$$U^{\lambda+\mu} = U^\lambda U^\mu. \qquad (10.16)$$

Let $f(x)$ be such that

$$\varphi(t) = M_x [(U^t f(x)) \bar{f}(x)] \qquad (10.17)$$

exists for every t. Clearly, by the Schwarz inequality,

$$\varphi(t) \leq [M_x(|U^t f(x)|^2) \, M_x(|f(x)|^2)]^{1/2}$$

$$= \left[\varphi(0) \lim_{A\to\infty} \frac{1}{2A} \int_{-\infty}^{\infty} |s_A U^t f(x)|^2 \, dx\right]^{1/2}$$

$$= \left[\varphi(0) \lim_{A\to\infty} \frac{1}{2A} \int_{-\infty}^{\infty} |U^t s_A f(x)|^2 \, dx\right]^{1/2}$$

$$= \varphi(0). \qquad (10.18)$$

It follows that $\varphi(t)$ is a bounded function, and that

$$S(u_1, u_2) = \underset{B\to\infty}{\text{l.i.m.}} \frac{1}{\pi} \int_{-B}^{B} \varphi(t) \frac{\sin(u_2 - u_1)t/2}{t} e^{i\left(\frac{u_1+u_2}{2}\right)t} dt \tag{10.19}$$

exists when $u_2 - u_1$ is given as a quadratically summable function of $u_2 + u_1$. For this one need only apply Plancherel's theorem.

Let us put

$$\varphi_A(t) = \frac{1}{2A} \int_{-\infty}^{\infty} [U^t s_A f(x)] \overline{s_A f(x)} \, dx. \tag{10.20}$$

If condition (10.17) holds for every t, we have

$$\varphi(t) = \lim_{A\to\infty} \frac{1}{2A} \int_{-\infty}^{\infty} [s_A U^t f(x)] \overline{s_A f(x)} \, dx, \tag{10.21}$$

which we may readily reduce to the form

$$\varphi(t) = \lim_{A\to\infty} \varphi_A(t) \tag{10.22}$$

by means of (10.08). We may easily prove $\varphi_A(t)$ not to exceed $\varphi_A(0)$ and hence to be uniformly bounded in A and t, for

$$\lim_{A\to\infty} \varphi_A(0) = \varphi(0). \tag{10.23}$$

Let us now introduce a new assumption concerning U^x. *Let the transformation W taking $f(x)$ into $U^x f(a)$ (a fixed) preserve* (10.02) *invariant.* Than if

$$\psi(x) = W s_A f(x) \tag{10.24}$$

we have, by the new assumption and (10.20),

$$\begin{aligned}\varphi_A(t) &= \frac{1}{2A} \int_{-\infty}^{\infty} U^i (U^x s_A f(a)) \overline{U^x s_A f(a)} \, dx \\ &= \frac{1}{2A} \int_{-\infty}^{\infty} \psi(x+t) \overline{\psi}(x) \, dx.\end{aligned} \tag{10.25}$$

As we may readily see (cf. (1.29)) $\varphi_A(t)$ is absolutely integrable over $(-\infty, \infty)$, and

$$P_A(u) = \frac{1}{\pi} \int_{-\infty}^{\infty} \varphi_A(t)\, e^{iut}\, dt \tag{10. 26}$$

exists, and is real and positive. Indeed, we might have replaced our new assumption by the assumption that

$$\frac{1}{\pi} \int_{-\infty}^{\infty} e^{iut}\, dt \int_{-\infty}^{\infty} [U^t f(x)] f(x)\, dx \tag{10. 27}$$

is positive-definite, and exists for every quadratically summable $f(x)$.

Thus it appears that

$$S_A(u_1, u_2) = \frac{2}{\pi} \int_{-\infty}^{\infty} \varphi_A(t) \frac{\sin (u_2 - u_1) t/2}{t}\, e^{i\left(\frac{u_1+u_2}{2}\right)t}\, dt \tag{10. 28}$$

exists, is monotone in u_1 and u_2, and has the property that

$$S_A(u_1, u_2) + S_A(u_2, u_3) = S_A(u_1, u_3). \tag{10. 29}$$

To see this, we need only reflect that

$$S_A(u_1, u_2) = \int_{u_1}^{u_2} P_A(u)\, du. \tag{10. 30}$$

Now

$$\varphi(t) \frac{\sin (u_2 - u_1) t/2}{t} = \underset{A\to\infty}{\text{l.i.m.}}\ \varphi_A(t) \frac{\sin (u_2 - u_1) t/2}{t}, \tag{10. 31}$$

so that

$$S(u_1, u_2) = \underset{A\to\infty}{\text{l.i.m.}}\ S_A(u_1, u_2). \tag{10. 32}$$

From this we may readily conclude that

$$S(u_1, u_2) + S(u_2, u_3) = S(u_1, u_3), \tag{10. 33}$$

and that $S(u_1, u_2)$ may be so defined as to be monotone in both arguments and increasing in u_2. $S(u_1, u_2)$ is the analogue to $S(u_2) - S(u_1)$ in our earlier sections.

We have

$$\frac{1}{u}\int_0^u S(-u, u)\,du = \frac{1}{\pi u}\int_{-\infty}^{\infty} \varphi(t)\frac{1-\cos ut}{t^2}\,dt. \tag{10.34}$$

Hence

$$S(-\infty, \infty) = \lim_{u\to\infty} \frac{1}{\pi u}\int_{-\infty}^{\infty} \varphi(t)\frac{1-\cos ut}{t^2}\,dt. \tag{10.35}$$

As in (5.47)

$$S(-\infty, \infty) = \lim_{\varepsilon\to 0} \frac{1}{2\varepsilon}\int_{-\varepsilon}^{\varepsilon} \varphi(t)\,dt. \tag{10.36}$$

We thus have arrived at a spectrum theory closely paralleling the theory here developed for trigonometric expansions. Thus for the general case of harmonic analysis, it is the group theory of transformations satisfying (10.08) and (10.24) which is important, rather than the recognized theory of unitary transformations.

Transformations U^t with the properties demanded in this section make their appearance in physics in connection with the Schrödinger equation, which often has its Eigenfunktionen also Eigenfunktionen of an operator analogous to U^t. A more specific instance of $U^t f(x)$ is

$$U^t f(x) = e^{iAt} f(x+t). \tag{10.37}$$

CHAPTER IV.

11. Examples of functions with continuous spectra.

The theory of harmonic analysis which we have so far developed has as one of its purposes the analysis of phenomena such as white light, involving continuous spectra. We have not yet proved our theory to be adequate to this purpose, for we have not yet given a single instance of a continuous spectrum. This lacuna it is the purpose of the present section to fill. To this end, we shall only consider functions $f(t)$ which over every interval $(n, n+1)$, n being an integer, assume one of the two values, 1 and -1. For such a function, the existence of

$$\varphi(x) = \lim_{T\to\infty} \frac{1}{2T} \int_{-T}^{T} f(t+x) f(t)\, dt$$

for all arguments will follow from its existence for all integral arguments, inasmuch as, if x lies between n and $n+1$,

$$\frac{1}{2T} \int_{-T}^{T} f(t+x) f(t)\, dt = \frac{x-n}{2T} \int_{-T}^{T} f(t+n+1) f(t)\, dt$$

$$+ \frac{n+1-x}{2T} \int_{-T}^{T} f(t+n) f(t)\, dt, \qquad (11.01)$$

so that

$$\varphi(x) = \lim_{T\to\infty} \frac{1}{2T} \int_{-T}^{T} f(t+x) f(t)\, dt = (x-n)\,\varphi(n+1) + (n+1-x)\,\varphi(n). \qquad (11.02)$$

An example of a function of this sort is given below, where the sequence of signs represents the signs of $f(t)$ over the intervals both to the right and the left of the zero point:

$+, -;$

$+, +; +, -; -, +; -, -$ repeated twice

$+, +, +; +, +, -; +, -, +; +, -, -; -, +, +; -, +, -;$
$-, -, +; -, -, -$ repeated four times

$+, +, +, +; +, +, +, -; +, +, -, +;$ etc. repeated eight times

. . . . (11.03)

Each repetition of a row is here counted as a separate row. In each row, all the possible arrangements of n symbols which may be either + or — occur arranged in a regular order. Thus in each row, the possible arrangements of p pluses or minuses occur equally often, except for the modification incurred by the possibility that such an arrangement may overlap one of the major divisions indicated by a semicolon in the above table. These semicolons become more and more infrequent as we proceed to later rows in our table, and do not affect the asymptotic distribution of sequence of p signs.

Thus the possible sequences of p signs occur with a probability approaching $1/2^p$ at the end of a comported row. However, the ratio of the number of terms in a row to that in all previous rows approaches zero, so that the effect of stopping at some intermediate point in a row becomes negligible. In other words,

$$\lim \frac{(\text{number of repetitions of a particular sequence of } p \text{ terms in first } n)}{n} = 1/2^p. \qquad (11.04)$$

Hence

$$\lim_{T\to\infty} \frac{1}{2T} \int_{-T}^{T} f(t+m) f(t)\, dt = 0. \qquad [m = \text{an integer} \neq 0] \quad (11.05)$$

Inasmuch as obviously

$$\lim_{T\to\infty} \frac{1}{2T} \int_{-T}^{T} [f(t)]^2\, dt = 1, \qquad (11.06)$$

we see that

$$\varphi(\tau) = \begin{cases} 1 - |\tau|; & [|\tau| \leq 1] \\ 0. & [|\tau| > 1] \end{cases} \qquad (11.07)$$

It follows that

$$S(u) = \frac{1}{2\pi} \int_{-\infty}^{\infty} \varphi(x) \frac{e^{iux} - 1}{ix}\, dx = \frac{1}{\pi} \int_0^1 (1-x) \frac{\sin ux}{x}\, dx = \frac{1}{\pi} \int_0^u \frac{1 - \cos v}{v^2}\, dv. \qquad (11.08)$$

Thus the function $f(t)$ has a continuous spectrum with spectral density $\frac{1}{\pi} \frac{1 - \cos u}{u^2}$. This fact that the spectrum has a spectral density is an even more restrictive condition than the condition that it should be continuous.

Every monotone function is known to be the sum of three parts: a step function with a denumerable set of steps; a function which is the integral of its derivative; and a continuous function with a derivative almost everywhere zero. This latter type of function has been ignored as a possibility in spectrum analysis. With both line and continuous spectra we are familiar, but the physicists have not considered the possibility of a spectrum in which the total energy of a region tends to zero as the width of the region decreases, but not proportionally in the limit to the width of the region. Nevertheless, functions with a spectrum of this type do exist, as Mr. Kurt Mahler has proved. I am incorporating into this paper an extremely ingenious note of Mr. Mahler, already published in the Journal of Mathematics and Physics of the Massachusetts Institute of Technology, giving an example of this kind.

Let ξ be a simple q-th root of unity, q being any positive integer greater than 1. Let $\bar{\xi}$ be the conjugate complex number, so that

$$\xi\bar{\xi} = 1. \tag{11.09}$$

We define the arithmetical function $\varrho(n)$ by the functional equations

$$\left.\begin{aligned} &\varrho(0) = 1; \\ &\varrho(qn + l) = \xi^l \varrho(n) \text{ for } \begin{bmatrix} l = 0, 1, 2, \ldots, q-1 \\ n = 0, 1, 2, \ldots \end{bmatrix} \end{aligned}\right\} \tag{11.10}$$

We have thus defined $\varrho(n)$ unambiguously for every positive integer n. We may write

$$\varrho(n) = \xi^{q(n)} \tag{11.11}$$

where $q(n)$ is the sum of the digits of n in the q-ary system of notation.

Our problem here is to give an asymptotic evaluation of

$$S_k(n) = \sum_{l=0}^{n-1} \varrho(l)\,\bar{\varrho}(l+k), \tag{11.12}$$

for arbitrary positive integral values of k and large values of n. If $k=0$, we have the obvious formula

$$S_0(n) = n. \tag{11.13}$$

We shall use this as a basis on which to determine

$$S_1(n) = \sum_{l=0}^{n-1} \varrho(l)\,\bar{\varrho}(l+1). \tag{11.14}$$

We may deduce at once from our fundamental equation (11.10) the functional equations of $S_1(n)$, namely

$$\begin{aligned} &S_1(0)=0, \\ &S_1(qn+l)=\bar{\xi}\, S_1(n)+((q-1)n+l)\,\bar{\xi}. \qquad [l=0, 1, \ldots, q-1] \end{aligned} \tag{11.15}$$

As is obvious, these equations determine $S_1(n)$ unambiguously.

We now see, however, that the series

$$\Sigma_1(n) = \bar{\xi}\left\{[n] - \left[\frac{n}{q}\right]\right\} + \bar{\xi}^2\left\{\left[\frac{n}{q}\right] - \left[\frac{n}{q^2}\right]\right\} + \bar{\xi}^3\left\{\left[\frac{n}{q^2}\right] - \left[\frac{n}{q^3}\right]\right\} + \cdots \qquad (11.\ 16)$$

satisfies the same functional equations (11. 15) as $S_1(n)$ and hence is identical with $S_1(n)$. We thus have

$$S_1(n) = \bar{\xi}\left\{[n] - \left[\frac{n}{q}\right]\right\} + \bar{\xi}^2\left\{\left[\frac{n}{q}\right] - \left[\frac{n}{q^2}\right]\right\} + \bar{\xi}^3\left\{\left[\frac{n}{q^2}\right] - \left[\frac{n}{q^3}\right]\right\} + \cdots. \qquad (11.\ 17)$$

Now let

$$q^r \leq n < q^{r+1}. \qquad (11.\ 18)$$

We see that

$$\begin{aligned} S_1(n) &= \bar{\xi}\left\{[n] - \left[\frac{n}{q}\right]\right\} + \bar{\xi}^2\left\{\left[\frac{n}{q}\right] - \left[\frac{n}{q^2}\right]\right\} + \cdots + \bar{\xi}^{r+1}\left\{\left[\frac{n}{q^r}\right] - \left[\frac{n}{q^{r+1}}\right]\right\} \\ &= n\bar{\xi}\left(1 - \frac{1}{q}\right)\left(1 + \frac{\bar{\xi}}{q} + \frac{\bar{\xi}^2}{q^2} + \cdots + \frac{\bar{\xi}^r}{q^r}\right) + O(r) \\ &= n\bar{\xi}\left(1 - \frac{1}{q}\right)\frac{1}{1 - \frac{\bar{\xi}}{q}} + O(1) + O(r), \end{aligned} \qquad (11.\ 19)$$

or by (11. 18)

$$S_1(n) = \frac{q-1}{q\xi - 1}\, n + O\,(\log n). \qquad (11.\ 20)$$

Formulae (11. 13) and (11. 20) are only special cases of the corresponding formula for arbitrary k. We obtain this in the following manner.

Since

$$S_k(qn+l) = S_k(qn) + O(1), \qquad [l = 0,\ 1,\ 2,\ \ldots,\ q-1] \qquad (11.\ 21)$$

we need only consider $S_k(qn)$. For this we have the formula

$$S_{qK+\lambda}(qn)\, \bar{\xi}^\lambda \{(q-\lambda)\, S_K(n) + \lambda\, S_{K+1}(n)\}. \qquad (11.\ 22)$$

We define a sequence $\sigma(k)$ by the functional equations

$$\begin{aligned} &\sigma(0) = 1 \\ &\sigma(qK+\lambda) = \bar{\xi}^\lambda\left(\frac{q-\lambda}{q}\,\sigma(K) + \frac{\lambda}{q}\,\sigma(K+1)\right). \end{aligned} \qquad (11.\ 23)$$

Then it is always true that

$$S_k(n) = \sigma(k)n + O(\log n). \tag{11. 24}$$

To begin with, we have proved this theorem for $k=0$ and $k=1$. Formula (11. 23) shows, however, that we may prove (11. 24) in general by a mathematical induction with respect to k.

$\sigma(k)$ is a very complicated arithmetical function. For small values of its argument ($K=0, 1, \cdots, q-1$, $\lambda=0, 1, \cdots, q-1$) we have

$$\left.\begin{aligned} \sigma(\lambda) &= \bar{\xi}^{\lambda}\,\frac{(q-\lambda)+(\lambda-1)\bar{\xi}}{q-\bar{\xi}}, \\ \sigma(Kq+\lambda) &= \bar{\xi}^{K+\lambda}\,\frac{(q+K)(q-\lambda)+((K-1)(q-\lambda)+(q-K-1)\lambda)\,\bar{\xi}+K\lambda\bar{\xi}^2}{q(q-\bar{\xi})}. \end{aligned}\right\} \tag{11. 25}$$

It is natural to extend our definition of $\sigma(k)$ to negative values of k by the formula

$$\sigma(-k) = \overline{\sigma(k)}. \tag{11. 26}$$

Formula (11. 24) is then true for negative as well as for positive arguments.

It is natural to investigate the functions

$$T_k(n) = \sum_{0}^{n-1} \sigma(l)\,\bar{\sigma}(l+k) \tag{11. 27}$$

which arise from σ in the same fashion as S_k arises from ϱ. We shall confine ourselves to the case

$$q=2;\ \xi=\bar{\xi}=-1.$$

We have here the equations

$$\left.\begin{aligned} \sigma(k) &= \bar{\sigma}(k); \\ \sigma(2k) &= \sigma(k); \\ \sigma(2k+1) &= -\frac{\sigma(k)+\sigma(k+1)}{2}. \end{aligned}\right\} \tag{11. 28}$$

Hence we have the following formulae:

$$T_{2k}(2n) = \sum_{m=0}^{n-1} (\sigma(2m)\,\sigma(2m+2k) + \sigma(2m+1)\,\sigma(2m+2k+1)) \tag{11.29}$$

$$= \sum_{m=0}^{n-1} \left(\sigma(m)\,\sigma(m+k) + \frac{(\sigma(m)+\sigma(m+1))(\sigma(m+k)+\sigma(m+k+1))}{4}\right) \tag{11.30}$$

or

$$\left| T_{2k}(2n) - \frac{3}{2} T_k(n) - \frac{1}{4} T_{k-1}(n) - \frac{1}{4} T_{k+1}(n) \right| < \text{const.} \tag{11.31}$$

and further

$$T_{2k+1}(2n) = \sum_{m=0}^{n-1} (\sigma(2m)\,\sigma(2m+2k+1) + \sigma(2m+1)\,\sigma(2m+2k+2))$$

$$= -\sum_{m=0}^{n-1} \left(\sigma(m)\frac{\sigma(m+k)+\sigma(m+k+1)}{2} + \frac{\sigma(m)+\sigma(m+1)}{2}\sigma(m+k+1)\right) \tag{11.32}$$

or

$$| T_{2k+1}(2n) + T_k(n) + T_{k+1}(n) | < \text{const.} \tag{11.33}$$

The array

$$\{\ldots, \varrho(n), \ldots, \varrho(1), \varrho(0), \varrho(1), \ldots, \varrho(n), \ldots\} = \{\ldots, a_{-n}, \ldots, a_{-1}, a_0, a_1, \ldots, a_n, \ldots\}$$

defines a function

$$f(t) = \begin{cases} a_n \text{ if } |t-n| \leq 1/8 \\ 0 \text{ if for no } n, |t-n| \leq 1/8 \end{cases} \tag{11.34}$$

and

$$\varphi(\tau) = \lim_{T\to\infty} \frac{1}{2T} \int_{-T}^{T} f(t+\tau)\bar{f}(t)\,dt = \frac{1}{4}\sigma\left(\left[\tau + \frac{1}{4}\right]\right) Q\left(\tau - \left[\tau + \frac{1}{4}\right]\right), \tag{11.35}$$

where

$$Q(x) - \begin{cases} 1 - 4|x|; & \left[-\frac{1}{4} < x < \frac{1}{4}\right] \\ 0. & \left[\frac{1}{4} \leq x < \frac{3}{4}\right] \end{cases}$$

We have

$$S(u) = \frac{1}{2\pi} \int_{-\infty}^{\infty} \varphi(\tau) \frac{e^{iu\tau} - 1}{i\tau}\,d\tau, \tag{11.36}$$

where $S(u)$ is of limited total variation. Then

$$S(u+\varepsilon) - S(u-\varepsilon) = \frac{1}{\pi} \int_{-\infty}^{\infty} \varphi(\tau) \frac{\sin \varepsilon\tau}{\tau} e^{iu\tau}\,d\tau. \tag{11.37}$$

Hence by (6. 15)

$$\lim_{\varepsilon\to 0}\frac{1}{2\varepsilon}\int_{-\infty}^{\infty} e^{-i\nu u}\,|\,S(u+\varepsilon)-S(u-\varepsilon)\,|^2\,du=\lim_{T\to\infty}\frac{1}{2T}\int_{-T}^{T}\varphi(x+\nu)\,\overline{\varphi}(x)\,dx. \qquad (11.\ 38)$$

Hence if the finite or denumerable set of discontinuities of $S(u)$ are at $u_1, u_2, \ldots$ and have values $A_1, A_2, \ldots$, respectively

$$\sum_1^\infty |\,A_k\,|^2=\lim_{\varepsilon\to 0}\frac{1}{2\varepsilon}\int_{-\infty}^{\infty} e^{-i\nu u}\,|\,S(u+\varepsilon)-S(u-\varepsilon)\,|^2\,du=$$

$$=\lim_{T\to\infty}\frac{1}{2T}\int_{-T}^{T}\varphi(x+\nu)\,\overline{\varphi}(x)\,dx \qquad (11.\ 39)$$

and exists for every ν. However,

$$\lim_{T\to\infty}\frac{1}{2T}\int_{-T}^{T}\varphi(x+\nu)\,\overline{\varphi}(x)\,dx=\frac{1}{16}\lim_{n\to\infty}\frac{T_{\left[\nu+\frac{1}{2}\right]}(n)}{n}R\left(\nu-\left[\nu+\frac{1}{2}\right]\right), \qquad (11.\ 40)$$

where

$$R(x)=\int_{-\frac{1}{2}}^{\frac{1}{2}} Q(y)\,Q(x+y)\,dy, \qquad (11.\ 41)$$

so that

$$\lim_{n\to\infty}\frac{T_k(n)}{n}$$

exists for every k. If we put

$$\varpi(k)=\lim_{n\to\infty}\frac{T_k(n)}{n}=\lim_{n\to\infty}\frac{1}{n}\sum_{l=0}^{n-1}\sigma(l)\,\sigma(l+k) \qquad (11.\ 42)$$

we may conclude from our equations for T_k that

$$\left.\begin{aligned}\varpi(2k)&=\frac{\varpi(k-1)+6\,\varpi(k)+\varpi(k+1)}{8};\\ \varpi(2k+1)&=-\frac{\varpi(k)+\varpi(k+1)}{2}.\end{aligned}\right\} \qquad (11.\ 43)$$

It follows that if

$$\varpi(0) = 0, \quad \varpi(1) = 0 \tag{11.44}$$

then for every k

$$\varpi(k) = 0. \tag{11.45}$$

We now put $k=0$ in (11.43), remembering that

$$\varpi(-k) = \varpi(k). \tag{11.46}$$

We obtain

$$\varpi(0) = \varpi(1) = 3\,\varpi(1). \tag{11.47}$$

Hence $\varpi(0)=\varpi(1)=0$ and $\varpi(k)$ is identically 0. In other words,

$$T_k(n) = o(n). \tag{11.48}$$

As $\sigma(2k)=\sigma(k)$, $\sigma(1)=-1/3$, we see that we cannot have

$$\lim_{k\to\infty} \sigma(k) = 0, \tag{11.49}$$

and hence we cannot have

$$\lim_{\tau\to\infty} \varphi(\tau) = 0. \tag{11.50}$$

It is thus impossible that

$$S(u) = \int_{-\infty}^{u} S'(v)\,dv + S(-\infty), \tag{11.51}$$

for then we should have by (5.43), (5.46)

$$\lim_{\tau\to\infty} \varphi(\tau) = \lim_{\tau\to\infty} \int_{-\infty}^{\infty} S'(u)\,e^{-iu\tau}\,du = 0. \tag{11.52}$$

On the other hand, as

$$\varpi(0) = 0,$$

we must have

$$\lim_{\varepsilon\to 0} \frac{1}{2\varepsilon} \int_{-\infty}^{\infty} |S(u+\varepsilon) - S(u-\varepsilon)|^2\,du = 0. \tag{11.53}$$

It follows that $S(u)$ is a continuous function which we have already seen not to be the integral of its derivative. This theorem of Mahler thus leads to a new type of spectrum.

27 — 29764. *Acta mathematica.* 55. Imprimé le 15 avril 1930.

12. Spectra depending on an infinite sequence of choices.

In the last section we gave concrete, well defined examples of functions with various continuous types of spectra. The present and the following sections are devoted to the generation of such functions by methods which instead of always yielding functions with continuous spectra, *almost* always yield such functions. The distinction is precisely analogous to that between rational mechanics of the classical kind and statistical mechanics. Theoretically, all the molecules of gas in a vessel might group themselves in a specified half of its volume; practically, such a contingency is unthinkably improbable.

The notion of probability is a new element not occurring in classical mechanics, but essential in statistical mechanics. It applies to a class of contingent situations, and has the essential properties of a measure. So too the idea of »almost always» in harmonic analysis depends on some more or less concealed notion of measure. In the present and the ensuing sections, we shall assimilate this notion of measure to that of Lebesgue, so that »almost always» will translate into »except for a set of Lebesgue measure zero».

Consider a sequence of independent tosses of a coin. By a sequence, we mean a record such as, »heads, tails, heads, heads, tails.» For such a finite sequence, the probability is 2^{-n}, where n is the number of tosses. That is, it is the same as the measure of the set of all the points on $(0, 1)$ with coordinates whose binary expansion begins $.10110$. This mapping immediately suggests a definition of probability for infinite sequences of tosses. The probability of any set of sequences of tosses is defined as the Lebesgue measure of the set of points whose binary representations correspond to sequences of tosses in the set, in such a manner that 1 corresponds to »heads» and 0 to »tails».[5] We can even represent sequences infinite in both directions by binary numbers in such a way as to define the probability of a set of sequences, by having recourse to some definite enumerative rearrangement of such a sequence.

If we have made »probability» a mere translation of »measure», »average» becomes the equivalent of »integral». We are accordingly able to use the entire body of theorems concerning the Lebesgue integral in the service of the calculus of probabilities.

We have not yet, however, correlated with our sequence of throws functions susceptible of a harmonic analysis. To do this, we take a certain zero point on a doubly infinite line to correspond with the zero point of our doubly infinite sequence of tosses, and if the nth toss is a head, we define $f(t)$ to be 1 for $n<t<n+1$; if a tail, to be -1. The question we wish to ask is: what is the probability distribution of spectral functions $S(u)$ for these functions $f(t)$?

We have, taking $f(t)=a_n$ for $n<t<n+1$,

$$\frac{1}{2n}\int_{-n}^{n} f(t+m)f(t)\,dt = \frac{1}{2n}(a_{m-n}a_{-n} + a_{m-n+1}a_{-n+1} + \cdots + a_{m+n-1}a_{n-1}). \quad (12.01)$$

Since the distribution of each a_n between negative and positive values is symmetrical and independent of that of every other,

$$\text{Average } \frac{1}{2n}\int_{-n}^{n} f(t+m)f(t)\,dt = \begin{cases} 0 \text{ if } m \neq 0; \\ 1 \text{ if } m = 0. \end{cases} \quad (12.02)$$

When $m=0$, this average is indeed identically 1. In every other case, when m is an integer,

$$\begin{aligned} &\text{Average } \left\{\frac{1}{2n}\int_{-n}^{n} f(t+m)f(t)\,dt\right\}^2 \\ &= \text{Average } \frac{1}{4n^2}(a_{m-n}a_{-n} + \cdots + a_{m+n-1}a_{n-1})^2 \\ &= \text{Average } \frac{1}{4n^2}(a_{m-n}^2 a_{-n}^2 + \cdots + a_{m+n-1}^2 a_{n-1}^2) \\ &= 1/2n, \end{aligned} \quad (12.03)$$

since the averages of all the non-square terms vanish. Hence

$$\frac{1}{2n^2}\int_{-n^2}^{n^2} f(t+m)f(t)\,dt > A \quad (12.04)$$

over a set of values of $f(t)$ with total probability

$$\leq \frac{A}{2n^2}. \quad (12.05)$$

Since the sum of these latter quantities forms a convergent series, with remainder after n terms tending to 0 as n increases, we must have

$$\lim_{n\to\infty} \frac{1}{2n^2} \int_{-n^2}^{n^2} f(t+m) f(t)\, dt < A, \tag{12.06}$$

except in a set of cases of arbitrarily small and hence of zero probability. Hence except in a set of cases of zero probability,

$$\lim_{n\to\infty} \frac{1}{2n^2} \int_{-n^2}^{n^2} f(t+m) f(t)\, dt = 0. \tag{12.07}$$

Here the procedure to the limit runs through integral values of n. This can be generalized at once. Let P be bounded, and let

$$n^2 < T < (n+1)^2.$$

Then

$$\begin{aligned} \frac{1}{2T}\int_{-T}^{T} P\,d\lambda &= \frac{n^2}{T}\frac{1}{2n^2}\int_{-n^2}^{n^2} P\,d\lambda + \frac{1}{2T}\left[\int_{n^2}^{T} + \int_{-T}^{-n^2}\right] P\,d\lambda \\ &= \frac{n^2}{T}\frac{1}{2n^2}\int_{-n^2}^{n^2} P\,d\lambda + \frac{\vartheta}{2T}\left[\int_{-(n+1)^2}^{(n+1)^2} P\,d\lambda - \int_{-n^2}^{n^2} P\,d\lambda\right] \qquad [0 \le \vartheta \le 1] \\ &= \frac{n^2}{T}\frac{1}{2n^2}\int_{-n^2}^{n^2} P\,d\lambda + \vartheta\left[\frac{(n+1)^2}{T}\cdot\frac{1}{2(n+1)^2}\int_{-(n+1)^2}^{(n+1)^2} P\,d\lambda - \frac{n^2}{T}\cdot\frac{1}{2n^2}\int_{-n^2}^{n^2} P\,d\lambda\right]. \end{aligned} \tag{12.08}$$

Thus if

$$\lim_{n\to\infty} \frac{1}{2n^2}\int_{-n^2}^{n^2} P\,d\lambda = L,$$

then

$$\lim_{T\to\infty} \frac{1}{2T}\int_{-T}^{T} P\,d\lambda = L. \tag{12.09}$$

We thus see that in case m is an integer other than zero, we almost always have

$$\lim_{T\to\infty}\frac{1}{2T}\int_{-T}^{T} f(t+m)f(t)\,dt = 0. \tag{12.10}$$

As in (11.07), we may conclude that

$$\varphi(x) = \begin{cases} 0; & (x > 1) \\ 1-x; & (0 < x < 1) \\ 1+x; & (-1 < x < 0) \\ 0; & (-1 < x) \end{cases} \tag{12.11}$$

and that

$$S(u) = \frac{1}{2\pi}\int_{-\infty}^{\infty} \varphi(x)\frac{e^{iux}-1}{ix}\,dx = \frac{1}{\pi}\int_{0}^{1}(1-x)\frac{\sin ux}{x}\,dx = \frac{1}{\pi}\int_{0}^{u}\frac{1-\cos v}{v^2}\,dv. \tag{12.12}$$

These propositions are true, not always, but almost always. Thus a haphazard sequence of positive and negative rectangular impulses almost always has the spectral intensity

$$\frac{1}{\pi}\frac{1-\cos v}{v^2} \tag{12.13}$$

which is numerically identical to the square of the Fourier transform of a single rectangular impulse. To see this, we need only reflect that the Fourier transform of such an impulse is

$$\frac{1}{\sqrt{2\pi}}\int_{-\frac{1}{2}}^{\frac{1}{2}} e^{iux}\,dx = \sqrt{\frac{2}{\pi}}\,\frac{\sin u/2}{u} = \sqrt{\frac{1-\cos u}{\pi u^2}}. \tag{12.14}$$

It would not be a difficult task to generalize this remark to impulses of other than rectangular shape. The essential generalization to make this fact of physical interest is, however, to eliminate the equal spacing of the individual impulses, to reduce the sequence of impulses to such an irregularity as is found in the Brownian motion. This is the problem of our next section. The principal difficulty is that the fundamental Lebesgue measure to which we reduce our probabilities is not so obviously at hand. There is no continuous infinity of choices which bears an obvious analogy to that involved in building up a binary fraction. Nevertheless, the distribution involved in the time paths of particles subject to the Brownian motion can be reduced to a Lebesgue measure, certain functions connected with these paths can almost always be analysed harmonically, and their spectra will almost always have a certain fixed distribution of energy if frequency. In other words, the properties of the distributions and functions of this section furnish an excellent working model for those to be expected of the functions discussed in the next section.

13. Spectra and integration in function-space.

From the very beginning, spectrum theory and statistics have joined hands in the theory of white light. The apparent contradiction between the additive character of electromagnetic displacement in the Maxwell theory and the observed additive character of the quadratic light-intensities is on the surface of things irreconcilable. The credit for resolving this antinomy is due to Lord Rayleigh. He considers the resultant of a large number of vibrations of arbitrary phase, and shows that the mean intensity of their sum is actually additive. He says, »It is with this mean intensity only that we are concerned in ordinary photometry. A source of light, such as a candle or even a soda flame, may be regarded as composed of a very large number of luminous centres disposed throughout a very sensible space; and even though it be true that the intensity at a particular point of a screen illuminated by it and at a particular moment of time is a matter of chance, further processes of averaging must be gone through before anything is arrived at of which our senses could ordinarily take cognizance. In the smallest interval of time during which the eye could be impressed, there would be opportunity for any number of rearrangements of phase, due either to motions of the particles or to irregularities in their modes of vibration. And even if we suppose that each luminous centre was fixed, and emitted perfectly regular vibrations, the manner of composition and consequent intensity would vary rapidly from point to point of the screen, and in ordinary cases the mean illumination over the smallest appreciable area would correspond to a thorough averaging of the phase-relationships. In this way the idea of the intensity of a luminous source, independently of any questions of phase, is seen to be justified, and we may properly say that two candles are twice as bright as one.»

Thus Rayleigh's statistics of light is a statistics in which the quantities distributed are amplitudes of sinusoidal vibrations. Such a theory involves a preliminary harmonic analysis, perhaps of a somewhat vague nature, but definite enough to be useful in the hands of a competent physicist. There is an alternative approach to spectrum statistics, which is of a more direct nature. Imagine a resonator — say a sea-shell — struck by a purely chaotic sequence of acoustical impulses. It will yield a response which still has a statistical element in it, but in which the selective properties of the resonator will have accentuated certain frequencies at the expense of others. It seems on the surface of it highly plausible that the outputs of two such resonators will almost always be additive as to intensities rather than merely as to amplitudes.

»Chaos» and »almost always» — there lies an entire statistical theory behind these terms. The simplest phenomenon to which the name »chaos» may be applied with any propriety is that of the Brownian motion. Here a small particle is impelled by atomic collisions in such a way that its future is entirely independent of its past. If we consider the X-coordinate of such a particle, the probability that this should alter a given amount in a given time is independent (1) of the entire past history of the particle; (2) of the instant from which the given interval is measured; (3) whether we are considering changes that increase or changes that decrease it. If we make the assumption that the distribution of the changes of x over a given interval of time is Gaussian, it follows as Einstein has pointed out that the probability that after a time t the change in x should lie between x_1 and x_1+dx_1 is

$$\frac{1}{\sqrt{\pi c t}} e^{-x_1^2/ct}\, dx_1. \tag{13.01}$$

Here c is a constant which we may and shall reduce to 1 by a proper choice of units. The particular manner in which t enters results from the fact that

$$\frac{1}{\sqrt{\pi(t_1+t_2)}} e^{-\frac{x_1^2}{t_1+t_2}} = \frac{1}{\pi\sqrt{t_1 t_2}} \int_{-\infty}^{\infty} e^{-\frac{x^2}{t_1}} e^{-\frac{(x_1-x)^2}{t_2}}\, dx. \tag{13.02}$$

This fundamental identity is tantamount to the statement that the probability that x should have changed by an amount lying between x_1 and x_1+dx_1 after a time t_1+t_2 is the total compound probability that the change of x over time

t_1 should be anything at all, and that it should then migrate in a subsequent interval of length t_2 to a position between x_1 and $x_1 + dx_1$.

A quantity x whose changes are distributed after the manner just mentioned is said to have them normally or chaotically distributed. Of course, what really is distributed is the function $x(t)$ representing the successive values of x. (There is no essential restriction in supposing $x(0)=0$.) Thus the conception of a purely chaotic distribution really introduces a certain system of measure and consequently of integration into function-space. This gives us the clue to the statistical study of spectra. We determine the response of a resonator in terms of functionals of the incoming chaotic disturbance. We then ask, »What is the distribution of various quantities connected with the spectrum of this response, as determined by integrating these quantities over function-space with respect to the original chaos?» Let it be remarked that the theory of integration appropriate to this problem is that developed by the author in his »Differential-space», and not the earlier theory of Gâteaux, which forms the starting point of most researches in this direction.

Before we can attack these more difficult problems we must establish out theory of integration on a firm basis. To do this, we shall establish a correspondence between the set of all functions and the points on a segment of a line AB of unit length, and shall use this correspondence to define integration over function-space in terms of Lebesgue integration over the segment. Let me say in passing that in my previous attacks on this problem, I have made use of a somewhat generalized form of integration due to P. J. Daniell. This form of integration, at least in all cases yet given, may be mapped into a Lebesgue integration over a one-dimensional manifold by a transformation retaining measure invariant. In as much as the literature contains a much greater wealth of proved theorems concerning the Lebesgue integral than of theorems concerning the Daniell integral — although the latter are not particularly difficult to establish — it has seemed to me more desirable to employ the method of mapping. This method of mapping is an extension to infinitely many dimensions of a method due to Radon.

The method of mapping consists in making certain sets of functions $x(t)$, which we shall call »quasi-intervals», correspond to certain intervals of AB. Our quasi-intervals will be sets of all functions $x(t)$ defined for $0 \leq t \leq 1$ such that

$$\begin{aligned} x(0) &= 0; \\ x_{11} \le x(t_1) &\le x_{12}; \\ x_{21} \le x(t_2) &\le x_{22}; \\ x_{31} \le x(t_3) &\le x_{32}; \\ &\cdots\cdots \\ x_{n1} \le x(t_n) &\le x_{n2}. \end{aligned}$$

$$(0 \le t_1 \le t_2 \le t_3 \le \cdots \le t_n \le 1). \tag{13.03}$$

By our definition of probability, the probability that $x(t)$ should lie in this quasi-interval is

$$\pi^{-n/2}[t_1(t_2-t_1)(t_3-t_2)\cdots(t_n-t_{n-1})]^{-1/2}\int_{x_{11}}^{x_{12}} d\xi_1 \cdots \int_{x_{n1}}^{x_{n2}} d\xi_n \cdot$$

$$\cdot \exp\left[-\xi_1^2 t_1^{-1} - \sum_2^n (\xi_k - \xi_{k-1})^2 (t_k - t_{k-1})^{-1}\right]. \tag{13.04}$$

Clearly, if the class of all functions $x(t)$ be divided into a finite number of quasi-intervals — some of which then must contain infinite values of x_{h1} or x_{h2} — the sum of their probabilities will be 1.

The quasi-intervals with which we shall be more specially concerned are the quasi-intervals $I(n; k_1, k_2, \ldots, k_{2^n})$ for which

$$\left.\begin{aligned} t_h &= h\,2^{-n}; \qquad (1 \le h \le 2^n) \\ x_{h1} &= \tan\,(k_h \pi 2^{-n}); \\ x_{h2} &= \tan\,((k_h+1)\pi 2^{-n}); \end{aligned}\right\} \tag{13.05}$$

where k_h is some integer between -2^{n-1} and $2^{n-1}-1$, inclusive. For the probability that $x(t)$ should lie in this interval let us write

$$P\{I(n; k_1, \ldots, k_{2^n})\}.$$

Let us notice that $I(n; k_1, \ldots, k_{2^n})$ is made of a finite number of quasi-intervals $I(n; l_1, \ldots, l_{2^{n+1}})$, and that the sum of the probabilities belonging to the latter gives the probability belonging to the former.

Let us now map the four quasi-intervals $I(1; k_1, k_2)$ on the segment AB, assigning to each in order an interval with length equal to its probability. Let us map the quasi-intervals $I(2; k_1, k_2, k_3, k_4)$ into intervals of the segment AB, translating probability into length, and in such a manner that if $I(2; l_1, l_2, l_3, l_4)$ forms a portion of $I(1; k_1, k_2)$, their images stand in the same relation. If we keep this process up indefinitely, we shall have mapped all the quasi-intervals $I(n; k_1, k_2, \ldots, k_{2^n})$ into intervals of AB in such a way that probability is translated into length, and that the end-points of the intervals of AB form an everywhere dense set.

Up to this point, our mapping has mapped quasi-intervals on intervals. We wish to deduce from it a mapping of functions on points. As a lemma for this purpose, we shall show that the functions $x(t)$ for which for any t_1 and t_2 that are terminating binaries,

$$|x(t_1)-x(t_2)| \geq 40\,h\,|t_1-t_2|^{1/4} \tag{13.06}$$

may be enclosed in a denumerable set of quasi-intervals such that the sum of the probabilities of these quasi-intervals is $O(h^{-n})$ for any n.

To show this, let us represent t_1 as the binary fraction

$$0 \cdot a_1\, a_2\, a_3 \cdots a_n \cdots$$

and t_2 as the binary fraction

$$0 \cdot b_1\, b_2\, b_3 \cdots b_n \cdots.$$

Let t_3 be a number whose binary expansion may be made to agree with that of t_1 up to and including a_j and with that of t_2 up to and including b_j. We shall choose t_3 so that j is as large as possible, even though this may necessitate the use of an expression for t_3 ending in $111 \cdots$ to agree with the smaller of the quantities t_1 and t_2 and of a terminating expression for t_3 to agree with the larger. The interval from t_1 to t_3 will then be expressible in the form

$$0 \cdot 00 \cdots 0 c_{j+1}\, c_{j+2} \cdots,$$

where there are j consecutive 0's after the final point, and every c is 0 or 1. The interval from t_2 to t_3 may be expressed in a similar manner. In other words, the interval from t_1 to t_2 may be reduced to the sum of a denumerable set of intervals from terminating binaries to adjacent terminating binaries of the same number of figures, such that there are not more than two intervals in the set of magnitude 2^{-j-k} where k is any positive integer, and such that every interval is of one of these sizes.

Now clearly,

$$|t_2 - t_1| \geq 2^{-1-j}. \tag{13.07}$$

Hence, if it is for particular values of t_2 and t_1 in question that (13.06) is fulfilled,

$$\begin{aligned} |x(t_1) - x(t_2)| &\geq 40\,h \cdot 2^{-(1+j)/4} \\ &> h \cdot 2^{1-(1+j)/4}/(1-2^{-1/4}) \\ &= 2\,h \sum_{k=j+1}^{\infty} 2^{-l/4}. \end{aligned} \tag{13.08}$$

If we now appeal to our analysis of the interval (t_1, t_2), we see that for *some* interval from $m \cdot 2^{-j-k}$ to $(m+1) \cdot 2^{-j-k}$, where m and k are integers and $0 \leq m < 2^{j+k}$, we shall have

$$\left| x\left(\frac{m}{2^{j+k}}\right) - x\left(\frac{m+1}{2^{j+k}}\right) \right| > h \cdot 2^{-(j+k)/4}. \tag{13.09}$$

Thus if for any pair of values t_1 and t_2 that are terminating binary fractions,

$$|x(t_1) - x(t_2)| \geq 40\,h\,|t_1 - t_2|^{1/4}, \tag{13.10}$$

then for some integers m and i $(m < 2^i)$

$$|x(m\,2^{-i}) - x((m+1)\,2^{-i})| > h \cdot 2^{-i/4}. \tag{13.11}$$

It merely remains to determine the measure of a denumerable set of our quasi-intervals $I(n; k_1, \ldots, k_{2^n})$ containing all the functions $x(t)$ for which, for some m and i, (13.11) holds.

To begin with, let m and i be fixed. Since our selected quasi-intervals ultimately divide the range of values of $x(m\,2^{-i})$ and of $x((m+1)2^{-i})$ to an arbitrary degree of fineness, there is no trouble in proving that the functions satisfying (13.11) may be enclosed in a finite set of selected quasi-intervals of total probability not exceeding

$$\varepsilon + \frac{2}{\sqrt{\pi 2^{-i}}} \int_{h \cdot 2^{-i/4}}^{\infty} e^{-\frac{x^2}{2^{-i}}}\,dx = \varepsilon + \frac{2}{\sqrt{\pi}} \int_{h \cdot 2^{i/4}}^{\infty} e^{-x^2}\,dx, \tag{13.12}$$

where ε is arbitrarily small. If we sum for m and i, we get as the sum of the probabilities of all our enclosing sets a quantity not exceeding

$$\eta + \sum_{i=1}^{\infty} \frac{2^{i+1}}{\sqrt{\pi}} \int_{h \cdot 2^{i/4}}^{\infty} e^{-x^2}\, dx$$

$$< \sum_{i=1}^{\infty} \frac{2^{i+1}}{\sqrt{\pi}} e^{-h \cdot 2^{i/4}}$$

$$< \sum_{i=1}^{\infty} \frac{2^{i+1-ni} h^{-4n}}{\sqrt{\pi}} \tag{13. 13}$$

for sufficiently large h. As usual η represents an arbitrarily small quantity. Expression (13. 13) clearly can be made to vanish more rapidly than any given negative power of h as h becomes infinite.

Let us now reconsider our mapping. If we leave out the ends of our intervals, which form a denumerable set of measure 0, every point on AB is uniquely characterized by and uniquely characterizes an infinite sequence of intervals, each containing the next, and tending to 0 in length. If we reject a denumerable set of quasi-intervals of arbitrarily small total probability, the remaining quasi-intervals and portions of quasi-intervals all contain functions satisfying the condition of equicontinuity

$$|x(t_1)-x(t_2)| < 40\, h\, |t_1-t_2|^{1/4}, \tag{13. 14}$$

so that if we modify AB by the removal of a set of points of arbitrarily small outer measure, as well as by the removal of the end-points of our intervals, every point of AB is characterized by a sequence of intervals closing down on it, by the succession of corresponding quasi-intervals, and by the uniquely determined function $x(t)$ common to this sequence of quasi-intervals and satisfying (13. 14). It follows at once that except for a set of points of zero measure, we have determined a unique mapping of the points of AB by functions satisfying (13. 14) for *some* h. Thus any functional of the latter functions determines a function on the line, which may be summable Lebesgue. In the latter case, we shall define the average of the functional as the Lebesgue integral of the corresponding function on AB.

Among the summable functionals are the expressions

$$P(x(t_1), x(t_2), \ldots, x(t_n)),$$

where P stands for a polynomial. This is readily seen to be the case when the expressions $t_1, \ldots, t_n$ are terminating binaries, and the extension to other values follows from the equicontinuity conditions we have already laid down. To see this, let us note that we have already given information enough to prove that the upper average (corresponding to upper integral) of

$$[\max |x(t)|]^n$$

is finite. This functional will, however, simultaneously dominate

$$P(x(t_1), x(t_2), \ldots, x(t_n))$$

in which we suppose P of the nth degree, and the set of approximating functionals

$$P(x(t_{11}), x(t_{12}), \ldots, x(t_{1n}));$$
$$\cdots\cdots\cdots$$
$$P(x(t_{m1}), x(t_{m2}), \ldots, x(t_{mn}));$$
$$\cdots\cdots\cdots$$

in which $t_{11}, \ldots, t_{mn}, \ldots$ are terminating binaries, and $\lim_{m\to\infty} t_{mn}-t_n$. That these functionals are actually approximating functionals results from the fact that

$$P(y_1, \ldots, y_n)$$

is continuous, and that $x(t)$ is almost always continuous. Now, there is a theorem to the effect that if a sequence of functions uniformly dominated by a Lebesgue summable function converges for each argument to a limit, and if the Lebesgue integrals of these functions converge to a limit, this limit is the Lebesgue integral of the limit function. This proves our theorem.

In case $t_1 \leq t_2 \leq \cdots \leq t_n$, the average of $P(x(t_1), \ldots, x(t_n))$ is readily seen to be

$$\pi^{-\frac{n}{2}}[t_1(t_2-t_1)\cdots(t_n-t_{n-1})]^{-1/2}\int_{-\infty}^{\infty} d\xi_1 \cdots \int_{-\infty}^{\infty} d\xi_n\, P(\xi_1, \ldots, \xi_n) \cdot$$
$$\cdot \exp\left[-\xi_1^2\, t_1^{-1} - \sum_2^n (\xi_k-\xi_{k-1})^2 (t_k-t_{k-1})^{-1}\right]. \qquad (13.15)$$

In particular, if $t_1 \leq t_2$,

$$\text{Average } (x(t_1)\,x(t_2)) = \frac{1}{\sqrt{\pi t_1 (t_2 - t_1)}} \int_{-\infty}^{\infty} d\xi_1 \int_{-\infty}^{\infty} \xi_1 \xi_2 e^{-\frac{\xi_1^2}{t_1} - \frac{(\xi_2-\xi_1)^2}{t_2-t_1}} d\xi_2$$

$$= t_1/2. \tag{13.16}$$

and if $t_1 \leqq t_2 \leqq t_3 \leq t_4$,

$$\text{Average } (x(t_1)\,x(t_2)\,x(t_3)\,x(t_4)) = \frac{1}{\pi^2 \sqrt{t_1(t_2-t_1)(t_3-t_2)(t_4-t_3)}} \cdot$$

$$\cdot \int_{-\infty}^{\infty} d\xi_1 \int_{-\infty}^{\infty} d\xi_2 \int_{-\infty}^{\infty} d\xi_3 \int_{-\infty}^{\infty} d\xi_4 \, \xi_1 \xi_2 \xi_3 \xi_4 \exp\left(-\frac{\xi_1^2}{t_1} - \frac{(\xi_2-\xi_1)^2}{t_2-t_1} - \frac{(\xi_3-\xi_2)^2}{t_3-t_2} - \frac{(\xi_4-\xi_3)^2}{t_4-t_3}\right)$$

$$= \frac{t_1 t_2}{2} + \frac{t_1 t_3}{4}. \tag{13.17}$$

The expressions just given are absolutely summable. Accordingly, by the familiar rules for inverting the order of integration, if $\alpha_1(t)$, $\alpha_2(t)$, $\alpha_3(t)$, $\alpha_4(t)$ are of limited total variation over $(0, 1)$,

$$\text{Average} \int_0^1 x(t)\, d\alpha_1(t) \int_0^1 x(t)\, d\alpha_2(t) = \int_0^1 \frac{t}{2}\, d\alpha_1(t) \int_t^1 d\alpha_2(s) + \int_0^1 \frac{t}{2}\, d\alpha_2(t) \int_t^1 d\alpha_1(s)$$

$$= \int_0^1 \frac{t}{2} (\alpha_2(1) - \alpha_2(t))\, d\alpha_1(t) + \int_0^1 \frac{t}{2} (\alpha_1(1) - \alpha_1(t))\, d\alpha_2(t)$$

$$= - \int_0^1 \frac{t}{2}\, d[\alpha_1(1) - \alpha_1(t)][\alpha_2(1) - \alpha_2(t)]$$

$$= \frac{1}{2} \int_0^1 [\alpha_1(1) - \alpha_1(t)][\alpha_2(1) - \alpha_2(t)]\, dt; \tag{13.18}$$

$$\text{Average} \int_0^1 x(t)\, d\alpha_1(t) \int_0^1 x(t)\, d\alpha_2(t) \int_0^1 x(t)\, d\alpha_3(t) \int_0^1 x(t)\, d\alpha_4(t)$$

$$= \int_0^1 \frac{t}{2}\, d\alpha_1(t) \int_t^1 d\alpha_2(s) \int_s^1 \left(s + \frac{u}{2}\right) d\alpha_3(u) \int_u^1 d\alpha_4(v)$$

$$+\int_0^1 \frac{t}{2}\, d\alpha_1(t) \int_t^1 d\alpha_2(s) \int_s^1 \left(s+\frac{u}{2}\right) d\alpha_4(u) \int_u^1 d\alpha_3(v)$$

$+22$ other terms representing different orders of $\alpha_1, \alpha_2, \alpha_3, \alpha_4$

$$=\frac{1}{2}\int_0^1 \frac{t}{2}\, d\alpha_1(t) \int_t^1 d\alpha_2(s) \int_s^1 \frac{u}{2}\, d\alpha_3(u) \int_u^1 d\alpha_4(v)$$

$$+\frac{1}{2}\int_0^1 \frac{t}{2}\, d\alpha_1(t) \int_t^1 \frac{u}{2}\, d\alpha_3(u) \int_u^1 d\alpha_2(s) \int_s^1 d\alpha_4(v)$$

$$+\frac{1}{2}\int_0^1 \frac{u}{2}\, d\alpha_3(u) \int_u^1 \frac{t}{2}\, d\alpha_1(t) \int_t^1 d\alpha_2(s) \int_s^1 d\alpha_4(v)$$

$$+\frac{1}{2}\int_0^1 \frac{u}{2}\, d\alpha_3(u) \int_u^1 d\alpha_4(v) \int_v^1 \frac{t}{2}\, d\alpha_1(t) \int_t^1 d\alpha_2(s)$$

$$+\frac{1}{2}\int_0^1 \frac{u}{2}\, d\alpha_3(u) \int_u^1 \frac{t}{2}\, d\alpha_1(t) \int_t^1 d\alpha_4(v) \int_v^1 d\alpha_2(s)$$

$$+\frac{1}{2}\int_0^1 \frac{t}{2}\, d\alpha_1(t) \int_t^1 \frac{u}{2}\, d\alpha_3(u) \int_u^1 d\alpha_4(v) \int_v^1 d\alpha_2(s)$$

$+$ all other terms representing different orders of $\alpha_1, \alpha_2, \alpha_3, \alpha_4$

$$=\frac{1}{2}\int_0^1 \frac{t}{2}\, d\alpha_1(t) \int_t^1 d\alpha_2(s) \int_0^1 \frac{u}{2}\, d\alpha_3(u) \int_u^1 d\alpha_4(v) + \text{etc.}$$

$$=\frac{1}{4}\int_0^1 [\alpha_1(1)-\alpha_1(t)][\alpha_2(1)-\alpha_2(t)]\,dt \int_0^1 [\alpha_3(1)-\alpha_3(s)][\alpha_4(1)-\alpha_4(s)]\,ds$$

$$+\frac{1}{4}\int_0^1 [\alpha_1(1)-\alpha_1(t)][\alpha_3(1)-\alpha_3(t)]\,dt \int_0^1 [\alpha_2(1)-\alpha_2(s)][\alpha_4(1)-\alpha_4(s)]\,ds$$

$$+\frac{1}{4}\int_0^1 [\alpha_1(1)-\alpha_1(t)][\alpha_4(1)-\alpha_4(t)]\,dt \int_0^1 [\alpha_2(1)-\alpha_2(s)][\alpha_3(1)-\alpha_3(s)]\,ds. \quad (13.19)$$

The point of this last argument is that

$$\frac{t}{2}\left(s+\frac{u}{2}\right)$$

may be written

$$\frac{t}{2}\frac{s}{2}+\frac{s}{2}\frac{t}{2}+\frac{t}{2}\frac{u}{2}$$

and that we may then take advantage of the existence in our expression of all permutations of $\alpha_1, \alpha_2, \alpha_3, \alpha_4$ to relabel our variables s, t, u, v so as to add the terms of our expression together again in a more symmetrical way, and represent it as a sum of three products of integrals such as we have already evaluated.

Up to the present, we have been considering probabilities depending on a basis function $x(t)$ defined over (0, 1). For the purposes of harmonic analysis, we wish to replace this by a basis function defined over $(-\infty, \infty)$. We may do this as follows: Let

$$\xi(\tau)=\sqrt{\pi}\left[\int_{\frac{1}{2}}^{\frac{1}{\pi}\cot^{-1}(-\tau)} x(t)\,d\csc\pi t - x\left(\frac{1}{\pi}\cot^{-1}(-\tau)\right)\sqrt{1+\tau^2}+x(1/2)\right]. \qquad (13.20)$$

As $x(t)$ is almost always bounded, this will almost always exist. Then

$$\text{Average } (\xi(\beta)-\xi(\alpha))^2=\frac{\pi}{2}\int_{\frac{1}{\pi}\cot^{-1}(-\alpha)}^{\frac{1}{\pi}\cot^{-1}(-\beta)}\csc^2\pi t\,dt=\frac{\beta-\alpha}{2}, \qquad (13.21)$$

in the case that $\beta>\alpha$. This is merely a particular case of (13.18). In the case that (α, β) and (γ, δ) do not overlap, a similar argument will show that

$$\text{Average } (\xi(\beta)-\xi(\alpha))(\xi(\gamma)-\xi(\delta))=0. \qquad (13.22)$$

Thus $\xi(\tau)$ has essentially the same distribution properties as $x(t)$, but over $(-\infty, \infty)$ instead of (0, 1).

We might, of course, have defined our distribution of $\xi(\tau)$ originally, without any recourse to that of $x(t)$. In any case, we should have had to make use of the fact that this distribution has certain equicontinuity properties, and these are somewhat easier to develop over a finite than over an infinite interval. The function $\xi(\tau)$ represents the result of a haphazard sequence of impulses, uniformly distributed in time, extending from $-\infty$ to ∞, in such a way that their zero value is taken to be at $\tau=0$. It is consequently immediately available for a harmonic analysis such as we discuss in this paper, while $x(t)$ is itself immediately adapted only for a Fourier series analysis.

Now let $\vartheta(\tau)$ represent the characteristic response in time of some resonator, the so-called indicial admittance. It may be real or complex, but we shall assume $\vartheta(\tau)\sqrt{1+\tau^2}$ to be of limited total variation over $(-\infty, \infty)$ and ϑ to be quadratically summable. As an immediate consequence of these assumptions,

$$\vartheta(-\infty)=\vartheta(\infty)=0. \tag{13.23}$$

We have

$$\int_{-\infty}^{\infty}\xi(\tau)\,d\vartheta(\tau)=\sqrt{\pi}\int_{-\infty}^{\infty}d\vartheta(\tau)\left[\int_{\frac{1}{2}}^{\frac{1}{\pi}\cot^{-1}(-\tau)}x(t)\,d\csc t-x\left(\frac{1}{\pi}\cot^{-1}(-\tau)\right)\sqrt{1+\tau^2}+x(1/2)\right]$$

$$=\sqrt{\pi}\int_0^1 x(t)[\vartheta(-\cot\pi t)\,d\csc\pi t+\csc\pi t\,d\vartheta(-\cot\pi t)]$$

$$=\sqrt{\pi}\int_0^1 x(t)\,d[\vartheta(-\cot\pi t)\csc\pi t]. \tag{13.24}$$

Hence if $\vartheta_1(\tau)$, $\vartheta_2(\tau)$, $\vartheta_3(\tau)$, and $\vartheta_4(\tau)$ satisfy the conditions we have laid down for $\vartheta(\tau)$,

$$\text{Average}\int_{-\infty}^{\infty}\xi(\tau)\,d\vartheta_1(\tau)\int_{-\infty}^{\infty}\xi(\tau)\,d\vartheta_2(\tau)=\frac{1}{2}\int_0^1\pi\csc^2\pi t\,\vartheta_1(-\cot\pi t)\,\vartheta_2(-\cot\pi t)\,dt$$

$$=\frac{1}{2}\int_{-\infty}^{\infty}\vartheta_1(\tau)\,\vartheta_2(\tau)\,d\tau; \tag{13.25}$$

29—29764. *Acta mathematica.* 55. Imprimé le 29 avril 1930.

$$\text{Average} \int_{-\infty}^{\infty} \xi(\tau)\, d\vartheta_1(\tau) \int_{-\infty}^{\infty} \xi(\tau)\, d\vartheta_2(\tau) \int_{-\infty}^{\infty} \xi(\tau)\, d\vartheta_3(\tau) \int_{-\infty}^{\infty} \xi(\tau)\, d\vartheta_4(\tau)$$

$$= \frac{1}{4} \int_{-\infty}^{\infty} \vartheta_1(\tau)\vartheta_2(\tau)\, d\tau \int_{-\infty}^{\infty} \vartheta_3(\tau)\vartheta_4(\tau)\, d\tau$$

$$+ \frac{1}{4} \int_{-\infty}^{\infty} \vartheta_1(\tau)\vartheta_3(\tau)\, d\tau \int_{-\infty}^{\infty} \vartheta_2(\tau)\vartheta_4(\tau)\, d\tau$$

$$+ \frac{1}{4} \int_{-\infty}^{\infty} \vartheta_1(\tau)\vartheta_4(\tau)\, d\tau \int_{-\infty}^{\infty} \vartheta_2(\tau)\vartheta_3(\tau)\, d\tau. \tag{13. 26}$$

We thus have succeeded in generalizing our theorems concerning the averages of products of linear functionals to the case where the basis function has an infinite range.

We wish to apply these results to the harmonic analysis of an expression $\int_{-\infty}^{\infty} \xi(\tau)\, d\vartheta(\tau+\lambda)$. To do this, we must evaluate the following averages:

$$\text{Average} \frac{1}{2T} \int_{-T}^{T} d\lambda \int_{-\infty}^{\infty} \xi(\tau)\, d\vartheta(\tau+\lambda) \int_{-\infty}^{\infty} \xi(\sigma)\, d\bar{\vartheta}(\sigma+\lambda)$$

$$= \frac{1}{2T} \int_{-T}^{T} d\lambda \;\text{Average} \left[\int_{-\infty}^{\infty} \xi(\tau)\, d\vartheta(\tau+\lambda) \int_{-\infty}^{\infty} \xi(\sigma)\, d\bar{\vartheta}(\sigma+\lambda) \right]$$

$$= \frac{1}{4T} \int_{-T}^{T} d\lambda \int_{-\infty}^{\infty} |\vartheta(\tau+\lambda)|^2\, d\tau$$

$$= \frac{1}{2} \int_{-\infty}^{\infty} |\vartheta(\tau)|^2\, d\tau. \tag{13. 27}$$

(Here as in what follows, the inversion of the operators $\frac{1}{2T}\int_{-T}^{T} d\tau$ and Average is permissable, since the integral to which our mapping process leads us is absolutely convergent.)

$$\text{Average}\left[\frac{1}{2T}\int_{-T}^{T}d\lambda\int_{-\infty}^{\infty}\xi(\tau)\,d\vartheta(\tau+\lambda)\int_{-\infty}^{\infty}\xi(\sigma)\,d\bar{\vartheta}(\sigma+\lambda)-\frac{1}{2}\int_{-\infty}^{\infty}|\vartheta(\tau)|^2\,d\tau\right]^2$$

$$=\text{Average}\,\frac{1}{4T^2}\int_{-T}^{T}d\lambda\int_{-T}^{T}d\mu\int_{-\infty}^{\infty}\xi(\tau)\,d\vartheta(\tau+\lambda)\int_{-\infty}^{\infty}\xi(\sigma)\,d\bar{\vartheta}(\sigma+\lambda)\int_{-\infty}^{\infty}\xi(\alpha)\,d\vartheta(\alpha+\mu)\cdot$$

$$\int_{-\infty}^{\infty}\xi(\beta)\,d\bar{\vartheta}(\beta+\mu)-\frac{1}{4}\left[\int_{-\infty}^{\infty}|\vartheta(\tau)|^2\,d\tau\right]^2$$

$$=\frac{1}{4T^2}\int_{-T}^{T}d\lambda\int_{-T}^{T}d\mu\,\frac{1}{4}\left\{\left|\int_{-\infty}^{\infty}\vartheta(\tau+\lambda)\bar{\vartheta}(\tau+\mu)\,d\tau\right|^2+\left|\int_{-\infty}^{\infty}\vartheta(\tau+\lambda)\vartheta(\tau+\mu)d\tau\right|^2\right\}$$

$$\leq\frac{1}{32T^2}\int_{-2T}^{2T}d\lambda\int_{-2T}^{2T}du\left\{\left|\int_{-\infty}^{\infty}\vartheta(\tau)\bar{\vartheta}(\tau+u)\,d\tau\right|^2+\left|\int_{-\infty}^{\infty}\vartheta(\tau)\vartheta(\tau+u)\,d\tau\right|^2\right\}$$

$$\leq\frac{1}{8T}\int_{-\infty}^{\infty}du\left\{\left|\int_{-\infty}^{\infty}\vartheta(\tau)\bar{\vartheta}(\tau+u)\,d\tau\right|^2+\left|\int_{-\infty}^{\infty}\vartheta(\tau)\vartheta(\tau+u)\,d\tau\right|^2\right\}. \tag{13.28}$$

The function in the bracket in the last expression will be summable since $\vartheta(\tau)$ is, as we see from (1.28).

It follows that for any positive number A,

$$\left|\frac{1}{2T}\int_{-T}^{T}d\lambda\left|\int_{-\infty}^{\infty}\xi(\tau)\,d\vartheta(\tau+\lambda)\right|^2-\frac{1}{2}\int_{-\infty}^{\infty}|\vartheta(\tau)|^2\,d\tau\right|<A \tag{13.29}$$

except for a set of values of $x(t)$ not exceeding

$$\frac{1}{8TA^2}\int_{-\infty}^{\infty}du\left\{\left|\int_{-\infty}^{\infty}\vartheta(\tau)\bar{\vartheta}(\tau+u)\,d\tau\right|^2+\left|\int_{-\infty}^{\infty}\vartheta(\tau)\vartheta(\tau+u)\,d\tau\right|^2\right\} \tag{13.30}$$

in outer measure. Let T now assume the successive values 1, 4, 9, ... Then the probability that (13.29) fails to be satisfied for some T from $1/n^2$ on does not exceed the remainder of the convergent series

$$\frac{1}{8A^2}\int_{-\infty}^{\infty} du \left\{\left|\int_{-\infty}^{\infty} \vartheta(\tau)\bar{\vartheta}(\tau+u)\,d\tau\right|^2 + \left|\int_{-\infty}^{\infty} \vartheta(\tau)\vartheta(\tau+u)\,d\tau\right|^2\right\}\left[1+\frac{1}{4}+\frac{1}{9}+\cdots\right]. \quad (13.31)$$

Inasmuch as this remainder is arbitrarily small, we almost always have

$$\overline{\lim_{n\to\infty}}\left|\frac{1}{2n^2}\int_{-n^2}^{n^2} d\lambda \left|\int_{-\infty}^{\infty} \xi(\tau)\,d\vartheta(\tau+\lambda)\right|^2 - \frac{1}{2}\int_{-\infty}^{\infty} |\vartheta(\tau)|^2\,d\tau\right| \leq A. \quad (13.32)$$

Since, however, A is an arbitrary positive quantity,

$$\lim_{n\to\infty}\left|\frac{1}{2n^2}\int_{-n^2}^{n^2} d\lambda \left|\int_{-\infty}^{\infty} \xi(\tau)\,d\vartheta(\tau+\lambda)\right|^2 - \frac{1}{2}\int_{-\infty}^{\infty} |\vartheta(\tau)|^2\,d\tau\right| = 0. \quad (13.33)$$

As in the preceding section, we may conclude that

$$\lim_{T\to\infty}\frac{1}{2T}\int_{-T}^{T} d\lambda \left|\int_{-\infty}^{\infty} \xi(\tau)\,d\vartheta(\tau+\lambda)\right|^2 = \frac{1}{2}\int_{-\infty}^{\infty} |\vartheta(\tau)|^2\,d\tau, \quad (13.34)$$

except in a set of cases of zero probability.

Let us now consider

$$\int_{-A}^{A} d\lambda \int_{-\infty}^{\infty} \xi(\tau)\,d\vartheta(\tau+\lambda) = \int_{-\infty}^{\infty} \xi(\tau)\,d\int_{-A}^{A} \vartheta(\tau+\lambda)\,d\lambda, \quad (13.35)$$

for rational values of A, $\vartheta(\tau)$ being subject to the conditions already laid down. Let us put

$$\int_{-\infty}^{\infty} \xi(\tau)\,d\vartheta(\tau+\lambda) = f(\lambda). \quad (13.36)$$

We have almost always, for any denumerable set of values of A, as for example, for all rational values of A,

$$\lim_{T\to\infty} \frac{1}{2T}\int_{-T}^{T} d\lambda \left| \int_{-A}^{A} f(\lambda+\mu)\,d\mu \right|^2 = \frac{1}{2}\int_{-\infty}^{\infty} d\tau \left| \int_{-A}^{A} \vartheta(\tau+\mu)\,d\mu \right|^2. \tag{13.37}$$

This results from the fact that the sum of a denumerable set of null sets is a null set. As before, let us put

$$s(u) = \frac{1}{2\pi}\int_{-1}^{1} f(x)\frac{e^{ixu}-1}{ix}\,dx + \frac{1}{2\pi}\underset{M\to\infty}{\text{l.i.m.}}\left[\int_{1}^{M} + \int_{-M}^{-1}\right] f(x)\frac{e^{ixu}}{ix}\,dx. \tag{13.38}$$

It then follows from (13.34) and (13.37), with the help of (6.23), that we shall almost always have

$$\lim_{\varepsilon\to 0}\frac{1}{2\varepsilon}\int_{-\infty}^{\infty} |s(u+\varepsilon)-s(u-\varepsilon)|^2\,du = \frac{1}{2}\int_{-\infty}^{\infty} |\vartheta(\tau)|^2\,d\tau, \tag{13.39}$$

and (for all rational A)

$$\lim_{\varepsilon\to 0}\frac{1}{2\varepsilon}\int_{-\infty}^{\infty}\frac{4\sin^2 Au}{u^2}|s(u+\varepsilon)-s(u-\varepsilon)|^2\,du = \frac{1}{2}\int_{-\infty}^{\infty} d\tau\left|\int_{-A}^{A}\vartheta(\tau+\mu)\,d\mu\right|^2. \tag{13.40}$$

Now let us put

$$\vartheta(\tau) = \frac{1}{\sqrt{2\pi}}\underset{M\to\infty}{\text{l.i.m.}}\int_{-M}^{M}\psi(u)e^{iu\tau}\,du. \tag{13.41}$$

Then

$$\int_{-A}^{A}\vartheta(\tau+\mu)\,d\mu = \sqrt{\frac{2}{\pi}}\int_{-\infty}^{\infty}\psi(u)\frac{\sin Au}{u}e^{iu\tau}\,du. \tag{13.42}$$

Thus if $\vartheta(\tau)\sqrt{1+\tau^2}$ is of limited total variation and $\vartheta(\tau)$ is quadratically summable, we almost always have for all rational A

$$\int_{-\infty}^{\infty} d\tau\left|\int_{-A}^{A}\vartheta(\tau+\mu)\,d\mu\right|^2 = \int_{-\infty}^{\infty}|\psi(u)|^2\frac{4\sin^2 Au}{u^2}\,du. \tag{13.43}$$

In other words, we almost always have for all rational A,

$$\lim_{\varepsilon\to 0}\int_{-\infty}^{\infty}\frac{\sin^2 Au}{u^2}\left\{\frac{1}{\varepsilon}|s(u+\varepsilon)-s(u-\varepsilon)|^2-|\psi(u)|^2\right\}du=0, \tag{13.44}$$

and

$$\lim_{\varepsilon\to 0}\int_{-\infty}^{\infty}\left\{\frac{1}{\varepsilon}|s(u+\varepsilon)-s(u-\varepsilon)|^2-|\psi(u)|^2\right\}du=0. \tag{13.45}$$

Thus we almost always have

$$\lim_{\varepsilon\to 0}\int_{-\infty}^{\infty}P(u)\left\{\frac{1}{\varepsilon}|s(u+\varepsilon)-s(u-\varepsilon)|^2-|\psi(u)|^2\right\}du=0, \tag{13.46}$$

in case

$$P(u)=\sum_1^n\frac{A_k\sin^2 N_k(u)}{u^2} \tag{13.47}$$

it follows from (13.45) that we may even replace $P(u)$ by

$$Q(u)=\underset{n\to\infty}{\text{uniform limit}}\ P_n(u) \tag{13.48}$$

where $P_n(u)$ is of the form given above for $P(u)$. Thus by the Weierstrass theorem, $Q(u)$ may be the quotient by u^2 of any continuous periodic function with any period. Since we can approximate by such a function Q to any continuous function vanishing at $\pm\infty$, our sole condition on Q may be replaced by

$$Q(u)=o(1)\quad\text{at}\quad u=\pm\infty. \tag{13.49}$$

Even this does not represent the utmost extension of our theorem. It follows at once by subtracting from 1 a Q vanishing outside of a finite range that

$$\lim_{N\to\infty}\ \lim_{\varepsilon\to 0}\left[\overline{\int_N^{\infty}}+\overline{\int_{-\infty}^{-N}}\right]\left\{\frac{1}{\varepsilon}|s(u+\varepsilon)-s(u-\varepsilon)|^2-|\psi(u)|^2\right\}du=0. \tag{13.50}$$

Thus a bounded modification of Q for large arguments produces a decreasing effect as the range of modification recedes to infinity, and we have as our sole condition to be imposed on the continuous function Q that

$$Q(u)=O(1)\quad\text{at}\quad u=\pm\infty. \tag{13.51}$$

A case of peculiar importance is where $e^{-ivu}=Q(u)$. Here

$$\lim_{\varepsilon\to 0}\frac{1}{2\,\varepsilon}\int_{-\infty}^{\infty} e^{-i\nu u}\,|\,s(u+\varepsilon)-s(u-\varepsilon)\,|^2\,du=\frac{1}{2}\int_{-\infty}^{\infty} e^{-i\nu u}\,|\,\psi(u)\,|^2\,du. \tag{13.52}$$

This exists for *every* ν for almost all $x(t)$. Thus by (6.15),

$$\varphi(\nu)=\frac{1}{2}\int_{-\infty}^{\infty} e^{-i\nu u}\,|\,\psi(u)\,|^2\,du, \tag{13.53}$$

and $\varphi(\nu)$ exists for *every* ν for almost every $x(t)$. As in section 3, let us put

$$S(u)=\frac{1}{2\,\pi}\int_{-\infty}^{\infty}\varphi(x)\,\frac{e^{iux}-1}{ix}\,dx, \tag{13.54}$$

so that we get by (5.42), (5.45)

$$\varphi(x)=\int_{-\infty}^{\infty} e^{-ixu}\,dS(u). \tag{13.55}$$

We have already shown φ and S to exist on the assumption that $\varphi(\nu)$ exists for every ν. We have, by (13.53) and (13.55), for almost all $x(t)$

$$\int_{-\infty}^{\infty} e^{-i\nu u}\,d\left[S(u)-\frac{1}{2}\int_{-\infty}^{u}|\,\psi(v)\,|^2\,dv\right]\equiv 0. \tag{13.56}$$

By processes now familiar (cf. (13.46)), we can replace $e^{-i\nu u}$ by functions Q which are merely continuous and bounded. We here make use of the absolute convergence of (13.56). Hence the average of $S(u)-\frac{1}{2}\int_{-\infty}^{u}|\,\psi(v)\,|^2\,dv$ vanishes over every interval, and

$$S(u)-\frac{1}{2}\int_{-\infty}^{u}|\,\psi(v)\,|^2\,dv\equiv 0; \tag{13.57}$$

and consequently

$$S'(u)=\frac{1}{2}\,|\,\psi(u)\,|^2, \tag{13.58}$$

except for a set of values of a zero measure. Inasmuch as $S'(u)$ is the spectral density of $f(x)$, we see that as a consequence of our assumptions that $\vartheta(\tau)$ is quadratically summable and of limited total variation when multiplied by $\sqrt{1+\tau^2}$ *the spectral density of* $\int_{-\infty}^{\infty} \xi(\tau) d\vartheta(\tau+\lambda)$ *is half the square of the modulus of the Fourier transform of* ϑ. Another way of phrasing this fact is: *if a linear resonator is stimulated by a uniformly haphazard sequence of impulses, each frequency responds with an amplitude proportional to that which it would have if stimulated by an impulse of that frequency and of unit energy.* An even simpler statement is: *the energy of a haphazard sequence of impulses is uniformly distributed in frequency.* This law of distribution bears a curious analogy to that predicted for white light by the incorrect Boltzmann law of radiation. The physical conditions which lead to this law of distribution of energy in frequency are that the sequence of impulses in question should be distributed over every interval of time in a Gaussian manner, that their past should not influence their future, that very many should occur over the smallest period of time to be investigated, and that the modulus of the Gaussian distribution of these impulses for a given time interval should depend only on the length of this interval. These conditions are approximately realized in the case of the Schroteffekt, where an electrical resonating circuit is set in vibration by the irregularities in the stream of electrons across a vacuum tube. It might also be realized in the ease of an acoustical system set in oscillation by such a noise as that of a sand blast. Theoretically this equipartition of energy might be used in the absolute calibration of acoustical instruments.

Just as the average of an expression depending on a single function $x(t)$ may be reduced a Lebesgue single integral, so a similar average depending on two independent functions $x(t)$ and $y(t)$ may be reduced to a Lebesgue double integral. On the assumption that $\vartheta_1(\tau)$ and $\vartheta_2(\tau)$ satisfy the conditions we have already laid down for $\vartheta(\tau)$ and that

$$\left.\begin{aligned} \xi(\tau) &= \sqrt{\pi} \int_{1/2}^{\frac{1}{\pi}\cot^{-1}(-\tau)} x(t)\, d \csc \pi t - x\left(\frac{1}{\pi}\cot^{-1}(-\tau)\right) + x(1/2); \\ \eta(\tau) &= \sqrt{\pi} \int_{1/2}^{\frac{1}{\pi}\cot^{-1}(-\tau)} y(t)\, d \csc \pi t - y\left(\frac{1}{\pi}\cot^{-1}(-\tau)\right) + y(1/2); \end{aligned}\right\} \quad (13.59)$$

$$\vartheta_{1,2}(\tau) = \frac{1}{\sqrt{2\pi}} \operatorname*{l.i.m.}_{M\to\infty} \int_{-M}^{M} \psi_{1,2}(u) e^{iu\tau} du \tag{13.60}$$

it is easy to prove by methods substantially identical with those already employed that

$$\lim_{T\to\infty} \frac{1}{2T} \int_{-T}^{T} \left| \int_{-\infty}^{\infty} \xi(\tau) d\vartheta_1(\tau+\lambda) + \int_{-\infty}^{\infty} \eta(\tau) d\vartheta_2(\tau+\lambda) \right|^2 d\lambda \tag{13.61}$$

almost always has a certain definite value. Inasmuch as a normal distribution for $x(t)$ implies the same for $-x(t)$,

$$\lim_{T\to\infty} \frac{1}{2T} \int_{-T}^{T} \left| -\int_{-\infty}^{\infty} \xi(\tau) d\vartheta_1(\tau+\lambda) + \int_{-\infty}^{\infty} \eta(\tau) d\vartheta_2(\tau+\lambda) \right|^2 d\lambda \tag{13.62}$$

almost always has the same value. Subtracting, we almost always have

$$\lim_{T\to\infty} \frac{1}{2T} \int_{-T}^{T} \Re \left[\int_{-\infty}^{\infty} \xi(\tau) d\vartheta_1(\tau+\lambda) \int_{-\infty}^{\infty} \eta(\tau) d\overline{\vartheta}_2(\tau+\lambda) \right] d\lambda = 0. \tag{13.63}$$

If we work in a similar manner with

$$\lim_{T\to\infty} \frac{1}{2T} \int_{-T}^{T} \left| \int_{-\infty}^{\infty} \xi(\tau) d\vartheta_1(\tau+\lambda) \pm i \int_{-\infty}^{\infty} \eta(\tau) d\vartheta_2(\tau+\lambda) \right|^2 d\lambda, \tag{13.64}$$

we see that almost always

$$\lim_{T\to\infty} \frac{1}{2T} \int_{-T}^{T} \Im \left[\int_{-\infty}^{\infty} \xi(\tau) d\vartheta_1(\tau+\lambda) \int_{-\infty}^{\infty} \eta(\tau) d\overline{\vartheta}_2(\tau+\lambda) \right] d\lambda = 0. \tag{13.65}$$

Hence almost always

$$\lim_{T\to\infty} \frac{1}{2T} \int_{-T}^{T} \left[\int_{-\infty}^{\infty} \xi(\tau) d\vartheta_1(\tau+\lambda) \int_{-\infty}^{\infty} \eta(\tau) d\overline{\vartheta}_2(\tau+\lambda) \right] d\lambda = 0 \tag{13.66}$$

30—29764. *Acta mathematica.* 55. Imprimé le 29 avril 1930.

and the coherency matrix of $\int_{-\infty}^{\infty} \xi(\tau) d\vartheta_1(\tau+\lambda)$ and $\int_{-\infty}^{\infty} \eta(\tau) d\vartheta_2(\tau+\lambda)$ is almost always

$$\left\| \begin{matrix} \frac{1}{2} |\psi_1(u)|^2 & 0 \\ 0 & \frac{1}{2} |\psi_2(u)|^2 \end{matrix} \right\|. \tag{13.67}$$

As a direct consequence, if the motion of a particle is independently haphazard in two directions at right angles, and if this motion influences a resonator with the same characteristics in the two directions, the coherency matrix of the motion of the resonator is unpolarized.

In the opinion of the author, the chief importance of this section is in showing in a systematic manner how the Lebesgue integral may be adapted to the needs of statistical mechanics. It is no new observation that sets of zero measure and sets of phenomena, not necessarily impossible, of probability zero, are in essence the same sort of thing. It is not, however, a particularly easy matter to translate any specific problem in statistical mechanics into its precise counterpart in the theory of integration. The author feels confident that methods closely resembling those here developed are destined to play a part in the statistical mechanics of the future, in such regions as those now being invaded by the theory of quanta.

CHAPTER V.

14. The spectrum of an almost periodic function.

The last paragraph was exclusively devoted to functions with continuous spectra; we now come to the most important known class of functions with spectra that are discrete. This is the class of almost periodic functions, the discovery of which is due to Harald Bohr. Let $f(x)$ be a continuous function, not necessarily real, defined for all real values of x between $-\infty$ and ∞. If ε is any positive quantity, Bohr defines τ_ε to be a translation number of $f(x)$ belonging to ε, in case for every real x,

$$|f(x+\tau_\varepsilon) - f(x)| \leq \varepsilon. \tag{14.01}$$

In case, whenever, ε is given, a quantity L_ε can be assigned, such that no interval $(a, a+L_\varepsilon)$ is free of translation numbers τ_ε belonging to ε, $f(x)$ is said to be *almost periodic.* Bohr's most fundamental theorem is: *the necessary and sufficient condition for a function $f(x)$ to be almost periodic is that for any positive quantity ε, there exist a finite set of complex numbers $A_1, A_2, \ldots, A_n$ and a set of real numbers $\Lambda_1, \Lambda_2, \ldots, \Lambda_n$, such that for all x*

$$\left| f(x) - \sum_1^n A_k e^{i\Lambda_k x} \right| < \varepsilon. \tag{14.02}$$

The next few sections of this paper are devoted to the proof of this theorem. In this proof we shall avail ourselves of the following theorems of Bohr concerning almost periodic functions, which are susceptible of a completely elementary proof:

Any finite set of almost periodic functions is *simultaneously* almost periodic, in the sense that for any ε, L_ε may be assigned for the whole set at once, in such a manner that in any interval $(a, a+L_\varepsilon)$, there is at least one translation number τ_ε, such that for every function $f(t)$ in the set, and every t,

$$|f(t+\tau_\varepsilon) - f(t)| \leq \varepsilon. \tag{14.03}$$

Hence any continuous function of a finite number of almost periodic functions yields an almost periodic function, as for example the sum or the product of a finite number of almost periodic functions. The limit of a uniformly convergent series or sequence of almost periodic functions is almost periodic. Every function that is periodic in the classical sense is almost periodic, and the same is true of

$$\Sigma A_n e^{i\Lambda_n t}, \tag{14.04}$$

in case $\Sigma |A_n|$ converges. Every almost periodic function is uniformly continuous. If $f(t)$ is almost periodic,

$$M\{f\} = \lim_{T\to\infty} \frac{1}{T} \int_z^{z+T} f(t)\,dt \tag{14.05}$$

exists as a uniform limit in z. If $f(t)$ is almost periodic, so is

$$\varphi(t) = M_x\{f(x+t)\bar{f}(x)\}. \tag{14.06}$$

(Here and later the symbol under the M indicates the variable on which the averaging is being done.) If $f(t)$ is a real non-negative almost periodic function, and $M\{f\}=0$, $f(t)$ is identically zero.

If $f(t)$ is almost periodic, then since $\varphi(t)$ is also almost periodic, it is continuous. Let us form

$$S(u)=\frac{1}{2\pi}\int_{-\infty}^{\infty}\varphi(x)\frac{e^{iux}-1}{ix}\,dx. \tag{14. 07}$$

From theorems already established

$$\varphi(0)=S(\infty)-S(-\infty). \tag{14. 08}$$

Let the discontinuities of $S(u)$ be at $u=\lambda_1, \lambda_2, \ldots$ These form a denumerable set, as $S(u)$ is of limited total variation, and indeed monotone. Let

$$a_n=S(\lambda_n+0)-S(\lambda_n-0). \tag{14. 09}$$

All these coefficients a_n are positive, and

$$\sum_1^{\infty} a_n \leq S(\infty)-S(-\infty)=M\{|f|^2\}. \tag{14. 10}$$

Let us form the function

$$\gamma(t)=\varphi(t)-\sum_0^{\infty} a_k e^{-i\lambda_k t}, \tag{14. 11}$$

As a simple computation will show (cf. (4. 05)),

$$\frac{1}{2\pi}\int_{-\infty}^{\infty}\gamma(x)\frac{e^{iux}-1}{ix}\,dx=S(u)-S_1(u), \tag{14. 12}$$

where $S_1(u)$ consists of the sum of all the jumps of $S(u)$ with abscissae less than u, together with half the jump (if any) with abscissa u. Hence

$$S_2(u)=\frac{1}{2\pi}\int_{-\infty}^{\infty}\gamma(x)\frac{e^{iux}-1}{ix}\,dx \tag{14. 13}$$

is a continuous function of limited total variation, say V. Let the total variation of $S_2(u)$ over the ranges $(-\infty, -B)$ and (B, ∞) be $V(B)$.

We have

$$\frac{1}{2\varepsilon}\int_{-\infty}^{\infty} |S_2(u+\varepsilon)-S_2(u-\varepsilon)|^2\,du$$

$$= \frac{1}{2\varepsilon}\left[\int_A^{\infty} + \int_{-\infty}^{-A}\right] |S_2(u+\varepsilon)-S_2(u-\varepsilon)|^2\,du + \frac{1}{2\varepsilon}\int_{-A}^{A} |S_2(u+\varepsilon)-S_2(u-\varepsilon)|^2\,du$$

$$\leq \frac{\max|S_2(v)|}{\varepsilon}\left[\int_A^{\infty} + \int_{-\infty}^{-A}\right] |S_2(u+\varepsilon)-S_2(u-\varepsilon)|\,du$$

$$+ \max_{-A\leq v\leq A} |S_2(v+\varepsilon)-S_2(v-\varepsilon)|\frac{1}{2\varepsilon}\int_{-A}^{A} |S_2(u+\varepsilon)-S_2(u-\varepsilon)|\,du$$

$$\leq \frac{\max|S_2(v)|}{\varepsilon}\left[\int_{A-\varepsilon}^{A+\varepsilon} + \int_{A+\varepsilon}^{A+3\varepsilon} + \int_{A+3\varepsilon}^{A+5\varepsilon} + \cdots + \int_{-A-\varepsilon}^{-A+\varepsilon} + \int_{-A-3\varepsilon}^{-A-\varepsilon}\right.$$

$$\left. + \int_{-A-5\varepsilon}^{-A-3\varepsilon} + \cdots\right] |S_2(u+\varepsilon)-S_2(u-\varepsilon)|\,du + \max_{-A\leq v\leq A} |S_2(v+\varepsilon)-S_2(v-\varepsilon)|\frac{1}{2\varepsilon}$$

$$\left[\int_{-A-\varepsilon}^{-A+\varepsilon} + \int_{-A+\varepsilon}^{-A+3\varepsilon} + \cdots + \int_{-A+\left(\left[\frac{2A}{\varepsilon}\right]-2\right)\varepsilon}^{-A+\left[\frac{2A}{\varepsilon}\right]\varepsilon}\right] |S_2(u+\varepsilon)-S_2(u-\varepsilon)|\,du$$

$$= \frac{\max|S_2(v)|}{\varepsilon}\int_{-\varepsilon}^{\varepsilon}\{|S_2(u+A+\varepsilon)-S_2(u+A-\varepsilon)| + |S_2(u+A+3\varepsilon)$$

$$-S_2(u+A+\varepsilon)| + \cdots + |S_2(u-A+\varepsilon)-S_2(u-A+\varepsilon)| + |S_2(u-A-\varepsilon)$$

$$-S_2(u-A-3\varepsilon)| + \cdots\}\,du + \max_{-A\leq v\leq A}|S_2(v+\varepsilon)-S_2(v-\varepsilon)|$$

$$\cdot\frac{1}{2\varepsilon}\int_{-\varepsilon}^{\varepsilon}\Big\{|S_2(u-A+\varepsilon)-S_2(u-A-\varepsilon)| + |S_2(u-A+3\varepsilon)-S_2(u-A+\varepsilon)|$$

$$+ \cdots + \left| S_2\left(u - A + \left[\frac{2A}{\varepsilon}\right]\varepsilon\right) - S_2\left(u - A + \left[\frac{2A}{\varepsilon}\right]\varepsilon - 2\varepsilon\right)\right|\Bigg\} du$$

$$\leq 2\, V(A - 2\varepsilon) \max |S_2(v)| + V \max_{-A \leq v \leq A} |S_2(v+\varepsilon) - S_2(v-\varepsilon)|. \tag{14. 14}$$

Since the function $S(u)$ is uniformly continuous over any finite range, this gives us

$$\lim_{\varepsilon \to 0} \frac{1}{2\varepsilon} \int_{-\infty}^{\infty} |S_2(u+\varepsilon) - S_2(u-\varepsilon)|^2\, du \leq 2\, V(A-\eta) \max |S_2(v)|. \tag{14. 15}$$

However, $V(A-\eta)$ tends to zero as A tends to infinity, and may be arbitrarily small. Hence

$$\lim_{\varepsilon \to 0} \frac{1}{2\varepsilon} \int_{-\infty}^{\infty} |S_2(u+\varepsilon) - S_2(u-\varepsilon)|^2\, du = 0. \tag{14. 16}$$

Applying (5. 53), we get

$$\lim_{T \to \infty} \frac{1}{2T} \int_{-T}^{T} |\gamma(t)|^2\, dt = 0. \tag{14. 17}$$

Since, however, $\gamma(t)$ and $|\gamma(t)|^2$ are almost periodic, we must have

$$\gamma(t) = 0, \tag{14. 18}$$

which yields us

$$S(u) = S_1(u) \tag{14. 19}$$

and

$$\varphi(t) = \sum_{1}^{\infty} a_k e^{-i\lambda_k t}. \tag{14. 20}$$

Thus $S(u)$ is a step function, and the spectrum of an almost periodic function is a pure line spectrum.

15. The Parseval theorem for almost periodic functions.

A further result is

$$M\{\varphi(t)\, e^{i\lambda_k t}\} = a_k. \tag{15. 01}$$

If we remember the uniformity properties of the means of almost periodic functions, this yields

$$a_k = \lim_{T\to\infty} \frac{1}{2T} \int_{-T}^{T} e^{i\lambda_k t}\, dt \lim_{U\to\infty} \frac{1}{2U} \int_{-U}^{U} f(x+t)\bar{f}(x)\, dx$$

$$= \lim_{U\to\infty} \frac{1}{2U} \int_{-U}^{U} \bar{f}(x) e^{-i\lambda_k x}\, dx \lim_{T\to\infty} \frac{1}{2T} \int_{-T}^{T} f(x+t) e^{i\lambda_k (x+t)}\, dt$$

$$= \lim_{U\to\infty} \frac{1}{2U} \int_{-U}^{U} \bar{f}(x) e^{-i\lambda_k x}\, dx \lim_{T\to\infty} \frac{1}{2T} \int_{-T}^{T} f(y) e^{i\lambda_k y}\, dy$$

$$= |M\{f(x) e^{i\lambda_k x}\}|^2. \qquad (15.02)$$

Hence

$$M\{|f|^2\} = \varphi(0) = S(\infty) - S(-\infty)$$

$$= S_1(\infty) - S_1(-\infty)$$

$$= \sum_{0}^{\infty} |M\{f(x) e^{i\lambda_k x}\}|^2. \qquad (15.03)$$

This is a precise analogue to the Hurwitz-Parseval theorem for periodic functions, and is the well-known fundamental theorem of Bohr.

16. The Weierstrass theorem for almost periodic functions.

The present section 16 is devoted to the proof of the approximation theorem for almost periodic functions, which tells us that the necessary and sufficient condition for a function to be almost periodic is that it can be expressed as the uniform limit of a trigonometrical polynomial. The main idea of the present proof is due to Weyl, although the form of the argument is much changed from that on his paper. The essence of the proof is that harmonic analysis is not applied directly to the almost periodic function discussed, but to certain related functions derived from what Bochner calls the Verschiebungsfunktion of the given function. In the discussion of the many different extensions of almost periodic functions, there is a function in each case analogous to this Verschiebungsfunktion which is almost periodic in the strict Bohr sense. As we shall see in the next section, this enables us to carry over to these more general cases practically the entire Bohr approximation theorem redefined to suit each particular definition.

Let $f(t)$ be almost periodic. Consider

$$g(x) = \max_t |f(x+t) - f(t)|. \tag{16.01}$$

We have

$$\begin{aligned} |g(x+\tau) - g(x)| &\leq \max_t \big||f(x+t+\tau) - f(t)| - |f(x+t) - f(t)|\big| \\ &\leq \max_t |f(x+t+\tau) - f(x+t)| \\ &= \max_t |f(t+\tau) - f(t)|. \end{aligned} \tag{16.02}$$

Hence any translation number for $f(t)$ pertaining to ε is a translation number for $g(t)$ pertaining to ε, and $g(t)$ is almost periodic. It is this function which Bochner calls the Verschiebungsfunktion of $f(t)$.

We have already indicated the fact that any continuous function of an almost periodic function is almost periodic. Let $H_\varepsilon(U)$ be befined as follows:

$$H_\varepsilon(U) = \begin{cases} 1; & [\,0 \leq U \leq \varepsilon/2\,] \\ 2 - \dfrac{2U}{\varepsilon}; & [\,\varepsilon/2 \leq U \leq \varepsilon\,] \\ 0. & [\,\varepsilon \leq U\,] \end{cases} \tag{16.03}$$

Let

$$\psi_\varepsilon(x) = \frac{H_\varepsilon[g(x)]}{M_x H_\varepsilon[g(x)]}. \tag{16.04}$$

Since $H_\varepsilon[g(x)]$ is somewhere positive, and it is everywhere non negative and almost periodic, $M_x H_\varepsilon[g(x)]$ cannot vanish. Hence $\psi_\varepsilon(x)$ exists and is almost periodic.

Let

$$f_\varepsilon(x) = M_t\{f(t)\,\psi_\varepsilon(x-t)\}. \tag{16.05}$$

The existence and almost periodic character of $f_\varepsilon(x)$ are proved without difficulty. The definition of ψ_ε ensures that

$$|f(x) - f(t)| \leq \varepsilon \tag{16.06}$$

if $\psi_\varepsilon(x-t) \neq 0$. Hence, since $f_\varepsilon(x)$ is a mean of these values of $f(t)$,

$$\max_x |f(x) - f_\varepsilon(x)| \leq \varepsilon. \tag{16.07}$$

Similarly, if

$$f^{(\varepsilon)}(x) = M_t\{f_\varepsilon(t)\,\psi_\varepsilon(x-t)\}, \tag{16.08}$$

$f^{(\varepsilon)}(x)$ exists and is almost periodic, and

$$\max_x |f^{(\varepsilon)}(x) - f_\varepsilon(x)| \leq \varepsilon. \tag{16.09}$$

Hence, by (16.07) and (16.09),

$$\max_x |f^{(\varepsilon)}(x) - f(x)| \leq 2\,\varepsilon. \tag{16.10}$$

We have

$$f^{(\varepsilon)}(x) = M_t\{\psi_\varepsilon(x-t)\, M_\tau\{f(\tau)\,\psi_\varepsilon(t-\tau)\}\}. \tag{16.11}$$

Bearing in mind the uniformity properties of almost-periodic functions, we have

$$f^{(\varepsilon)}(x) = M_\tau\{f(\tau)\, M_t\{\psi_\varepsilon(x-t)\,\psi_\varepsilon(t-\tau)\}\}. \tag{16.12}$$

However, by (14.20),

$$M_t\{\psi_\varepsilon(x-t)\,\psi_\varepsilon(t-\tau)\} = \sum_1^\infty a_k e^{i\lambda_k(x-\tau)} \tag{16.13}$$

where all the coefficients a_k are positive, and $\sum_1^\infty a_k$ converges. Hence

$$f^{(\varepsilon)}(x) = M_\tau\left\{f(\tau)\sum_1^\infty a_k e^{i\lambda_k(x-\tau)}\right\}. \tag{16.14}$$

Since

$$\left|\frac{1}{2\,T}\int_{-T}^{T} f(\tau)\, e^{i\lambda_k(x-\tau)}\, d\tau\right| \leq \max|f(\tau)|, \tag{16.15}$$

it follows that we can invert the order of M and Σ, and that

$$f^{(\varepsilon)}(x) = \sum_1^\infty a_k\, e^{i\lambda_k x}\, M\{f(\tau)\, e^{-i\lambda_k \tau}\}. \tag{16.16}$$

Inasmuch as

$$|M\{f(\tau)\, e^{-i\lambda_k\tau}\}| \leq \max|f(\tau)|, \tag{16.17}$$

$f^{(\varepsilon)}(x)$ is the sum of a uniformly convergent series of trigonometric terms. That is to say, we can choose N so large that

31—29764. *Acta mathematica*. 55. Imprimé le 29 avril 1930.

$$\max_x \left| f^{(\varepsilon)}(x) - \sum_1^N a_k e^{i\lambda_k x} M\{f(\tau)e^{-i\lambda_k \tau}\} \right| \leq \varepsilon, \tag{16. 18}$$

and hence that

$$\max_x \left| f(x) - \sum_1^N a_k e^{i\lambda_k x} M\{f(\tau)e^{-i\lambda_k \tau}\} \right| \leq 3\,\varepsilon. \tag{16. 19}$$

In other words, we have proved Bohr's approximation theorem, to the effect that it is possible to approximate uniformly to any desired degree of accuracy to an almost periodic function by means of trigonometrical polynomials.

17. Certain generalizations of almost periodic functions.

It will be noticed that in the proof of the Weierstrass theorem for almost periodic functions, the spectrum of the function to be analyzed was not directly introduced, but rather that of the auxiliary function $\psi_\varepsilon(t)$. In many cases, when the function $f(t)$ is not almost periodic in the classical sense, an auxiliary function $\psi_\varepsilon(t)$ may be defined, which will be almost periodic in the classical sense, and which may be employed to establish the approximation theorem for $f(t)$, in whatever. sense this theorem may hold. It would be possible in this manner to establish the approximation theorems for the almost periodic functions of the generalized types of Weyl, Besicovitch, Stepanoff, and others, but one example will suffice to show the power of the method, and to this we shall confine ourselves. This example, which is due to Mr. C. F. Muckenhoupt, is that of functions almost periodic in the mean.

We shall confine our attention to functions $f(x, t)$ defined over the range $(-\infty < t < \infty,\ x_0 \leq x \leq x_1)$, quadratically summable in x, and continuous in the mean in t in the sense that

$$\lim_{\tau \to 0} \int_{x_0}^{x_1} |f(x, t) - f(x, t+\tau)|^2 dx = 0. \tag{17. 01}$$

We shall say that τ_ε is a *translation number* of $f(x, t)$ pertaining to ε in case for all t,

$$\int_{x_0}^{x_1} |f(x, t+\tau_\varepsilon) - f(x, t)|^2 dx < \varepsilon^2. \tag{17. 02}$$

In case, given ε, we can always assign a finite quantity L_ε, such that each interval $(A, A+L_\varepsilon)$ contains at least one translation number τ_ε pertaining to ε, then $f(x, t)$ is said to be *almost periodic in the mean*. I case $f(x, t)$ is almost periodic in the mean, Mr. Muckenhoupt's theorem is:

Given any positive quantity ε *there can be assigned a trigonometrical polynomial*

$$P_\varepsilon(x, t) = \sum_1^N A_n(x)\, e^{i\Lambda_n t}, \tag{17.03}$$

such that

$$A_1(x),\ A_2(x),\ \ldots,\ A_n(x) \tag{17.04}$$

are all quadratically summable, and for all t,

$$\int_{x_0}^{x_1} |f(x, t) - P_\varepsilon(x, t)|^2\, dx < \varepsilon^2. \tag{17.05}$$

There are a number of elementary theorems which Mr. Muckenhoupt proves along lines not differing in any essential way from those followed by Bohr in the proof of the corresponding theorems for functions almost periodic in the original sense. Thus every function almost periodic in the mean is bounded in the mean, in the sense that

$$\int_{x_0}^{x_1} |f(x, t)|^2\, dx \tag{17.06}$$

is bounded; and is uniformly continuous in the mean, in the sense that

$$\lim_{\varepsilon \to 0} \max_{\tau < \varepsilon} \int_{x_0}^{x_1} |f(x, t) - f(x, t+\tau)|^2\, dx = 0. \tag{17.07}$$

In this and subsequent formulas, the maximum value indicated by »max» need not be actually attained. Any finite set of functions almost periodic in the mean are *simultaneously* almost periodic in the mean, in the sense that, given ε, an L_ε may be assigned in such a manner that every interval $(A, A+L_\varepsilon)$ contains at least one τ_ε which is a translation number pertaining to ε of all the functions of the set. Hence the sum of two or more functions almost periodic in the mean is almost periodic in the mean. Similarly, the product in the ordinary sense is almost periodic in the mean. The uniform limit in the mean of a set of functions almost periodic in the mean is itself almost periodic in the mean. If $f(x, t)$ is almost periodic in the mean,

$$\underset{T\to\infty}{\text{l.i.m.}} \frac{1}{T}\int_a^{a+T} f(x, t)\, dt \tag{17.08}$$

exists as a uniform limit in the mean in a, and is independent of a. We shall represent it by the symbol

$$M\{f(x, t)\}. \tag{17.09}$$

Mr. Muckenhoupt now puts

$$g(t) = \max_z \int_{x_0}^{x_1} |f(x, t+z) - f(x, z)|^2\, dx. \tag{17.10}$$

Clearly

$$\begin{aligned} |g(t+\tau) - g(t)| &\leq \max_z \int_{x_0}^{x_1} \big||f(x, t+\tau+z) - f(x, z)|^2 - |f(x, t+z) - f(x, z)|^2\big|\, dx \\ &\leq \max_z \left\{ \int_{x_0}^{x_1} \big||f(x, t+\tau+z) - f(x, z)| + |f(x, t+z) - f(x, z)|\big|^2\, dx \right\}^{1/2} \\ &\quad \cdot \max_z \left\{ \int_{x_0}^{x_1} \big||f(x, t+\tau+z) - f(x, z)| - |f(x, t+z) - f(x, z)|\big|^2\, dx \right\}^{1/2}. \end{aligned} \tag{17.11}$$

To evaluate this, let us consider the maximum of each of the integrals under the radical sign separately. The first does not exceed

$$16 \max_z \int_{x_0}^{x_1} |f(x, z)|^2\, dx; \tag{17.12}$$

the second does not exceed

$$\max_z \int_{x_0}^{x_1} |f(x, \tau+z) - f(x, z)|^2\, dx. \tag{17.13}$$

Hence we may write

$$|g(t+\tau)-g(t)| \leq 16\left[\max_z \int_{x_0}^{x_1} |f(x, z+\tau)-f(x, z)|^2\, dx\right]^{1/2}$$

$$\cdot\left[\max_z \int_{x_0}^{x_1} |f(x, z)|^2 dx\right]^{1/2}. \qquad (17.\ 14)$$

If τ_ε is a translation number of $f(x, t)$ pertaining to ε, we have

$$|g(t+\tau_\varepsilon)-g(t)| \leq 16\,\varepsilon\left[\max_z \int_{x_0}^{x_1} |f(x, z)|^2 dx\right]^{1/2}, \qquad (17.\ 15)$$

so that any translation number of $f(x, t)$ pertaining to ε is a translation number of $g(t)$ pertaining to

$$16\,\varepsilon\left[\max_z \int_{x_0}^{x_1} |f(x, z)|^2 dx\right]^{1/2}. \qquad (17.\ 16)$$

Thus $g(t)$ is almost periodic in the classical sense, and is distinct from 0 unless $f(x, t)$ is independent of t, in the sense that $f(x, t_1)=f(x, t_2)$ almost everywhere.

As in the last section, let

$$H_\varepsilon(U)=\begin{cases} 1 & ; \quad [0 \leq U \leq \varepsilon/2] \\ 2-\dfrac{2\,U}{\varepsilon} & ; \quad [\varepsilon/2 \leq U \leq \varepsilon] \\ 0 & ; \quad [\varepsilon \leq U] \end{cases} \qquad (17.\ 17)$$

and let

$$\psi_\varepsilon(t)=\frac{H_\varepsilon[g(t)]}{M H_\varepsilon[g(t)]}\cdot \qquad (17.\ 18)$$

As before, $H_\varepsilon[g(t)]$ is dictinct from 0, and $\psi_\varepsilon(x)$ exists, and is almost periodic. As before, we put

$$f_\varepsilon(x, t)=M_\tau[f(x, \tau)\,\psi_\varepsilon(t-\tau)], \qquad (17.\ 19)$$

and

$$f^{(\varepsilon)}(x, t)=M_\tau[f(x, \tau)\,\psi_\varepsilon(t-\tau)]. \qquad (17.\ 20)$$

A proof precisely parallel to that of (16. 10) and (16. 12) shows that

$$\max_t \int_{x_0}^{x_1} |f(x,t) - f^{(\varepsilon)}(x,t)|^2 \, dx \leq 4\varepsilon. \tag{17.21}$$

and that

$$f^{(\varepsilon)}(x,t) = M_\sigma[f(x,\sigma) M_\tau[\psi_\varepsilon(t-\tau)\psi_\varepsilon(\tau-\sigma)]]. \tag{17.22}$$

As before,

$$M_\tau[\psi_\varepsilon(t-\tau)\psi_\varepsilon(\tau-\sigma)] = \sum_0^\infty a_k e^{i\lambda_k(t-\sigma)} \tag{17.23}$$

where all the a_k's are positive, and $\sum_0^\infty a_k$ converges. Hence

$$f^{(\varepsilon)}(x,t) = M_\sigma\left[f(x,\sigma)\sum_0^\infty a_k e^{i\lambda_k(t-\sigma)}\right]. \tag{17.24}$$

We have

$$\lim_{N\to\infty} \int_{x_0}^{x_1} \left| M_\tau\left[f(x,\sigma)\sum_N^\infty a_k e^{i\lambda_k(t-\sigma)}\right]\right|^2 dx$$

$$\leq \lim_{N\to\infty}\left[\sum_N^\infty a_k\right]^2 \max_\sigma \int_{x_0}^{x_1} |f(x,\sigma)|^2 \, dx = 0. \tag{17.25}$$

Hence since

$$\sum_0^\infty a_k$$

converges, we can invert the order of M and Σ in (17.25), and get

$$f^{(\varepsilon)}(x,t) = \underset{N\to\infty}{\text{l.i.m.}} \sum_0^n a_k e^{i\lambda_k t} \underset{\sigma}{M}[f(x,\sigma)e^{-i\lambda_k\sigma}]. \tag{17.26}$$

This convergence in the mean is uniform with respect to t. Combining (17.25) and (17.26), our theorem is proved.

This theorem has an interesting dynamical application. Really significant dynamical applications of almost periodic functions have been rather scarce, as no one has yet produced an example of an almost periodic function entering into a dynamical system with a finite number of degrees of freedom in which the frequencies or exponents are not linearly dependent (with rational coefficients) on a finite set of quantities. However, dynamical systems with an infinite number of degress of freedom are familiar enough in connection with boundary value problems, and in these, it is well known that the solution may involve an infinite linearly independent set of time frequencies. Mr. Muckenhoupt has succeeded in showing, under certain very general conditions, that the solution of such a problem is almost periodic in the mean with respect to the time, the space variables playing the rôle above assigned to x. In this proof, the existence of an integral invariant such as the energy is of the utmost importance, as is also the condition that when all the coordinates and velocities of the system are less in value than some given constant, the energy is also necessarily less than some constant.

Let us consider as an example a vibrating string, whose density and tensions are functions of position, but not of time. Let the mass density be $\mu(x)$ and the tension $T(x)$. The equation of motion is then

$$\frac{\partial}{\partial x}\left[T(x)\frac{\partial y}{\partial x}\right] = \mu(x)\frac{\partial^2 y}{\partial t^2}. \tag{17. 27}$$

We consider the ends to be fixed, giving us

$$y(x_0) = y(x_1) = 0, \tag{17. 28}$$

and we take T and μ, as is always physically the case, finite and positive. We shall also suppose them to have bounded derivatives of all orders.

Thus the total energy of the system is

$$E_0 = \frac{1}{2}\int_{x_0}^{x_1}\left[\mu(x)\left(\frac{\partial y}{\partial t}\right)^2 + T(x)\left(\frac{\partial y}{\partial x}\right)^2\right]dx. \tag{17. 29}$$

If we assume density and tension independent of the time, we have

$$\frac{\partial E_0}{\partial t} = \int_{x_0}^{x_1}\left[\mu(x)\frac{\partial y}{\partial t}\frac{\partial^2 y}{\partial t^2} + T(x)\frac{\partial y}{\partial x}\frac{\partial^2 y}{\partial x\,\partial t}\right]dx, \tag{17. 30}$$

or by (17. 27),

$$\begin{aligned}\frac{\partial E_0}{\partial t} &= \int_{x_0}^{x_1}\left\{\frac{\partial y}{\partial t}\frac{\partial}{\partial x}\left[T(x)\frac{\partial y}{\partial x}\right] + T(x)\frac{\partial y}{\partial x}\frac{\partial}{\partial x}\left(\frac{\partial y}{\partial t}\right)\right\}dx \\ &= \left[T(x)\frac{\partial y}{\partial x}\frac{\partial y}{\partial t}\right]_{x_0}^{x_1} = 0. \end{aligned} \tag{17. 31}$$

Thus the total energy is invariant, as was to be expected from physical considerations.

Inasmuch as $\partial^n y/\partial t^n$ also satisfies equation (17. 27) and boundary conditions (17. 28), we see that all the expressions

$$E_n = \frac{1}{2} \int_{x_0}^{x_1} \left[\mu(x) \left(\frac{\partial^{n+1} y}{\partial t^{n+1}} \right)^2 + T(x) \left(\frac{\partial^{n+1} y}{\partial t\, \partial x^n} \right)^2 \right] dx \tag{17. 32}$$

are invariants, at least if $y(x, t)$ is sufficiently often differentiable. We shall term E_n the $(n+1)$st energy of the system.

Let us now take $\partial y/\partial t$ and $\partial y/\partial x$ to be continuous, and let

$$E_0 \leq E; \quad \mu(x) \geq M; \quad T(x) \geq T.$$

Clearly $E_0 \geq 0$, and

$$\int_{x_0}^{x_1} \left(\frac{\partial y}{\partial t} \right)^2 dx \leq \frac{2\, E_0}{M} \leq \frac{2\, E}{M}. \tag{17. 33}$$

Similarly, $\int_{x_0}^{x_1} \left(\frac{\partial y}{\partial x} \right)^2 dx$ is bounded, provided only the first energy E_0 is finite. Furthermore, since

$$y^2 = \left[\int_{x_0}^{x} \frac{\partial y}{\partial x}\, dx \right]^2 \leq \int_{x_0}^{x} dx \int_{x_0}^{x} \left(\frac{\partial y}{\partial x} \right)^2 dx, \tag{17. 34}$$

by the Schwarz inequality, y is bounded. Similarly, if E_1 is also bounded,

$$\int_{x_0}^{x_1} \left(\frac{\partial^2 y}{\partial t^2} \right)^2 dx, \quad \int_{x_0}^{x_1} \left(\frac{\partial^2 y}{\partial x\, \partial t} \right)^2 dx, \quad \text{and} \quad \int_{x_0}^{x_1} \left(\frac{\partial^2 y}{\partial x^2} \right)^2 dx \tag{17. 35}$$

will be likewise; in the last case, as a result of (17. 27); if E_2 is also bounded,

$$\int_{x_0}^{x_1} \left(\frac{\partial^3 y}{\partial t^3} \right)^2 dx, \quad \int_{x_0}^{x_1} \left(\frac{\partial^3 y}{\partial t^2\, \partial x} \right)^2 dx, \quad \int_{x_0}^{x_1} \left(\frac{\partial^3 y}{\partial t\, \partial x^2} \right)^2 dx, \quad \text{and} \quad \int_{x_0}^{x_1} \left(\frac{\partial^3 y}{\partial x^3} \right)^2 dx \tag{17. 36}$$

will be, and so on indefinitely.

Let us now introduce

$$(y, y_1) = \left\{ \int_{x_0}^{x_1} \left[|y - y_1|^2 + \left| \frac{\partial y}{\partial x} - \frac{\partial y_1}{\partial x} \right|^2 + \left| \frac{\partial y}{\partial t} - \frac{\partial y_1}{\partial t} \right|^2 \right] dx \right\}^{1/2} \tag{17.37}$$

as the distance between two functions, $y(x, t)$ and $y_1(x, t)$. If we write

$$y \sim \sum_{-\infty}^{\infty} A_n(t) e^{\frac{2\pi i n x}{x_1 - x_0}}; \quad y_1 \sim \sum_{-\infty}^{\infty} B_n(t) e^{\frac{2\pi i n x}{x_1 - x_0}};$$

we may approximate *uniformly* to (y, y_1) by

$$\int_{x_0}^{x_1} \left[\left| \sum_{-N}^{N} (A_n - B_n) e^{\frac{2\pi i n x}{x_1 - x_0}} \right|^2 \left(1 + \frac{4\pi n^2}{(x_1 - x_0)^2} \right) + \left| \sum_{-N}^{N} (A'_n - B'_n) e^{\frac{2\pi i n x}{x_1 - x_0}} \right|^2 dx \right] \tag{17.38}$$

for all functions y and y_1 for which E_0 and E_1 are finite, since then

$$\left.\begin{aligned} \sum_{-\infty}^{\infty} n^4 |A_n(t) - B_n(t)|^2 &= \frac{x_1 - x_0}{(2\pi)^4} \int_{x_0}^{x_1} \left(\frac{\partial^2 y}{\partial t^2} \right)^2 dx; \\ \sum_{-\infty}^{\infty} n^2 |A'_n(t) - B'_n(t)|^2 &= \frac{x_1 - x_0}{(2\pi)^2} \int_{x_0}^{x_1} \left(\frac{\partial^2 y}{\partial x \partial t} \right)^2 dx \end{aligned}\right\} \tag{17.39}$$

are uniformly bounded. Now, a bounded region in space of m dimensions may be divided into a finite number of compartments such that the distance between two points in the same compartment does not exceed ε. Hence we can divide the entire class of functions $y(x, t)$ for which E_0 and E_1 are finite into a finite number of classes such that the distance between two functions in the same class does not exceed ε.

Let us do this, and let us discard every class which is not actually represented by $y(x, t)$ for some value of t. Then we may assign a time-interval L'_ε within which $y(x, t)$ enters every class that it ever enters. Then, whatever τ may be, we may determine τ_1 between 0 and L'_ε such that

$$(y(x, \tau), \quad y(x, \tau_1))^2 < \varepsilon. \tag{17.40}$$

Since

$$y(x, t) - y(x, t + \tau - \tau_1)$$

satisfies the differential equation (17. 27),

$$\mathfrak{E} = \frac{1}{2}\int_{x_0}^{x_1}\Big[\mu(x)\left(\frac{\partial y(x,t)}{\partial t} - \frac{\partial y(x,t+\tau-\tau_1)}{\partial t}\right)^2$$

$$+ T(x)\left(\frac{\partial y(x,t)}{\partial x} - \frac{\partial y(x,t+\tau-\tau_1)}{\partial x}\right)^2\Big]dx \qquad (17.\ 41)$$

is invariant, and since for $t=\tau_1$,

$$\mathfrak{E} \leq (\max \mu + \max T)\,(y(x,\tau_1),\ y(x,\tau))^2$$

$$< (\max \mu + \max T)\,\varepsilon, \qquad (17.\ 42)$$

it follows that for all t,

$$\mathfrak{E} < (\max \mu + \max T)\varepsilon, \qquad (17.\ 43)$$

and hence by (17. 34) and (17. 37)

$$|y(x,t)-y(x,t+\tau-\tau_1)| < \sqrt{2\,\frac{x_1-x_0}{M}(\max \mu + \max T)\varepsilon}\,. \qquad (17.\ 44)$$

Since for every τ, there is a value of τ_1 between 0 and L'_ε, there is a value of $\tau-\tau_1$ over every interval of length L'_ε. Thus $y(x,t)$ is an a almost periodic function taken with respect to the time, uniformly in x, and is *a fortiori* almost periodic in the mean, in case E_0 and E_1 are finite. It follows that we may so determine $A_1(x), \ldots, A_n(x); \Lambda_1, \ldots, \Lambda_n$ that for all t,

$$\int_{x_0}^{x_1}\left|y(x,t) - \sum_1^n A_k(x)\,e^{i\Lambda_k(t)}\right|^2 dx < \varepsilon. \qquad (17.\ 45)$$

It is possible to go further than this, as Mr. Muckenhoupt has done, and to show that the method we have given for obtaining $f_\varepsilon(x,t)$, $f^{(\varepsilon)}(x,t)$, and $A_k(x)\,e^{i\Lambda_k(t)}$ assures us that all the functions

$$A_k(x)\,e^{i\Lambda_k t}$$

are solutions of the original differential equation, or that the functions

$$A_k(x)$$

are all solutions of the ordinary differential equation

$$\frac{d}{dx}(T(x)\,A'_k(x)) + \Lambda_k^2\,\mu(x)\,A_k(x) = 0; \qquad (17.\ 46)$$

— that is, are what is known as Eigenfunktionen of the dynamical problem. This proof rests on the fact that each one of these functions may be obtained from its predecessor, and ultimately from $f(x, t)$, by a process of weighted averaging in the variable t which transforms every solution of a linear differential equation with coefficients constant with respect to the time into another solution of the same equation, or at least of the corresponding integral equation. Hence, if $y(x, 0) = F(x)$; $\frac{\partial y(x, 0)}{\partial t} = 0$ is a possible set of initial conditions for the motion of the vibrating string. We may write

$$F(x) = \underset{n \to \infty}{\text{l.i.m}} \sum_{1}^{n} A_k(x), \tag{17.47}$$

where the $A_k(x)$ are in general Eigenfunktionen of the problem that depend on n. Thus if the set of possible initial conditions of the string is closed, as we may show to be the case by direct methods, every quadratically summable function may be expanded in terms of a denumerable set of Eigenfunktionen, and the Eigenfunktionen may be shown to be a denumerable closed set.

The methods of Mr. Muckenhoupt are susceptible of extension to the treatment of a much wider class of Eigenfunktion problems, in any finite number of dimensions. The detail of this extension awaits further investigation.

Bibliography.

The works and papers covering the various theories belonging to general harmonic analysis fall into several imperfectly related categories. Among these are:

(1) The various papers written from the physical standpoint, with the explicit purpose of clearing up obscure points in the theories of interference, of coherency, and of polarization.

(2) Directly related to these, the various memoirs connecting with the Schuster theory of the periodogram.

(3) A group of memoirs preceding the Bohr theory of almost periodic functions, applying various extensions of the notion of periodicity in celestial mechanics and other similar fields.

(4) Papers written from the point of view of the mathematician, and dealing with trigonometric series not proceeding according to integral multiples of the argument.

(5) The Bohr theory of almost periodic functions, and papers directly inspired by it.

(6) Papers dealing with haphazard motion, and using ideas directly pertinent to generalized harmonic analysis.

(7) The Hahn direction of work, treating generalized harmonic analysis from the standpoint of ordinary convergence, rather than from that of convergence in the mean.

(8) The papers assuming essentially the standpoint of the present author, in whose work the generalizations of the Parseval theorem play the central rôle.

(9) Papers dealing rather with the rigorous theory of the Fourier integral itself than with its generalizations.

(10) Papers not dealing directly with generalized harmonic analysis, which it is desirable to cite for one reason or another.

In citing any paper, it will be indicated to which of these categories it belongs. Each paper will furthermore be quoted in the footnotes by the name of its author, together with an index number given in the bibliography.

A. C. Berry 1. Doctoral dissertation, Harvard, 1929. Unpublished. (8)

—— 2. The Fourier transform theorem. Jour. Math. and Phys. Mass. Inst. Technology 8, 106—118. (1929). (8)

A. Besicovitch 1. Sur quelques points de la théorie des fonctions presque périodiques. C.R. 180, 394—397. (1925). (5)

—— 2. On Parseval's theorem for Dirichlet series. Proc. Lond. Math. Soc. 25, 25—34. (1926). (5)

—— 3. On generalized almost periodic functions. Proc. Lond. Math. Soc. 24, 495—512. (1926). (5)

A. Besicovitch and H. Bohr 1. Some remarks on generalizations of almost periodic functions. Danske Vidensk. Selskab. 8, No. 5. (1927). (5)

—— 2. On generalized almost periodic functions. Journ. Lond. Math. Soc. 3, 172—176. (1928). (5)

G. D. Birkhoff. Dynamical systems. Am. Math. Society, New York, 1927. Pp. 218—220. (5)

S. Bochner 1. Properties of Fourier series of almost periodic functions. Proc. Lond. Math. Soc. 26, 433. (1925). (5)

—— 2. Sur les fonctions presque-périodiques de Bohr. C.R. 180, 1156. (1925). (5)

—— 3. Beiträge zur Theorie der fastperiodischen Funktionen. Math. Ann. 96, 119—147. (1926). (5)

—— 4. Beiträge zur Theorie der fastperiodischen Funktionen. Math. Ann. 96, 383—409. (1926). (5)

—— 5. Über Fourierreihen fastperiodischer Funktionen. Berliner Sitzungsberichte, 26. (1926). (5)

—— 6. Konvergenzsätze für Fourierreihen grenzperiodischer Funktionen. Math. Zeitschr. 27, 187—211. (1927). (5)

—— 7. Darstellung reellvariabler und analytischer Funktionen durch verallgemeinerte Fourier- und Laplace-Integrale. Math. Ann. 97, 632—674. (1926—7). (8)

S. BOCHNER 8. Über gewisse Differential- und allgemeinere Gleichungen, deren Lösungen fastperiodisch sind. I. Teil. Der Existenzsatz. Math. Ann. 102, 489—504. (1929). (5)

S. BOCHNER and G. H. HARDY. Note on two theorems of Norbert Wiener. Jour. Lond. Math. Soc. 1, 240—242. (1926). (10)

P. BOHL. Über die Darstellung von Funktionen einer Variabeln durch trigonometrische Reihen mit mehreren einer Variabeln proportionalen Argumenten. (Dorpat, 1893). (3)

H. BOHR 1. Sur les fonctions presque périodiques. C.R. Oct. 22, 1923. (5)

—— 2. Sur l'approximation des fonctions presque périodiques par des sommes trigonométriques. C.R. Nov. 26, 1923. (5)

—— 3. Über eine quasi-periodische Eigenschaft Dirichletscher Reihen mit Anwendung auf die Dirichletschen L Funktionen. Math. Ann. 85, 115—122. (4)

—— 4. Zur Theorie der fastperiodischen Funktionen. Acta Math. 45, 29—127. (124). (5)

—— 5. Zur Theorie der fastperiodischen Funktionen. II. Acta Math. 46, 101—214. (1925). (5)

—— 6. Zur Theorie der fastperiodischen Funktionen. III. Acta Math. 47, 237—281. (1926). (5)

—— 7. Einige Sätze über Fourierreihen fastperiodischer Funktionen. Math. Ztschr. 23, 38—44. (1925). (5)

8. Sur une classe de transcendantes entières. C.R. 181, 766. (1925). (5)

—— 9. Sur le théorème d'unicité dans la théorie des fonctions presque-périodiques. Bull. Sci. Math. 50, 1—7. (1926). (5)

10. On the explicit determination of the upper limit of an almost periodic function. Jour. Lond. Math. Soc. 1. (1926). (5)

—— 11. Ein Satz über analytische Fortsetzung fastperiodischer Funktionen. Crelle, 157, 61—65. (5)

—— 12. En Klasse hele transcendente Funktioner. Mat. Tidsskrift, B. Aarg. 1926, 41—45. (5)

—— 13. Allgemeine Fourier- und Dirichlet-Entwicklungen. Abh. aus dem math. Sem. d. Hamburgischen Univ. 4, 366—374. (1926). (5)

14. Fastperiodische Funktionen. Jahresb. d. D.M.V. 33, 25—41. (1925). (5)

—— 15. En Sætning om Fourierrækker for næstenperiodiske Funktioner. Mat. Tid. B. Aarg. 1924, 31—37. (5)

—— 16. Über die Verallgemeinerungen fastperiodischer Funktionen. Math. Ann. 99, 357—366. (1928). (5)

H. BOHR and O. NEUGEBAUER. Über lineare Differentialgleichungen mit konstanten Koeffizienten und fastperiodischer rechter Seite. Gött. Sitzb. 1926, 1—13. (5)

E. BOREL. Les probabilités dénombrables et leurs applications arithmétiques. Rend. di Palermo 27, 247—271. (1909). (10)

J. C. BURKILL 1. The expression in Stieltjes integrals of the inversion formulae of Fourier and Hankel. Proc. Lond. Math. Soc. 25, 513—524. (1926). (7)

J. C. BURKILL 2. On Mellin's inversion formula. Proc. Camb. Phil. Soc. 23, 356—360. (1926). (7)

V. BUSH, F. D. GAGE, and H. R. STEWART. A continuous integraph. Jour. Franklin Institute. 63—84. (1927). (10)

V. BUSH and H. L. HAZEN. Integraph solution of differential equations. 575—615. (1927). (10)

CARSE and SHEARER. A course in Fourier's analysis and periodogram analysis for the mathematical laboratory. London, 1915. (2)

P. J. DANIELL 1. A general form of integral. Annals of Math. 19, 279—294. (1918). (10)

—— 2. Integrals in an infinite number of dimensions. Ann. of Math. 20, 281—288. (1919). (10)

—— 3. Further properties of the general integral. Ann. of Math. 21, 203—220. (1920). (10)

W. DORN. Fouriersche Integrale als Grenzwerte Fourierscher Reihen. Wiener Sitzungsber., 1926, 127—147. (7)

A. EINSTEIN. Zur Theorie der Brownschen Bewegung. Ann. der Phys. (4) 19, 372—381. (1906). (10)

E. ESCLANGON 1. Sur une extension de la notion de périodicité. C.R. 135, 891—894. (1902). (3)

—— 2. Sur les fonctions quasi périodiques moyennes, déduites d'une fonction quasi périodique. C.R. 157, 1389—1392. (1913). (3)

—— 3. Sur les intégrales quasi périodiques d'une équation différentielle linéaire. C.R. 160, 652—653. (1915). (3)

—— 4. Sur les intégrales quasi périodiques d'une équation différentielle linéaire. C.R. 161, 488—489. (1915). (3)

J. FAVARD 1. Sur les fonctions harmoniques presque-périodiques. C.R. 182, 757. (1926). (5)

—— 2. Sur les equations différentielles à coefficients presque-periodiques. Acta Math. 51, 31—81. (1927). (5)

P. FRANKLIN 1. Almost periodic recurrent motions. Math. Ztschr. 30, 325—331. (1929). (5)

—— 2. The elementary theory of almost periodic functions of two variables. Jour. Math. Phys. M.I. T. 5, 40—55. (1925). (5)

—— 3. The fundamental theorem of almost periodic functions of two variables. Jur. Math. Phys. M.I. T. 5, 201—237. (1926). (5)

—— 4. Classes of functions orthogonal on an infinite interval, having the power of the continuum. Jour. Math. Phys. M.I. T. 8, 74—79. (1929). (5)

—— 5. Approximation theorems for generalized almost periodic functions. Math. Ztschr. 29, 70—87. (1928). (5)

G. L. GOUY. Sur le mouvement lumineux. Journal de physique, 5, 354—362. (1886). (1)

H. HAHN 1. Über die Verallgemeinerung der Fourierschen Integralformel. Acta Math. 49, 301—353. (1926). (7)

H. Hahn 2. Über die Methode der arithmetischen Mittel in der Theorie der verallgemeinerten Fourierintegrale. Wiener Sitzungsber., 1925, 449—470. (7)

G. H. Hardy, A. E. Ingham, and G. Pólya. Notes on moduli and mean values. Proc. Lond. Math. Soc. 27, 401—409. (1928). (10)

E. W. Hobson. The theory of functions of a real variable and the theory of Fourier series. Vol. 2, second edition. Cambridge, 1927. (9)

S. Izumi 1. Über die Summierbarkeit der Fourierschen Integralformel. Tôhoku Journal 30, 96—110. (1929). (7)

—— 2. On the Cahen-Mellin's Inversion Formula. Tôhoku Journal 30, 111—114. (1929). (7)

M. Jacob 1. Über ein Theorem von Bochner-Hardy-Wiener. Jour. Lond. Math. Soc. 3, 182—187. (1928). (10)

—— 2. Über den Eindeutigkeitssatz in der Theorie der verallgemeinerten trigonometrischen Integrale. Math. Ann. 100, 279—294. (1928). (7)

—— 3. Über den Eindeutigkeitssatz in der Theorie der trigonometrischen Integrale. Math. Ann. 97, 663—674. (1927). (9)

G. W. Kenrick 1. Doctoral dissertation, Mass. Inst. Technology, 1927. (8)

—— 2. The analysis of irregular motions with applications to the energy-frequency spectrum of static and of telegraph signals. Phil. Mag. (7) 7, 176—196. (1929). (8)

E. H. Linfoot. Generalization of two theorems of H. Bohr. Jour. Lond. Math. Soc. 3, 177—182. (1928). (5)

S. B. Littauer. On a theorem of Jacob. Jour. Lond. Math. Soc. 4, 226—231. (1929). (10)

K. Mahler. On the translation properties of a simple class of arithmetical functions. J. Math. Phys. Mass. Inst. Technology 6, 158—164. (1927). (8)

C. F. Muckenhoupt. Almost periodic functions and vibrating systems. Doctoral dissertation, Mass. Inst. Technology, 1929. J. Math. Phys. Mass. Inst. Technology 8, 163—200. (1929). (5)

M. Plancherel 1. Contribution à l'étude de la répresentation d'une fonction arbitraire par des intégrales définies. Rendiconti di Palermo, 30, 289—335. (1910). (9)

—— 2. Sur la représentation d'une fonction arbitraire par une intégrale définie. C.R. 150, 318—321. (1910). (9)

—— 3. Sur la convergence et sur la sommation par les moyennes de Cesàro de

$$\lim_{z=\infty}\int_a^z f(x)\cos xy\,dx.$$ Math. Ann. 76, 315—326. (1915). (9)

H. Poincaré. Leçons sur la théorie mathématique de la lumière. Paris, 1889. (1)

S. Pollard 1. The summation of a Fourier integral. Proc. Camb. Phil. Soc. 23, 373—382. (1926). (9)

—— 2. On Fourier's integral. Proc. Lond. Math. Soc. 26, 12—24. (1927). (9)

S. POLLARD 3. Identification of the coefficients in a trigonometrical integral. Proc. Lond. Math. Soc. 25, 451—468. (1926). (9)

A. PRINGSHEIM. Über neue Gültigkeitsbedingungen der Fourierschen Integralformel. Math. Ann. 68, 367—408. (1910). (9)

J. RADON, Theorie und Anwendungen der absolutadditiven Mengenfunktionen. Wien. Ber. 122, 1295—1438. (1913). (10)

Lord RAYLEIGH 1. On the resultant of a large number of vibrations of the same pitch and of arbitrary phase. Phil. Mag. 10, 73—78. (1880). (6)

—— 2. Wave theory of light. Ency. Britt., 1888. Cf. especially § 4. (1)

—— 3. On the character of the complete radiation at a given temperature. Phil. Mag. 27, 460. (1889). (1)

—— 4. Röntgen rays and ordinary light. Nature, 57, 607. (1898). (1)

—— 5. On the spectrum of an irregular disturbance. Phil. Mag. 5, 238—243. (1903). (6)

—— 6. Remarks concerning Fourier's theorem as applied to physical problems. Phil. Mag. 24, 864—869. (1912). (1)

—— 7. On the problem of random vibrations, and of random flights in one, two, or three dimensions. Phil. Mag. 37, 321—347. (1919). (6)

—— 8. On the resultant of a number of unit vibrations, whose phases are at random over a range not limited to an integral number of periods. Phil. Mag. 37, 498—515. (1919). (6)

F. RIESZ. Sur la formule d'inversion de Fourier. Acta litt. ac Sci. Univ. Hung. 3, 235—241. (1927). (9)

H. L. RIETZ. Mathematical statistics. Chicago, 1927. (10)

R. SCHMIDT 1. Über divergente Folgen und lineare Mittelbildungen. Math. Ztschr. 22, 89—152. (1925). (10)

—— 2. Über das Borelsche Summierungsverfahren. Schriften der Köningsberger gelehrten Gesellschaft, 1, 202—256. (1925). (10)

—— 3. Die trigonometrische Approximation für eine Klasse von verallgemeinerten fastperiodischen Funktionen. Math. Ann. 100, 334—356. (1928). (5)

I. SCHOENBERG 1. Über total monotone Folgen mit stetiger Belegungsfunktion. Mat. Ztschr. 30, 761—768. (1929). (8)

—— 2. Über die asymptotische Verteilung reeller Zahlen mod 1. Mat. Ztschr. 28, 177—200. (1928). (8)

A. SCHUSTER 1. On interference phenomena. Phil. Mag. 37, 509—545. (1894). (1)

—— 2. The periodogram of magnetic declination. Camb. Phil. Trans. 18, 108. (1899). (2)

—— 3. The periodogram and its optical analogy. Proc. Roy. Soc. 77, 136—140. (1906). (2)

—— 4. The theory of optics. London, 1904. (1)

—— 5. On lunar und solar periodicities of earthquakes. Proc. Roy. Soc. London 61, 455—465. (1897). (2)

—— 6. On hidden periodicities. Terrestrial Magnetism 3, 13. (1897). (2)

—— 7. The periodogram of magnetic declination. Trans. Camb. Phil. Soc. 18, 107—135. (1900). (2)

H. STEINHAUS. Les probabilités dénombrables et leur rapport à la théorie de la mésure. Fund. Math. 3, 286—310. (1923). (10)

W. STEPANOFF. Über einige Verallgemeinerungen der fastperiodischen Funktionen. Math. Ann. 90, 473—492. (1925). (5)

G. SZEGÖ. Zur Theorie der fastperiodischen Funktionen. Math. Ann. 96, 378—382. (1926). (5)

G. I. TAYLOR. Diffusion by continuous movements. Proc. Lond. Math. Soc. 20, 196—212. (1920). (6)

E. C. TITCHMARSH 1. A contribution to the theory of Fourier transforms. Proc. Lond. Math. Soc. 23, 279—289. (1924). (9)

—— 2. Recent advances in science mathematics. Science progress 91, 372—386. (1929). (5)

C. DE LA VALLÉE POUSSIN 1. Sur les fonctions presque périodiques de H. Bohr. Annales de la Société Scientifique de Bruxelles, A, 47, 141. (1927). (5)

—— 2. Sur les fonctions presque périodiques de H. Bohr. Note complementaire et explicative. A.S.S. Bruxelles, A, 48, 56—57. (1928). (5)

T. VIJAYARAGHAVAN 1. A Tauberian theorem. Jour. Lond. Math. Soc. 1, 113—120. (1926). (10)

—— 2. A theorem concerning the summability of series by Borel's method. Proc. Lond. Math. Soc. 27, 316—326. (1928). (10)

V. VOLTERRA. Leçons sur les fonctions de lignes. Paris, 1913. (10)

J. D. WALSH. A generalization of the Fourier cosine series. Trans. Am. Math. Soc. 21, 101—116. (1920). (4)

H. WEYL 1. Integralgleichungen und fastperiodische Funktionen. Math. Ann. 97, 338—356. (1926). (5)

—— 2. Beweis des Fundamentalsatzes in der Theorie der fastperiodischen Funktionen. Berliner Sitzungsber. 1926, 211—214. (5)

—— 3. Quantenmechanik und Gruppentheorie. Ztschr. f. Physik 46, 1—46. (1927). (10)

N. WIENER 1. On the representation of functions by trigonometrical integrals. Math. Ztschr. 24, 575—617. (1925). (8)

—— 2. The harmonic analysis of irregular motion. J. Math. Phys. Mass. Inst. Technology 5, 99—122. (1925). (8)

—— 3. The harmonic analysis of irregular motion II. J.M.P.M.I.T. 5, 158—191. (1926). (8)

—— 4. The spectrum of an arbitrary function. Proc. Lond. Math. Soc. 27, 487—496. (1928). (8)

—— 5. Coherency matrices and quantum theory. J.M.P.M.I.T. 7, 109—125. (1928). (8)

—— 6. Harmonic analysis and the quantum theory. Jour. of Franklin Institute 207, 525—534. (1929). (8)

—— 7. The average of an analytic functional. Proc. Nat. Acad. Sci. 7, 253—260. (1921). (6)

—— 8. The average of an analytic functional and the Brownian motion. Proc. Nat. Acad. Sci. 7, 294—298. (1921). (6)

33—29764. *Acta mathematica.* 55. Imprimé le 9 mai 1930.

N. WIENER 9. The quadratic variation of a function and its Fourier coefficients. J.M.P.M.I.T. 3, 72—94. (1924). (8)

—— 10. Differential-space. J.M.P.M.I.T. 2, 131—174. (1923). (6)

—— 11. The average value of a functional. Proc. Lond. Math. Soc. 22, 454—467. (1922). (6)

—— 12. On a theorem of Bochner and Hardy. Jour. Lond. Math. Soc. 2, 118—123. (1927). (10)

—— 13. A new method in Tauberian theorems. J.M.P.M.I.T 7, 161—184. (1928). (10)

—— 14. The spectrum of an array and its application to the study of the translation properties of a simple class of arithmetical functions. J.M.P.M.I.T. 6, 145—157. (1927). (8)

—— 15. Harmonic analysis and group theory. J.M.P.M.I.T. 8, 148—154. (1929). (8)

—— 16. The operational calculus. Math. Ann. 95, 557—584. (1926). (8)

—— 17. Verallgemeinerte trigonometrische Entwicklungen. Gött. Nachrichten, 1925, 151—158. (8)

A. WINTNER. Spektraltheorie der unendlichen Matrizen. Leipzig, 1929. (5)

W. H. YOUNG. On non-harmonic Fourier series. Proc. Lond. Math. Soc. 18, 307—335. (1919). (4)

Footnotes.

1. In accordance with the listing adopted in the bibliography, the following references will give the background of the corresponding sections.

1. PLANCHEREL 1, 2, 3; and in general, papers under rubric (9).
2. Papers under rubric (2).
3 and 4. WIENER 1, 2, 3, 4, 17.
5. SCHMIDT 1, 2; VIJAYARAGHAVAN 1, 2; WIENER 13; JACOB 1.
6. BOCHNER 7.
7. HAHN 1. 2; DORN; JACOB 2, 3.
8. BERRY 1, 2.
9. POINCARÉ; WIENER 5, 6; WEYL 3; RIETZ.
10. WIENER 15.
11. MAHLER; WIENER 14.
12. WIENER 14.
13. WIENER 2, 3, 7, 8, 10, 11; EINSTEIN; TAYLOR; RAYLEIGH 1, 2, 3, 4, 5, 7, 8.
14, 15, 16. Papers under rubric (5).
17. MUCKENHOUPT.

2. SCHUSTER 5., p. 464.
3. HOBSON, § 492.
4. WEYL 3.
5. VOLTERRA, Ch. 7.
6. BOREL; STEINHAUS.

TAUBERIAN THEOREMS.*

By Norbert Wiener.

INTRODUCTION.

Numerous important branches of mathematics and physics concern themselves with the asymptotic behavior of functions for very large or very

* Received July 20, 1931.

Reprinted from *Ann. of Math.*, *33*, 1932, pp. 1–100. (Courtesy of the *Annals of Mathematics*.)

small values of their arguments, or of certain parameters. Statistical mechanics is that branch of mechanics which concerns itself, not with an individual dynamical system of a finite number of degrees of freedom, but with the asymptotic behavior of dynamical systems as the number of degrees of freedom increases without limit. The analytical theory of numbers is likewise concerned with the behavior of assemblages of whole numbers as the size of these assemblages increases. To this domain of ideas belong asymptotic series in analysis, and a whole order of concepts clustering about the operational calculus of Heaviside and the Fourier series and integrals.

The Fourier integral is of peculiar interest in the study of asymptotic problems. If $f(x)$ is a function of Lebesgue class L_2, by a theorem of Plancherel,[1] there is related to it another function $g(u)$, also of L_2, defined by the equation

$$g(u) = \underset{A \to \infty}{\text{l. i. m.}} \frac{1}{\sqrt{2\pi}} \int_{-A}^{A} f(x)\, e^{iux}\, dx \tag{0.01}$$

(where l. i. m. stands for "limit in the mean"), such that

$$\int_{-\infty}^{\infty} |g(u)|^2\, du = \int_{-\infty}^{\infty} |f(x)|^2\, dx, \tag{0.02}$$

and

$$f(x) = \underset{A \to \infty}{\text{l. i. m.}} \frac{1}{\sqrt{2\pi}} \int_{-A}^{A} g(u)\, e^{-iux}\, du. \tag{0.03}$$

The functions $f(x)$ and $g(u)$ have the reciprocal relation, that asymptotic properties of each correspond to local properties of the other. They are known as Fourier transforms of one another. It is easy to see that $f(x+\lambda)$ and $g(u)\,e^{-iu\lambda}$ are likewise Fourier transforms of one another.

There is a large class of asymptotic problems in which the asymptotic property to be investigated is connected in some obvious and simple way with the entire class of functions $f(x+\lambda)$. Since the Fourier transforms of the functions of this class only differ by factors $e^{-iu\lambda}$, independent of the particular function $f(x)$, Fourier transformation is here a peculiarly useful tool. An example of such a problem is the investigation of the asymptotic behavior of the integral

$$\int_{-\infty}^{\infty} k(\lambda)\, f(x+\lambda)\, d\lambda. \tag{0.04}$$

Many problems which on first investigation do not appear to be concerned with integrals of the above sort may be put into such a form by an elementary transformation of variables. For example, if we put

[1] Cf. Plancherel (1), Titchmarsh (3), Mellin (1), Wiener (7). Cf. Bibliography on p. 94.

$$(0.05) \qquad x = \log \xi, \quad \lambda = \log \mu, \quad f(x) = \varphi(\xi), \quad \mu k(\lambda) = Q(\mu),$$

the integral in question assumes the form

$$(0.06) \qquad \int_0^\infty Q(\mu)\, \varphi(\xi\mu)\, d\mu.$$

Thus the study of the asymptotic properties of this integral also fall under those accessible through Fourier developments.

In 1925 a paper by Robert Schmidt[2] appeared in which the class of theorems known as Tauberian was brought into relation with the asymptotic properties of integrals of this type. Tauberian theorems gain their name from a theorem published by A. Tauber[3] in 1897, to the effect that if

$$(0.07) \qquad \lim_{x \to 1-0} \sum_0^\infty a_n x^n = A,$$

and

$$(0.08) \qquad a_n = o(1/n),$$

then

$$(0.09) \qquad \sum_0^\infty a_n = A.$$

This is a conditioned converse of Abel's theorem, which stated that (0.07) follows from (0.09), without the mediation of any hypothesis such as (0.08). Such conditioned inverses of Abel's theorem, and of other analogous theorems which assert that the convergence of a series implies its summability by a certain method to the same sum, have been especially studied by G. H. Hardy and J. E. Littlewood, and have been termed by them *Tauberian*.

It is the service of Hardy and Littlewood[4] to have replaced hypothesis (0.08) by hypotheses of the form

$$(0.10) \qquad a_n = O\left(\frac{1}{n}\right),$$

or even of the form

$$(0.11) \qquad n a_n > -K.$$

The importance of these generalizations is scarcely to be exaggerated. They far exceed in significance Tauber's original theorem. The work of Hardy and Littlewood, unlike that of Tauber, makes very appreciable demands on analytical technique, and is capable, among other things, of supplying the gaps in Poisson's imperfect discussion of the convergence

[2] Schmidt (2).

[3] Tauber (1).

[4] Hardy (2), Littlewood (1), Hardy and Littlewood (4), (10), (13), (18) etc.

of the Fourier series. For these reasons, I feel that it would be far more appropriate to term these theorems Hardy-Littlewood theorems, were it not that usage has sanctioned the other appelation.

As we said, Tauberian theorems reduce to theorems on the conditioned equivalence of the asymptotic values of certain averages. The auxiliary conditions of Hardy and Littlewood, together with certain more extended conditions given by Schmidt[5] and discussed by Vijayaraghavan[6] and Szász,[7] become restrictions on the magnitude of the function or mass distribution of which the average is to be taken. An even further reduction eliminates this function from all consideration, and finds the essence of the Tauberian theorem in the linear properties of the weighting functions used in determining the average in question. The linear properties which are of importance concern the closure of the set of functions derivable from a given weighting function by translation, and the first chapter of this monograph is devoted to the study of such closure properties. The second chapter is occupied with the formulation and proof of the fundamental Tauberian theorems concerning averages, and the third with the transformation of these theorems into a recognized Tauberian form.

Some of the most interesting applications of Tauberian theorems have been in the analytic theory of numbers. Here two different avenues of approach are possible. Hardy and Littlewood[8] have shown that the prime number theorem, to the effect that the number of primes less than n is asymptotically $n/\log n$, is entirely equivalent to the Tauberian theorem concerning Lambert series, that if

$$\lim_{x\to 1-0}(1-x)\sum_{0}^{\infty}\frac{n\,a_n\,x^n}{1-x^n}=A \tag{0.12}$$

and

$$n\,a_n > -K, \tag{0.11}$$

then

$$\sum_{0}^{\infty} a_n = A. \tag{0.09}$$

This theorem falls under the category of theorems demonstrable by the methods of the present paper, and involves no further information concerning the Riemann zeta function $\zeta(\sigma+\tau i)$ than that this must be free of zeros on the line $\sigma=1$. This theorem was already contained in the author's first note on his method, dated 1928, and establishes the usefulness of Lambert series in the proof of the prime number theorem, which had

[5] Schmidt (1) and (2).

[6] Vijayaraghavan (1) and (2).

[7] Szász (1) and (2).

[8] Hardy and Littlewood (11).

frequently been questioned[9]. The proof was however needlessly complicated by the fact that after a good deal of labor was spent in transforming the author's average theorem into recognizable Tauberian form, this form did not appear to be especially direct in its relation to the prime number theorem. The present memoir gives a much more direct proof, starting from a theorem of the average type.

The other approach to the prime number theorem employs a theorem of S. Ikehara[10], which is itself a generalization of a theorem of Landau. Landau[11] showed that if $\sum a_n n^{-x}$ is a Dirichlet series with positive coefficients, representing a function $\psi(x)$ analytic on and to the right of $\Re(x) = 1$, except for a pole of order 1 at $x = 1$, for which the residue is A, and if the function in question is $O(|x|^k)$ for $\Re(x) \geqq 1$, then

$$\lim_{n\to\infty} \frac{1}{n} \sum_0^n a_l = A. \tag{0.13}$$

Ikehara proved by a Tauberian theorem that the condition that $\psi(x) = O(|x|^k)$ is inessential. By applying the theorem to $\psi(x) = -\zeta'(x)/\zeta(x)$ it is at once seen that the prime number theorem again follows from the fact that the Riemann zeta function has no zeros on the line $\Re(x) = 1$.

Chapter V is devoted to miscellaneous applications of Tauberian theorems. Among these perhaps the most important are to trigonometric developments and their summability. Here a theorem of Ramanujan[12] on Fourier transforms is of service. A particular application of Tauberian theorems to the criterion of Young[13] for the convergence of a Fourier series is due to Dr. Littauer[14], and is here discussed. We also take up the application of Tauberian theorems to asymptotic series and Wintner's[15] work.

There is a further field of application for Tauberian theorems where we need to introduce something like Robert Schmidt's[16] notion of "gestrahlte Matrizen". This we discuss in Chapter VI. In this category we find certain theorems of Hardy and Littlewood[17] related to Abel summation, as also the Tauberian theorems that concern themselves with Borel summation[18].

[9] Wiener (4).
[10] Ikehara (1).
[11] Landau (3).
[12] Hardy and Littlewood (8).
[13] Young (1).
[14] Littauer (2).
[15] Wintner (1).
[16] Schmidt (2).
[17] Cf. Hardy and Littlewood (10), (13), (18); Karamata (1), (2), (3), (4); Szász (1), (2); Doetsch (3); etc.
[18] Cf. Hardy and Littlewood (2), (3); Doetsch (4); Schmidt (1); Vijayaraghavan (1); Wiener (4).

In Chapter VII, we discuss certain theorems not strictly of a Tauberian nature, in that they involve no positiveness or boundedness condition. We here follow an earlier paper[19] in which the author proved the Hardy-Littlewood necessary and sufficient condition for the summability of a Fourier series. There our result was not a "best possible" one, as the work of Bosanquet[20] and Paley has demonstrated. At present, our method has succeeded in yielding their strict theorem.

Chapter VIII is devoted to the development of certain Tauberian theorems intimately connected with our generalized harmonic analysis[21]. These suggest certain new definitions of summability, and we discuss definitions of this sort.

The genesis of the present paper may deserve a word of comment. In the preparation of a previous investigation on generalized harmonic analysis, the author found himself obliged to make use of certain theorems communicated to him by Mr. A. E. Ingham and of a Tauberian nature. A correspondence arose with Professor Hardy, finally resulting in a number of papers of a Tauberian character. While interested in theorems of this type, the author had the work of Dr. Robert Schmidt brought to his notice, and at Dr. Schmidt's suggestion, began to search for a method of combining the trigonometric attack of his own earlier papers with the generality of the Schmidt standpoint. It was only by a radical change in the manner in which trigonometric methods were applied that this attempt succeeded. This change consisted in a logarithmic change of base before the introduction of the harmonic analysis. It was quite along lines contemplated by Schmidt, who correlated with his Tauberian theorems a certain moment problem. Schmidt, however, found in the problem of uniqueness the significant aspect of this moment problem. The present author looked for it rather in the problem of the existence of a solution together with certain associated problems of approximation. It was Dr. Schmidt himself who furnished the experimentum crucis which established the greater scope of the methods of this paper. He suggested that the Tauberian theorem for Lambert series had resisted his methods of proof, and that it might be desirable to try on it the edge of any new method.

The origination of a new method is but the prelude to a large amount of detailed application before its power can be judged or its limitations defined. In this task the author has had the valuable aid of his students, Drs. S. B. Littauer and S. Ikehara. He also wishes to express his gratitude for the criticisms of Professors Hardy and Tamarkin at more than one stage of his work.

[19] Cf. Hardy and Littlewood (12); Wiener (5).

[20] Bosanquet (1); Paley (1).

[21] Wiener (7).

CHAPTER I.

THE CLOSURE OF THE SET OF TRANSLATIONS OF A GIVEN FUNCTION.

1. **Closure in class L_2.** Let $f(x)$ be a function, real or complex, defined for all real arguments over $(-\infty, \infty)$. Let it belong to Lebesgue class L_2—that is, let it be measurable, and let

$$\int_{-\infty}^{\infty} |f(x)|^2 \, dx \tag{1.01}$$

be finite. We shall term the class of all functions $f(x+\lambda)$ for all real values of λ the class of *translations of* $f(x)$. We wish to know when this class is closed or complete—that is, when it is possible, whenever a function $F(x)$ from L_2 is given, and any positive quantity ε, to find a function $F_1(x)$ of the form

$$F_1(x) = \sum_1^n A_k f(x+\lambda_k), \tag{1.02}$$

such that

$$\int_{-\infty}^{\infty} | F(x) - F_1(x) |^2 \, dx \leqq \varepsilon. \tag{1.03}$$

The situation is governed by the following theorem:

THEOREM I. *The necessary and sufficient condition for the set of all translations of $f(x)$ to be closed is that the real zeros of its Fourier transform*

$$g(u) = \underset{A\to\infty}{\text{l. i. m.}} \frac{1}{\sqrt{2\pi}} \int_{-A}^{A} f(x)\, e^{iux} \, dx \tag{0.01}$$

should form a set of zero measure.

The existence of this Fourier transform results from a familiar theorem of Plancherel.[1] This theorem further requires that

$$\int_{-\infty}^{\infty} |g(u)|^2 \, du = \int_{-\infty}^{\infty} |f(x)|^2 \, dx \tag{0.02}$$

and that

$$f(x) = \underset{A\to\infty}{\text{l. i. m.}} \frac{1}{\sqrt{2\pi}} \int_{-A}^{A} g(u)\, e^{-iux} \, du. \tag{0.03}$$

The Fourier transform of $f(x+\lambda)$ is $e^{-iu\lambda} g(u)$.

On account of (0.02), properties of convergence in the mean are conserved under a Fourier transformation, for these concern merely the integral of the square of the modulus of the difference of two functions, which is equal to the integral of the square of the difference of their Fourier transforms. Accordingly, Theorem I is a corollary of:

LEMMA Ia. *If $g(u)$ is a function belonging to L_2, then the set of functions $e^{-iu\lambda}g(u)$ is closed when and only when the zeros of $g(u)$ form a set of zero measure.*

Since every zero of $g(u)$ is also a zero of $e^{-iu\lambda}g(u)$, the necessity of this condition is at once obvious, for if

$$G_1(u) = \sum_1^n A_k e^{-iu\lambda_k} g(u) \tag{1.04}$$

is any linear combination of the functions $e^{-iu\lambda_k}g(u)$, then

$$\int_A^B |G_1(u) - 1|^2 du \tag{1.05}$$

must equal or exceed the measure of the set of the points between A and B where $g(u)$ vanishes. That is, if the measure of this set is not zero, the function which is 1 between A and B and zero elsewhere cannot be a limit in the mean of functions $G_1(u)$.

Now as to the sufficiency of the condition in Lemma Ia: let

$$\varphi(u) = \begin{cases} 1 - \frac{|u|}{A}, & [|u| \leqq A]; \\ 0, & [|u| > A]. \end{cases} \tag{1.06}$$

Then

$$\varphi(u) = \frac{1}{\pi A}\int_{-\infty}^{\infty} e^{-iu\mu}\frac{1-\cos A\mu}{\mu^2}\,d\mu. \tag{1.07}$$

This is an absolutely convergent integral, and hence a limit of a sequence of polynomials

$$\sum A_n e^{-iu\lambda_n} \tag{1.08}$$

converging boundedly, and uniformly over every finite interval. Hence

$$e^{-i\frac{un\pi}{A}}\varphi(u)g(u) \tag{1.09}$$

is a limit in the mean of such functions as those in (1.04).

Now let $\psi(u)$ be a quadratically summable function satisfying the conditions

$$\int_{-A}^{A} \psi(u)\varphi(u)g(u)e^{\frac{-iun\pi}{A}}\,du = 0. \tag{1.10}$$

$$[n = \cdots, -2, -1, 0, 1, 2, \cdots]$$

Since $\psi(u)\varphi(u)g(u)$ is absolutely summable,[22] we shall have

$$\psi(u)\varphi(u)g(u) = 0 \tag{1.11}$$

[22] This follows from the Parseval theorem for Fourier integrals.

except over a set of measure zero.[23] If $g(u)$ does not vanish except over a set of measure zero, since $\varphi(u)$ has no zeros, $\psi(u)$ vanishes over $(-A, A)$ except over a set of measure zero. Thus if $g(u) \neq 0$, there is no quadratically summable function not almost everywhere zero, orthogonal to every function $e^{\frac{-iun\pi}{A}} \varphi(u) g(u)$ over $(-A, A)$. This means that every function of L_2 vanishing everywhere outside $(-A, A)$ is a limit in the mean of polynomials[24] in $e^{\frac{-iun\pi}{A}} \varphi(u) g(u)$ and hence in $e^{-iu\lambda_n} g(u)$. Every function of L_2 is however the limit in the mean of a sequence of functions of L_2 vanishing for large arguments. Thus Lemma Ia and Theorem I are established.

2. **Closure in class L_1.** The Lebesgue class L_1 consists of all measurable functions $f(x)$ which are absolutely integrable over $(-\infty, \infty)$. If $f(x)$ is such a function, its Fourier transform

$$g(u) = \frac{1}{\sqrt{2\pi}} \int_{-\infty}^{\infty} f(x) e^{iux} dx \tag{2.01}$$

of course exists, and is bounded and continuous. As before, the Fourier transform of $f(x+\lambda)$ is $e^{-iu\lambda} g(u)$.

We shall say that a class C of functions $\varphi_\lambda(x)$ of L_1 is closed L_1 if, whenever $F(x)$ is a function of L_1 and ε is a positive number, there is a polynomial

$$F_1(x) = \sum_1^n A_k \varphi_{\lambda_k}(x) \tag{2.02}$$

of functions of C, such that

$$\int_{-\infty}^{\infty} |F(x) - F_1(x)| \, dx \leqq \varepsilon. \tag{2.03}$$

We shall prove the following theorem:

THEOREM II. *If $f(x)$ is a function of L_1, a necessary and sufficient condition for the set of all translations of $f(x)$ to be closed L_1 is that its Fourier transform*

$$g(u) = \frac{1}{\sqrt{2\pi}} \int_{-\infty}^{\infty} f(x) e^{iux} dx \tag{2.01}$$

should have no real zeros.

[23] This follows from the Riemann theorem on the unicity of the function with a given Fourier series.

[24] Here we make use of the familiar theorem that if any linear class of functions of L_2 is given, every function of L_2 is the sum of a function expressible in the mean by the functions of the class and a function orthogonal to all functions of the class.

The necessity of this condition is again obvious. If $g(u_1) = 0$, the same is true for the argument u_1 of the Fourier transform of

$$F_1(x) = \sum_1^n A_k f(x+\lambda_k). \tag{1.02}$$

Let $f_1(x)$ be a function of L_1 with a Fourier transform $g_1(x)$ for which

$$g_1(u) \neq 0.$$

Then

$$\begin{aligned}\int_{-\infty}^{\infty} |F_1(x)-f_1(x)|\, dx &\geqq \left|\int_{-\infty}^{\infty} (f_1(x)-F_1(x))\, e^{iu_1 x}\, dx\right| \\ &= \sqrt{2\pi}\; |g_1(u_1)| > 0.\end{aligned} \tag{2.04}$$

This proves that f_1 is not a limit in the mean (L_1) of any sequence of F_1's.

To prove the sufficiency of the condition of Theorem II, we shall have recourse to a sequence of lemmas.

LEMMA IIa. *If*

$$f_1(x) = \sum_{-\infty}^{\infty} a_n e^{inx}, \qquad f_2(x) = \sum_{-\infty}^{\infty} b_n e^{inx}, \tag{2.05}$$

and

$$\sum_{-\infty}^{\infty} |a_n| = A < \infty, \qquad \sum_{-\infty}^{\infty} |b_n| = B < \infty, \tag{2.06}$$

and if

$$f_1(x)\, f_2(x) \sim \sum_{-\infty}^{\infty} c_n e^{inx} \tag{2.07}$$

then

$$\sum_{-\infty}^{\infty} |c_n| \leqq AB. \tag{2.08}$$

The proof is immediate, for

$$\begin{aligned}\sum_{-\infty}^{\infty} |c_n| &= \sum_{n=-\infty}^{\infty} \left|\sum_{k=-\infty}^{\infty} a_k b_{n-k}\right| \leqq \sum_{n=-\infty}^{\infty} \sum_{k=-\infty}^{\infty} |a_k|\; |b_{n-k}| \\ &= \sum_{-\infty}^{\infty} |a_n| \sum_{-\infty}^{\infty} |b_n| = AB.\end{aligned} \tag{2.09}$$

LEMMA IIb. *If $f(x)$ is a function of period 2π, and if at every point y, there is a function $f_y(x)$, coincident with $f(x)$ over some interval $(y-\varepsilon_y, y+\varepsilon_y)$, of period 2π, and such that its Fourier series converges absolutely, then the Fourier series of $f(x)$ converges absolutely*[25].

[25] We shall say that a Fourier series converges absolutely when the sum of the moduli of the coefficients converges, the series being represented in complex exponential form. This is of course equivalent to the absolute convergence of the series in this form for any single real argument.

To prove this, let us reflect that by the Heine-Borel theorem, any period of $f(x)$ may be covered by a finite number of overlapping intervals $(y-\epsilon_y, y+\epsilon_y)$. Let such an interval be (A, B), let its neighbor to the left have its right-hand end-point at C, and let its neighbor to the right have its left-hand end-point at D. There will be no loss in generality if we suppose that $A < C < D < B$. Let us define $\varphi_y(x)$ by

$$\varphi_y(x) = \begin{cases} 0\ ; & [-\pi \leqq x < A] \\ \dfrac{x-A}{C-A}; & [\ A \leqq x < C] \\ 1\ ; & [\ C \leqq x < D] \\ \dfrac{B-x}{B-D}; & [\ D \leqq x < B] \\ 0 & [\ B \leqq x < \pi]. \end{cases} \tag{2.10}$$

It is easy to show that the Fourier series of $\varphi_y(x)$ converges absolutely. Thus the Fourier series of

$$\varphi_y(x)\, f(x) = \varphi_y(x)\, f_y(x) \tag{2.11}$$

converges absolutely, by Lemma IIa. However, if we add together the functions $\varphi_y(x)$ for a set of intervals completely covering a period of $f(x)$, the sum will be $f(x)$ itself. Thus $f(x)$ will be the sum of a finite number of functions with absolutely convergent Fourier series and must itself have an absolutely convergent Fourier series.

LEMMA IIc. *If*

$$f(x) = \sum_{-\infty}^{\infty} a_n e^{inx} \tag{2.12}$$

and

$$|a_0| > \sum_{-\infty}^{-1} |a_n| + \sum_{1}^{\infty} |a_n|, \tag{2.13}$$

then the Fourier series of $1/f(x)$ *converges absolutely.*

We may write

$$\begin{aligned} \frac{1}{f(x)} &= \frac{1}{a_0 + \sum_{-\infty}^{-1} a_n e^{inx} + \sum_{1}^{\infty} a_n e^{inx}} \\ &= \frac{1}{a_0}\Big\{1 - \frac{1}{a_0}\Big[\sum_{-\infty}^{-1} a_n e^{inx} + \sum_{1}^{\infty} a_n e^{inx}\Big] \\ &\quad + \frac{1}{a_0^2}\Big[\sum_{-\infty}^{-1} a_n e^{inx} + \sum_{1}^{\infty} a_n e^{inx}\Big]^2 - \cdots\Big\}. \end{aligned} \tag{2.14}$$

That this is more than formally true, and that $1/f(x)$ is continuous, results from the fact that the geometrical progression

$$(2.15)\ \frac{1}{|a_0|}\left\{1+\frac{1}{|a_0|}\left[\sum_{-\infty}^{-1}|a_n|+\sum_{1}^{\infty}|a_n|\right]+\frac{1}{a_0^2}\left[\sum_{-\infty}^{-1}|a_n|+\sum_{1}^{\infty}|a_n|\right]^2+\cdots\right\}$$

converges. Term by term, series (2.15) is greater than or equal to the sum of the sums of the moduli of the coefficients of the Fourier series of the successive terms of (2.14), and hence is greater than or equal to the sum of the moduli of the coefficients of the Fourier series of $1/f(x)$.

LEMMA IId. *Let*

$$\sum_{-\infty}^{\infty}|a_n| = A < \infty, \tag{2.16}$$

and let

$$f(x) = \sum_{-\infty}^{\infty} a_n e^{inx}. \tag{2.17}$$

Let

$$f(y) \neq 0.$$

Then there is a neighborhood $(y-\varepsilon,\ y+\varepsilon)$ *of* y *and a function* $f_y(x)$ *with absolutely convergent Fourier series such that over* $(y-\varepsilon,\ y+\varepsilon)$,

$$f_y(x) = f(x) \tag{2.18}$$

and that the Fourier series of $f_y(x)$, *let us say*

$$\sum_{-\infty}^{\infty} c_n e^{inx} \tag{2.19}$$

has the property that

$$|c_0| > \sum_{-\infty}^{-1}|c_n| + \sum_{1}^{\infty}|c_n|. \tag{2.20}$$

There is manifestly no restriction in taking y to be 0. Let us introduce the auxiliary function

$$\varphi_\varepsilon(x) = \begin{cases} 1 & ;\ [|x|<\varepsilon] \\ 2-\dfrac{|x|}{\varepsilon} & ;\ [\varepsilon \leqq |x| < 2\varepsilon] \\ 0 & .\ [2\varepsilon \leqq |x|]. \end{cases} \tag{2.201}$$

This will have the Fourier series

$$\frac{3\varepsilon}{2\pi}+\sum_{1}^{\infty}\frac{e^{inx}+e^{-inx}}{\pi n^2 \varepsilon}(\cos \varepsilon n - \cos 2\varepsilon n). \tag{2.202}$$

If we represent the function $\{f(x)\,\varphi_\varepsilon(x)+f(0)\,[1-\varphi_\varepsilon(x)]\}$ by the Fourier series

$$\sum_{-\infty}^{\infty} c_n e^{inx},$$

we shall have

$$c_n = \frac{3\varepsilon}{2\pi} a_n + \sum_{k=1}^{\infty} \frac{a_{n-k} + a_{n+k}}{\pi k^2 \varepsilon} (\cos \varepsilon k - \cos 2\varepsilon k)$$

$$- \sum_{-\infty}^{\infty} a_k \frac{\cos \varepsilon n - \cos 2\varepsilon n}{\pi n^2 \varepsilon}; \qquad [n \neq 0],$$

(2.203)

$$c_0 = a_0 + \sum_{1}^{\infty} (a_{-k} + a_k) \left[1 + \frac{\cos \varepsilon k - \cos 2\varepsilon k}{\pi k^2 \varepsilon} - \frac{3\varepsilon}{2\pi} \right].$$

Hence

$$\lim_{\varepsilon \to 0} |c_0| = \left| \sum_{-\infty}^{\infty} a_k \right| = |f(0)| > 0, \tag{2.204}$$

and

$$\sum_{1}^{\infty} [|c_n| + |c_{-n}|]$$

$$\leq \sum_{m=-\infty}^{\infty} |a_m| \left\{ \sum_{n=-\infty}^{\infty}{}'' \left| \frac{\cos \varepsilon n - \cos 2\varepsilon n}{\pi n^2 \varepsilon} - \frac{\cos \varepsilon (m-n) - \cos 2\varepsilon (m-n)}{\pi (m-n)^2 \varepsilon} \right| \right.$$

(2.205)

$$\left. + \left| \frac{\cos \varepsilon m - \cos 2\varepsilon m}{\pi m^2 \varepsilon} - \frac{3\varepsilon}{2\pi} \right| \right\}$$

$$= \sum_{-\infty}^{\infty} |a_m| A_m.$$

Here $\sum''$ means that the values 0 and m are omitted.

We have

$$\left| \frac{\cos \varepsilon n - \cos 2\varepsilon n)}{\pi n^2 \varepsilon} \right| = \left| \frac{2 \sin \frac{\varepsilon n}{2} \sin \frac{3\varepsilon n}{2}}{\pi n^2 \varepsilon} \right| \leq \begin{cases} \frac{3\varepsilon}{2\pi}; \\ \frac{2}{\pi n^2 \varepsilon}; \end{cases} \tag{2.206}$$

and

$$\left| \frac{\cos \varepsilon n - \cos 2\varepsilon n}{\pi n^2 \varepsilon} - \frac{\cos \varepsilon (m-n) - \cos 2\varepsilon (m-n)}{\pi (m-n)^2 \varepsilon} \right|$$

$$= \left| \int_{n-m}^{n} \frac{d}{dx} \left(\frac{\cos \varepsilon x - \cos 2\varepsilon x}{\pi x^2 \varepsilon} \right) dx \right| \tag{2.21}$$

$$\leq \frac{\varepsilon^2}{\pi} \left| \int_{n-m}^{n} \right| \max_y \frac{d}{dy} \frac{\cos y - \cos 2y}{y^2} \left| dx \right| \leq |m| \varepsilon^2 c.$$

This formula may be applied in the limiting form whenever $m = n$ or $m = 0$. It results from (2.205) and (2.206) that for all sufficiently small ε

$$A_m \leq 4 \left[\frac{1}{\varepsilon} + 1 \right] \frac{3\varepsilon}{2\pi} + \frac{8}{\pi\varepsilon} \sum_{[1/\varepsilon]}^{\infty} n^{-2} < \frac{15}{\pi}, \tag{2.22}$$

and from (2.205) and (2.21) that for all sufficiently small ε

$$A_m \leqq 2[\varepsilon^{-3/2}+1]\,|m|\,\varepsilon^2 c + \frac{8}{\pi\varepsilon}\sum_{[\varepsilon^{-3/2}]}^{\infty} n^{-2} \tag{2.23}$$
$$< \varepsilon^{1/2}[2\,|m|\,c + 9/\pi].$$

Combining (2.205), (2.22) and (2.23), we get

$$\lim_{\varepsilon \to 0} \sum_{-\infty}^{\infty} |a_m|\, A_m = 0, \tag{2.24}$$

which with (2.204) yields us

$$\sum_{1}^{\infty} [|c_n| + |c_{-n}|] < |c_0|. \tag{2.25}$$

If we take

$$f_y(x) = f(x)\,\varphi_\varepsilon(x) + f(0)\,(1-\varphi_\varepsilon(x)), \tag{2.26}$$

Lemma IId is established.

As a direct consequence of Lemmas IIb, IIc, and IId we obtain:

LEMMA IIe. *If $f(x)$ is a function with an absolutely convergent Fourier series, which nowhere vanishes for real arguments, $1/f(x)$ has an absolutely convergent Fourier series.*

To prove this let us note that, in every neighborhood, by IId, $f(x)$ coincides with a function which by IIc has a reciprocal with an absolutely convergent Fourier series. By IIa, since in every neighborhood $1/f(x)$ coincides with a function with an absolutely convergent Fourier series, its Fourier series converges absolutely.

LEMMA IIf. *Let $f(x)$ be a function with an absolutely convergent Fourier series over the interval $(-\pi, \pi)$, and vanishing over neighborhoods including π and $-\pi$. Let $f(x)$ vanish everywhere outside $(-\pi, \pi)$. Let*

$$g(u) = \frac{1}{\sqrt{2\pi}} \int_{-\pi}^{\pi} f(x)\, e^{iux}\, dx. \tag{2.27}$$

Then the integral

$$\int_{-\infty}^{\infty} |g(u)|\, du \tag{2.28}$$

will converge. Conversely, if $f(x)$ vanishes over the region indicated, and (2.28) converges, its Fourier series will converge absolutely.

Let ε be such that $f(x)$ has no non-zero, values outside $(-\pi+2\varepsilon, \pi-2\varepsilon)$. Let us define $\varphi(x)$ by:

$$\varphi(x) = \begin{cases} 1 & ; \ [|x| < \pi - \varepsilon] \\ \dfrac{\pi - |x|}{\varepsilon}; & [\pi - \varepsilon \leqq |x| < \pi] \\ 0 & . \ [\pi \leqq |x|] \end{cases} \tag{2.29}$$

Let

$$f(x) = \sum_{-\infty}^{\infty} a_n e^{inx}. \tag{2.17}$$

We shall have

$$\varphi(x) = \frac{1}{\pi} \int_{-\infty}^{\infty} e^{-iux} \frac{\cos u(\pi - \varepsilon) - \cos u\pi}{u^2 \varepsilon} du, \tag{2.30}$$

and hence

$$\begin{aligned} f(x) = f(x)\varphi(x) &= \frac{1}{\pi} \int_{-\infty}^{\infty} du \sum_{-\infty}^{\infty} a_n e^{-i(u-n)x} \frac{\cos u(\pi - \varepsilon) - \cos u\pi}{u^2 \varepsilon} \\ &= \frac{1}{\pi} \int_{-\infty}^{\infty} e^{iux} \left[\sum_{-\infty}^{\infty} a_n \frac{\cos(u+n)(\pi-\varepsilon) - \cos(u+n)\pi}{(u+n)^2 \varepsilon} \right] du, \end{aligned} \tag{2.31}$$

where the rearrangement is possible since both integral and sum form part of an absolutely convergent double summation process. Then

$$g(u) = \sqrt{\frac{2}{\pi}} \sum_{-\infty}^{\infty} a_n \frac{\cos(u+n)(\pi - \varepsilon) - \cos(u+n)\pi}{(u+n)^2 \varepsilon} \tag{2.32}$$

will satisfy the condition that (2.28) shall be finite, and equation (2.27).

To prove the converse part of Lemma IIf, let us notice that as a result of the convergence of (2.28),

$$\sum_{-\infty}^{\infty} \left| \int_{n-1/2}^{n+1/2} g(u)\, du \right| \tag{2.33}$$

converges. Now,

$$\begin{aligned} \int_{n-1/2}^{n+1/2} g(u)\, du &= \frac{1}{\sqrt{2\pi}} \int_{-\pi}^{\pi} f(x)\, dx \int_{n-1/2}^{n+1/2} e^{iux}\, du \\ &= \frac{1}{\sqrt{2\pi}} \int_{-\pi}^{\pi} f(x) \frac{2 \sin x/2}{x} e^{inx}\, dx. \end{aligned} \tag{2.34}$$

Thus the Fourier series of

$$f(x) \frac{2 \sin x/2}{x} \tag{2.35}$$

converges absolutely. Since the same is true of the Fourier series of

$$\frac{x}{2 \sin x/2} \varphi(x) \tag{2.36}$$

as may be determined by a simple computation, it follows from Lemma IIa that the same is true for the Fourier series of $f(x)$.

LEMMA IIg. *Let $f(x)$ be a function of class L_1. Then*

$$\lim_{\varepsilon \to 0} \int_{-\infty}^{\infty} |f(x+\varepsilon)-f(x)|\, dx = 0. \tag{2.37}$$

This is a well-known theorem and is proved in Hobson's *Theory of Functions of a Real Variable.*

LEMMA IIh. *Let $f(x)$ be a function of class L_1. Then if $\Phi(y)$ is any function of class L_1, differing from 0 only over $(-\varepsilon, \varepsilon)$,*

$$\begin{aligned}\int_{-\infty}^{\infty} \left| f(x) \int_{-\infty}^{\infty} \Phi(y)\, dy - \int_{-\infty}^{\infty} f(x+y)\, \Phi(y)\, dy \right| dx \\ \leq \int_{-\infty}^{\infty} |\Phi(y)|\, dy \max_{|z| \leq \varepsilon} \int_{-\infty}^{\infty} |f(x+z)-f(x)|\, dx.\end{aligned} \tag{2.38}$$

The proof of this lemma merely depends on the inversion of the order of integration in an absolutely convergent double integral.

LEMMA IIi. *Let $f(x)$ be a function of class L_1. Then*

$$\lim_{n \to \infty} \int_{-\infty}^{\infty} \left| f(x) - \frac{1}{\pi n} \int_{-\infty}^{\infty} f(x+y) \frac{\sin^2 n y}{y^2}\, dy \right| dx = 0. \tag{2.39}$$

This is a theorem of the Fejér type. It depends on the splitting of

$$\Phi_0(y) = \frac{\sin^2 n y}{y^2}$$

into the sum $\Phi_1(y) + \Phi_2(y)$ where

$$\Phi_1(y) = \begin{cases} \Phi_0(y)\,(1 - |y|\, n^{1/2}); & [|y| \leq n^{-1/2}] \\ 0; & [|y| > n^{-1/2}]. \end{cases} \tag{2.40}$$

Lemmas IIg and IIh enable us to show that

$$\lim_{n \to \infty} \int_{-\infty}^{\infty} \left| f(x) - \frac{1}{\pi n} \int_{-\infty}^{\infty} f(x+y)\, \Phi_1(y)\, dy \right| dx = 0 \tag{2.41}$$

while

$$\begin{aligned}\lim_{n \to \infty} \int_{-\infty}^{\infty} \left| \frac{1}{\pi n} \int_{-\infty}^{\infty} f(x+y)\, \Phi_2(y)\, dy \right| dx \\ \leq \int_{-\infty}^{\infty} |f(x)|\, dx \lim_{n \to \infty} \frac{1}{\pi n} \int_{-\infty}^{\infty} \Phi_2(y)\, dy.\end{aligned} \tag{2.42}$$

Now

$$\frac{1}{\pi n}\int_{-\infty}^{\infty}\Phi_2(y)\,dy$$

$$= \frac{1}{\pi n}\int_{-\infty}^{\infty}\frac{\sin^2 ny}{y^2}\,dy - \frac{1}{\pi n}\int_{-n^{-1/2}}^{n^{-1/2}}(1-|y|\,n^{1/2})\frac{\sin^2 ny}{y^2}\,dy$$

$$(2.421)\qquad = \frac{1}{\pi}\int_{-\infty}^{\infty}\frac{\sin^2 z}{z^2}\,dz - \frac{1}{\pi}\int_{-n^{1/2}}^{n^{1/2}}(1-|z|\,n^{-1/2})\frac{\sin^2 z}{z^2}\,dz$$

$$= \frac{2}{\pi}\int_{n^{1/2}}^{\infty}\frac{\sin^2 z}{z^2}\,dz + \frac{2}{\pi n^{1/2}}\int_{0}^{n^{1/2}}\frac{\sin^2 z}{z}\,dz$$

$$= O(n^{-1/2}\log n).$$

Hence

$$(2.422)\qquad \lim_{n\to\infty}\int_{-\infty}^{\infty}\left|\frac{1}{\pi n}\int_{-\infty}^{\infty}f(x+y)\,\Phi_2(y)\,dy\right|dx = 0,$$

and Lemma IIi is established.

We are now in a position to proceed to the proof of Theorem II. Let $F(x)$ be any function of L_1. By Lemma IIi, we can find a function $F_\eta(x)$ of the form

$$(2.43)\qquad \frac{1}{\pi n}\int_{-\infty}^{\infty}F(x+y)\,\frac{\sin^2 ny}{y^2}\,dy,$$

such that

$$(2.44)\qquad \int_{-\infty}^{\infty}|F(x)-F_\eta(x)|\,dx<\eta.$$

The Fourier transform $H_1(u)$ of this function $F_\eta(x)$ is

$$\frac{1}{\sqrt{2\pi}}\int_{-\infty}^{\infty}e^{iux}\,dx\,\frac{1}{\pi n}\int_{-\infty}^{\infty}F(x+y)\,\frac{\sin^2 ny}{y^2}\,dy$$

$$(2.45)\qquad = \frac{1}{\sqrt{2\pi}}\int_{-\infty}^{\infty}F(x)\,e^{iux}\,dx\,\frac{1}{\pi n}\int_{-\infty}^{\infty}\frac{\sin^2 ny}{y^2}\,e^{-iuy}\,dy$$

$$= \begin{cases}\left(1-\dfrac{|u|}{2n}\right)\dfrac{1}{\sqrt{2\pi}}\displaystyle\int_{-\infty}^{\infty}F(x)\,e^{iux}\,dx; & [|u|<2n]\\ 0\quad ; & [|u|\geq 2n]\end{cases}$$

and will vanish for all values of its argument u larger than $2n$ in modulus. The same will be the case with the Fourier transform $H_2(u)$ of

$$(2.46)\qquad \frac{1}{2\pi n}\int_{-\infty}^{\infty}f(x+y)\,\frac{\sin^2 2ny}{y^2}\,dy$$

which will vanish for values of its argument greater in modulus than $4n$, and only for such values. If we expand these Fourier transforms in Fourier series over $(-8n, 8n)$, by Lemma IIf, these series converge absolutely.

By an argument substantially identical with that used in proving Lemma IIb, we see that in the neighborhood of every point of $(-3n, 3n)$, there is a function coinciding locally with $1/H_2(u)$, and with an absolutely convergent Fourier series. It follows from Lemma IIa that the same is true of $H_1(u)/H_2(u)$ over the same interval. Since 0 enjoys this property also, the function which is $H_1(u)/H_2(u)$ over $(-3n, 3n)$ and 0 over the rest of a period $(-8n, 8n)$ has an absolutely convergent Fourier series, and hence, by Lemma IIf, an absolutely convergent Fourier integral. That is, we may write

$$H_1(u) = H_2(u)\, H_3(u) \tag{2.47}$$

where $H_3(u)$ is of the form

$$H_3(u) = \int_{-\infty}^{\infty} \Phi(x)\, e^{iux}\, dx \tag{2.48}$$

and

$$\int_{-\infty}^{\infty} |\Phi(x)|\, dx \tag{2.49}$$

converges. We shall then have

$$\begin{aligned}
&\frac{1}{\sqrt{2\pi}} \int_{-\infty}^{\infty} F_\eta(x)\, e^{iux}\, dx \\
&= \frac{1}{\sqrt{2\pi}} \int_{-\infty}^{\infty} \frac{e^{iux}}{2\pi n}\, dx \int_{-\infty}^{\infty} f(x+y) \frac{\sin^2 2ny}{y^2}\, dy \int_{-\infty}^{\infty} \Phi(z)\, e^{iuz}\, dz \\
&= \frac{1}{\sqrt{2\pi}} \int_{-\infty}^{\infty} e^{iux}\, dx \\
&\quad \times \left[\frac{1}{2\pi n} \int_{-\infty}^{\infty} f(x+\omega)\, d\omega \int_{-\infty}^{\infty} \frac{\sin^2 2ny}{y^2}\, \Phi(y-\omega)\, dy\right],
\end{aligned} \tag{2.50}$$

or

$$\begin{aligned}
&\int_{-\infty}^{\infty} e^{iux}\, dx \\
&\left[F_\eta(x) - \frac{1}{2\pi n} \int_{-\infty}^{\infty} f(x+\omega)\, d\omega \int_{-\infty}^{\infty} \frac{\sin^2 2ny}{y^2}\, \Phi(y-\omega)\, dy\right] = 0.
\end{aligned} \tag{2.51}$$

Here we may replace e^{iux} by any function with an absolutely convergent Fourier integral, and hence by functions positive over an arbitrarily small region and zero everywhere else. This can only be if

$$
(2.52)\quad \begin{aligned} F_\eta(x) &= \frac{1}{2\pi n}\int_{-\infty}^{\infty} f(x+\omega)\,d\omega \int_{-\infty}^{\infty} \frac{\sin^2 2ny}{y^2}\,\Phi(y-\omega)\,dy \\ &= \int_{-\infty}^{\infty} f(x+\omega)\,\Psi(\omega)\,d(\omega). \end{aligned}
$$

Here $\Psi(\omega)$ is absolutely integrable.

By Lemma IIh and Lemma IIg,

$$
(2.53)\quad \begin{aligned} \lim_{n\to\infty} \int_{-\infty}^{\infty} \left| \int_{-\infty}^{\infty} f(x+\omega)\,\Psi(\omega)\,d\omega - \sum_{k=-n^2}^{n^2-1} f\left(x+\frac{k}{n}\right) \int_{\frac{k}{n}}^{\frac{k+1}{n}} \Psi(\omega)\,d\omega \right| dx \\ = 0. \end{aligned}
$$

Combining this with (2.52) and Lemma IIi, we obtain Theorem II.

With only a slight modification of detail, we have proved:

THEOREM III. *Let $f_1(x)$ and $f_2(x)$ be two functions of L_1. Let*

$$
(2.54)\quad g_1(u) = \frac{1}{\sqrt{2\pi}} \int_{-\infty}^{\infty} f_1(x)\,e^{iux}\,dx
$$

and

$$
(2.55)\quad g_2(u) = \frac{1}{\sqrt{2\pi}} \int_{-\infty}^{\infty} f_2(x)\,e^{iux}\,dx.
$$

Let the set of points where $g_2(u) \neq 0$ consist only of inner points of the set where $g_1(u) \neq 0$. Then if $\varepsilon > 0$ there is a polynomial

$$
(2.56)\quad f_3(x) = \sum_{1}^{N} a_n f_1(x+\lambda_n)
$$

such that

$$
(2.57)\quad \int_{-\infty}^{\infty} |f_2(x) - f_3(x)|\,dx < \varepsilon.
$$

On the basis of Theorem III, we are in a position to prove the general:

THEOREM IV. *Let Σ be a class of functions of L_1. Then a necessary and sufficient condition for the class Σ_1, containing all functions $f(x+\lambda)$ whenever $f(x)$ belongs to Σ, to be closed L_1, is that there should be no real zero common to all the Fourier transforms of functions of Σ.*

The proof of the necessity of the condition of Theorem IV does not differ from the similar proof in the case of Theorem II. As to the sufficiency, it follows at once from Theorem III and the fact that the Fourier transform of a function of L_1 differs from 0 on an open set, that for every u, an ε may be assigned, such that

$$
(2.58)\quad e^{iux}\frac{\sin^2 \varepsilon x}{x^2}
$$

is a function approximable L_1 by polynomials in functions of Σ_1. Let $(-U, U)$ be any interval of frequencies. By the Heine-Borel theorem, this may be overlaid with a finite number of overlapping intervals $U_n - \varepsilon_n$, $U_n + \varepsilon_n$. Let us form

$$\sum e^{iU_n x} \frac{\sin^2 \varepsilon_n x}{x^2} \tag{2.59}$$

for this set of intervals. This will be a function of L_1, approximable L_1 by polynomials in the function of Σ_1, whose Fourier transform may be shown to be

$$\sum \frac{\varepsilon_n}{4}\left(1-2\left|\frac{U_n+u}{\varepsilon_n}\right|\right)\left[\operatorname{sgn}\left(1-2\left|\frac{U_n+u}{\varepsilon_n}\right|\right)+1\right] \tag{2.60}$$

and to have no zeros over $(-U, U)$. Thus by Theorem III, every function of L_1 with a Fourier transform vanishing outside $(-U+\overset{\varepsilon}{n}, U-\overset{\varepsilon}{n})$ will be approximable L_1 by polynomials in the functions of Σ_1. At this stage, the introduction of Lemma IIi serves to complete the proof of Theorem IV.

In Theorems II and IV, the necessary part is nearly trivial and all the difficulty resides in the proof of sufficiency. There are certain applications, however, where it is precisely the necessary part of the theorem that is significant. We may prove a function to have no zeros if it is the Fourier transform of a function of L_1 the set of whose translations is closed L_1. To establish this closure is indeed more than is needed: it is enough to produce for each u among the class of functions to which we may approximate L_1 by polynomials in the translations of our given functions, at least one function whose Fourier transform does not vanish for the argument u.

It may be shown if $F(x)$ is a non-negative function for which

$$\int_{-\infty}^{\infty} F(x)\, dx > A \tag{2.61}$$

and

$$\left[\int_B^{\infty} + \int_{-\infty}^{-B}\right] F(x)\, dx < \varepsilon A, \tag{2.62}$$

then

$$\left|\int_{-\infty}^{\infty} F(x)\, e^{iux}\, dx\right| > \left(\frac{1}{2} - \frac{3\varepsilon}{2}\right) A \qquad \left[|u| \leqq \frac{\pi}{3B}\right]. \tag{2.63}$$

This results from the fact that for $-B \leqq \lambda \leqq B$,

$$\Re(e^{iux}) = \cos ux \geqq \cos\frac{\pi}{3} = \frac{1}{2}.$$

Accordingly, if for every value of B and ε, we may find a function of this sort which may be approximated L_1 by polynomials in the translations of $f(x)$, the Fourier transform of $f(x)$ has no zeros. As in the case of Theorem IV, the function $f(x)$ may be replaced by the class Σ, and the class of translations of $f(x)$ by Σ_1. Then the condition that the Fourier transform of $f(x)$ shall have no zeros is to be replaced by the condition that there shall be no zero common to all the Fourier transforms of functions of Σ.

This method is a conceivable one for the investigation of the zeros of functions with known Fourier transforms, and might be applied to the study of the zeros of the Riemann zeta function. So far it has yielded no results. Under certain conditions, the Taylor series may be introduced to reduce the question of the closure of the set of translations of $f(x)$ to that of the closure of the set of functions

$$f^{(n)}(x). \tag{2.65}$$

This matter has not yet, however, been subjected to an adequate investigation.

3. A sub-class of L_1. In connection with Stieltjes distributions of mass, we shall have to consider "kernels" of class L_1, fulfilling a certain more restrictive condition. This condition on a measurable function $f(x)$ is that the series

$$\sum_{k=-\infty}^{\infty} \overline{\lim_{kA+B\leq x\leq (k+1)A+B}} |f(x)| \tag{3.01}$$

shall converge. We may easily verify that this condition is in fact independent of A and B, and that it can only be fulfilled by functions of L_1. With it goes a certain definition of the "distance" between two functions $f(x)$ and $g(x)$, this "distance" being defined as

$$(f(x), g(x)) = \sum_{-\infty}^{\infty} \overline{\lim_{k\leq x\leq k+1}} |f(x)-g(x)|. \tag{3.02}$$

This notion of "distance" has a very important difference from

$$\int_{-\infty}^{\infty} |f(x)-g(x)|^2 dx \tag{3.03}$$

and

$$\int_{-\infty}^{\infty} |f(x)-g(x)| dx \tag{3.04}$$

in that it is not true that if (3.01) is finite, then

$$\lim_{\varepsilon\to 0} \sum_{k=-\infty}^{\infty} \overline{\lim_{k\leq x\leq k+1}} |f(x+\varepsilon)-f(x)| = 0. \tag{3.05}$$

Accordingly, if we are to prove theorems analogous to those in the last paragraph without a radical change in method of proof, we must introduce a restriction which will make (3.05) hold. Otherwise we shall be unable to establish the analogue of Lemma IIg.

A condition of the desired sort is that $f(x)$ shall be continuous. It will then be uniformly continuous over any finite interval, and we shall be able, first to choose A so large that

$$\sum_{k=0}^{\infty}\left\{\max_{A+k\leq x\leq A+k+1}|f(x)|+\max_{-A-k-1\leq x\leq -A-k}|f(x)|\right\}<\varepsilon \tag{3.06}$$

and then to choose η so small that for $-A-1\leqq x\leqq A+1$, $0<\eta_1<\eta$,

$$\max|f(x+\eta_1)-f(x)|<\frac{\varepsilon}{2A+2}. \tag{3.07}$$

It will follow that for $0<\eta_1<\eta$,

$$\sum_{k=-\infty}^{\infty}\max_{k\leq x\leq k+1}|f(x+\varepsilon)-f(x)|<3\varepsilon. \tag{3.08}$$

We shall call the class of all continuous functions for which (3.01) is finite, the class M_1, and we shall say that a class Σ of functions of M_1 is *closed* M_1, if whenever $F(x)$ belongs to M_1 and $\varepsilon>0$, there is a polynomial

$$F_1(x)=\sum_{h=1}^{n}A_h f_h(x) \tag{3.09}$$

in functions of the class Σ such that

$$\sum_{k=-\infty}^{\infty}\max_{k\leq x\leq k+1}|F(x)-F_1(x)|<\varepsilon. \tag{3.10}$$

The following theorem is valid:

THEOREM V. *Let $f(x)$ belong to L_1. Then a necessary and sufficient condition that whenever $f_1(x)$ is a function of M_1 and $\varepsilon>0$, there shall exist a function $\alpha(x)$ of M_1 such that*

$$\sum_{k=-\infty}^{\infty}\max_{k\leq x\leq k+1}\left|f_1(x)-\int_{-\infty}^{\infty}f(y)\,\alpha(x-y)\,dy\right|<\varepsilon \tag{3.11}$$

is that for no real u

$$\frac{1}{\sqrt{2\pi}}\int_{-\infty}^{\infty}f(x)\,e^{iux}\,dx=0. \tag{3.12}$$

The proof differs in no essential respect from that of Theorem II. Let us note that if $f_2(x)$ belongs to L_1 and

$$(3.13)\qquad \frac{1}{\sqrt{2\pi}}\int_{-\infty}^{\infty} f_2(x)\, e^{iux}\, dx = 0, \qquad [|u| > A]$$

then $f_2(x)$ belongs to M_1. To see this, let us put

$$(3.14)\qquad \frac{1}{\sqrt{2\pi}}\int_{-\infty}^{\infty} f_2(x)\, e^{-iux}\, dx = g_2(u);$$

$g_2(u)$ is bounded and differs from 0 only over a bounded range, and hence belongs to L_2. Thus

$$(3.15)\qquad \underset{B\to\infty}{\text{l.i.m.}}\ \frac{1}{\sqrt{2\pi}}\int_{-B}^{B} g_2(u)\, e^{iux}\, du = f_3(x)$$

exists as a function of L_2. Further

$$(3.16)\qquad g_2(u) = \underset{B\to\infty}{\text{l.i.m.}}\ \frac{1}{\sqrt{2\pi}}\int_{-B}^{B} f_3(x)\, e^{-iux}\, dx$$

exists and belongs to L_2. If we put

$$(3.17)\qquad F(x) = \int_{-\infty}^{\infty} G(u)\, e^{iux}\, du$$

where G belongs both to L_2 and L_1, it follows readily from (3.14) and (3.16) that

$$(3.18)\qquad \int_{-\infty}^{\infty} F(x)\,[f_2(x) - f_3(x)]\, dx = 0.$$

We can choose a set of F's vanishing outside a given finite interval and closed L_2 and hence L_1 over this interval. Then over this interval, which is arbitrary, f_2 and f_3 can differ at most on a null set. Thus f_2 belongs to L_2.

Now

$$(3.19)\qquad g_2(u) = g_2(u)\, W(u)$$

where

$$(3.20)\qquad W(u) = \begin{cases} 0\ ; & [|u| > 2A] \\ 2 - \dfrac{|u|}{A}\ ; & [A < |u| < 2A] \\ 1\ . & [|u| < A] \end{cases}$$

Thus by the Parseval theorem,

$$(3.21)\qquad \begin{aligned} f_2(x) &= \frac{1}{2\pi}\int_{-\infty}^{\infty} f_2(y)\, dy \int_{-\infty}^{\infty} W(u)\, e^{iu(x-y)}\, du \\ &= \frac{1}{\pi A}\int_{-\infty}^{\infty} f_2(y)\, \frac{\cos A(x-y) - \cos 2A(x-y)}{(x-y)^2}\, dy. \end{aligned}$$

From this it follows that

$$(3.22)\quad \sum_{k=-\infty}^{\infty} \max_{k \leq x \leq k+1} |f_2(x)| \leq \frac{1}{\pi A} \int_{-\infty}^{\infty} |f_2(y)|\, dy \sum_{-\infty}^{\infty} O\left(\frac{1}{k-m}\right)^2 \leq O(1) \int_{-\infty}^{\infty} |f_2(y)|\, dy$$

and that $f_2(x)$, which obviously can be modified so as to be continuous, belongs to M_1.

It only remains to prove the analogue of Lemma IIi, and to show that every function of class M_1 may be approximated with any degree of accuracy in the M_1 sense by functions with Fourier transforms vanishing for large arguments. This follows exactly the lines suggested by IIi. We wish to show, that is, that if $f(x)$ belongs to M_1,

$$(3.221)\quad \lim_{N\to\infty} \left(f(x), \frac{1}{\pi N} \int_{-\infty}^{\infty} f(x+y) \frac{\sin^2 Ny}{y^2}\, dy \right) \equiv 0.$$

As before, we put

$$(3.222)\quad \Phi_1(y) = \begin{cases} \dfrac{\sin^2 Ny}{y^2}(1 - |y|\, N^{1/2}); & [|y| \leq N^{-1/2}] \\ 0 & [y > N^{-1/2}] \end{cases}$$

$$(3.223)\quad \Phi_2(y) = \frac{\sin^2 Ny}{y^2} - \Phi_1(y).$$

The proof that

$$(3.224)\quad \lim_{N\to\infty} \left(f(x), \frac{1}{\pi N} \int_{-\infty}^{\infty} f(x+y)\, \Phi_1(y)\, dx \right) = 0$$

follows exactly the lines of that of (2.41), for the analogue of Lemma IIg has already been shown to be true, while that of Lemma IIh is proved by the same arguments of absolute convergence as the lemma itself. Again,

$$(3.225)\quad \lim_{N\to\infty} \sum_{k=-\infty}^{\infty} \max_{k \leq x \leq k+1} \left| \frac{1}{\pi N} \int_{-\infty}^{\infty} f(x+y)\, \Phi_2(y)\, dy \right| \leq 2 \sum_{k=-\infty}^{\infty} \max_{k \leq x \leq k+1} |f(x)| \lim_{N\to\infty} \frac{1}{\pi N} \int_{-\infty}^{\infty} \Phi_2(y)\, dy,$$

and this, by (2.421), is zero. This completes the proof of (3.221).

An extension of Theorem V in the direction of Theorem IV is the following:

THEOREM VI. *Let Σ be a class of functions of L_1. Let Σ_1 be the class of all functions of the form*

$$\sum_{1}^{N} \int_{-\infty}^{\infty} f_n(y)\, \alpha_n(x+y)\, dy \tag{3.23}$$

where $f_1, \cdots, f_N$ belong to Σ and $\alpha_1, \cdots, \alpha_N$ to M_1. Then Σ_1 is closed M_1 when and only when there is no real value of u which is a zero common to all the functions

$$\frac{1}{\sqrt{2\pi}} \int_{-\infty}^{\infty} f_n(x)\, e^{iux}\, dx \tag{3.24}$$

corresponding to functions $f_n(x)$ of class Σ.

As an easy corollary of Theorem V, using (3.10), we have:

THEOREM VII. *If $f(x)$ is a function of M_1, a necessary and sufficient condition for the set of its translations to be closed M_1 is that for no u should we have*

$$\frac{1}{\sqrt{2\pi}} \int_{-\infty}^{\infty} f(x)\, e^{iux}\, dx = 0. \tag{3.25}$$

CHAPTER II.

ASYMPTOTIC PROPERTIES OF AVERAGES[26].

4. Averages of bounded functions. Our fundamental theorem is:

THEOREM VIII. *Let $f(x)$ be a bounded measurable function, defined over $(-\infty, \infty)$. Let $K_1(x)$ be a function in L_1, and let*

$$\frac{1}{\sqrt{2\pi}} \int_{-\infty}^{\infty} K_1(x)\, e^{iux}\, dx \neq 0 \tag{4.01}$$

for every real u. Let

$$\lim_{x\to\infty} \int_{-\infty}^{\infty} f(\xi)\, K_1(\xi - x)\, d\xi = A \int_{-\infty}^{\infty} K_1(\xi)\, d\xi. \tag{4.02}$$

Then if $K_2(x)$ is any function in L_1,

$$\lim_{x\to\infty} \int_{-\infty}^{\infty} f(\xi)\, K_2(\xi - x)\, d\xi = A \int_{-\infty}^{\infty} K_2(\xi)\, d\xi. \tag{4.03}$$

Conversely, let $K_1(\xi)$ be a function of L_1, and let $\int_{-\infty}^{\infty} K_1(\xi)\, d\xi \neq 0$. Let (4.02) *imply* (4.03) *whenever $K_2(x)$ belongs to L_1 and $f(x)$ is bounded. Then* (4.01) *holds.*

---

[26] The emphasis here placed on averages is in the same order of ideas as was first introduced into the theory by Schmidt (1) (2).

As to the first part of Theorem VIII, it is clear that it is valid whenever $K_2(x)$ is of the form $\sum_{k=1}^{n} A_k K_1(x+\lambda_k)$. It is also clear that if

$$|f(x)| \leqq B; \quad \int_{-\infty}^{\infty} |K_2(x)-K_3(x)|\, dx \leqq \varepsilon, \tag{4.04}$$

then

$$\left| \int_{-\infty}^{\infty} f(\xi)\, K_2(\xi - x)\, d\xi - \int_{-\infty}^{\infty} f(\xi)\, K_3(\xi - x)\, d\xi \right| \leqq B\varepsilon.$$

An application of Theorem II completes the proof.

As to the second part of the proof, it is merely necessary to suppose that

$$\frac{1}{\sqrt{2\pi}} \int_{-\infty}^{\infty} K_1(x)\, e^{iu_1 x}\, dx = 0 \tag{4.05}$$

and to take

$$f(x) = e^{iu_1 x} \tag{4.06}$$

to obtain a contradiction.

Theorem VIII has an extension in the sense of Theorem III, in which $K_1(x)$ is replaced by a class of functions Σ_1 (4.02) holds for every function of the class, and there is no u for which for *every* function $K_1(x)$ of Σ_1

$$\frac{1}{\sqrt{2\pi}} \int_{-\infty}^{\infty} K_1(x)\, e^{iux}\, dx = 0. \tag{4.07}$$

The converse of this extended theorem is also valid.

5. Averages of bounded Stieltjes distributions. Here our theorem is:

THEOREM IX. *Let $f(x)$ be a function of limited total variation over every finite range, and let*

$$\int_{y}^{y+1} |d f(x)| \tag{5.01}$$

be bounded in y. Let $K_1(x)$ be a continuous function of L_1, and let

$$\sum_{k=-\infty}^{\infty} \overline{\lim_{k \leq x \leq k+1}} |K_1(x)| \tag{5.02}$$

converge. Let

$$\frac{1}{\sqrt{2\pi}} \int_{-\infty}^{\infty} K_1(x) e^{iux}\, dx \neq 0 \qquad [-\infty < u < \infty] \tag{4.01}$$

and let

$$\lim_{x \to \infty} \int_{-\infty}^{\infty} K_1(\xi - x)\, df(\xi) = A \int_{-\infty}^{\infty} K_1(\xi)\, d\xi. \tag{5.03}$$

Then if $K_2(x)$ is any function of M_1,

$$\lim_{x \to \infty} \int_{-\infty}^{\infty} K_2(\xi - x)\, df(\xi) = A \int_{-\infty}^{\infty} K_2(\xi)\, d\xi. \tag{5.04}$$

Conversely, let (5.02) *converge. Let* $\int_{-\infty}^{\infty} K_1(x)\,dx \neq 0$. *Let* (5.03) *imply* (5.04) *for every function of* K_2 *of* M_1 *and every* $f(x)$ *for which* (5.01) *is bounded. Then* (4.01) *holds.*

To prove this theorem we need the following elementary:

LEMMA IXa. *Let*

$$\int_{y-0}^{y+1+0} |df(x)| \leqq B \qquad [-\infty < y < \infty] \tag{5.05}$$

and let

$$\sum_{k=-\infty}^{\infty} \overline{\lim_{k \leqq x \leqq k+1}} |K_2(x) - K_3(x)| \leqq \epsilon. \tag{5.06}$$

Then

$$\left| \int_{-\infty}^{\infty} K_2(\xi - x)\,df(x) - \int_{-\infty}^{\infty} K_3(\xi - x)\,df(x) \right| \leqq B\epsilon. \tag{5.061}$$

With this lemma at our disposal, the proof of Theorem IX does not differ in any important respect from that of Theorem VIII. Of course, as in Theorem VI, we must replace the polynomials in translations of K_1 that figure in Theorem VIII by absolutely convergent integrals in these translations. The proof of the lemma itself is immediate.

The modification of Theorem IX with a hypothesis involving a whole class of kernels will be used later, so we shall formulate it as a separate theorem:

THEOREM X. *Let* $f(x)$ *be a function of limited total variation over every finite range, and let*

$$\int_{y}^{y+1} |df(x)| \tag{5.01}$$

be bounded in y. *Let* Σ *be a class of continuous functions of* L_1, *each one of which, for example* $K_1(x)$, *has the properties that*

$$\sum_{k=-\infty}^{\infty} \overline{\lim_{k \leqq x \leqq k+1}} |K_1(x)| \tag{5.02}$$

converges, and that

$$\lim_{x \to \infty} \int_{-\infty}^{\infty} K_1(\xi - x)\,df(\xi) = A \int_{-\infty}^{\infty} K_1(\xi)\,d\xi. \tag{5.03}$$

Let there be no u *which is a real zero for all the functions*

$$\frac{1}{\sqrt{2\pi}} \int_{-\infty}^{\infty} K_1(x)\,e^{iux}\,dx \tag{5.07}$$

for which $K_1(x)$ *belongs to* Σ. *Then if* $K_2(x)$ *is any function belonging to* M_1,

$$\lim_{x\to\infty}\int_{-\infty}^{\infty} K_2(\xi - x)\, df(\xi) = A\int_{-\infty}^{\infty} K_2(\xi)\, d\xi. \tag{5.04}$$

Conversely, if the class Σ *contains at least one function with non zero integral and has the property that each member of it is a member of* L_1 *and satisfies the condition that* (5.02) *shall converge, and if, whenever* $f(x)$ *satisfies the condition that* (5.01) *be bounded, and* (5.03) *for every member of* Σ, *then* (5.04) *holds for every function* $K_2(x)$ *belonging to* M_1, *then there is no u such that* (4.07) *holds for every function* $K_1(x)$ *belonging to* Σ.

6. **Averages of unilaterally bounded distributions and functions.** Let $K_1(x)$ not be equivalent to 0, and let it be a continuous function (or a function continuous except for a finite number of finite jumps) not identically zero and such that

$$K_1(x) \geqq 0, \quad \sum_{k=-\infty}^{\infty} \max_{k \leqq x \leqq k+1} K_1(x) < p \qquad (-\infty < x < \infty). \tag{6.01}$$

We shall say that the mass-distribution determined by $f(x)$ is *bounded below* when

$$\int_y^{y+1} |df(x)| - f(y+1) + f(y) \leqq N \qquad (-\infty < x < \infty) \tag{6.02}$$

and that it is *bounded above* when $-f(x)$ determines a mass-distribution bounded below. We shall prove:

LEMMA XIa. *Let the distribution corresponding to* $f(x)$ *be bounded below (or above). Let* (6.01) *hold, and let*

$$\overline{\int_{-\infty}^{\infty}} K_1(\xi - x)\, df(\xi) \leqq M; \quad (\geqq M); \qquad (-\infty < x < \infty). \tag{6.03}$$

Then there is a Q such that

$$\int_y^{y+1} |df(x)| \leqq Q \qquad (-\infty < y < \infty). \tag{6.04}$$

To prove this, let us notice that

$$\begin{aligned} \overline{\int_{-\infty}^{\infty}} K_1(\xi - x)\, d\int_{\xi_0}^{\xi} |df(y)| &\leqq \overline{\int_{-\infty}^{\infty}} K_1(\xi - x)\, d\left[\int_{\xi_0}^{\xi} |df(y)| - f(\xi)\right] \\ &+ \overline{\int_{-\infty}^{\infty}} K_1(\xi - x)\, df(\xi) \leqq N \sum_{k=-\infty}^{\infty} \max_{k \leqq x \leqq k+1} K(\xi) + M. \end{aligned} \tag{6.05}$$

Since K is "stückweise stetig", B, a and b exist such that

$$K_1(u) > B > 0 \qquad (a \leqq u \leqq b). \tag{6.06}$$

Hence

$$M + N \sum_{k=-\infty}^{\infty} \max_{k \leqq x \leqq k+1} K_1(\xi) \geqq B \int_{a+x}^{b+x} d\int_{\xi_0}^{\xi} |df(y)|, \tag{6.07}$$

or in other terms

$$\frac{1}{B}\left\{\left[\frac{1}{b-a}\right]+1\right\}\left\{M+N\sum_{k=-\infty}^{\infty}\max_{k\leq x\leq k+1}K_1(\xi)\right\} \geqq \int_y^{y+1}|df(x)| \qquad (-\infty<y<\infty). \tag{6.08}$$

If we combine this lemma with Theorem X, we get:

THEOREM XI. *Let $f(x)$ be a function of limited total variation over every finite interval, for which*

$$\int_y^{y+1}|df(x)|-f(y+1)+f(y)\leqq N \qquad (-\infty<y<\infty). \tag{6.02}$$

Let Σ be a class of continuous functions K_1, for each of which

$$\sum_{k=-\infty}^{k=\infty}\max_{k\leq x\leq k+1}|K_1(x)| \tag{5.02}$$

converges, and let

$$\lim_{x\to\infty}\int_{-\infty}^{\infty}K_1(\xi-x)\,df(\xi) = A\int_{-\infty}^{\infty}K_1(\xi)\,d\xi \tag{5.03}$$

for every K_1 belonging to Σ. Let there be no real u for which every

$$\frac{1}{\sqrt{2\pi}}\int_{-\infty}^{\infty}K_1(x)\,e^{iux}\,dx \tag{5.07}$$

vanishes, for which K_1 belongs to Σ. Let $Q(x)$ be a continuous function belonging to Σ, for which

$$Q(x)\geq 0, \qquad \left|\int_{-\infty}^{\infty}Q(\xi-x)\,df(x)\right|\leqq T \qquad (-\infty<x<\infty). \tag{6.09}$$

Then if $K_2(-x)$ is any function belonging to M_1,

$$\lim_{x\to\infty}\int_{-\infty}^{\infty}K_2(\xi-x)\,df(\xi) = A\int_{-\infty}^{\infty}K_2(\xi)\,d\xi. \tag{5.04}$$

We shall have frequent occasion to use Theorems X and XI in a form in which the infinite range of x is mapped on a semi-infinite range by an exponential transformation. This may happen in two ways: either 0 or ∞ for the new argument may correspond to $+\infty$ for x. In the first case, let us write:

$$\xi=-\log\lambda,\quad f(\xi)=\int\lambda^{-1}\,d\varphi(\lambda),\quad K_{1,2}(\xi)=\lambda N_{1,2}(\lambda),\quad Q(\xi)=\lambda M(\lambda) \tag{6.10}$$

and in the second,

$$\xi=\log\lambda,\quad f(\xi)=\int\lambda^{-1}\,d\varphi(\lambda),\quad K_{1,2}(\xi)=\lambda N_{1,2}(\lambda),\quad Q(\xi)=\lambda M(\lambda). \tag{6.11}$$

Then Theorem XI yields:

THEOREM XI′. *Let $\varphi(\lambda)$ be a function of limited total variation over every interval $(\varepsilon, 1/\varepsilon)$ where $0<\varepsilon<1$. Let $\varphi(0) = 0$ and let*

$$\int_u^{2u} \lambda^{-1} \, |d\varphi(\lambda)| - \int_u^{2u} \lambda^{-1} \, d\varphi(\lambda) \leqq N \qquad (0<u<\infty). \tag{6.12}$$

Let Σ be a class of continuous functions $N_1(\lambda)$ for each of which

$$\sum_{k=-\infty}^{\infty} \max_{2^k \leqq \lambda \leqq 2^{k+1}} \lambda \, |N_1(\lambda)| \tag{6.13}$$

converges, and let

$$\lim_{\lambda\to 0\,(\infty)} \frac{1}{\lambda} \int_0^\infty N_1\left(\frac{\mu}{\lambda}\right) d\varphi(\mu) = A \int_0^\infty N_1(\mu)\, d\mu \tag{6.14}$$

for every $N_1(\lambda)$ belonging to Σ. Let there be no real u for which every

$$\int_0^\infty N_1(\lambda)\, \lambda^{iu}\, d\lambda \tag{6.15}$$

vanishes, for which N_1 belongs to Σ. Let $M(\lambda)$ be a function belonging to Σ, for which

$$M(\lambda) \geqq 0, \quad \left| \frac{1}{\lambda} \int_0^\infty M\left(\frac{\mu}{\lambda}\right) d\varphi(\mu) \right| \leqq \text{const.}\ \textit{for}\ 0 \leqq \lambda < \infty. \tag{6.16}$$

Then if $N_2(\lambda)$ is any continuous function for which

$$\sum_{k=-\infty}^{\infty} \max_{2^k \leqq \lambda \leqq 2^{k+1}} \lambda \, |N_2(\lambda)| \tag{6.17}$$

converges,

$$\lim_{\lambda\to 0\,(\infty)} \frac{1}{\lambda} \int_0^\infty N_2\left(\frac{\mu}{\lambda}\right) d\varphi(\mu) = A \int_0^\infty N_2(\mu)\, d\mu. \tag{6.18}$$

In the case where

$$\varphi(\lambda) = \int_{\lambda_0}^{\lambda} f(x)\, dx, \tag{6.19}$$

(6.12) will be satisfied if

$$f(\lambda) > -\frac{N}{2 \log 2}. \tag{6.20}$$

If the other hypotheses of Theorem XI′ are satisfied, (6.18) will become

$$\lim_{\lambda\to 0(\infty)} \frac{1}{\lambda} \int_0^\infty N_2\left(\frac{\mu}{\lambda}\right) f(\mu)\, d\mu = A \int_0^\infty N_2(\mu)\, d\mu. \tag{6.21}$$

As a particular admissible N_2, we may take

$$(6.22)\qquad N_2(\mu) = \begin{cases} 1 & ; \quad [0 \leqq \mu < 1] \\ 1 - \dfrac{\mu - 1}{\varepsilon} & ; \quad [1 \leqq \mu < 1 + \varepsilon] \\ 0 & . \quad [1 + \varepsilon \leqq \mu] \end{cases}$$

Thus (6.21) will assume the form

$$(6.23)\qquad \lim_{\lambda \to 0(\infty)} \frac{1}{\lambda} \left[\int_0^\lambda f(\mu)\, d\mu + \int_\lambda^{\lambda(1+\varepsilon)} \left(1 - \frac{\mu - \lambda}{\lambda \varepsilon}\right) f(\mu)\, d\mu \right] = A \left(1 + \frac{\varepsilon}{2}\right).$$

Now, by (6.20)

$$(6.24)\qquad \frac{1}{\lambda} \int_\lambda^{\lambda(1+\varepsilon)} \left(1 - \frac{\mu - \lambda}{\lambda \varepsilon}\right) f(\mu)\, d\mu \geqq \frac{1}{\lambda} \int_\lambda^{\lambda(1+\varepsilon)} \left(1 - \frac{\mu - \lambda}{\lambda \varepsilon}\right) \left(\frac{-N}{2 \log 2}\right) d\mu$$
$$= \frac{-N\varepsilon}{4 \log 2}.$$

Hence by (6.23)

$$(6.25)\qquad \overline{\lim_{\lambda \to 0(\infty)}} \frac{1}{\lambda} \int_0^\lambda f(\mu)\, d\mu \leqq A \left(1 + \frac{\varepsilon}{2}\right) + \frac{N\varepsilon}{4 \log 2}.$$

Since ε is arbitrarily small,

$$(6.25)\qquad \overline{\lim_{\lambda \to 0(\infty)}} \frac{1}{\lambda} \int_0^\lambda f(\mu)\, d\mu \leqq A.$$

Again, we may write (6.21) in the form

$$(6.26)\qquad \lim_{\lambda \to 0(\infty)} \frac{1}{\lambda(1+\varepsilon)} \left[\int_0^{\lambda(1+\varepsilon)} f(\mu)\, d\mu - \int_\lambda^{\lambda(1+\varepsilon)} \frac{\mu - \lambda}{\lambda \varepsilon} f(\mu)\, d\mu \right] = \frac{A(1 + \varepsilon/2)}{1 + \varepsilon}$$

from which we may conclude as above that

$$(6.27)\qquad \underline{\lim_{\lambda \to 0(\infty)}} \frac{1}{\lambda} \int_0^\lambda f(\mu)\, d\mu \geqq A.$$

Combining (6.25) and (6.27), we see that

$$(6.28)\qquad \lim_{\lambda \to 0(\infty)} \frac{1}{\lambda} \int_0^\lambda f(\mu)\, d\mu = A.$$

We thus get

THEOREM XI″. *Let $f(x)$ be a function bounded over every interval $(\varepsilon, 1/\varepsilon)$, where $0 < \varepsilon < 1$. Let*

$$(6.29)\qquad f(\lambda) > -K \ (\textit{or}\ f(\lambda) < K)$$

for every argument. Let Σ be a class of continuous functions $N_1(\lambda)$ for each of which (6.13) *converges, and let*

$$\text{(6.30)} \qquad \lim_{\lambda\to 0(\infty)} \frac{1}{\lambda}\int_0^\infty N_1\left(\frac{\mu}{\lambda}\right) f(\mu)\,d\mu = A\int_0^\infty N_1(\mu)\,d\mu$$

for every $N_1(\lambda)$ *belonging to* Σ. *Let there be no real* u *for which every expression* (6.15) *vanishes with* N_1 *belonging to* Σ. *Let* $M(\lambda)$ *belong to* Σ, *and let*

$$\text{(6.31)} \quad M(\lambda) \geqq 0, \frac{1}{\lambda}\int_0^\infty M\left(\frac{\mu}{\lambda}\right) f(\mu)\,d\mu \leqq \text{const. } for\ 0 \leqq \lambda < \infty.$$

Then (6.28) *is true.*

Another case where a stricter conclusion than (6.18) may be drawn is where $\varphi(\lambda)$ is monotone. In this case, (6.12) is automatically satisfied. If we take $N_2(\mu)$ as in (6.22), we shall be able to write (6.18) in the form

$$\text{(6.32)} \quad \lim_{\lambda\to 0(\infty)} \frac{1}{\lambda}\left[\varphi(\lambda) + \int_\lambda^{\lambda(1+\varepsilon)}\left(1 - \frac{\mu-\lambda}{\lambda\varepsilon}\right) d\varphi(\mu)\right] = A\left(1+\frac{\varepsilon}{2}\right)$$

and hence

$$\text{(6.33)} \qquad \overline{\lim_{\lambda\to 0(\infty)}} \frac{\varphi(\lambda)}{\lambda} \leqq A.$$

From the analogue of (6.26) it follows that

$$\text{(6.34)} \qquad \underline{\lim_{\lambda\to 0(\infty)}} \frac{\varphi(\lambda)}{\lambda} \geqq A,$$

and hence that

$$\text{(6.35)} \qquad \varphi(\lambda) \sim \lambda A.$$

We thus obtain

THEOREM XI‴. *If in the hypothesis of Theorem* XI′, (6.12) *is replaced by the condition that* $\varphi(\lambda)$ *is monotone. Then* (6.35) *follows.*

It is even possible to weaken the hypothesis here given, and to replace (6.12) by the condition that

$$\text{(6.36)} \quad \lim_{\varepsilon\to+0} \overline{\lim_{-\infty<\mu<\infty}} \left[\int_\mu^{\mu(1+\varepsilon)} \lambda^{-1}\,|d\varphi(\lambda)| - \int_\mu^{\mu(1+\varepsilon)} \lambda^{-1}\,d\varphi(\lambda)\right] = 0$$

to establish (6.25).

CHAPTER III.

TAUBERIAN THEOREMS AND THE CONVERGENCE OF SERIES AND INTEGRALS.

7. The Hardy-Littlewood condition. We now enter upon the realm of ideas which has longest been associated with Tauberian theorems.[27] The class Σ of the last paragraph now consists of all functions of the form

[27] Hardy (2); Hardy and Littlewood (4), (10), (13) etc.; Littlewood (1); Landau (9), (4).

$$M(x) - M(x+\lambda) \tag{7.01}$$

for which we have

$$M(-\infty) = 1, \quad M(\infty) = 0, \quad \int_{-\infty}^{\infty} M(\xi - x)\, df(\xi) \quad \text{bounded.} \tag{7.02}$$

Condition (5.02) now asserts the convergence of

$$\sum_{k=-\infty}^{\infty} \max_{k \leq x \leq k+1} |M(x) - M(x+\lambda)| \tag{7.03}$$

which will be automatically fulfilled if $M(x)$ is monotone and satisfies (7.02). We shall also suppose $M(x)$ continuous. Condition (5.07) becomes the condition that

$$\frac{1}{\sqrt{2\pi}} \int_{-\infty}^{\infty} [M(x) - M(x+\lambda)]\, e^{iux}\, dx \tag{7.04}$$

shall not vanish for any real u for every λ. Let us suppose M to vanish exponentially at $+\infty$, and let

$$\frac{1}{\sqrt{2\pi}} \int_{-\infty}^{\infty} M(x)\, e^{xz}\, dx = \varphi(z). \tag{7.05}$$

$\varphi(z)$ will clearly be analytic over some vertical strip to the right of the origin. We shall obviously have

$$\frac{1}{\sqrt{2\pi}} \int_{-\infty}^{\infty} [M(x) - M(x+\lambda)]\, e^{xz}\, dz = \varphi(z)\,(1 - e^{-\lambda z}) \tag{7.06}$$

over this strip, and by analytic continuation, which we assume to be possible, the non-vanishing of (7.04) becomes the non-vanishing of $\varphi(z)$ over the imaginary axis.

If

$$\lim_{x \to \infty} \int_{-\infty}^{\infty} M(\xi - x)\, df(\xi) = B, \tag{7.07}$$

(5.03) is satisfied for $A = 0$. As is obviously permissible, let us put

$$K_2(\xi) = \begin{cases} M(\xi) - 1 \quad ; & (-\infty < \xi < 0) \\ M(\xi) - 1 + \dfrac{\xi}{\varepsilon}; & (0 \leq \xi < \varepsilon) \\ M(\xi) \quad . & (\varepsilon \leq \xi < \infty) \end{cases} \tag{7.08}$$

Then (5.04) assumes the form

$$\lim_{x \to \infty} \int_{-\infty}^{\infty} K_2(\xi - x)\, df(\xi) = 0, \tag{7.09}$$

which becomes

$$\lim_{x \to \infty} \left[\int_{-\infty}^{x} df(\xi) + \int_{x}^{x+\varepsilon} \left(1 - \frac{\xi - x}{\varepsilon}\right) df(\xi) \right] = B. \tag{7.10}$$

If we put

$$f(-\infty) = 0 \tag{7.11}$$

(5.04) assumes the form

$$\lim_{x\to\infty} \frac{1}{\varepsilon} \int_x^{x+\varepsilon} f(\xi)\, d\xi = B. \tag{7.12}$$

Restating Theorem XI in this new form, we get

THEOREM XII. *Let $f(x)$ be a function of limited total variation over every range $(-\infty, A)$, let $f(-\infty) = 0$, and let*

$$\int_y^{y+1} |df(x)| - f(y+1) + f(y) \leq N \qquad (-\infty < y < \infty). \tag{6.02}$$

Let $M(x)$ be a monotonely decreasing continuous function, such that

$$M(-\infty) = 1, \quad M(x) = O(e^{-ux}) \text{ at } \infty \qquad (u > 0). \tag{7.13}$$

Let

$$\frac{1}{\sqrt{2\pi}} \int_{-\infty}^{\infty} M(x)\, e^{xz}\, dx = \varphi(z) \tag{7.05}$$

over a strip to the right of the origin, and let $\varphi(z)$, continued analytically on to the imaginary axis, have no zeros there. (We assume the possibility of this continuation.) *Let*

$$\int_{-\infty}^{\infty} M(\xi - x)\, df(\xi) \tag{7.14}$$

be bounded, and let

$$\lim_{x\to\infty} \int_{-\infty}^{\infty} M(\xi - x)\, df(\xi) = B. \tag{7.07}$$

Then

$$\lim_{x\to\infty} \frac{1}{\varepsilon} \int_x^{x+\varepsilon} f(\xi)\, d\xi = B. \tag{7.12}$$

Let us put

$$M(\log x) = N(x); \quad f(\log x) = F(x).$$

Theorem XII then becomes:

THEOREM XIII. *Let $F(x)$ be a function vanishing at the origin and of limited total variation over any finite range (including the origin), and let for some $\Lambda > 1$,*

$$\int_y^{\Lambda y} |dF(x)| - F(\Lambda y) + F(y) \leq N \qquad (0 < y < \infty). \tag{7.16}$$

Let $N(x)$ be a monotonely decreasing continuous function, such that

$$N(0) = 1, \qquad N(x) = O(x^{-\lambda}) \text{ at } \infty \qquad [\lambda > 0]. \tag{7.17}$$

Let

$$\frac{1}{\sqrt{2\pi}} \int_0^{\infty} N(x)\, x^{z-1}\, dx = \varphi(z) \tag{7.18}$$

over a strip to the right of the origin, and let $\varphi(z)$, *continued analytically on to the imaginary axis, as we assume to be possible for all points but the origin, have no zeros there. Let*

$$\int_0^\infty N\left(\frac{\xi}{x}\right) dF(\xi) \tag{7.19}$$

be bounded, and let

$$\lim_{x\to\infty} \int_0^\infty N\left(\frac{\xi}{x}\right) dF(\xi) = B. \tag{7.20}$$

Then if $A_1 > 1$,

$$\lim_{x\to\infty} \frac{1}{\log A_1} \int_x^{xA_1} F(\xi)\frac{d\xi}{\xi} = B. \tag{7.21}$$

A case of particular interest and importance is where

$$F(x) = \sum_{n=0}^{[x]} a_n. \tag{7.22}$$

Here

$$\begin{aligned} \int_y^{Ay} |dF(x)| - F(Ay) + F(y) &= \sum_{n=[y]+1}^{[Ay]} (-a_n + |a_n|) \\ &\leq \begin{cases} \max 2na_n; \ (n \leq [Ay]), \\ 0. \end{cases} \end{aligned} \tag{7.23}$$

if A is sufficiently near to 1. (7.23) is to be interpreted as meaning that the expression on the left hand side is less than or equal to the greater of the two expressions in brackets.

We thus obtain the

COROLLARY. *If* $N(x)$ *is subject to the conditions of Theorem* XIII, *if*

$$\int_0^\infty N\left(\frac{\xi}{x}\right) dF(\xi) = \sum_{n=0}^{\infty} a_n N\left(\frac{n}{x}\right) \tag{7.24}$$

is bounded, if

$$na_n < K \qquad (\text{or } na_n > -K) \tag{7.25}$$

for all n, *and if*

$$\lim_{x\to\infty} \sum_{n=0}^{\infty} a_n N\left(\frac{n}{x}\right) = B \tag{7.26}$$

then

$$\sum_0^\infty a_n = B. \tag{7.27}$$

Here instead of (7.27), what we directly prove is

$$\lim_{x\to\infty} \frac{1}{\log A_1} \int_x^{xA_1} \sum_{n=0}^{[\xi]} a_n \frac{d\xi}{\xi} = B. \tag{7.28}$$

(7.27) may however be written in the form

$$\lim_{x\to\infty} \frac{1}{\log A_1} \int_x^{xA_1} \sum_{n=0}^{[x]} a_n \frac{d\xi}{\xi} = B. \tag{6.281}$$

Our conclusion (7.27) will then follow if we can show that we can so choose A_1, that

$$\begin{aligned} &\overline{\lim_{x\to\infty}} \frac{1}{\log A_1} \int_x^{xA_1} \sum_{n=[x]+1}^{[\xi]} a_n \frac{d\xi}{\xi} < \varepsilon, \\ &\overline{\lim_{x\to\infty}} \frac{1}{\log A_1} \int_x^{xA_1} \sum_{n=[\xi]+1}^{[xA_1]} a_n \frac{d\xi}{\xi} < \varepsilon. \end{aligned} \tag{7.29}$$

However,

$$\frac{1}{\log A_1} \int_x^{xA_1} \sum_{n=[x]+1}^{[\xi]} a_n \frac{d\xi}{\xi} \leq \frac{1}{\log A_1} \int_x^{xA_1} \frac{K}{x} (xA_1 - x) \frac{d\xi}{\xi} = K(A_1 - 1). \tag{7.30}$$

Again,

$$\frac{1}{\log A_1} \int_x^{xA_1} \sum_{n=[\xi]+1}^{[A_1 x]} a_n \frac{d\xi}{\xi} \leq K(A_1 - 1). \tag{7.31}$$

Since we may take A_1 as near to 1 as we wish, (7.29) is established, and (7.27) follows at once. This type of argument is to be found in the work of Szász.

8. **The Schmidt condition.**[28] This last corollary covers the work of Hardy and Littlewood on Tauberian theorems. An extension of their conditions is due to Robert Schmidt. To arrive at theorems of his type, let M be subject to the conditions stated in the hypothesis to Theorem XII; let us put

$$g(x) = f(x+\alpha) - f(x) \tag{8.01}$$

and let us assume that

$$-f(x) + |f(x)| < A(x + |x|) + c \qquad (A, c > 0). \tag{8.02}$$

As the integrals in question converge absolutely by (8.02) and (7.13), whenever they exist we may invert our order of integration and write:

$$\begin{aligned} \int_y^{y+\alpha} dx \int_{-\infty}^{\infty} M(\xi - x)\, df(\xi) &= \int_y^{y+\alpha} dx \int_{-\infty}^{\infty} M(\xi)\, df(\xi + x) \\ &= \int_{-\infty}^{\infty} M(\xi)\bigl(f(\xi+y+\alpha) - f(\xi+y)\bigr)\, d\xi \\ &= \int_{-\infty}^{\infty} M(\xi - y)\bigl(f(\xi+\alpha) - f(\xi)\bigr)\, d\xi \\ &= \int_{-\infty}^{\infty} M(\xi - y)\, g(\xi)\, d\xi. \end{aligned} \tag{8.03}$$

[28] Schmidt (1), (2). Throughout this section, the author has been strongly influenced by the methods of Vijayaraghavan and Szász.

Thus if $M(x)$ is subject to the conditions in the hypothesis to Theorem XII, if it is bounded above or below, and (7.07) is valid, and (7.14) bounded, whether (6.02) is true or not, it follows from Theorem XII itself on replacing $f(x)$ by $\frac{1}{\alpha}\int_{-\infty}^{x} g(\xi)\,d\xi$ that

$$\begin{aligned} B &= \lim_{y\to\infty} \frac{1}{\alpha}\int_{y}^{y+\alpha} \frac{dx}{\alpha}\int_{-\infty}^{x} g(\xi)\,d\xi \\ &= \lim_{y\to\infty} \frac{1}{\alpha}\int_{y}^{y+\alpha} \frac{dx}{\alpha}\int_{x}^{x+\alpha} f(\xi)\,d\xi \qquad (8.04) \\ &= \lim_{y\to\infty} \frac{1}{\alpha^2}\int_{-\alpha}^{\alpha} f(\xi+\alpha+y)\,(\alpha-|\xi|)\,d\xi. \end{aligned}$$

Let us now introduce Schmidt's notion of a "slowly decreasing" function. Let $f(u)$ have the property that when u and v run through a sequence of pairs of values for which

$$v \geqq u, \qquad v-u \to 0 \tag{8.05}$$

then

$$\underline{\lim}_{u\to\infty} (f(v)-f(u)) \geqq 0; \tag{8.06}$$

we shall call $f(u)$ *slowly decreasing*. (Schmidt treats sequences instead of functions, and on another scale, but the difference is unimportant.) Schmidt shows that we may write

$$f(u) = f_1(u)+f_2(u) \tag{8.07}$$

where $f_2(u)$ is monotone increasing, and $f_1(u)$ satisfies the condition that whenever (8.05) is fulfilled, then

$$\lim_{u\to\infty} (f_1(v)-f_1(u)) = 0. \tag{8.08}$$

He proves that there exists a function $T(\alpha)$ such that

$$\overline{\lim_{\substack{u\to\infty \\ u\leqq v\leqq u+\alpha}}} |f_1(v)-f_1(u)| = T(\alpha) \tag{8.09}$$

and

$$\lim_{\alpha\to 0} T(\alpha) = 0. \tag{8.10}$$

He proves that a number T exists, such that

$$|f_1(v)-f_1(u)| \leqq T\cdot(v-u) \qquad (v>u+\alpha). \tag{8.11}$$

From this, it readily follows that if f is bounded near $-\infty$, (8.02) is valid and $g(x)$ is bounded below. Furthermore,

$$\overline{\lim_{x\to\infty}} \left(f(x)-\frac{1}{\alpha^2}\int_{-\alpha}^{\alpha} f(\xi+\alpha+x)\,(\alpha-|\xi|)\,d\xi \right) \leqq T(2\alpha) \tag{8.12}$$

and

(8.13) $$\varliminf_{x\to\infty}\left(-\frac{1}{\alpha^2}\int_{-\alpha}^{\alpha} f(\xi+\alpha+x)\,(\alpha-|\xi|)\,d\xi+f(x+2\alpha)\right)\geqq -T(2\alpha);$$

going to the limit

(8.14) $$\varliminf_{x\to\infty} f(x)\geqq -T(2\alpha)+B$$

and

(8.15) $$\varlimsup_{x\to\infty} f(x)\leqq B+T(2\alpha).$$

Then by (8.10), it follows that

(8.16) $$\lim_{x\to\infty} f(x) = B.$$

This yields

THEOREM XIV. *In the hypothesis of Theorem* XII, *condition* (6.02) *may be replaced by the assumption that* $f(x)$ *is of limited total variation over any finite range, and is slowly decreasing, and the conclusion* (7.12) *may be replaced by* (8.16). Again, Theorem XIII becomes

THEOREM XV. *In the hypothesis to Theorem* XIII, *condition* (7.16) *may be replaced by the condition that if* u *and* v *run through a sequence for which*

(8.17) $$v\geqq u,\quad \frac{v}{u}\to 1$$

then

(8.18) $$\varliminf_{u\to\infty}\,(F(v)-F(u))\geqq 0.$$

In the conclusion to Theorem XIII, (7.21) *may then be replaced by*

(8.161) $$\lim_{x\to\infty} F(x) = B.$$

In the corollary to Theorem XIII, (7.25) *may be replaced by the hypothesis that if*

(8.19) $$s_n = a_0+a_1+\cdots+a_n$$

and $p\to\infty$, $q\to\infty$ *in such a manner that*

(8.20) $$q>p,\quad \frac{q}{p}\to 1$$

then

(8.21) $$\varliminf_{p\to\infty}\,(s_q-s_p)\geqq 0.$$

The conclusion of this corollary remains unchanged.

The hypothesis that (8.20) shall imply (8.21) is Schmidt's hypothesis concerning slowly decreasing sequences in unaltered form.

It will be noted that the proof here given for the Schmidt theorems does not essentially differ from the proofs given by Vijayaraghavan[29] and Szász[30] for theorems of these types. In both cases the transition is made to a Tauberian theorem of more standard type in which s_n is itself averaged, instead of appearing in the conclusion as the average of a unilaterally or bilaterally bounded mass distribution. In both cases, moreover, the Schmidt condition is again used to give a bilateral or unilateral estimate of the difference between s_n and its average.

CHAPTER IV.

TAUBERIAN THEOREMS AND PRIME NUMBER THEORY

9. **Tauberian theorems and Lambert series.** One of the most important applications of Tauberian theorems is to the proof of the prime number theorem of Hadamard[31] and de la Vallée Poussin,[32] to the effect that the number of primes less than N is asymptotically $N/\log N$. The prime numbers bear an exceedingly close relation to series of the form

$$\sum a_n \frac{x^n}{1-x^n} \tag{9.01}$$

known as Lambert series, after their eighteenth century discoverer. Until recent times, however, all attempts to employ Lambert series effectively in the study of prime numbers had proved a failure, and indeed Knopp[33] has characterized one of these directions of attack as "verführerisch". Hardy and Littlewood[34] finally showed that the prime number theorem was equivalent to a Tauberian theorem concerning Lambert series, but did not succeed in establishing an autonomous proof of this theorem.

Our general Tauberian theorems suffice to furnish this autonomous proof, and indeed, the Tauberian theorem which we shall find it easiest to establish directly leads more directly to the prime number theorem than does the theorem of Hardy and Littlewood. The latter is also directly demonstrable by our methods. In both, the cardinal point in the proof is that the Riemann zeta function, $\zeta(x+iy)$, has no zeros on the line $x=1$. The proof of this goes back to the first proofs of the prime number theorem, and has always been recognized as that property of the Riemann zeta

[29] Vijayaraghavan (1), (2).
[30] Szász (1), (2).
[31] Hadamard (1).
[32] de la Vallée Poussin (1).
[33] Knopp (3).
[34] Hardy and Littlewood (11); Hardy (6); Ananda-Rau (2).

function which is most central in the proofs of this theorem, but all earlier proofs had made some use of the behavior of the zeta function at infinity. These further properties now appear as inessential, and the non-vanishing of the function becomes the only non-elementary feature of the zeta function in question.

For example, the usual proof of the prime number theorem employs a lemma of Landau[35] to the following effect:

LANDAU'S LEMMA. *Let*

$$F(x) = \sum_{n=1}^{\infty} a_n n^{-x} \qquad [\Re(x) > 1] \tag{9.02}$$

and let

$$a_n \geqq 0 \qquad [n = 1, 2, \cdots]. \tag{9.03}$$

Let $F(x)$ when analytically continued be without singularities on $\Re(x) = 1$, except for a pole of order one at $x = 1$, with principal part $A/(x-1)$. Let there be some α for which

$$F(x) = O(|x|^{\alpha}) \tag{9.04}$$

in the right half-plane. Then

$$A = \lim_{n\to\infty} \frac{1}{n} \sum_{1}^{n} a_k. \tag{9.05}$$

In the following section, we shall show—following Ikehara— that condition (9.04) is not needed. When the lemma is used in the proof of the prime number theorem,

$$a_n = \Lambda(n) \text{ (to be defined immediately)}; \quad F(x) = -\zeta'(x)/\zeta(x) \tag{9.06}$$

and it will be seen that here too the only non-elementary property of the zeta function which we use is that it does not vanish on the line $\Re(x) = 1$.

To return to Lambert series, let $\Lambda(n)$ be the number-theoretic function determined by:

$\Lambda(p^k) = \log p$ if p is a prime and k is a positive integer;

$\Lambda(n) = 0$ if n is not of the form p^k.

Let $0 \leqq x < 1$; then

$$\sum_{1}^{\infty} \log m\, x^m = \sum_{m=1}^{\infty} x^m \sum_{n/m} \Lambda(n) = \sum_{n=1}^{\infty} \Lambda(n) \sum_{n/m} x^m = \sum_{n=1}^{\infty} \Lambda(n) \frac{x^n}{1-x^n}. \tag{9.07}$$

[35] Landau (1).

Hence

$$(9.08)\quad \begin{aligned}\sum_1^\infty \Lambda(n)\frac{x^n}{1-x^n} &= \sum_1^\infty \log m\,\frac{x^m - x^{m+1}}{1-x}\\ &= \frac{x}{1-x}\sum_1^\infty \log\left(1+\frac{1}{m}\right)x^m\\ &= \frac{x}{1-x}\sum_1^\infty\left[\frac{1}{m}+O\left(\frac{1}{m^2}\right)\right]x^m\\ &= \frac{x}{1-x}\left[\log\frac{1}{1-x}+\sum_1^\infty O\left(\frac{1}{m^2}\right)x^m\right].\end{aligned}$$

If we put $x = e^{-\xi}$ and multiply by ξ we have

$$(9.09)\quad \sum_1^\infty \Lambda(n)\frac{\xi e^{-n\xi}}{1-e^{-n\xi}} = \frac{\xi e^{-\xi}}{1-e^{-\xi}}\left[\sum_1^\infty O\left(\frac{1}{m^2}\right)e^{-m\xi} - \log(1-e^{-\xi})\right].$$

For $\xi > 0$, the derived series on both sides converge absolutely and uniformly, and we may differentiate (9.09) term by term. On multiplication by ξ this yields us

$$(9.10)\quad \begin{aligned}&\xi\sum_1^\infty \Lambda(n)\frac{d}{dn\xi}\left(\frac{n\xi e^{-n\xi}}{e^{-n\xi}-1}\right)\\ &\quad = \frac{\xi e^{-\xi}(1-\xi-e^{-\xi})}{(1-e^{-\xi})^2}\left[\log(1-e^{-\xi})-\sum_1^\infty O\left(\frac{1}{m^2}\right)e^{-m\xi}\right]\\ &\quad\quad + \frac{\xi^2 e^{-\xi}}{1-e^{-\xi}}\left[\frac{e^{-\xi}}{1-e^{-\xi}}-\sum_1^\infty O\left(\frac{1}{m}\right)e^{-m\xi}\right]\\ &\quad = O(\xi)\,[O(\log\xi)-O(1)]+[\xi+O(\xi^2)]\left[\frac{1}{\xi}+O(\log\xi)\right]\\ &\quad = 1+O(\xi\log\xi)\end{aligned}$$

as $\xi\to 0$. If we put

$$(9.11)\quad N_1(u) = \frac{d}{du}\left(\frac{ue^{-u}}{e^{-u}-1}\right)$$

this leads us to

$$(9.12)\quad \lim_{\xi\to 0}\xi\sum_1^\infty \Lambda(n)\,N_1(n\xi) = 1.$$

This is a particular case of (6.14), which forms part of the hypothesis to Theorem XI'''. $\Lambda(n)$ is positive, as the hypothesis of that theorem further demands. The function $N_1(u)$ may be written

$$(9.13)\quad N_1(u) = \frac{d}{du}\left(\frac{u}{1-e^u}\right) = \frac{1+(u-1)e^u}{(1-e^u)^2} = \begin{cases} O(1) & \text{as } u\to 0;\\ O(ue^{-u}) & \text{as } u\to\infty.\end{cases}$$

Thus of that part of the hypothesis of Theorem XI''' containing N_1, which is now both M and the whole class Σ, it only remains for us to verify the boundedness of

$$\xi \sum_{1}^{\infty} \Lambda(n) N_1(n\xi) \tag{9.14}$$

which follows from the fact that there is a K such that for $\xi > \xi_0 > 0$

$$\begin{aligned} \left| \xi \sum_{1}^{\infty} \Lambda(n) N_1(n\xi) \right| &\leq K \sum_{1}^{\infty} n(n\xi e^{-n\xi}) \\ &= K\xi^2 \frac{d^2}{d\xi^2} \frac{e^{-\xi}}{1-e^{-\xi}} \to 0 \quad \text{as} \quad \xi \to 0 \end{aligned} \tag{9.15}$$

and from (9.12), and to verify further that

$$\int_0^\infty N_1(u)\, u^{ix}\, du \neq 0. \tag{9.16}$$

Now

$$\begin{aligned} &\int_0^\infty N_1(u)\, u^{ix}\, du \\ &= \int_0^\infty u^{ix}\, d\left(\frac{u}{1-e^u}\right) \\ &= \lim_{\lambda\to 0} \int_0^\infty u^{ix+\lambda}\, d\left(\frac{u}{1-e^u}\right) \\ &= -\lim_{\lambda\to 0} \int_0^\infty \frac{u}{1-e^u}\, d(u^{ix+\lambda}) \\ &= -\lim_{\lambda\to 0} (ix+\lambda) \int_0^\infty u^{ix+\lambda} \frac{du}{1-e^u} \qquad (9.17) \\ &= \lim_{\lambda\to 0} (ix+\lambda) \int_0^\infty u^{ix+\lambda} (e^{-u} + e^{-2u} + \cdots)\, du \\ &= \lim_{\lambda\to 0} (ix+\lambda)\, \Gamma(ix+\lambda+1)\, (1 + 2^{-(ix+\lambda+1)} + 3^{-(ix+\lambda+1)} + \cdots) \\ &= \lim_{\lambda\to 0} (ix+\lambda)\, \Gamma(ix+\lambda+1)\, \zeta(ix+\lambda+1) \\ &= ix\, \Gamma(ix+1)\, \zeta(ix+1). \end{aligned}$$

Let us take over from the theory of the Riemann zeta function the following facts:

(a) that the Riemann zeta function $\zeta(z)$ is analytic on the line with real part 1, except for a pole of the first order with principal part $1/(z-1)$;

(b) that the Riemann zeta function has no zeros with real part 1.

It then follows that

$$\int_0^\infty N_1(u)\, u^{ix}\, du \neq 0 \tag{9.18}$$

for any real x, and that

$$\int_0^\infty N_1(u)\,du = \lim_{\lambda\to 0} \lambda\,\Gamma(\lambda+1)\,\zeta(\lambda+1) = 1. \tag{9.19}$$

Thus it follows from (6.35) that

$$\lim_{N\to\infty} \frac{1}{N}\sum_{n=1}^{N} \Lambda(n) = 1. \tag{9.22}$$

Let us put $\pi(u)$ for the number of primes less than or equal to u, and let us write

$$\varpi(u) = \pi(u) + \tfrac{1}{2}\pi(u^{1/2}) + \tfrac{1}{3}\pi(u^{1/3}) + \cdots. \tag{9.23}$$

It is easy to prove by elementary means that

$$\lim_{u\to\infty} \frac{\pi(u)}{u} = 0. \tag{9.24}$$

Moreover,

$$\begin{aligned} |\varpi(u) - \pi(u)| &= \left| \tfrac{1}{2}\pi(u^{1/2}) + \cdots + \frac{\pi\left(u^{\frac{1}{[\log u/\log 2+1]}}\right)}{[\log u/\log 2]+1} \right| \\ &\leq u^{1/2}([\log u/\log 2]+1) \\ &= O(u^{1/2}\log u). \end{aligned} \tag{9.25}$$

Thus

$$\lim_{u\to\infty} \frac{\varpi(u)}{u} = 0. \tag{9.26}$$

We may write (9.22) in the form

$$\lim_{N\to\infty} \frac{1}{N}\int_0^N \log\, u\, d\varpi(u) = 1. \tag{9.27}$$

On integrating by parts, we see that

$$\lim_{N\to\infty} \left\{ \frac{\varpi(N)\log N}{N} - \frac{1}{N}\int_0^N \frac{\varpi(u)\,du}{u} \right\} = 1. \tag{9.28}$$

Now

$$\lim_{N\to\infty} \frac{1}{N}\int_0^N \frac{\varpi(u)}{u}\,du = 0 \tag{9.29}$$

because of (9.26). Thus

$$\lim_{N\to\infty} \frac{\varpi(N)\log N}{N} = 1 \tag{9.30}$$

or

$$\varpi(N) \sim \frac{N}{\log N}. \tag{9.31}$$

From (9.25) it follows that

$$\pi(N) \sim \frac{N}{\log N}. \tag{9.32}$$

This is the famous prime number theorem of de la Valleé Poussin and Hadamard.

10. Ikehara's Theorem. The Landau Theorem (XVI) received several successive generalizations at the hands of Landau himself, and of Hardy and Littlewood, perhaps the most general of which was indicated by Hardy and Littlewood[36] to be the following:

THEOREM XVI. *Let:*

(i) *the series* $\Sigma a_n \lambda_n^{-s}$ *be absolutely convergent for* $\Re(s) > \sigma_0 > 0$;

(ii) *the function* $F(s)$ *defined by the series be regular for* $\Re(s) > c$ *where* $0 < c \leqq \sigma_0$ *and continuous for* $\Re(s) \geqq c$, *except for a simple pole with residue* g *at* $s = c$;

(iii)

$$F(s) = O(e^{C|t|}) \tag{10.01}$$

for some finite C, *uniformly for* $\sigma \geqq c$;

(iv)

$$\frac{\lambda_n}{\lambda_{n-1}} \to 1; \tag{10.02}$$

(v) a_n *be real, and satisfy one of the inequalities*

$$a_n > -K\lambda_n^{c-1}(\lambda_n - \lambda_{n-1}), \quad a_n < K\lambda_n^{c-1}(\lambda_n - \lambda_{n-1}) \tag{10.03}$$

or complex, and of the form

$$O\{\lambda_n^{c-1}(\lambda_n - \lambda_{n-1})\}. \tag{10.04}$$

Then

$$A_n = a_1 + a_2 + \cdots + a_n \sim \frac{g\lambda_n^c}{c}. \tag{10.05}$$

The vital change between the Landau and the Hardy-Littlewood theorem is the looser form of (iii), which replaces a restriction of the form

$$F(s) = O(|t|^k). \tag{9.04}$$

Both these restrictions are inessential, and the true theorem is that of Ikehara[37], which reads as follows:

THEOREM XVII. *Let* $\alpha(x)$ *be a monotone increasing function, and let*

$$\int_{1+0}^{\infty} x^{-u}\, d\alpha(x) = f(u) \tag{10.06}$$

---

[36] Hardy and Littlewood (8).

[37] Ikehara (1).

converge for $\Re(u) > 1$. *Let*

$$f(u) - \frac{A}{u-1} = g(u) \tag{10.07}$$

converge uniformly to a finite limit as

$$\Re(u) \to 1 \tag{10.08}$$

over any finite interval of the line $\Re(u) = 1$. *Then*

$$A = \lim_{N\to\infty} \frac{\alpha(N)}{N}. \tag{10.09}$$

To prove this, let us put

$$\beta(\xi) = \alpha(e^\xi)\, e^{-\xi} + \int_0^\xi e^{-\xi}\, \alpha(e^\xi)\, d\xi - A\xi. \quad (\xi > 0). \tag{10.10}$$

Thus

$$d\beta(\xi) = e^{-\xi}\, d\alpha(e^\xi) - A\, d\xi. \tag{10.11}$$

Let us assume—what is no essential restriction—that

$$\beta(x) = \beta(+0), \quad (-\infty \leqq x \leqq 0). \tag{10.12}$$

Then (10.07) becomes

$$g(u) = \int_{-\infty}^{\infty} e^{(1-u)\xi}\, d\beta(\xi), \quad [\Re(u) > 1]. \tag{10.13}$$

What we wish to prove is that

$$\lim_{\eta\to\infty} \int_{-\infty}^{\eta} e^{\xi-\eta}\, d\beta(\xi) = 0 \tag{10.14}$$

which is equivalent to (10.09).

If $\varepsilon > 0$ and η is real, we have

$$\begin{aligned} &\int_{-B}^{B} \left(1 - \frac{|u|}{B}\right) g(iu + \varepsilon + 1)\, e^{iu\eta}\, du \\ &\quad = -\int_{-B}^{B} \left(1 - \frac{|u|}{B}\right) du \int_{-\infty}^{\infty} e^{iu(\eta-\xi)}\, e^{-\varepsilon\xi}\, d\beta(\xi). \end{aligned} \tag{10.15}$$

As this double integral is absolutely convergent, it becomes

$$\begin{aligned} &\int_{-\infty}^{\infty} e^{-\varepsilon\xi}\, d\beta(\xi) \int_{-B}^{B} \left(1 - \frac{|u|}{B}\right) e^{iu(\eta-\xi)}\, du \\ &\quad = -\int_{-\infty}^{\infty} \frac{2(\cos B(\eta-\xi) - 1)}{B(\eta-\xi)^2}\, e^{-\varepsilon\xi}\, d\beta(\xi). \end{aligned} \tag{10.16}$$

Now,

$$\begin{aligned} &\lim_{\varepsilon\to 0} \int_{-B}^{B} \left(1 - \frac{|u|}{B}\right) g(iu + \varepsilon + 1)\, e^{iu\eta}\, d\eta \\ &\quad = \int_{-B}^{B} \left(1 - \frac{|u|}{B}\right) g(iu + 1)\, e^{iu\eta}\, d\eta. \end{aligned} \tag{10.17}$$

Thus

$$\lim_{\varepsilon \to 0} \int_{-\infty}^{\infty} \frac{2(\cos B(\eta - \xi) - 1)\, e^{-\varepsilon \xi}}{B(\eta - \xi)^2}\, d\beta(\xi) \tag{10.18}$$

exists. Remembering that

$$\lim_{\varepsilon \to 0} \int_0^{\infty} \frac{2(\cos B(\eta - \xi) - 1)}{B(\eta - \xi)^2}\, e^{-\varepsilon \xi}\, d\xi = \int_0^{\infty} \frac{2(\cos B(\eta - \xi) - 1)}{B(\eta - \xi)^2}\, d\xi, \tag{10.19}$$

and that

$$\frac{2(\cos B(\eta - \xi) - 1)}{B(\eta - \xi)^2}\, e^{(-1-\varepsilon)\xi}\, d\alpha(e^{\xi}) \tag{10.20}$$

is a non-positive integrand, we have, by a theorem due to Bray[38] and fundamental in the theory of the Stieltjes integral

$$\lim_{\varepsilon \to 0} \int_0^{\infty} \frac{2(\cos B(\eta - \xi) - 1)}{B(\eta - \xi)^2}\, e^{(-1-\varepsilon)\xi}\, d\alpha(e^{\xi}) = \int_0^{\infty} \frac{2(\cos B(\eta - \xi) - 1)}{B(\eta - \xi)^2}\, e^{-\xi}\, d\alpha(e^{\xi}) \tag{10.21}$$

and hence

$$\lim_{\varepsilon \to 0} \int_{-\infty}^{\infty} \frac{2(\cos B(\eta - \xi) - 1)}{B(\eta - \xi)^2}\, e^{-\varepsilon \xi}\, d\beta(\xi) = \int_{-\infty}^{\infty} \frac{2(\cos B(\eta - \xi) - 1)}{B(\eta - \xi)^2}\, d\beta(\xi). \tag{10.22}$$

Moreover,

$$-\lim_{\eta \to \infty} \int_{-\infty}^{\infty} \frac{2(\cos B(\eta - \xi) - 1)}{B(\eta - \xi)^2}\, d\beta(\xi) = \lim_{\eta \to \infty} \int_{-B}^{B} \left(1 - \frac{|u|}{B}\right) g(iu+1)\, e^{iu\eta}\, du = 0 \tag{10.23}$$

because $\left(1 - \frac{|u|}{B}\right) g(iu+1)$ is summable over $(-B, B)$. Similarly

$$\int_{-\infty}^{\infty} \frac{2(\cos B(\eta - \xi) - 1)}{B(\eta - \xi)^2}\, d\beta(\xi) \tag{10.24}$$

may be proved to be bounded because of (10.23) and (10.12).

We know that

$$\int_n^{n+1} d\beta(\xi) > -A \tag{10.25}$$

[38] Bray (1).

and that all the other conditions of Theorem XI are satisfied, with the possible exception of the non-vanishing of (6.15). To see that this is also satisfied, we need only reflect that

$$(10.26)\quad \int_{-\infty}^{\infty} \frac{2(\cos B\eta - 1)}{B\eta^2} e^{iu\eta}\, d\eta = -\frac{1}{2\pi}\left(1 - \frac{|u|}{B}\right) \qquad [|u| < B].$$

Thus there are no zeros common to all these functions for all values of B, and the non-vanishing of (6.15) follows.

It will be observed that the full force of our Tauberian method is scarcely needed for this theorem. In Theorem XI, which is the critical part of the proof, the difficulty of proof is considerably lessened if $K_1(x)$ assumes such a special form as $2(\cos Bx - 1)/Bx^2$.

As a corollary of Theorem XVII, we may prove:

THEOREM XVIII. *Let* $\gamma(x)$ *be a monotone increasing function, and let*

$$(10.27)\quad \int_{1+0}^{\infty} x^{-u}\, d\gamma(x) = \varphi(u)$$

converge for $\Re(u) > 1$. *Let*

$$(10.28)\quad F(u) = e^{\varphi(u)}(u-1)^A \qquad [0 < A < \tfrac{4}{3}]$$

when continued analytically, be regular for $\Re(u) = 1$, *and let it not vanish for* $u = 1$. *Then,*

$$(10.29)\quad A = \lim_{N\to\infty} \frac{1}{N} \int_{1+0}^{N} \log x\, d\gamma(x).$$

To prove this, let us put

$$(10.30)\quad \begin{cases} \int_{1+0}^{x} \log \xi\, d\gamma(\xi) = \alpha(x); \\ -\varphi'(u) = f(u). \end{cases}$$

The theorem then reduces itself to Theorem XVII, provided we can establish that (10.07) approaches a finite limit as $\Re(u) \to 1$. From the regularity of (10.28), it follows that

$$(10.31)\quad \varphi(u) + A \log(u-1)$$

is regular for $\Re(u) = 1$, except for logarithmic singularities with *negative* infinities. It is also clear that there is no singularity for $u = 1$.

Now,

$$(10.32)\quad \Re(\varphi(u)) = \int_{1+0}^{\infty} x^{-\Re(u)} \cos(\Im(u)\log x)\, d\gamma(x)$$

so that

$$(10.33)\quad \Re(\varphi(1+\epsilon+iv)) = \int_{1+0}^{\infty} x^{-(1+\epsilon)} \cos(v \log x)\, d\gamma(x);$$

$$\Re(\varphi(1+\varepsilon+2iv)) = \int_{1+0}^{\infty} x^{-(1+\varepsilon)} \cos(2v \log x)\, d\gamma(x); \tag{10.34}$$

and

$$\varphi(1+\varepsilon) = \int_{1+0}^{\infty} x^{-(1+\varepsilon)}\, d\gamma(x). \tag{10.35}$$

Thus

$$\begin{aligned} &3\varphi(1+\varepsilon) + 4\Re(\varphi(1+\varepsilon+iv)) + \Re(\varphi(1+\varepsilon+2iv)) \\ &= \int_{1+0}^{\infty} x^{-(1+\varepsilon)} (3 + 4\cos v \log x + \cos(2v \log x))\, d\gamma(x) \end{aligned} \tag{10.36}$$

and since

$$\begin{aligned} 3 + 4\cos\varphi + \cos 2\varphi &= 3 + 4\cos\varphi + 2\cos^2\varphi - 1 \\ &= 2(1+\cos\varphi)^2 \\ &\geqq 0 \end{aligned} \tag{10.37}$$

it follows that

$$3\varphi(1+\varepsilon) + 4\Re(\varphi(1+\varepsilon+iv)) + \Re(\varphi(1+\varepsilon+2iv)) \geqq 0, \tag{10.38}$$

or

$$\Re(\varphi(1+\varepsilon+iv)) \geqq -\tfrac{3}{4}\varphi(1+\varepsilon) - \tfrac{1}{4}\Re(\varphi(1+\varepsilon+2iv)). \tag{10.39}$$

Thus

$$\begin{aligned} &\overline{\lim_{\varepsilon\to 0}} \frac{\Re(\varphi(1+\varepsilon+iv))}{\log\varepsilon} \\ &\leqq -\tfrac{3}{4}\lim_{\varepsilon\to 0}\frac{\varphi(1+\varepsilon)}{\log\varepsilon} - \tfrac{1}{4}\underline{\lim_{\varepsilon\to 0}}\frac{\Re(\varphi(1+\varepsilon+2iv))}{\log\varepsilon} \leqq \frac{3A}{4} < 1. \end{aligned} \tag{10.40}$$

On the other hand, if $1+iv$ is a logarithmic singularity of φ with a negative coefficient, it is a zero of $F(u)$ of integral order n, and

$$\overline{\lim_{\varepsilon\to 0}} \frac{\Re(\varphi(1+\varepsilon+iv))}{\log\varepsilon} = n > 1. \tag{10.41}$$

Thus $\varphi(u) + A\log(u-1)$ has no logarithmic singularities with negative coefficients, for $\Re(u) = 1$, and hence is analytic throughout this whole line. Thus, by differentiation,

$$f(u) - \frac{A}{u-1} \tag{10.42}$$

is analytic on the line in question, and the finiteness of the limit of (10.07) is established.

It will be observed that our proof, which completes the demonstration of Theorem XVIII, follows closely the lines of Landau's proof[39] that the Riemann zeta function has no zeros on the line $\Re(u) = 1$, and includes it

[39] Landau (2).

as a particular case. The prime number theorem itself is a particular case of Theorem XVIII. Let us put

$$\gamma(x) = \varpi(x). \tag{10.43}$$

Then

$$e^{\varphi(u)} = \zeta(u) \tag{10.44}$$

and the hypothesis of Theorem XVIII is manifestly satisfied. (10.29) then becomes

$$1 = \lim_{N \to \infty} \frac{1}{N} \int_{1+0}^{N} \log x \, d\varpi(x) \tag{9.27}$$

which we have shown to be equivalent to the prime number theorem.

Thus we have repeatedly shown that the prime number theorem is basically Tauberian in nature. It might consequently be expected that the more refined theorems as to the distribution of the primes, based on Riemann's hypothesis as to the distribution of the zeros of the zeta function, and established by Hardy, Littlewood and others, might be easy to establish on a Tauberian basis. Such a formula as

$$\lim_{\xi \to 0} \xi \sum_{1}^{\infty} (\Lambda(n) - 1)\, n^{\alpha} \,(n^{-\alpha} \xi^{-\alpha} N_1(n\xi)) = 0 \quad \left(0 < \alpha < \frac{1}{2}\right) \tag{10.45}$$

which follows from (9.10) much as does (9.12), appears to lend a certain color to this view. Here the theorem to which this seems to lead is

$$\lim_{N \to \infty} \frac{1}{N} \sum_{1}^{N} (\Lambda(n) - 1)\, n^{\alpha} = 0, \tag{10.46}$$

and the condition of non-vanishing on the Fourier transform becomes the Riemann condition

$$\zeta(ix + 1 - \alpha) \neq 0. \tag{10.47}$$

The author considers these hopes illusory and deceptive. Let it be noted that (10.45) does not form a satisfactory hypothesis to a Tauberian theorem until we have some hold on the boundedness of the mass distribution whose integral is

$$M(u) = \sum_{1}^{[u]} (\Lambda(n) - 1)\, n^{\alpha}. \tag{10.48}$$

Such information would already presuppose as much information as (10.45) can yield concerning all smaller values of α than occur in (10.46). In other words, Tauberian theorems merely transform a O into a o.

Another way of stating the same thing is to say that a Tauberian theorem always operates in the neighborhood of a single ordinate in the plane of the zeta function. This is because it depends on a division of the range of this function into near and remote parts, and because this division has validity in the theory of functions of a real variable, not in the theory of functions of a complex variable. On the other hand, the more refined properties of the distribution of primes depend on the behavior of the zeta function in the entire strip between ordinates $\frac{1}{2}$ and 1, inclusive, and can only be discussed with the aid of Cauchy's theorem.

Of course, no proof of the limitations of so vague a thing as a method has real mathematical cogency. At any time some super-Tauberian theorem may come to light and prove to be central in the utmost refinements of prime number theory. For the present, however, Tauberian theorems do not seem to lie on the main avenue of progress.

CHAPTER V.

SPECIAL APPLICATIONS OF TAUBERIAN THEOREMS.

11. **On the proof of special Tauberian theorems.** In the sequel, we shall show that the greater part of all known Tauberian theorems may be proved without great difficulty on the basis of the general theorems of the present paper. However, most of the particular theorems were proved in the first instance by entirely different methods. In individual cases, these methods are simpler and more direct than the general method here indicated. This is especially noticeable in the case of Karamata's[40] proof of the original Abel-Tauber theorem.

Being in possession of a general method, we may consider with advantage the particular methods and why they function. All Tauberian theorems of the type discussed in this paper are intimately related to the solution of an integral equation of the form

$$\int_{-\infty}^{\infty} F(y)\, G(x-y)\, dy = H(x). \tag{11.01}$$

The most direct and general method of solving such an equation is by the use of Fourier transforms. Nevertheless, there are many cases in which a repeated differentiation will reduce such an equation to a linear differential equation of finite order, and many more where the same repeated differentiation will lead to a differential equation of infinite order, but of manageable form. Thus it is appropriate in many cases to employ a technique

[40] Karamata (2), (3), (4).

of repeated differentiation, and this has been done by Hardy, Littlewood[41], and Vijayaraghavan,[42] though scarcely from an explicit consideration of the integral equations in question. So far, the successes of this method have been confined to cases where the analytic properties of the Fourier transform of F are extremely simple, and it has failed to throw any light on the Tauberian theorems of prime number theory.

The methods of Robert Schmidt[43] lie more along the lines of the present paper, in as much as he has seen the essential role played by the integral

$$\int_0^y K(x)\, x^u\, dx = R(u) \tag{11.02}$$

in the study of the kernel of a Tauberian theorem. However, he has devoted his attention to $R(u)$ for real integral arguments instead of for complex general arguments. Furthermore, Schmidt's general theorem concerns the unicity of the solution of his moment problem rather than its existence theory. As a consequence there is a wide gap between his general moment theorem and the particular Tauberian theorems which he obtains as corollaries. This gap he actually fills in in two cases, that of the Abel-Tauber theorem and that of the Borel-Tauber theorem, but he gives no general method by which it may be filled in in a new case. His actual procedure is closely allied to that of Hardy, Littlewood and Vijayaraghavan. Schmidt's chief service to the subject is in his great improvement of what may be called the auxiliary apparatus of the theory of Tauberian theorems, through his invention of the notions of "langsam abfallende Funktionen" and "gestrahlte Matrizen".

Karamata's elegant method leads to the study of the closure of a set of translations of a given function, and thus most closely approximates to that developed here. His function is

$$f(x) = e^x e^{-e^x} \tag{11.03}$$

and the problem is solved through Weierstrass' theory of polynomial approximation. The translations considered are accordingly those of the form

$$f(x + \log \mu). \tag{11.04}$$

Szász has carried further the study of this particular set of translations of a given function. This is a far more difficult study than that of the

[41] Cf. Littlewood (1).
[42] Vijayaraghavan (1), (2).
[43] Schmidt (1), (2).

closure of the complete set of translations, and here the general solution of the closure problem is not yet known to me.

12. Examples of kernels for which Tauberian theorems hold. Among the kernels admissible in the role of the $N(x)$ of Theorem XIII are:

(1) The Riesz kernels[44]

$$N(x) = \begin{cases} (1-x)^{\lambda}; & [0<x<1] \\ 0 \quad ; & [1<x] \end{cases} \quad [\lambda>0]. \tag{12.01}$$

(2) The Abel kernel

$$N(x) = e^{-x}; \tag{12.02}$$

(3) The kernel

$$N(x) = \frac{1}{1+x^2}, \tag{12.03}$$

corresponding to the method of summation of the series Σa_n, which gives as its partial sum the Abel average

$$\lambda \int_0^\infty f(x)\, e^{-\lambda x}\, dx \tag{12.04}$$

of the cosine series

$$f(x) = \sum a_n \cos nx. \tag{12.05}$$

In the respective cases we have

(1)

$$\int_0^\infty N(x)\, x^{z-1}\, dx = \frac{\Gamma(\lambda+1)\,\Gamma(z)}{\Gamma(\lambda+z+1)}; \tag{12.06}$$

(2)

$$\int_0^\infty N(x)\, x^{z-1}\, dx = \Gamma(z); \tag{12.07}$$

(3)

$$\int_0^\infty N(x)\, x^{z-1}\, dx = \frac{\pi}{2 \sin \frac{\pi z}{2}}. \tag{12.08}$$

None of these functions vanishes for purely imaginary values of z.

A possible $N_1(\mu)$ of Theorem XI‴ is

$$N_1(\mu) = \begin{cases} (\lambda+1)(1-\mu)^{\lambda}; & [0<\mu<1] \\ 0 & [1<\mu] \end{cases} \quad [\lambda \geqq 0]. \tag{12.09}$$

Here

$$\int_0^\infty N_1(\mu)\, \mu^{iu}\, d\mu = \frac{\Gamma(\lambda+2)\,\Gamma(iu+1)}{\Gamma(iu+\lambda+2)} \neq 0 \qquad \text{if } u \text{ is real.} \tag{12.10}$$

[44] Riesz (1).

13. A theorem of Ramanujan. As a lemma in our further work, we shall find it convenient to introduce a theorem formulated by Ramanujan[45] and first proved by Hardy and Titchmarsh,[46] although with a formulation somewhat different from that here given. The theorem is intrinsically interesting, and is perhaps worth presenting in some detail.

Let $f(x)$ be a function of L_2, defined over $(0, \infty)$. Then $e^{x/2}f(e^x)$ will also belong to L_2 over $(-\infty, \infty)$; for

$$\int_{-\infty}^{\infty} |e^{x/2}f(e^x)|^2\,dx = \int_{-0}^{\infty} |f(x)|^2\,dx. \tag{13.01}$$

The Fourier transform of $e^{x/2}f(e^x)$ will be

$$\begin{aligned} h(u) &= \sqrt{\frac{1}{2\pi}}\ \underset{A\to\infty}{\text{l. i. m.}} \int_{-A}^{A} e^{iux}\, e^{x/2} f(e^x)\,dx \\ &= \frac{1}{\sqrt{2\pi}}\ \underset{\varepsilon\to 0}{\text{l. i. m.}} \int_{\varepsilon}^{1/\varepsilon} f(x)\, x^{iu-1/2}\,dx. \end{aligned} \tag{13.02}$$

The sine transform of $f(x)$ will be

$$g_1(y) = \sqrt{\frac{2}{\pi}}\ \underset{B\to\infty}{\text{l. i. m.}} \int_0^B f(x)\sin xy\,dx, \tag{13.03}$$

and its cosine transform,

$$g_2(y) = \sqrt{\frac{2}{\pi}}\ \underset{B\to\infty}{\text{l. i. m.}} \int_0^B f(x)\cos xy\,dx. \tag{13.04}$$

Let us put

$$k_1(u) = \frac{1}{\sqrt{2\pi}}\ \underset{\varepsilon\to 0}{\text{l. i. m.}} \int_{\varepsilon}^{1/\varepsilon} g_1(y)\, y^{iu-1/2}\,dy, \tag{13.05}$$

and

$$k_2(u) = \frac{1}{\sqrt{2\pi}}\ \underset{\varepsilon\to 0}{\text{l. i. m.}} \int_{\varepsilon}^{1/\varepsilon} g_2(y)\, y^{iu-1/2}\,dy. \tag{13.06}$$

We have

$$\begin{aligned} \frac{1}{z}\int_0^z g_1(y)\,dy &= \sqrt{\frac{2}{\pi}} \int_0^{\infty} f(x)\,dx\, \frac{1}{z}\int_0^z \sin xy\,dy \\ &= \sqrt{\frac{2}{\pi}} \int_0^{\infty} f(x)\frac{1-\cos xz}{xz}\,dx. \end{aligned} \tag{13.07}$$

Similarly

$$\frac{1}{z}\int_0^z g_2(y)\,dy = \sqrt{\frac{2}{\pi}} \int_0^{\infty} f(x)\frac{\sin xz}{xz}\,dx. \tag{13.08}$$

[45] Hardy and Littlewood (8).

[46] Hardy and Titchmarsh (1).

However, on an exponential transformation, these become

$$(13.09)\quad \int_{-\infty}^{\zeta} e^{\frac{\eta-\zeta}{2}} (e^{\eta/2} g_2(e^{\eta}))\, d\eta = \sqrt{\frac{2}{\pi}} \int_{-\infty}^{\infty} \sin(e^{\xi+\zeta})\, e^{-\frac{\xi+\zeta}{2}} (e^{\xi/2} f(e^{\xi}))\, d\xi,$$

and

$$(13.10)\quad \int_{-\infty}^{\zeta} e^{\frac{\eta-\zeta}{2}} (e^{\eta/2} g_1(e^{\eta}))\, d\eta = \sqrt{\frac{2}{\pi}} \int_{-\infty}^{\infty} (1-\cos(e^{\xi+\zeta}))\, e^{-\frac{\xi+\zeta}{2}} (e^{\xi/2} f(e^{\xi}))\, d\xi.$$

If we now make a Fourier transformation and make use of the Parseval theorem, it appears that

$$(13.11)\quad k_1(u) \int_0^{\infty} e^{-\eta/2} e^{iu\eta}\, d\eta = \sqrt{\frac{2}{\pi}}\, h(-u) \int_{-\infty}^{\infty} (1-\cos e^{\xi})\, e^{-\xi/2} e^{iu\xi}\, d\xi,$$

and

$$(13.12)\quad k_2(u) \int_0^{\infty} e^{-\eta/2} e^{iu\eta}\, d\eta = \sqrt{\frac{2}{\pi}}\, h(-u) \int_{-\infty}^{\infty} \sin e^{\xi}\, e^{-\xi/2} e^{iu\xi}\, d\xi.$$

Thus

$$(13.13)\quad \begin{aligned} \frac{k_1(u)}{h(-u)} &= \frac{\sqrt{\frac{2}{\pi}} \int_0^{\infty} (1-\cos x)\, x^{iu-3/2}\, dx}{\int_0^{\infty} e^{-\eta(1/2-iu)}\, d\eta} \\ &= \sqrt{\frac{2}{\pi}} \sin\left(iu+\frac{1}{2}\right)\frac{\pi}{2}\, \Gamma\left(iu+\frac{1}{2}\right). \end{aligned}$$

Here we make use of the formulae:

$$(13.131)\quad \int_0^{\infty} x^{\nu} \sin x\, dx = \Gamma(\nu+1) \cos\frac{\nu\pi}{2} \qquad [-1 \geqq \Re(\nu) > -2]$$

and

$$(13.132)\quad \int_0^{\infty} x^{\nu} (1-\cos x)\, dx = \Gamma(\nu+1) \sin\frac{\nu\pi}{2} \qquad [-1 > \Re(\nu) > -3].$$

Similarly

$$(13.14)\quad \frac{k_2(u)}{h(-u)} = \sqrt{\frac{2}{\pi}} \cos\left(iu+\frac{1}{2}\right)\frac{\pi}{2}\, \Gamma\left(iu+\frac{1}{2}\right).$$

The duality of the relation between $k_1(u)$ or $k_2(u)$ and $h(u)$ is shown by the familiar formulae of the gamma function,

$$(13.15)\quad \begin{aligned} &\frac{2}{\pi} \sin\left(iu+\frac{1}{2}\right)\frac{\pi}{2} \sin\left(-iu+\frac{1}{2}\right)\frac{\pi}{2}\, \Gamma\left(iu+\frac{1}{2}\right) \Gamma\left(-iu+\frac{1}{2}\right) \\ &= \frac{2}{\pi} \cos\left(iu+\frac{1}{2}\right)\frac{\pi}{2} \cos\left(-iu+\frac{1}{2}\right)\frac{\pi}{2}\, \Gamma\left(iu+\frac{1}{2}\right) \Gamma\left(-iu+\frac{1}{2}\right) \\ &= \frac{1}{\pi} \cos iu\pi\, \Gamma\left(iu+\frac{1}{2}\right) \Gamma\left(-iu+\frac{1}{2}\right) = 1. \end{aligned}$$

Thus

$$\left|\frac{k_2(u)}{h(-u)}\right| = 1, \qquad \left|\frac{k_1(u)}{h(-u)}\right| = 1, \tag{13.16}$$

and the real zeros of $k_1(u)$, $k_2(u)$, and $h(-u)$ are the same. In other words, if the integrals in question exist, the zeros of

$$\begin{cases} \dfrac{1}{\sqrt{2\pi}} \displaystyle\int_0^\infty f(x)\, x^{-iu-1/2}\, dx, \\ \dfrac{1}{\sqrt{2\pi}} \displaystyle\int_0^\infty g_1(y)\, y^{iu-1/2}\, dy, \\ \text{and } \dfrac{1}{\sqrt{2\pi}} \displaystyle\int_0^\infty g_2(y)\, y^{iu-1/2}\, dy, \end{cases} \tag{13.17}$$

are the same. This is a very valuable way of determining new kernels whose Fourier transforms (on a logarithmic scale) have no zeros.

14. **The summation of trigonometrical developments.** Let $f(x)$ be an even function of class L_2, and, as above, let

$$g_2(y) = \sqrt{\frac{2}{\pi}} \underset{B\to\infty}{\text{l.i.m.}} \int_0^B f(x) \cos xy\, dx. \tag{13.04}$$

Let $K(x)$ belong to L_2 over $(0, \infty)$, and let us form

$$x \int_0^\infty g_2(w)\, K(wx)\, dw. \tag{14.01}$$

This will equal

$$\frac{1}{x} \int_0^\infty z f(z) \frac{x}{z}\, k\left(\frac{z}{x}\right) dz, \tag{14.02}$$

where

$$k(z) = \sqrt{\frac{2}{\pi}} \underset{B\to\infty}{\text{l.i.m.}} \int_0^B K(w) \cos wz\, dw; \tag{14.03}$$

for the cosine transform of $xK(wx)$ is $k\left(\dfrac{z}{x}\right)$. Thus an average of $g_2(w)$ at infinity will appear formally as an average of $z f(z)$ about the origin, and an average of $g_2(w)$ about the origin will appear as an average of $z f(z)$ at infinity. The kernels in the two cases will be $K(x)$ and $\dfrac{1}{x} k(x)$. Now by § 13, we have formally

$$\left|\sqrt{\frac{1}{2\pi}} \int_0^\infty \frac{k(x)}{x} x^\lambda\, dx\right| = \left|\sqrt{\frac{1}{2\pi}} \int_0^\infty K(x)\, x^{-\lambda}\, dx\right|, \tag{14.04}$$

when $\Re(\lambda) = \frac{1}{2}$. If

$$(14.05) \quad \sqrt{\frac{1}{2\pi}} \int_0^\infty \frac{k(x)}{x} x^\lambda \, dx \quad \text{and} \quad \sqrt{\frac{1}{2\pi}} \int_0^\infty K(x)\, x^{-\lambda}\, dx$$

converge for $0 < \Re(\lambda) \leq \frac{1}{2}$, it then follows that

$$(14.06) \quad \sqrt{\frac{1}{2\pi}} \int_0^\infty \frac{k(x)}{x} x^{iu}\, dx \neq 0 \qquad [-\infty < u < \infty]$$

when, and only when,

$$(14.07) \quad \frac{1}{\sqrt{2\pi}} \int_0^\infty K(x)\, x^{iu}\, dx \neq 0 \qquad [-\infty < u < \infty].$$

Thus if $k(x)/x$ in the role of $N_1(x)$, and $N_2(x)$ satisfy the other conditions specified in the hypothesis of Theorem XI, and (14.07) holds, it is possible to infer from

$$(14.08) \quad \lim_{x \to {0 \atop \infty}} x \int_0^\infty g_2(w)\, K(wx)\, dw = A \int_0^\infty K(w)\, dw$$

to the conclusion that

$$(14.09) \quad \lim_{x \to {0 \atop \infty}} x \int_0^\infty z f(z)\, N_2(zx)\, dz = A \int_0^\infty N_2(z)\, dz.$$

The condition that f belongs to L_2 may in many cases be very considerably altered and relaxed.

It is possible to read the relation between (14.01) and (14.02) in the reverse direction, and to infer from

$$(14.10) \quad \lim_{x \to \infty} \int_0^\infty f(z)\, k\left(\frac{z}{x}\right) dz = Ak(0)$$

and other conditions completing the hypothesis of Theorem XI, to

$$(14.11) \quad \lim_{x \to 0} x \int_0^\infty g_2(w)\, K_2(wx)\, dw = A \int_0^\infty K_2(w)\, dw.$$

The condition that

$$(14.12) \quad \frac{1}{\sqrt{2\pi}} \int_0^\infty k(x)\, x^{z-1}\, dx$$

should determine a function over a strip to the right of the origin, which, when continued has no zeros over the imaginary axis, becomes the same condition for

$$(14.13) \quad \frac{1}{\sqrt{2\pi}} \int_0^\infty K(x)\, x^{-z}\, dx.$$

We thus arrive at a whole class of theorems relating the partial sum of a Fourier development with the average of the function represented about some point. For example, it is easy to prove:

THEOREM XVIII. *If $f(x)$ is a non-negative function of class L_2 over $(-\pi, \pi)$, if it is summable at any point by a Riesz mean of any positive order, or by an Abel mean, to any value A, it is so summable to A by Riesz means of all positive orders, and by an Abel mean. A necessary and sufficient condition that this should take place at a given point x is that*

$$\lim_{\varepsilon \to 0} \frac{1}{2\varepsilon} \int_{x-\varepsilon}^{x+\varepsilon} f(x)\,dx = A. \tag{14.14}$$

This theorem is due to Hardy and Littlewood.

15. **Young's criterion for the convergence of a Fourier series.** We now come to a region in which S. B. Littauer has done work, which constitutes the theorems proved in the present section. Young has proved the following theorem: *for the Fourier series of the integrable function $f(u)$ to converge to s for $u = x$, it is sufficient that*

$$\varphi(t) \to 0 \tag{15.01}$$

and that

$$\int_0^t |d(u\,\varphi(u))| = O(t) \tag{15.02}$$

for small t, where

$$\varphi(t) = \frac{1}{2}\{f(x+t)+f(x-t)-2s\}. \tag{15.03}$$

Further work on this theorem has been done by Young, Pollard, Hardy and Littlewood. The chief theorem to which Hardy and Littlewood came was the following:

THEOREM XIX. *A necessary and sufficient condition that the Fourier series of the integrable function $f(u)$ be summable $(C, -1+\delta)$ for any positive δ is that*

$$\varphi_1(t) = o(t) \tag{15.04}$$

[*provided that*

$$\int_0^t |d(u\,\varphi(u))| = O(t), \qquad [0 \leq t < \pi]] \tag{15.05}$$

where

$$\varphi_1(t) = \int_0^t \varphi(u)\,du. \tag{15.06}$$

This theorem in its original form is not directly adaptable to proof by the methods of this paper, inasmuch as Cesàro summation of fractional order does not depend on a kernel of the form $N(nx)$. On the other hand, if the Cesàro summation is replaced by Riesz summation, the theorem is reduced to a particular case of theorems already proved. We shall here content ourselves with proving this related theorem, as applied to Fourier integrals rather than series, in the case where $f(x)$ differs from zero only over a finite range.[47] Let us write

$$t\,\varphi(t) = \psi(t). \tag{15.07}$$

Then (15.05) becomes

$$\int_0^t |d\,\psi(t)| = O(t), \qquad [0 \leqq t < \pi] \tag{15.08}$$

which yields the boundedness of

$$\int_\mu^{2\mu} \lambda^{-1} |d\,\psi(\lambda)| - \int_\mu^{2\mu} \lambda^{-1}\, d\,\psi(\lambda). \tag{15.09}$$

Furthermore

$$\begin{aligned} \varphi_1(t) &= \int_0^t \varphi(u)\,du = \int_0^t \psi(u)\,\frac{du}{u} \\ &= \int_0^t \frac{du}{u} \int_0^u d\,\psi(v) = \int_0^t d\,\psi(v) \int_v^t \frac{du}{u} \\ &= \int_0^t \log\frac{t}{v}\, d\,\psi(v). \end{aligned} \tag{15.10}$$

Thus (15.04) becomes

$$\lim_{t\to 0} \frac{1}{t} \int_0^t \log\frac{t}{v}\, d\,\psi(v) = 0. \tag{15.11}$$

We have

$$\int_0^1 \log\lambda\, \lambda^{iu}\, d\lambda = -(iu+1)^{-2} \neq 0. \tag{15.12}$$

This is similar in form to (6.15).

The Riesz sum of kth order of the Fourier series of $f(x) - s$ for $x = 0$ is

$$\begin{aligned} &\lim_{\varepsilon\to 0} \frac{1}{\varepsilon} \int_0^\infty \frac{\varphi(t)\,dt}{\Gamma(1+k)} \int_0^1 (1-\lambda)^k \cos\frac{\lambda t}{\varepsilon}\, d\lambda \\ &= \lim_{\varepsilon\to 0} \frac{1}{\varepsilon} \int_0^\infty \frac{dt}{t\,\Gamma(1+k)} \int_0^t d\psi(u) \int_0^1 (1-\lambda)^k \cos\frac{\lambda t}{\varepsilon}\, d\lambda \\ &= \lim_{\varepsilon\to 0} \frac{1}{\varepsilon} \int_0^\infty d\psi(u) \int_u^\infty \frac{dt}{t\,\Gamma(1+k)} \int_0^1 (1-\lambda)^k \cos\frac{\lambda t}{\varepsilon}\, d\lambda \\ &= \lim_{\varepsilon\to 0} \frac{1}{\varepsilon} \int_0^\infty d\psi(u) \int_{u/\varepsilon}^\infty \frac{dt}{t\,\Gamma(k+1)} \int_0^1 (1-\lambda)^k \cos\lambda t\, d\lambda. \end{aligned} \tag{15.13}$$

[47] It is possible to make the transition to Theorem XIX, but several of the steps require some consideration. In particular, the equivalence of Riesz and Cesàro summation for orders > -1 was proved by Riesz (Proc. Lond. Math. Soc., 2 (22), 412–419 (1924).

Here if $k > -1$,

$$\int_0^\infty \left| \int_0^1 (1-\lambda)^k \cos \lambda\tau \, d\lambda \right| d\tau \tag{15.14}$$

is finite, and the inversion of integration in (15.13) is hence applied to an absolutely convergent integral, in view of (15.08).

We have

$$\int_0^\infty \omega^{iu} \, d\omega \int_\omega^\infty \frac{d\tau}{\tau \Gamma(1+k)} \int_0^1 (1-\lambda)^k \cos \lambda\tau \, d\lambda$$

$$= \frac{1}{iu+1} \int_0^\infty \frac{\omega^{iu}}{\Gamma(1+k)} d\omega \int_0^1 (1-\lambda)^k \cos \lambda\omega \, d\lambda$$

$$= \frac{1}{iu+1} \int_0^\infty \frac{\omega^{iu}}{\Gamma(1+k)} d\omega \int_{\lambda=0}^{\lambda=1} (1-\lambda)^k \frac{d \sin \lambda\omega}{\omega}$$

$$= \frac{1}{(iu+1)\Gamma(k)} \int_0^\infty \omega^{iu-1} \, d\omega \int_0^1 (1-\lambda)^{k-1} \sin \lambda\omega \, d\lambda$$

$$= \frac{1}{(iu+1)\Gamma(k)} \int_0^1 (1-\lambda)^{k-1} \, d\lambda \int_0^\infty \omega^{iu-1} \sin \lambda\omega \, d\omega$$

$$= \frac{1}{(iu+1)\Gamma(k)} \int_0^1 (1-\lambda)^{k-1} \lambda^{-iu} \, d\lambda \int_0^\infty z^{iu-1} \sin z \, dz \tag{15.15}$$

$$= \frac{1}{(iu+1)\Gamma(k)} \frac{\Gamma(k)\Gamma(1-iu)}{\Gamma(k+1-iu)} \Gamma(iu) \cos (iu-1)\frac{\pi}{2}$$

$$= \frac{\Gamma(1-iu)\Gamma(iu)}{(iu+1)\Gamma(k+1-iu)} \sin \frac{\pi iu}{2}$$

$$= \frac{\pi \sin \frac{\pi iu}{2}}{(iu+1)\Gamma(k+1-iu) \sin \pi iu}$$

$$= \frac{\pi}{2(iu+1)\Gamma(k+1-iu) \cos \frac{\pi iu}{2}}$$

$$\neq 0.$$

In proving this, we have made use of (13.131) and (13.132).

The other conditions of the hypothesis of Theorem XI′ are readily proved to be satisfied, and we have already shown that if (15.05) is satisfied, (15.11) *and*

$$\lim_{\varepsilon \to 0} \frac{1}{\varepsilon} \int_0^\infty \frac{\varphi(t)\, dt}{\Gamma(1+k)} \int_0^1 (1-\lambda)^k \cos \frac{\lambda t}{\varepsilon} \, d\lambda = 0, \tag{15.16}$$

are equivalent. Thus if (15.05) *is satisfied,* (15.04) *is completely equivalent to the statement that the Fourier series of* $f(x)$ *is summable to s by Riesz (or Cesàro) sums of any given order exceeding* -1.

16. Tauberian theorems and asymptotic series. Certain asymptotic problems arise in the discussion of the behavior of an integral

$$\int_a^b (g(t))^x \Phi(t)\, dt \tag{16.01}$$

for large values of x. By a change of variable, this problem may be reduced to the consideration of

$$\int_0^\infty e^{-xt} \Phi(t)\, dt. \tag{16.02}$$

In particular, let us discuss the situation which arises when $\nu > 0$ and

$$\lim_{x\to\infty} x^\nu \int_0^\infty e^{-xt} \Phi(t)\, dt = A, \tag{16.03}$$

which we may write

$$\lim_{x\to\infty} x \int_0^\infty (x\,t)^{\nu-1} e^{-xt} \frac{\Phi(t)}{t^{\nu-1}}\, dt = A. \tag{16.04}$$

If

$$x \int_0^\infty (x\,t)^{\nu-1} e^{-xt} \frac{\Phi(t)}{t^{\nu-1}}\, dt \tag{16.05}$$

is bounded and $\Phi(t)$ is positive, it will follow by Theorem XI″ that

$$\lim_{x\to\infty} x \int_0^{1/x} \frac{\Phi(t)}{t^{\nu-1}}\, dt = \frac{A}{\int_0^\infty t^{\nu-1} e^{-t}\, dt} = \frac{A}{\Gamma(\nu)}, \tag{16.06}$$

or that

$$\lim_{\varepsilon\to 0} \frac{1}{\varepsilon} \int_0^\varepsilon \frac{\Phi(t)}{t^{\nu-1}}\, dt = \frac{A}{\Gamma(\nu)}. \tag{16.07}$$

Again, it will follow that

$$\lim_{x\to\infty} x \int_{1/x}^\infty \frac{\Phi(t)}{t^{\nu-1}} \frac{dt}{x^2 t^2} = \frac{A}{\Gamma(\nu)}, \tag{16.08}$$

which yields

$$\lim_{\varepsilon\to 0} \varepsilon \int_0^\varepsilon t^{\nu-1} \Phi\left(\frac{1}{t}\right) dt = \frac{A}{\Gamma(\nu)}. \tag{16.09}$$

The condition that

$$\Gamma(iu+\nu) = \int_0^\infty e^{-t}\, t^{iu+\nu-1}\, dt \neq 0 \tag{16.10}$$

is obviously satisfied. Indeed the Tauberian theorems of this section (which are due to Ikehara[48]) differ only from those of the last section in that x tends to infinity instead of to 0.

If $\frac{\Phi(t)}{t^{\nu-1}}$ is bounded and tends to B at the origin and $\nu > 0$,

$$\lim_{x\to\infty} x \int_0^\infty (x\,t)^{\nu-1} e^{-xt} \left[\frac{\Phi(t)}{t^{\nu-1}} - B\right] dt = 0. \tag{16.11}$$

Then

$$\lim_{x\to\infty} x \int_0^\infty (xt)^{\nu-1} e^{-xt} \frac{\Phi(t)}{t^{\nu-1}}\, dt = B\,x^\nu \int_0^\infty t^{\nu-1} e^{-xt}\, dt = B\,\Gamma(\nu). \tag{16.12}$$

Thus

$$\lim_{x\to\infty} x^\nu \int_0^{\infty} e^{-xt}\, \Phi(t)\, dt = B\,\Gamma(\nu). \tag{16.13}$$

In particular, let there be a neighborhood of the origin in which $\Phi(t)$ is $\nu-1$ times differentiable, with a bounded derivative of the $(\nu-1)$st order. Let

$$\Phi(0) = 0; \quad \Phi'(0) = 0; \quad \cdots \quad \Phi^{(\nu-2)}(0) = 0; \quad \Phi^{(\nu-1)}(0) = A \tag{16.14}$$

and let $\frac{\Phi(t)}{t^{\nu-1}}$ be bounded outside the neighborhood in question. Then

$$\lim_{x\to\infty} x^\nu \int_0^\infty e^{-xt}\, \Phi(t)\, dt = A. \tag{16.03}$$

If now $\Phi(t) - A\,t^{\nu-1}$ has a bounded derivative of order $\mu - 1 > \nu - 1$ at the origin, and vanishes there with all its derivatives of order less than $\mu - 1$, and $\Psi(t)$ is a function which equals $\Phi(t) - \frac{A\,t^{\nu-1}}{\Gamma(\nu)}$ in some neighborhood of the origin, and for which $\frac{\Psi(t)}{t^{\mu-1}}$ is bounded and tends to A at the origin, we have

$$\lim_{x\to\infty} x^\mu \int_0^\infty e^{-xt}\, \Psi(t)\, dt = A. \tag{16.15}$$

By a repetition of this process, it is easy to show that if

$$\Phi(t) = \frac{A_1\, t^{\nu_1-1}}{\Gamma(\nu_1)} + \frac{A_2\, t^{\nu_2-1}}{\Gamma(\nu_2)} + \cdots + \frac{A_k\, t^{\nu_k-1}}{\Gamma(\nu_k)} + \Psi(t), \tag{16.16}$$

if $\Psi(t)$ is bounded and $(\mu - 1)$ times differentiable with bounded $(\mu - 1)$st derivative in some neighborhood of the origin, if

---

[48] In a paper not yet published.

(16.17) $\Psi(0) = 0; \quad \Psi'(0) = 0; \quad \cdots \quad \Psi^{(\mu-1)}(0) = 0,$

and if $\mu > \nu_1$, $\mu > \nu_2$, $\cdots$, $\mu > \nu_k$, then

(16.18) $$\int_0^\infty e^{-xt}\,\Phi(t)\,dt = A_1 x^{-\nu_1} + A_2 x^{-\nu_2} + \cdots + A_k x^{-\nu_k} + O(x^{-\mu+\varepsilon})$$

as $x \to \infty$. This is an adequate basis for the theory of asymptotic expansions of Laplace integrals, and enables Wintner's work[49] on the subject to be greatly simplified.

CHAPTER VI.

KERNELS ALMOST OF THE CLOSED CYCLE.

17. The reduction of kernels almost of the closed cycle to kernels of the closed cycle. In the present section, we shall approach very close to the work of Robert Schmidt[50] in his discussion of "gestrahlte Mittelbildungen", although our terminology will be somewhat different. Up to this point, we have been discussing means of the form

(17.01) $$\int_{-\infty}^{\infty} K_1(\xi - x)\,f(\xi)\,d\xi$$

or

(17.02) $$\int_{-\infty}^{\infty} K_1(\xi - x)\,d\varphi(\xi),$$

or means only differing from these by a change of variable from ξ to a function of ξ. Let us now turn our attention to means of the form

(17.03) $$\int_{-\infty}^{\infty} K_1(\xi, x)\,d\varphi(\xi),$$

where

(17.04) $$K_1(\xi, x) = K_1(\xi - x) + K_1^*(\xi, x)$$

where $K_1(x)$ satisfies the condition that (5.02) converges and that (5.07) does not vanish, and where

(17.05) $$\lim_{x\to\infty} \sum_{-\infty}^{\infty} \max_{n \leq y \leq n+1} |K_1^*(y, x)| = 0.$$

Let $K_1(x, y)$ and $K_1(x)$ be continuous, and let

(17.06) $$\int_{-\infty}^{\infty} K_1(x, y)\,d\varphi(x) \quad \text{and} \quad \int_y^{y+1} |d\varphi(x)| - \varphi(y+1) + \varphi(y)$$

[49] Wintner (1).

[50] Schmidt (2).

be bounded. Let

$$\lim_{x\to\infty} \int_{-\infty}^{\infty} K_1(\xi, x)\, d\varphi(\xi) = A \int_{-\infty}^{\infty} K_1(\xi)\, d\xi. \tag{17.061}$$

The argument of Lemma XIa will need no substantial alteration to show that there is a Q such that

$$\int_y^{y+1} |d\varphi(x)| \leqq Q. \tag{6.04}$$

It will then follow from (17.05) that

$$\lim_{x\to\infty} \int_{-\infty}^{\infty} K_1(\xi - x)\, d\varphi(\xi) = A \int_{-\infty}^{\infty} K_1(\xi)\, d\xi \tag{5.03}$$

and that

$$\int_{-\infty}^{\infty} K_1(\xi - x)\, d\varphi(\xi)$$

is bounded. Hence, by Theorem X, if K_2 is any function belonging to N_1,

$$\lim_{x\to\infty} \int_{-\infty}^{\infty} K_2(\xi - x)\, d\varphi(\xi) = A \int_{-\infty}^{\infty} K_2(\xi)\, d\xi. \tag{5.04}$$

In Chapter III, the kernel $K(\xi - x)$ is replaced by a kernel of the form

$$M(\xi + \alpha - x) - M(\xi - x) \tag{17.07}$$

where $M(x)$ is a monotone function for which

$$M(-\infty) = 1, \qquad M(x) = O(e^{-\mu x}) \text{ at } +\infty; \qquad [\mu > 0] \tag{7.13}$$

and for which the function

$$\frac{1}{\sqrt{2\pi}} \int_{-\infty}^{\infty} M(x)\, e^{xz}\, dx \tag{7.05}$$

which is defined over a strip to the right of the origin, when continued analytically on to the imaginary axis, has no zeros there. We can replace the kernel $M(\xi - x)$ by a kernel of the form

$$M(\xi, x) = M_1(\xi - x) + M_2(\xi, x) \tag{17.08}$$

where $M_1(x)$ satisfies the conditions we have already laid down for $M(x)$, and where

$$\lim_{x\to\infty} \int_0^{\infty} |dM_2(\xi, x)| = 0 \tag{17.09}$$

and

$$M_2(\xi, x) = O(e^{-\mu\xi}) \text{ at } \infty \text{ uniformly.} \tag{17.10}$$

Under this change, Theorem XIV still remains valid.

18. **A Tauberian theorem of Hardy and Littlewood.**[51] Hardy and Littlewood have proved the following theorem:

THEOREM XX. *Let $f(x)=\sum a_n x^n$ be a power series with positive coefficients, and let*

$$(18.01)\qquad f(x)\sim \frac{A}{(1-x)^\alpha}\qquad (\alpha>0,\ x\to 1).$$

Then

$$(18.02)\qquad \sum_1^n a_k \sim \frac{n^\alpha A}{\Gamma(\alpha+1)}\qquad (n\sim\infty).$$

We may write (18.01)

$$(18.03)\qquad \sum a_n e^{-n\xi}\sim A\,\xi^{-\alpha}$$

or

$$(18.04)\qquad \lim_{\xi\to 0}\xi\sum \frac{a_n}{n^{\alpha-1}}(n\xi)^{\alpha-1}e^{-n\xi}=A.$$

This falls under (6.30), and

$$(18.05)\qquad \int_0^\infty \xi^{\alpha-1}e^{-\xi}\xi^{iu}\,d\xi=\Gamma(\alpha+iu)\neq 0.$$

Since $a_n/n^{\alpha-1}>0$,

$$(18.06)\qquad \lim_{\xi\to 0}\xi\sum_1^{[1/\xi]}\frac{a_n}{n^{\alpha-1}}(n\xi)^{\alpha-1}=A\frac{\int_0^1 \xi^{\alpha-1}\,d\xi}{\int_0^\infty \xi^{\alpha-1}e^{-\xi}\,d\xi}=\frac{A}{\Gamma(\alpha+1)}$$

because of Theorem XI‴. Writing m for $1/\xi$, and letting m become infinite through integral values, we get

$$(18.07)\qquad \lim_{m\to\infty}\frac{1}{m^\alpha}\sum_1^m a_n=\frac{A}{\Gamma(\alpha+1)},$$

which is only another way of writing (18.02).

This however is not the most general theorem proved by Hardy and Littlewood in this connection. They show that if

$$(18.08)\qquad f(x)\sim\frac{A}{(1-x)^\alpha}L\left(\frac{1}{1-x}\right),$$

where

$$(18.09)\qquad L(u)=(\log u)^{\alpha_1}(\log\log u)^{\alpha_2}\cdots(\log^{(n)}u)^{\alpha_n}$$

and

$$(18.10)\qquad u^\alpha L(u)\neq O(1)\ \text{at}\ \infty,$$

then

$$(18.11)\qquad \sum_1^n a_k\sim\frac{n^\alpha A\,L(n)}{\Gamma(\alpha+1)}.$$

[51] Hardy and Littlewood (4).

We may write our theorem to be proved in the form that if

$$\xi \sum_1^\infty \frac{a_n}{n^{\alpha-1}} (n\xi)^{\alpha-1} e^{-n\xi} \sim A L(\xi^{-1}), \tag{18.12}$$

then (18.11) follows. This we may again write

$$e^{-\mu} \sum_1^\infty \frac{a_n}{n^{\alpha-1}} e^{(\log n-\mu)(\alpha-1)} e^{-e^{\log n-\mu}} \sim A L(e^\mu) \tag{18.13}$$

as $\mu \to \infty$. Now, if $M(\xi)$ is a continuous function, defined over $(0, \infty)$, and asymptotic to $L(\xi)$ at ∞, we may show that:

$$\int_0^\infty e^{(u-\mu)\alpha} e^{-e^{(u-\mu)}} M(e^u)\,du \sim M(e^\mu) \int_{-\infty}^\infty e^u e^{-e^u}\,du \sim L(e^\mu). \tag{18.14}$$

Here $M(\xi)$ is introduced instead of $L(\xi)$ in order that we may have no trouble with finite singularities of $L(\xi)$.

To prove (18.14), let us reflect that

$$\begin{aligned}\int_0^\infty e^{(u-\mu)\alpha} e^{-e^{(u-\mu)}} M(e^u)\,du &= \int_0^{\mu-A} e^{(u-\mu)\alpha} e^{-e^{(u-\mu)}} M(e^u)\,du \\ &+ \int_{\mu-A}^{\mu+A} e^{(u-\mu)\alpha} e^{-e^{(u-\mu)}} M(e^u)\,du \\ &+ \int_{\mu+A}^{\infty} e^{(u-\mu)\alpha} e^{-e^{(u-\mu)}} M(e^u)\,du.\end{aligned} \tag{18.141}$$

Since $M(\xi)$ is asymptotic to $L(\xi)$, which is asymptotically increasing, the first integral is asymptotically less than

$$\begin{aligned}&L(e^u) \int_0^{\mu-A} e^{(u-\mu)\alpha} e^{-e^{(u-\mu)}}\,du \\ &\leqq L(e^\xi) \int_{-\infty}^{-A} e^u e^{-e^u}\,du \leqq L(e^\mu)\, e^{-A}.\end{aligned} \tag{18.142}$$

As to the second integral, it is asymptotically

$$L(e^\mu) \int_{-A}^{A} e^{\alpha u} e^{-e^u}\,du, \tag{18.143}$$

while the third one is ultimately less than

$$\begin{aligned}&e^{A^2} e^{-e^A} \int_{\mu+A}^\infty e^{-(u-\mu)} L(e^u)\,du \\ &\leqq e^{A^2} e^{-e^A} e^\mu \int_\mu^\infty L(e^u)\, d(-e^{-u}) \\ &= e^{A^2} e^{-e^A} \left[L(e^\mu) + e^\mu \int_\mu^\infty e^{-u} \frac{d}{du}(L(e^u))\,du\right] \\ &\leqq e^{A^2} e^{-e^A} \left[L(e^\mu) + C e^\mu \int_\mu^\infty e^{-u} u^{\alpha_1-1/2}\,du\right].\end{aligned} \tag{18.144}$$

5

Now,

$$\begin{aligned}\int_{\mu}^{\infty} e^{-u}\, u^{\alpha_1-1/2}\, du &= -\int_{\mu}^{\infty} u^{\alpha_1-1/2}\, d(e^{-u}) \\ &= e^{-\mu}\, \mu^{\alpha_1-1/2} + \int_{\mu}^{\infty} (\alpha_1-\tfrac{1}{2})\, e^{-u}\, u^{\alpha_1-3/2}\, du \\ &\leqq e^{-\mu}\left(\mu^{\alpha_1-1/2} + \int_{\mu}^{\infty} (\alpha_1-\tfrac{1}{2})\, u^{\alpha_1-3/2}\, du\right) \\ &\leqq 2e^{-\mu}\, \mu^{\alpha_1-1/2}.\end{aligned} \tag{18.145}$$

Thus by (18.144)

$$\int_{\mu+A}^{\infty} e^{(u-\mu)\alpha}\, e^{-e^{(u-\mu)}}\, M(e^u)\, du \leqq e^{A^2}\, e^{-e^A}\, [L(e^{\mu}) + o(L(e^{\mu}))]. \tag{18.146}$$

Combining this with (18.141), (18.142) and (18.143), we see that

$$\begin{aligned}\overline{\lim_{\mu\to\infty}} \left| \int_0^{\infty} e^{(u-\mu)\alpha}\, e^{-e^{(u-\mu)}}\, M(e^u)\, du - L(e^{\mu}) \int_{-\infty}^{\infty} e^{u}\, e^{-e^{u}}\, du \right| \Big/ L(e^{\mu}) \\ \leqq \varphi(A)\end{aligned} \tag{18.147}$$

where $\varphi(A)$ vanishes as A becomes infinite. Since A is arbitrarily large, (18.14) follows.

Consequently, by (18.14), if we put

$$d\psi(u) = \frac{1}{u}\, d \sum_{0}^{[u]} \frac{a_n}{n^{\alpha-1}} - A\, M(u), \tag{18.15}$$

(18.12) becomes

$$\int_0^{\infty} (w\,\xi)^{\alpha}\, e^{-w\xi}\, d\psi(w) \to 0. \tag{18.16}$$

On the other hand, we may write (18.11) in the form

$$\int_0^{1/\xi} (w\,\xi)^{\alpha}\, d\psi(w) \to 0. \tag{18.17}$$

Both integrals, (18.16) and (18.17), converge absolutely. We wish to make the transition from (18.16) to (18.17).

If $\alpha_1 \leqq 1$, $\psi(w)$ is "slowly decreasing" in the sense of Schmidt, we can apply Theorem XV, and it appears that $\psi(\lambda) \to 0$ as $\lambda \to \infty$. From this it follows that

$$\int_0^{1/\xi} (w\,\xi)^{\alpha}\, d\psi(w) = \psi\left(\frac{1}{\xi}\right) - \psi(0) - \alpha \int_0^{1/\xi} \xi^{\alpha}\, w^{\alpha-1}\, \psi(w)\, dw \to 0. \tag{18.18}$$

We now come to the more general case where $\alpha_1 > 1$. Our hypothesis becomes

$$e^{-\mu} \sum \frac{a_n}{n^{\alpha-1}(\log n)^{[\alpha_1]}}\, e^{(\log u-\mu)(\alpha-1)}\, e^{-e^{\log n-\mu}}\, \frac{\log n^{[\alpha_1]}}{\mu} \sim A\, \mathcal{A}(e^{\mu}), \tag{18.19}$$

where

(18.20) $$\Lambda(u) = (\log u)^{\alpha_1 - [\alpha_1]} (\log \log u)^{\alpha_2} \cdots (\log^{(n)} u)^{\alpha_n}.$$

The kernel of (18.19), in the sense of § 17, is

(18.21)
$$e^{(\alpha-1)(y-x)} e^{-e^{y-x}} \left\{\frac{y}{x}\right\}^{[\alpha_1]} = e^{(\alpha-1)(y-x)} e^{-e^{y-1}} \left\{1 - \frac{x-y}{x}\right\}^{[\alpha_1]}$$
$$+ \left\{-[\alpha_1]\frac{x-y}{x} + \frac{[\alpha_1]\{[\alpha_1]-1\}}{2} \frac{(x-y)^2}{x^2} + \cdots + \frac{(x-y)}{x^{[\alpha_1]}}\right\}.$$

It hence appears that this kernel is of the form indicated in (17.04) where K_1 and K_1^* satisfy the appropriate conditions. Thus if we put $M^*(\xi) \sim \Lambda(\xi)$ at ∞ and $M^*(\xi)$ is continuous, and if

(18.22) $$d\psi^*(u) = \frac{1}{u} d \sum_0^{[u]} \frac{a_n}{n^{\alpha-1} (\log n)^{[\alpha_1]}} - A M^*(u),$$

it follows just as in the case where $\alpha_1 \leqq 1$ that

(18.23) $$\int_0^{1/\xi} (w\xi)^\alpha \, d\psi^*(w) \to 0.$$

This however is only another way of writing (18.11).

19. The Tauberian theorem of Borel summation.[52] The Borel sum of the series with partial sums s_m is

(19.01) $$\lim_{x \to \infty} e^{-x} \sum_0^\infty \frac{s_n x^n}{n!} = s.$$

Let us put $n = [u^2]$ and

(19.02) $$\sum_1^n s_m (\sqrt{m} - \sqrt{m-1}) = f(u)$$

for $u > 0$. Let $f(u) = 0$ for negative u. Let us also put $x = y^2$. Then (19.01) becomes

(19.03) $$\lim_{y \to \infty} \int_0^\infty \frac{e^{-y^2} y^{2u^2}}{\Gamma(u^2+1)} \frac{df(u)}{u - \sqrt{u^2-1}} = s.$$

Now

(19.04)
$$\frac{e^{-y^2} y^{2u^2}}{\Gamma(u^2+1)(u - \sqrt{u^2-1})} = \sqrt{\frac{2}{\pi}} e^{u^2-y^2} \left(\frac{y}{u}\right)^{2u^2} \left(1 + O\left(\frac{1}{u^2}\right)\right)$$
$$= \sqrt{\frac{2}{\pi}} e^{-2(u-y)^2} e^{3u^2-4uy+y^2} \left(\frac{y}{u}\right)^{2u^2} \left(1 + O\left(\frac{1}{u^2}\right)\right).$$

[52] For this section, cf. Schmidt (2); Vijayaraghavan (1).

Thus over the range $y - y^{1/6} \leqq u \leqq y + y^{1/6}$,

$$(19.05)\quad \frac{e^{-y^2} y^{2u^2}}{\Gamma(u^2+1)(u-\sqrt{u^2-1})} - \sqrt{\frac{2}{\pi}}\, e^{-2(u-y)^2} = \sqrt{\frac{2}{\pi}}\, e^{-2(u-y)^2} O\left(\frac{1}{y^2}\right)$$

because

$$(19.06)\quad \begin{aligned} & e^{3u^2-4uy+y^2}\left(\frac{y}{u}\right)^{2u^2} \\ &= \exp\left\{3u^2 - 4uy + y^2 + 2u^2\left[\frac{y-u}{u} - \frac{(y-u)^2}{2u^2} + O\left(\frac{(y-u)^3}{u^3}\right)\right]\right\} \\ &= 1 + O\left(\frac{(y-u)^3}{u}\right) \\ &= 1 + O(y^{-1/2}). \end{aligned}$$

Moreover, it is always true that

$$(19.07)\quad \begin{aligned} \frac{e^{-y^2} y^{2u^2}}{\Gamma(u^2+1)(u-\sqrt{u^2-1})} &= \sqrt{\frac{2}{\pi}}\, e^{u^2-y^2}\left(\frac{y}{u}\right)^{2u^2}\left(1+O\left(\frac{1}{u^2}\right)\right) \\ &\leqq \sqrt{\frac{2}{\pi}}\, e^{u^2-y^2}\, e^{\left(\frac{y-u}{u}\right)2u^2}\left(1+O\left(\frac{1}{u^2}\right)\right) \\ &= \sqrt{\frac{2}{\pi}}\, e^{-(u-y)^2}\left(1+O\left(\frac{1}{u^2}\right)\right). \end{aligned}$$

Thus the conditions (17.05) and (17.06) are satisfied if we put

$$(19.08)\quad \frac{e^{-y^2} y^{2u^2}}{\Gamma(u^2+1)(u-\sqrt{u^2-1})} = K_1(u, y)$$

and

$$(19.09)\quad \sqrt{\frac{2}{\pi}}\, e^{-2(u-y)^2} = K_1(u-y).$$

It follows at once that

$$(19.10)\quad \lim_{y\to\infty} \sqrt{\frac{2}{\pi}} \int_{-\infty}^{\infty} e^{-2(u-y)^2}\, df(u) = s$$

and that

$$(19.101)\quad \sqrt{\frac{2}{\pi}} \int_{-\infty}^{\infty} e^{-2(u-y)^2}\, df(u)$$

is bounded.

We now introduce again in its appropriate form Robert Schmidt's definition of a "langsam abfallende Folge". The sequence $\{s_m\}$ has this property if, whenever $q = q(p)$ $(p = 0, 1, \cdots)$ runs through such a sequence of indices that

$$(19.11)\quad q \geqq p \quad \text{and} \quad \frac{q-p}{\sqrt{p}} \to 0$$

then

$$\underline{\lim}_{p\to\infty} (s_q - s_p) \geqq 0. \tag{19.12}$$

Schmidt proves by elementary methods that if $s_0, s_1, \cdots$ is such a sequence, then

$$C(\Lambda) = \underline{\lim}_{\substack{p\to\infty \\ p\leqq q\leqq p+\Lambda\sqrt{p}}} (s_q - s_p) \tag{19.13}$$

exists, and

$$\lim_{\Lambda\to 0} C(\Lambda) = 0. \tag{19.14}$$

Furthermore, he shows that there exists a constant K, such that

$$s_q - s_p \geqq -K(\sqrt{q} - \sqrt{p}) \tag{19.15}$$

for all $p = 0, 1, \cdots$ and all $q \geqq p + \sqrt{p}$. Thus he shows that a K exists such that

$$\frac{s_p}{\sqrt{p+1}} \geqq -K \qquad (p = 0, 1, \cdots). \tag{19.16}$$

Let the sequence s_m be "langsam abfallend" and let (19.01) hold. Still following Schmidt, we see that

$$\begin{aligned} e^{-x}\sum_0^\infty \frac{s_n x^n}{n!} &= \sum_0^{[x]} \frac{e^{-x} s_n x^n}{n!} + \sum_{[x]+1}^{[x+\sqrt{x}\,]} \frac{e^{-x} s_n x^n}{n!} + \sum_{[x+\sqrt{x}\,]+1}^{\infty} \frac{e^{-x} s_n x^n}{n!} \\ &\geqq -K\sum_0^\infty \frac{e^{-x}\sqrt{n+1}\,x^n}{n!} + s_{[x]}\sum_{[x]+1}^{[x+\sqrt{x}\,]} \frac{e^{-x}x^n}{n!} - C(1)\sum_{[x]+1}^{[x+\sqrt{x}\,]} \frac{e^{-x}x^n}{n!}. \end{aligned} \tag{19.17}$$

If we build up the appropriate $f(u)$ as in (19.02), and make the transformation which led from (19.01) to (19.03), we get:

$$e^{-x}\sum_0^\infty \frac{\sqrt{n+1}\,x^n}{n!} \leqq \int_0^\infty \frac{e^{-x}x^{u^2}}{\Gamma(u^2+1)} \frac{d(u^2)}{u - \sqrt{u^2-1}}. \tag{19.18}$$

Again, by (19.07)

$$e^{-x}\sum_0^\infty \frac{\sqrt{n+1}\,x^n}{n!} \leqq \int_{-\infty}^\infty e^{-(u-\sqrt{x})^2} O(u)\,du = O(\sqrt{x}). \tag{19.19}$$

Obviously

$$\sum_{[x]+1}^{[x+\sqrt{x}\,]} \frac{e^{-x}x^n}{n!} < 1. \tag{19.20}$$

On the other hand

$$\varliminf_{x\to\infty} \sum_{[x]+1}^{[x+\sqrt{x}]} \frac{e^{-x}x^n}{n!} > \varliminf_{x\to\infty} \frac{e^{-x}x^x}{e^{-x-\sqrt{x}}(x+\sqrt{x})^x\sqrt{2\pi(x+\sqrt{x})}}[\sqrt{x}+1]$$

$$(19.21) \qquad = \lim_{x\to\infty} \frac{e^{\sqrt{x}}}{\left(1+\frac{1}{\sqrt{x}}\right)^x \sqrt{2\pi}} = \left(\frac{e}{2\pi}\right)^{1/2}.$$

Thus (19.17) yields

$$(19.22) \quad \overline{\lim_{x\to\infty}}\, x^{-1/2} s_{[x]} \leqq \overline{\lim_{x\to\infty}} \sqrt{\frac{2\pi}{e}}\,[s x^{-1/2}+KO(1)+C(1)\,x^{-1/2}] = O(1),$$

which we may simplify and write

$$(19.23) \qquad s_{[x]} = O(x^{1/2}).$$

Let us now introduce the function

$$(19.24) \quad g_\alpha(x) = \frac{1}{\alpha^2}\left[\int_{x+\alpha}^{x+2\alpha} - \int_x^{x+\alpha}\right] f(y)\,dy.$$

By (19.02),

$$(19.25) \quad g_\alpha(x) = \frac{1}{\alpha^2}\left[\int_{(x+\alpha)^2}^{(x+2\alpha)^2} - \int_{x^2}^{(x+\alpha)^2}\right] \frac{dM}{2\sqrt{M}} \sum_1^{[M]} s_m(\sqrt{m}-\sqrt{m-1}).$$

Now,

$$\int_0^\nu \frac{dM}{2\sqrt{M}} \sum_1^{[M]} s_m(\sqrt{m}-\sqrt{m-1})$$

$$(19.26) \qquad = \sum_1^{[\nu]} s_m(\sqrt{m}-\sqrt{m-1}) \int_m^\nu \frac{dM}{2\sqrt{M}}$$

$$= \sum_1^{[\nu]} s_m(\sqrt{m}-\sqrt{m-1})(\sqrt{\nu}-\sqrt{m}).$$

Substituting this in (19.25), and writing S for an average of the s_n's for $x^2 < n < (x+2\alpha)^2+1$, we have

$$g_\alpha(x) = \frac{1}{\alpha^2} \sum_1^{[(x+2\alpha)^2]} s_m(\sqrt{m}-\sqrt{m-1})(x+2\alpha-\sqrt{m})$$

$$-\frac{2}{\alpha^2} \sum_1^{[(x+\alpha)^2]} s_m(\sqrt{m}-\sqrt{m-1})(x+\alpha-\sqrt{m})$$

$$+\frac{1}{\alpha^2} \sum_1^{[x^2]} s_m(\sqrt{m}-\sqrt{m-1})(x-\sqrt{m})$$

$$= \frac{1}{\alpha^2}\left\{ \sum_{[x^2]+1}^{[(x+\alpha)^2]} s_m(\sqrt{m}-\sqrt{m-1})(\sqrt{m}-x) \right.$$

$$\left. + \sum_{[(x+\alpha)^2]+1}^{[(x+2\alpha)^2]} s_m(\sqrt{m}-\sqrt{m-1})(x+2\alpha-\sqrt{m}) \right\}$$

$$(19.27) \quad = \frac{S}{\alpha^2}\left\{\sum_{[x^2]+1}^{[(x+\alpha)^2]}(\sqrt{m}-\sqrt{m-1})(\sqrt{m}-x) + \sum_{[(x+\alpha)^2]+1}^{[(x+2\alpha)^2]}(\sqrt{m}-\sqrt{m-1})(x+2\alpha-\sqrt{m})\right\}$$

$$= \frac{S}{\alpha^2}\left[\int_{x+\alpha}^{x+2\alpha} - \int_{x}^{x+\alpha}\right]\sqrt{[x^2]}\,dx$$

$$= S\left\{1+\frac{1}{\alpha^2}\left[\int_{x+\alpha}^{x+2\alpha} - \int_{x}^{x+\alpha}\right](\sqrt{[x^2]}-x)\,dx\right\}$$

$$= S\left\{1+O\left(\frac{1}{x^2}\right)+\frac{1}{\alpha^2}\left[\sum_{[x^2]}^{[(x+\alpha)^2]} - \sum_{[(x+\alpha)^2]}^{[(x+2\alpha)^2]}\right]\frac{(\sqrt{m+1}-\sqrt{m})^2}{2}\right\}$$

$$= S\left(1+O\left(\frac{1}{x^2}\right)\right).$$

It is indeed easy to show that as x increases, the weight of the later s_n's averaged in S increases at the expense of the weight of the earlier ones.

We may show that the condition that $\{s_m\}$ is "langsam abfallend" may be put in the form that if

$$(19.28) \qquad v \geqq u, \quad v-u\to 0$$

then

$$(19.29) \qquad \underline{\lim}_{u\to\infty}(s_{[v^2]}-s_{[u^2]}) = 0.$$

Hence by (19.27) and (19.23), we may readily show that

$$(19.30) \qquad \underline{\lim}_{u\to\infty}(g_\alpha(v)-g_\alpha(u)) \geqq 0.$$

Again remembering (19.23), which makes all the integrals in question absolutely convergent, we have

$$\lim_{y\to\infty}\sqrt{\frac{2}{\pi}}\int_{-\infty}^{\infty} e^{-2(u-y)^2} g_\alpha(u)\,du$$

$$(19.31) \quad = \lim_{y\to\infty}\sqrt{\frac{2}{\pi}}\int_{-\infty}^{\infty} e^{-2(u-y)^2}\,du\,\frac{1}{\alpha}\int_u^{u+\alpha}\frac{dv}{\alpha}\int_v^{v+\alpha} df(w)$$

$$= \lim_{y\to\infty}\frac{1}{\alpha^2}\sqrt{\frac{2}{\pi}}\int_y^{y+\alpha}dv\int_v^{v+\alpha}dw\int_{-\infty}^{\infty}e^{-2(u-w)^2}\,df(u)$$

$$= \lim_{y\to\infty}\sqrt{\frac{2}{\pi}}\int_{-\infty}^{\infty}e^{-2(u-y)^2}\,df(u) = s.$$

An integration by parts yields:

$$(19.32) \qquad s = \lim_{y\to\infty}\sqrt{\frac{2}{\pi}}\int_{-\infty}^{\infty}d\,g_\alpha(v)\int_v^{\infty}e^{-2(u-y)^2}\,du.$$

Similar methods establish the boundedness of

$$(19.33) \qquad \sqrt{\frac{2}{\pi}} \int_{-\infty}^{\infty} d\, g_\alpha(v) \int_{v}^{\infty} e^{-2(u-y)^2}\, du.$$

We may then apply Theorem XIV, taking

$$(19.34) \qquad M(x) = \sqrt{\frac{2}{\pi}} \int_{x}^{\infty} e^{-2u^2}\, du.$$

In this case

$$(19.35) \qquad \varphi(z) = \sqrt{\frac{1}{2\pi z}}\, e^{z^2/8}.$$

Hence

$$(19.36) \qquad \lim_{x \to \infty} g_\alpha(x) = s.$$

Let us now return to (19.27), making use of the existence of (19.13) and of (19.14). Then if x is sufficiently large,

$$(19.37) \qquad S - s_{[x^2]} \geqq C(5\alpha)$$

and

$$(19.38) \qquad s_{[(x+2\alpha)^2]} - S \geqq C(5\alpha).$$

It then follows from (19.27) and (19.36) that

$$(19.39) \qquad \overline{\lim_{n \to \infty}}\, s_n \leqq s - C(5\alpha)$$

and

$$(19.40) \qquad \underline{\lim_{n \to \infty}}\, s_n \geqq s + C(5\alpha).$$

Combining these, and remembering that $C(\Delta) \leqq 0$, we see that the following theorem holds:

THEOREM XXI. *Let* (19.01) *hold, and let* s_m *be "langsam abfallend". Then*

$$(19.41) \qquad \lim_{n \to \infty} s_n = s.$$

CHAPTER VII.

A QUASI-TAUBERIAN THEOREM.

20. **The quasi-Tauberian theorem.** Up to the present point, all the Tauberian theorems we have discussed have involved some auxiliary condition of boundedness or positiveness or slow decrease. In the present chapter, we shall discuss a theorem without any such auxiliary condition as to the function averaged. The type of theorem is so fundamentally

different from that already discussed that the nomenclature, "Tauberian", seems to the author unfortunate. The theorems now to be discussed are in essence much closer to Abel's theorem than to Tauber's theorem.

By an integration by parts, we establish the following:

LEMMA XXII a. *Let*

$$\int_B^C |F(x)|\,|df(x)| \tag{20.01}$$

exist for every $C > B$, *and let*

$$\int_B^\infty F(x)\,df(x) \tag{20.02}$$

exist as the limit of $\int_B^C F(x)\,df(x)$ *as* $C\to\infty$. *Let*

$$\int_B^C |G(x)|\,|df(x)| \tag{20.03}$$

exist for each $C > B$, *let* $G(x)$ *and* $F(x)$ *be continuous, and let*

$$\frac{G(x)}{F(x)} \to L \text{ as } x\to\infty; \qquad \int_D^\infty \left| d\,\frac{G(x)}{F(x)} \right| < \infty. \tag{20.04}$$

Then

$$\int_B^\infty G(x)\,df(x) \tag{20.05}$$

will exist, and

$$\begin{aligned} &\int_B^\infty G(x)\,df(x) \\ &= L\int_B^\infty F(x)\,df(x) - \int_B^\infty d\left(\frac{G(x)}{F(x)}\right)\int_B^x F(\xi)\,df(\xi). \end{aligned} \tag{20.06}$$

Now let us suppose that $f(x)$ is of limited total variation over every finite interval, that $K_1(x)$ is bounded and continuous, that

$$\int_{-\infty}^\infty |d(K_1(x)\,e^{-\lambda x})| < \text{const.}, \tag{20.061}$$

and that

$$\int_D^\infty K_1(-x)\,d\,f(x) \tag{20.07}$$

exists in the sense of (20.02). Let us suppose further that as $x\to-\infty$

$$K_1(x) \sim A_1\,e^{\lambda x}, \qquad (\lambda > 0) \tag{20.08}$$

where $A_1 \neq 0$. Then

$$\int_{-\infty}^{-D}\left|d\frac{e^{\lambda x}}{K_1(x)}\right|<\text{const.,}\tag{20.081}$$

if D is sufficiently large that for $x \leqq -D$,

$$|e^{-\lambda x}K_1(x)| \geqq \text{const.} > 0.\tag{20.082}$$

As a particular case of (20.06), if B is large enough,

$$\begin{aligned}&\int_B^\infty e^{-\lambda x}\,d\,f(x)\\ &= \frac{1}{A_1}\int_B^\infty K_1(-x)\,d\,f(x) - \int_B^\infty d\left(\frac{e^{-\lambda x}}{K_1(-x)}\right)\int_B^x K_1(-\xi)\,d\,f(\xi)\end{aligned}\tag{20.083}$$

and hence

$$\int_B^\infty e^{-\lambda x}\,d\,f(x)\tag{20.084}$$

is bounded. A further application of (20.06) yields us

$$\begin{aligned}&\int_B^\infty K_1(y-x)\,d\,f(x)\\ &= A_1 e^{\lambda y}\int_B^\infty e^{-\lambda x}\,d\,f(x) - \int_B^\infty d(K_1(y-x)\,e^{\lambda x})\int_B^x e^{-\lambda\xi}\,d\,f(\xi).\end{aligned}\tag{20.085}$$

From this we may conclude that

$$\int_0^\infty K_1(y-x)\,d\,f(x) = \begin{cases}O(e^{\lambda y}) & \text{as } y\to\infty,\\ O(1) & \text{as } y\to-\infty\end{cases}\tag{20.09}$$

and that

$$\lim_{B\to\infty} e^{-\lambda y}\int_B^\infty K_1(y-x)\,d\,f(x) = 0\tag{20.10}$$

uniformly in y.

Now let $R(z)$ be a function for which

$$\int_{-\infty}^\infty e^{-\lambda z}\,|dR(z)|\tag{20.11}$$

and

$$\int_{-\infty}^\infty |dR(z)|\tag{20.12}$$

are finite. It follows from (20.10) that

$$\begin{aligned}&\int_{-\infty}^\infty dR(y-z)\int_0^\infty K_1(z-x)\,d\,f(x)\\ &= \lim_{B\to\infty}\int_{-\infty}^\infty dR(y-z)\int_0^B K_1(z-x)\,d\,f(x)\\ &= \lim_{B\to\infty}\int_0^B d\,f(x)\int_{-\infty}^\infty K_1(z-x)\,d\,R(y-z)\\ &= \int_0^\infty d\,f(x)\int_{-\infty}^\infty K_1(y-x-z)\,d\,R(z).\end{aligned}\tag{20.13}$$

Thus if

$$\lim_{y\to\infty}\int_0^\infty K_1(y-x)\,d\,f(x) = A\int_{-\infty}^\infty K_1(x)\,dx \tag{20.14}$$

then

$$\int_0^\infty K_1(y-x)\,d\,f(x) \tag{20.15}$$

will be bounded, by (20.09), and it follows that

$$\begin{aligned}
&\lim_{y\to\infty}\int_0^\infty d\,f(x)\int_{-\infty}^\infty K_1(y-x-z)\,d\,R(z)\\
&\quad = \lim_{y\to\infty}\int_{-\infty}^\infty d\,R(y-z)\int_0^\infty K_1(z-x)\,d\,f(x)\\
&\quad = \int_{-\infty}^\infty d\,R(-z)\lim_{y\to\infty}\int_0^\infty K_1(z+y-x)\,d\,f(x)\\
&\quad = A\int_{-\infty}^\infty K_1(x)\,d\,x\int_{-\infty}^\infty d\,R(-z)\\
&\quad = A\int_{-\infty}^\infty d\,x\int_{-\infty}^\infty K_1(x-z)\,d\,R(z).
\end{aligned} \tag{20.16}$$

Hence if $K_1(x)$ is bounded and continuous, if $f(x)$ is of limited total variation over every finite interval, if (20.14), (20.061) *and* (20.08) *hold, if* (20.11) *and* (20.12) *are finite, and if*

$$K_2(x) = \int_{-\infty}^\infty K_1(x-z)\,d\,R(z), \tag{20.17}$$

it follows that

$$\lim_{y\to\infty}\int_0^\infty K_2(y-x)\,d\,f(x) = A\int_{-\infty}^\infty K_2(x)\,d\,x. \tag{20.18}$$

This is a sufficiently important theorem to dignify by a number; we shall call it Theorem XXII.

A closely related proposition is:

Theorem XXIII. *In the hypothesis of Theorem* XXII, *in case*

$$K_1(x) = 0, \qquad [x > 0], \tag{20.19}$$

we may replace the assumption of the finiteness of (20.11) *and* (20.12) *by that of the finiteness of*

$$\int_{-\infty}^\infty |\,d\,R(z)\,|. \tag{20.20}$$

The conclusion remains valid.

To prove this, let us reflect that if (20.08) holds as $x \to \infty$, since

$$\int_0^\infty K_1(y-x)\,df(x) = \int_y^\infty K_1(y-x)\,df(x), \tag{20.21}$$

it will follow from (20.06) that

$$\begin{aligned}\int_y^\infty e^{-\lambda x}\,df(x) &= \frac{e^{-\lambda y}}{A_1}\int_y^\infty K_1(y-x)\,df(x)\\ &\quad -\int_y^\infty d\left(\frac{e^{-\lambda x}}{K_1(y-x)}\right)\int_y^x K_1(y-\xi)\,df(\xi)\\ &= e^{-\lambda y}\,O(1).\end{aligned} \tag{20.22}$$

By a precisely similar argument,

$$\begin{aligned}&\int_B^\infty K_1(y-x)\,df(x)\\ &= A_1 e^{\lambda y}\int_B^\infty e^{-\lambda x}\,df(x) - \int_B^\infty d_x\,(e^{\lambda x} K_1(y-x))\int_B^\infty e^{-\lambda \xi}\,df(\xi)\\ &= A_1 e^{\lambda(y-B)}O(1) - \int_B^\infty d_x(e^{\lambda x} K_1(y-x))\,(e^{-\lambda B}O(1)+e^{-\lambda x}O(1))\\ &= O(e^{\lambda(y-B)}) \quad as \quad B\to\infty\end{aligned} \tag{20.23}$$

if $B > y$. Thus by (20.21), $\int_B^\infty K_1(y-x)\,df(x)$ is less than a function which is bounded and decreases monotonely to 0 as B becomes infinite. Consequently, by a theorem of Daniell on the Stieltjes integral,[53] in combination with the finiteness of (20.21),

$$\lim_{B\to\infty}\int_{-\infty}^\infty dR(y-z)\int_B^\infty K_1(z-x)\,df(x) = 0. \tag{20.24}$$

Thus the inversions of integration in (20.13) are again permissible.

Formally and heuristically, (20.17) is equivalent to

$$\frac{\int_{-\infty}^\infty K_2(x)\,e^{ux}\,dx}{\int_{-\infty}^\infty K_1(x)\,e^{ux}\,dx} = \int_{-\infty}^\infty e^{ux}\,dR(x). \tag{20.25}$$

If K_1 and K_2 are both $O(e^{-(u+\varepsilon)x})$ at $+\infty$ and $O(e^{+\varepsilon x})$ at $-\infty$,

$$k_2(u) = \int_{-\infty}^\infty K_2(x)\,e^{ux}\,dx \tag{20.26}$$

and

$$k_1(u) = \int_{-\infty}^\infty K_1(x)\,e^{ux}\,dx \tag{20.27}$$

[53] Daniell (1).

are both analytic over $-\varepsilon < \Re(u) < \varepsilon + \mu$. Formally,

$$R(x) = \frac{1}{2\pi i} \int_0^x d\xi \int_{-\infty}^{\infty} \frac{k_2(u)}{k_1(u)} e^{-u\xi} du, \tag{20.28}$$

the second integral being taken along any ordinate in the strip $-\varepsilon < \Re(u) < \varepsilon + \mu$.

Now let $K_1(x)$ belong to L_2. Let $k_2(u)/k_1(u)$ be analytic over $-\varepsilon \leqq \Re(u) \leqq \mu + \varepsilon$, and let it be quadratically summable over every ordinate in that strip. Then

$$e^{-\varepsilon x} \operatorname*{l.i.m.}_{A\to\infty} \int_{-A}^{A} \frac{k_2(iu)}{k_1(iu)} e^{-iux} du \tag{20.29}$$

and

$$e^{(\mu+\varepsilon)x} \operatorname*{l.i.m.}_{A\to\infty} \int_{-A}^{A} \frac{k_2(iu)}{k_1(iu)} e^{-iux} du \tag{20.30}$$

are quadratically summable. As a consequence,

$$\operatorname*{l.i.m.}_{A\to\infty} \int_{-A}^{A} \frac{k_2(iu)}{k_1(iu)} e^{-iux} du \tag{20.31}$$

and

$$e^{\mu x} \operatorname*{l.i.m.}_{A\to\infty} \int_{-A}^{A} \frac{k_2(iu)}{k_1(iu)} e^{-iux} du \tag{20.32}$$

are absolutely summable. If we assume R to be defined as in (20.28) and $\mu = \lambda$, we obtain:

THEOREM XXII′. *Let $K_1(x)$ be bounded and continuous. Let $f(x)$ be of limited total variation over every finite interval. Let* (20.14), (20.061) *and* (20.08) *hold. Let $k_2(u)$ and $k_1(u)$ be defined as in* (20.26) *and* (20.27), *respectively. Let $K_1(x)$ belong to L_2. Let $k_2(u)/k_1(u)$ be analytic over $-\varepsilon \leqq \Re(u) \leqq \lambda + \varepsilon$, and let it belong to L_2 over every ordinate in that strip. Then* (20.18) *follows.*

Similarly, we have:

THEOREM XXIII′. *In the hypothesis of Theorem* XXII′, *in case $K_1(x)$ vanishes for positive arguments, we may replace the strip $-\varepsilon \leqq \Re(u) \leqq \lambda + \varepsilon$ by the narrower strip $-\varepsilon \leqq \Re(u) \leqq \varepsilon$.*

21. Applications of the quasi-Tauberian theorem. If

$$K^{(m)}(x) = \begin{cases} (m+1)(1-e^x)^m e^x; & [x<0] \\ 0 & [x>0] \end{cases} \tag{21.01}$$

and

$$^{(m)}K(x) = \frac{2e^x}{\pi} \int_0^1 (1-z)^m \cos(ze^x)\, dz; \qquad [m>0] \tag{21.02}$$

then (20.08) is clearly satisfied for $\lambda = 1$, and

$$K_1(x) = K^{(m)}(x) \tag{21.03}$$

or

$$K_1(x) = {}^{(m)}K(x). \tag{21.04}$$

If we put

$$\tau_m(u) = \int_{-\infty}^{\infty} K^{(m)}(x)\, e^{ux}\, dx = \frac{\Gamma(m+2)\,\Gamma(u+1)}{\Gamma(u+m+2)}, \tag{21.05}$$

and

$$\sigma_m(u) = \int_{-\infty}^{\infty} {}^{(m)}K(x)\, e^{ux}\, dx = \frac{\Gamma(m+1)}{\Gamma(m+1-u)\cos\dfrac{\pi u}{2}}, \tag{21.06}$$

then as $|\mathfrak{I}(u)| \to \infty$

$$\left|\frac{\sigma_m(u)}{\tau_n(u)}\right| \sim \frac{2\,\Gamma(m+1)}{\sqrt{2\pi}\,\Gamma(n+2)\, e^{\Re(u)-m}}\,|\mathfrak{I}(u)|^{\Re(u)-m+n+1/2}. \tag{21.07}$$

Thus if $n > m$, then

$$\lim_{y\to\infty} \int_0^{\infty} {}^{(m)}K(y-x)\, df(x) = A \int_0^{\infty} {}^{(m)}K(x)\, dx \tag{21.08}$$

will imply

$$\lim_{y\to\infty} \int_0^{\infty} K^{(n)}(y-x)\, df(x) = A \int_{-\infty}^{\infty} K^{(n)}(x)\, dx, \tag{21.09}$$

while if $m > n+1$, (21.09) will imply (21.08).

With a little manipulation, already indicated by the authors in question, this result is seen to be equivalent to an important theorem of Hardy and Littlewood,[54] which gives the necessary and sufficient condition for the summability of Fourier series and integrals by Cesàro sums of some order. In the integral form, the theorem reads as follows: *Let $f(x)$ be a measurable function defined over $(-\infty, \infty)$, and zero outside $(-A, A)$. Let*

$$\varphi(y) = \tfrac{1}{2}\{f(x+y)+f(x-y)-2s\}. \tag{21.10}$$

Then if we write B_m for the proposition

$$\lim_{\lambda\to\infty} \lambda \lim_{\varepsilon\to 0} \int_{\varepsilon}^{1/\lambda} \varphi(y)\,(1-\lambda y)^{m-1}\, dy = 0, \tag{21.11}$$

and C_m for the proposition

$$\lim_{\lambda\to\infty} \lambda \lim_{\varepsilon\to 0} \int_{\varepsilon}^{\infty} \varphi(y)\, dy \int_0^1 (1-z)^m \cos\lambda yz\, dz = 0, \tag{21.12}$$

B_m implies $C_{m+\varepsilon}$ for $m \geqq 1$, while C_m implies $B_{m+1+\varepsilon}$ for $m > 0$.

[54] Hardy and Littlewood (12).

While the general method of the present section has been developed in an earlier paper of the author,[55] his final results were not stated correctly. The correct result is due to Bosanquet[56] and Paley, who have shown it to be a "best possible" result in both directions. Their theorem also applies to $m > -1$, and is otherwise somewhat more general.

CHAPTER VIII.

TAUBERIAN THEOREMS AND SPECTRA.

22. A further type of asymptotic behavior. The O and o symbols do not exhaust the possible terms in which we may describe the behavior of a function at infinity. A proposition which may be regarded as in some wise a generalization of

$$\lim_{x\to\infty} f(x) = A \tag{22.01}$$

is

$$\lim_{B\to\infty} \frac{1}{B}\int_0^B |f(x)-A|^2\,dx = 0. \tag{22.02}$$

If $f(x)$ is bounded and measurable, (22.01) clearly implies (22.02), while (22.02) does not imply (22.01). Proposition (22.02) has a certain analogy to the different types of "strong convergence" to a limit which a function may exhibit: namely

$$\int_0^\infty |f(x)-A|\,dx \text{ converges} \tag{22.03}$$

and

$$\int_0^\infty |f(x)-A|^2 dx \text{ converges.} \tag{22.04}$$

The series analogues of the latter

$$\sum_1^\infty |s_n - A| \text{ converges} \tag{22.05}$$

and

$$\sum_1^\infty |s_n - A|^2 \text{ converges} \tag{22.06}$$

imply the ordinary convergence of s_n to A, but are not implied by it. On the other hand, neither (22.03) nor (22.04) is implied by ordinary convergence, although (22.04) implies (22.02). In contrast with propositions

[55] Wiener (5).

[56] Bosanquet (1), (2); Paley (1).

(22.03)—(22.06), which represent various types of "strong convergence", we shall express (22.02) in the usual language by saying that $f(x)$ is *strongly summable*[57] to A as $x\to\infty$. We shall say that A is a *sublimit* of $f(x)$, and shall write

$$\operatorname*{slm}_{x\to\infty} f(x) = A. \tag{22.07}$$

The sublimit of a function $f(x)$ differs markedly from the ordinary limit in that neither its existence nor its value are invariant if we replace x by a monotone function of x becoming infinite with x. On the other hand, the sublimit of $f(x)$ has a relation to the harmonic analysis of $f(x)$ far closer than does the ordinary limit.

Closely related to the notion of sublimit is that of subboundedness, which bears to the ordinary notion of boundedness much the same relation which that of sublimit does to the ordinary notion of limit. A function $f(x)$ is said to be *subbounded* if

$$\frac{1}{B}\int_0^B |f(x)|^2\,dx \tag{22.08}$$

is bounded. This juxtaposition of a notion of limit and a notion of boundedness suggests a generalized form of Tauberian theorem. To be specific, let us ask what conditions beyond the subboundedness of $f(x)$ are sufficient to make

$$\operatorname*{slm}_{x\to\infty} \int_0^\infty K_1(x-\xi)f(\xi)\,d\xi = A\int_{-\infty}^\infty K_1(\xi)\,d\xi \tag{22.09}$$

imply

$$\operatorname*{slm}_{x\to\infty} \int_0^\infty K_2(x-\xi)f(\xi)\,d\xi = A\int_{-\infty}^\infty K_2(\xi)\,d\xi. \tag{22.10}$$

This problem belongs to the range of ideas treated by the author in his work on generalized harmonic analysis. It gives a clearer picture of the real significance of Tauberian theorems. On the other hand, it does not at present offer an alternative approach to them, since Tauberian theorems of the type already discussed in this paper play an essential role in the establishment of a theory of generalized harmonic analysis. Results involving these theorems will be applied in this section.

To return to (22.09) and (22.10), no generality is lost by taking $A = 0$, as this simply amounts to replacing $f(x)$ by $f(x)-A$. Let us then take $A = 0$, and let us put $f(x) = 0$ for negative arguments, which

[57] Hardy and Littlewood (15).

is clearly permissible. We then wish to find a set of conditions which in conjunction with the boundedness for large B of

(22.11) $$\frac{1}{2B}\int_{-B}^{B}|f(x)|^2\,dx$$

(or what is the same, with the fact that $\frac{f(x)}{1+|x|}$ belongs to L_2) and the proposition

(22.12) $$\lim_{B\to\infty}\frac{1}{2B}\int_{-B}^{B}\left|\int_{-\infty}^{\infty}K_1(x-\xi)f(\xi)\,d\xi\right|^2 dx = 0$$

are sufficient to imply

(22.13) $$\lim_{B\to\infty}\frac{1}{2B}\int_{-B}^{B}\left|\int_{-\infty}^{\infty}K_2(x-\xi)f(\xi)\,d\xi\right|^2 d\xi = 0.$$

We shall prove the following:

THEOREM XXIV. *Let $f(x)$ be measurable, and let (22.11) be bounded. Let (22.12) be valid. Let K_1 and K_2 be measurable, and let*

(22.14) $$\int_{-\infty}^{\infty}(1+|x|)^{3+\varepsilon}|K_1(x)|^2\,dx<\infty$$
$$\left(\text{and hence}\int_{-\infty}^{\infty}(1+|x|)\,|K_1(x)|\,dx<\infty\right),$$

and

(22.15) $$\int_{-\infty}^{\infty}(1+|x|)^{3+\varepsilon}|K_2(x)|^2\,dx<\infty$$
$$\left(\text{and hence}\int_{-\infty}^{\infty}(1+|x|)\,|K_2(x)|\,dx<\infty\right).$$

Furthermore, let

(22.16) $$\int_{-\infty}^{\infty}K_1(x)\,e^{iux}\,dx\neq 0$$

for all real u. Then (22.13) is true.

For this theorem we shall need the following:

LEMMA XXIVa. *Let $f(x, u, \lambda)$ belong to L_2 as a function of u and be measurable in (x, u). Let $\int_{-\infty}^{\infty}|f(x, u, \lambda)|^2\,du$ be bounded, and let*

(22.17) $$\int_{-\infty}^{\infty}|f(x, u, \lambda)-f(x, u)|^2\,du\to 0$$

uniformly over any finite range of x as $\lambda\to\infty$. Let K be measurable, and let

$$(22.18) \qquad \int_{-\infty}^{\infty} |K(x)|\, dx$$

be finite. Let

$$(22.19) \qquad \int_{-\infty}^{\infty} K(x) f(x, u, \lambda)\, dx$$

exist for all λ. *Then*

$$(22.20) \quad \int_{-\infty}^{\infty} \left| \int_{-\infty}^{\infty} K(x) f(x, u, \lambda)\, dx - \int_{-\infty}^{\infty} K(x) f(x, u)\, dx \right|^2 du \to 0 \text{ as } \lambda \to \infty.$$

To prove this lemma, let us reflect that

$$(22.21) \quad \begin{aligned} &\left[\int_A^B |K(x)|\, dx\right]^2 \max_x \int_{-\infty}^{\infty} |f(x, u, \lambda) - f(x, u)|^2\, du \\ &\geqq \int_A^B |K(x)|\, dx \int_A^B |K(y)|\, dy \int_{-\infty}^{\infty} du\, |f(x, u, \lambda) - f(x, u)| \\ &\qquad\qquad \times |f(y, u, \lambda) - f(y, u)| \\ &= \int_{-\infty}^{\infty} du \int_A^B |K(x)|\, dx \int_A^B |K(y)|\, dy\, |f(x, u, \lambda) - f(x, u)| \\ &\qquad\qquad \times |f(y, u, \lambda) - f(y, u)| \\ &\geqq \int_{-\infty}^{\infty} du \left| \int_A^B K(x)\, [f(x, u, \lambda) - f(x, u)]\, dx \right|^2. \end{aligned}$$

By taking A large and B infinite, or $-B$ large and $-A$ infinite, we may make the left-hand member uniformly small. Over any finite range (A, B) the left-hand member tends to 0 as $\lambda \to \infty$. By combining these facts the lemma follows. It will be noted that it involves the existence almost everywhere of

$$(22.22) \qquad \int_{-\infty}^{\infty} K(x) f(x, u)\, dx.$$

We may now return to the proof of our main theorem. In a previous paper[58], the author has shown that

$$(22.23) \quad s(u) = \frac{1}{2\pi} \int_{-1}^{1} f(x) \frac{e^{iux} - 1}{ix}\, dx + \operatorname*{l.i.m.}_{A \to \infty} \frac{1}{\pi} \left[\int_1^A + \int_A^{-1} \right] f(x) \frac{e^{iux}}{ix}\, dx$$

exists. Moreover,

$$(22.24) \quad \begin{aligned} \int_{-\infty}^{\infty} |K_1(x - \xi) f(\xi)|\, d\xi &\leqq \int_{-\infty}^{\infty} |K_1(\xi)|\, |f(x - \xi)|\, d\xi \\ &\leqq \left[\int_{-\infty}^{\infty} (1 + |\xi|)^2 |K_1(\xi)|^2\, d\xi \int_{-\infty}^{\infty} \frac{|f(x - \xi)|^2}{(1 + |\xi|)^2}\, d\xi \right]^{1/2} \\ &< \infty, \end{aligned}$$

58 Wiener (7).

so that

$$\frac{1}{2B}\int_{-B}^{B}\left|\int_{-\infty}^{\infty} K_1(x-\xi)f(\xi)\,d\xi\right|^2 dx$$

$$\leqq \frac{1}{2B}\int_{-B}^{B} dx \int_{-\infty}^{\infty} |K_1(\xi)|\,|f(x-\xi)|\,d\xi \int_{-\infty}^{\infty} |K_1(\eta)|\,|f(x-\eta)|\,d\eta$$

$$(22.25)\quad = \frac{1}{2B}\int_{-\infty}^{\infty} |K_1(\xi)|\,d\xi \int_{-\infty}^{\infty} |K_1(\eta)|\,d\eta \int_{-B}^{B} |f(x-\xi)|\,|f(x-\eta)|\,dx$$

$$\leqq \frac{1}{2B}\int_{-\infty}^{\infty} |K_1(\xi)|\,d\xi \int_{-\infty}^{\infty} |K_1(\eta)|\,d\eta \int_{-B-|\xi|-|\eta|}^{B+|\xi|+|\eta|} |f(x)|^2\,dx$$

$$\leqq \frac{\text{const.}}{B}\int_{-\infty}^{\infty} |K_1(\xi)|\,d\xi \int_{-\infty}^{\infty} |K_1(\eta)|\,d\eta\,[B+|\xi|+|\eta|]$$

$$< \text{const.}$$

Thus the function

$$(22.26)\qquad f_1(x) = \int_{-\infty}^{\infty} K_1(x-\xi)f(\xi)\,d\xi$$

and the analogous function

$$(22.27)\qquad f_2(x) = \int_{-\infty}^{\infty} K_2(x-\xi)f(\xi)\,d\xi$$

define functions $s_1(u)$ and $s_2(u)$, corresponding to them in the same way in which $s(u)$ corresponds to $f(x)$.

In his previous paper,[58] the author has shown that

$$(22.28)\quad s(u+\varepsilon)-s(u-\varepsilon) = \frac{1}{\pi}\ \underset{A\to\infty}{\text{l.i.m.}} \int_{-A}^{A} f(x)\frac{\sin\varepsilon x}{x} e^{iux}\,dx.$$

Similarly

$$(22.29)\quad s_1(u+\varepsilon)-s_1(u-\varepsilon) = \frac{1}{\pi}\ \underset{A\to\infty}{\text{l.i.m.}} \int_{-A}^{A} \frac{\sin\varepsilon x}{x} e^{iux}\,dx \int_{-\infty}^{\infty} K_1(\xi)f(x-\xi)\,d\xi.$$

There is no trouble in modifying (22.28) into

$$(22.30)\quad s(u+\varepsilon)-s(u-\varepsilon) = \frac{1}{\pi}\ \underset{A\to\infty}{\text{l.i.m.}} \int_{-A-\xi}^{A-\xi} f(x)\frac{\sin\varepsilon x}{x} e^{iux}\,dx.$$

Thus

$$s_1(u+\varepsilon)-s_1(u-\varepsilon)-(s(u+\varepsilon)-s(u-\varepsilon))\int_{-\infty}^{\infty} K_1(\xi)\,e^{iu\xi}\,d\xi$$

$$= \frac{1}{\pi}\Bigg[\underset{A\to\infty}{\text{l.i.m.}} \int_{-\infty}^{\infty} K_1(\xi)\,d\xi \int_{-A}^{A} f(x-\xi)\frac{\sin\varepsilon x}{x} e^{iux}\,dx$$

$$(22.31)\qquad - \int_{-\infty}^{\infty} K_1(\xi)\,d\xi\ \underset{A\to\infty}{\text{l.i.m.}} \int_{-A}^{A} f(x-\xi)\frac{\sin\varepsilon(x-\xi)}{x-\xi} e^{iux}\,dx\Bigg]$$

$$= \frac{1}{\pi}\ \underset{A\to\infty}{\text{l.i.m.}} \int_{-\infty}^{\infty} K_1(\xi)\,d\xi \int_{-A}^{A} f(x-\xi)\left[\frac{\sin\varepsilon x}{x} - \frac{\sin\varepsilon(x-\xi)}{x-\xi}\right] e^{iux}\,dx.$$

In the proof of this we have used our last lemma.

Clearly

$$\left| \frac{\sin \varepsilon x}{\varepsilon x} - \frac{\sin \varepsilon (x - \xi)}{\varepsilon (x - \xi)} \right| \leqq 2. \tag{22.32}$$

Again,

$$\left| \frac{\sin \varepsilon x}{\varepsilon x} - \frac{\sin \varepsilon (x-\xi)}{\varepsilon (x-\xi)} \right| = \left| \int_{x-\xi}^{x} \frac{d}{dw} \left(\frac{\sin \varepsilon w}{\varepsilon w} \right) dw \right|$$

$$= \left| \int_{x-\xi}^{x} \frac{\varepsilon^2 w \cos \varepsilon w - \varepsilon \sin \varepsilon w}{\varepsilon^2 w^2} dw \right|$$

$$\leqq \varepsilon |\xi| \max_{w \, \mathrm{in} (x-\xi, x)} \left| \frac{\cos \varepsilon w}{\varepsilon w} - \frac{\sin \varepsilon w}{\varepsilon^2 w^2} \right| \tag{22.321}$$

$$\leqq 2 \varepsilon |\xi| \max \frac{1}{|\varepsilon w|}$$

$$\leqq \left| \frac{8 \xi}{|x| + |\xi|} \right|.$$

Thus and hence by an application of (22.32)

$$\frac{1}{\varepsilon \xi} \int_{-A}^{A} f(x-\xi) \left[\frac{\sin \varepsilon x}{x} - \frac{\sin \varepsilon (x-\xi)}{x - \xi} \right] e^{iux} \, dx \tag{22.33}$$

converges in the mean uniformly in u as $A \to \infty$ over any finite range of ξ. Hence by our lemma, we may write

$$\begin{aligned} s_1(u+\varepsilon) - s_1(u-\varepsilon) - (s(u+\varepsilon) - s(u-\varepsilon)) \int_{-\infty}^{\infty} K_1(\xi) \, e^{iu\xi} \, d\xi \\ = \frac{1}{\pi} \int_{-\infty}^{\infty} K_1(\xi) \, d\xi \, \underset{A \to \infty}{\mathrm{l.\,i.\,m.}} \int_{-A}^{A} f(x-\xi) \left[\frac{\sin \varepsilon x}{x} - \frac{\sin \varepsilon (x-\xi)}{x-\xi} \right] e^{iux} \, dx. \end{aligned} \tag{22.34}$$

By a further use of (22.321) and our lemma, since

$$\underset{\varepsilon \to 0}{\mathrm{l.\,i.\,m.}} \frac{1}{\varepsilon^{1/2} \xi} \underset{A \to \infty}{\mathrm{l.\,i.\,m.}} \frac{1}{\pi} \int_{-A}^{A} f(x-\xi) \left[\frac{\sin \varepsilon x}{x} - \frac{\sin \varepsilon (x-\xi)}{x-\xi} \right] e^{iux} dx = 0 \tag{22.35}$$

uniformly in ξ, it follows that

$$\lim_{\varepsilon \to 0} \frac{1}{\varepsilon} \int_{-\infty}^{\infty} \left| s_1(u+\varepsilon) - s_1(u-\varepsilon) - (s(u+\varepsilon) - s(u-\varepsilon)) \int_{-\infty}^{\infty} K_1(\xi) \, e^{iu\xi} \, d\xi \right|^2 du = 0. \tag{22.36}$$

From (22.12) and a theorem of the previous paper,[59] it follows that

$$\lim_{\varepsilon \to 0} \frac{1}{\varepsilon} \int_{-\infty}^{\infty} |s_1(u+\varepsilon) - s_1(u-\varepsilon)|^2 \, du = 0. \tag{22.37}$$

[59] Wiener (7), (5.52).

Combining (22.36) and (22.37), we get

$$(22.38)\quad \lim_{\varepsilon\to 0}\frac{1}{\varepsilon}\int_{-\infty}^{\infty}\left|\int_{-\infty}^{\infty}K_1(\xi)\,e^{iu\xi}\,d\xi\right|^2|s(u+\varepsilon)-s(u-\varepsilon)|^2\,du=0.$$

Since over any finite range, by (22.16),

$$(22.39)\quad \left|\int_{-\infty}^{\infty}K_1(\xi)\,e^{iu\xi}\,d\xi\right|>\text{const.},$$

it results that for any finite C,

$$(22.40)\quad \lim_{\varepsilon\to 0}\frac{1}{\varepsilon}\int_{-C}^{C}|s(u+\varepsilon)-s(u-\varepsilon)|^2\,du=0.$$

As in (22.36), we have

$$(22.41)\quad \lim_{\varepsilon\to 0}\frac{1}{\varepsilon}\int_{-\infty}^{\infty}\left|s_2(u+\varepsilon)-s_2(u-\varepsilon)-(s(u+\varepsilon)-s(u-\varepsilon))\int_{-\infty}^{\infty}K_2(\xi)\,e^{iu\xi}\,d\xi\right|^2du=0.$$

This yields us

$$(22.42)\quad \lim_{\varepsilon\to 0}\frac{1}{\varepsilon}\int_{-C}^{C}\left|s_2(u+\varepsilon)-s_2(u-\varepsilon)-(s(u+\varepsilon)-s(u-\varepsilon))\int_{-\infty}^{\infty}K_2(\xi)\,e^{iu\xi}\,d\xi\right|^2du=0.$$

Combining this with (22.40), we get

$$(22.43)\quad \lim_{\varepsilon\to 0}\frac{1}{\varepsilon}\int_{-C}^{C}\left|s_2(u+\varepsilon)-s_2(u-\varepsilon)\right|^2du=0.$$

By our previous paper,[60]

$$(22.44)\quad \frac{1}{\pi\varepsilon}\int_{-\infty}^{\infty}|f_2(x)|^2\frac{\sin^2\varepsilon x}{x^2}\,dx=\frac{1}{2\varepsilon}\int_{-\infty}^{\infty}|s_2(u+\varepsilon)-s_2(u-\varepsilon)|^2\,du,$$

and

$$(22.45)\quad \frac{1}{\pi\varepsilon}\int_{-\infty}^{\infty}\frac{\sin^2\varepsilon x}{x^2}\,dx\left|\frac{1}{2\lambda}\int_{-\lambda}^{\lambda}f_2(x+y)\,dy\right|^2=\frac{1}{2\varepsilon}\int_{-\infty}^{\infty}\frac{\sin^2\lambda u}{\lambda^2u^2}|s_2(u+\varepsilon)-s_2(u-\varepsilon)|^2\,du.$$

---

[60] Wiener (7), (5.52).

Now

$$\overline{\lim_{\varepsilon\to 0}}\left|\frac{1}{\pi\varepsilon}\int_{-\infty}^{\infty}|f_2(x)|^2\frac{\sin^2\varepsilon x}{x^2}dx-\frac{1}{\pi\varepsilon}\int_{-\infty}^{\infty}\frac{\sin^2\varepsilon x}{x^2}dx\left|\frac{1}{2\lambda}\int_{-\lambda}^{\lambda}f_2(x+y)\,dy\right|^2\right|$$

$$\leqq \overline{\lim_{\varepsilon\to 0}}\frac{\text{const.}}{\varepsilon}\int_{-\infty}^{\infty}\left|f_2(x)-\frac{1}{2\lambda}\int_{-\lambda}^{\lambda}f_2(x+y)\,dy\right|^2\frac{\sin^2\varepsilon x}{x^2}dx$$

$$=\overline{\lim_{\varepsilon\to 0}}\frac{\text{const.}}{\varepsilon}\int_{-\infty}^{\infty}\left|\int_{-\infty}^{\infty}f(x-\xi)\right.$$

$$\left.\times\left[K_2(\xi)-\frac{1}{2\lambda}\int_{-\lambda}^{\lambda}K_2(\xi+y)\,dy\right]d\xi\right|^2\frac{\sin^2\varepsilon x}{x^2}dx$$

$$\leqq \overline{\lim_{B\to\infty}}\text{ const.}\frac{1}{2B}\int_{-B}^{B}\left|\int_{-\infty}^{\infty}f(x-\xi)\right. \tag{22.46}$$

$$\left.\times\left[K_2(\xi)-\frac{1}{2\lambda}\int_{-\lambda}^{\lambda}K_2(\xi+y)\,dy\right]d\xi\right|^2dx$$

$$\leqq \text{const.}\overline{\lim_{B\to\infty}}\frac{1}{B}\int_{-\infty}^{\infty}\left|K_2(\xi)-\frac{1}{2\lambda}\int_{-\lambda}^{\lambda}K_2(\xi+y)\,dy\right|d\xi$$

$$\times\int_{-\infty}^{\infty}\left|K_2(\eta)-\frac{1}{2\lambda}\int_{-\lambda}^{\lambda}K_2(\eta+z)\,dz\right|d\eta\,(B+|\xi|+|\eta|)$$

$$\leqq \text{const.}\overline{\lim_{B\to\infty}}\left[\int_{-\infty}^{\infty}\left|K_2(\xi)-\frac{1}{2\lambda}\int_{-\lambda}^{\lambda}K_2(\xi+y)\,dy\right|\left(1+\frac{|\xi|}{B}\right)d\xi\right]^2,$$

where

$$\begin{aligned}&\overline{\lim_{B\to\infty}}\int_{-\infty}^{\infty}\left|K_2(\xi)-\frac{1}{2\lambda}\int_{-\lambda}^{\lambda}K_2(\xi+y)\,dy\right|\left(1+\frac{|\xi|}{B}\right)d\xi\\&=\int_{-\infty}^{\infty}\left|K_2(\xi)-\frac{1}{2\lambda}\int_{-\lambda}^{\lambda}K_2(\xi+y)\,dy\right|d\xi.\end{aligned} \tag{22.47}$$

Hence

$$\overline{\lim_{\varepsilon\to 0}}\left|\frac{1}{\pi\varepsilon}\int_{-\infty}^{\infty}|f_2(x)|^2\frac{\sin^2\varepsilon x}{x^2}dx-\frac{1}{\pi\varepsilon}\int_{-\infty}^{\infty}\frac{\sin^2\varepsilon x}{x^2}dx\left|\frac{1}{2\lambda}\int_{-\lambda}^{\lambda}f_2(x+y)\,dy\right|^2\right|$$

$$\leqq \text{const.}\left[\int_{-\infty}^{\infty}\left|K_2(\xi)-\frac{1}{2\lambda}\int_{-\lambda}^{\lambda}K_2(\xi+y)\,dy\right|d\xi\right]^2. \tag{22.48}$$

As we have seen in section 1,

$$\lim_{\lambda\to 0}\int_{-\infty}^{\infty}\left|K_2(\xi)-\frac{1}{2\lambda}\int_{-\lambda}^{\lambda}K_2(\xi+y)\,dy\right|d\xi=0, \tag{22.49}$$

from which it immediately follows that

$$\lim_{\lambda\to 0}\overline{\lim_{\varepsilon\to 0}}\left|\frac{1}{\pi\varepsilon}\int_{-\infty}^{\infty}|f_2(x)|^2\frac{\sin^2\varepsilon x}{x^2}dx - \frac{1}{\pi\varepsilon}\int_{-\infty}^{\infty}\frac{\sin^2\varepsilon x}{x^2}dx\left|\frac{1}{2\lambda}\int_{-\lambda}^{\lambda}f_2(x+y)\,dy\right|^2\right| = 0. \tag{22.50}$$

In combination with (22.44) and (22.45) this yields us

$$\lim_{\lambda\to 0}\overline{\lim_{\varepsilon\to 0}}\frac{1}{2\varepsilon}\int_{-\infty}^{\infty}\left(1-\frac{\sin^2\lambda u}{\lambda^2 u^2}\right)|s_2(u+\varepsilon)-s_2(u-\varepsilon)|^2\,du = 0. \tag{22.51}$$

Since

$$1-\frac{\sin^2\lambda u}{\lambda^2u^2}>\frac{1}{2}\quad\text{if}\quad u>\frac{2}{\lambda} \tag{22.52}$$

it results at once that

$$\lim_{C\to\infty}\overline{\lim_{\varepsilon\to 0}}\frac{1}{2\varepsilon}\left[\int_{C}^{\infty}+\int_{-\infty}^{-C}\right]|s_2(u+\varepsilon)-s_2(u-\varepsilon)|^2\,du = 0. \tag{22.53}$$

Combining this with (22.43), we may readily conclude that

$$\lim_{\varepsilon\to 0}\frac{1}{2\varepsilon}\int_{-\infty}^{\infty}|s_2(u+\varepsilon)-s_2(u-\varepsilon)|^2\,du = 0, \tag{22.54}$$

or in view of a theorem of our previous paper, that

$$\lim_{T\to\infty}\frac{1}{2T}\int_{-T}^{T}|f_2(x)|^2\,dx = 0. \tag{22.13}$$

This establishes Theorem XXIV. It will be noted that this theorem has an immediate extension in the direction of Theorem XI′.

There is another Tauberian theorem concerning strong summability which we may discuss here. It is due to Hardy and Littlewood.[61] The particular case of it which most directly interests us is the following:

THEOREM XXV. *Let*

$$\frac{1}{\xi}\int_{0}^{\xi}|f(x)|^2\,dx \tag{22.55}$$

be bounded, and let

$$\lim_{\varepsilon\to 0}\frac{1}{\varepsilon}\int_{0}^{\varepsilon}f(x)\,dx = 0. \tag{22.56}$$

[61] Hardy and Littlewood (16).

Then if

$$s_n(0) = \frac{1}{\pi} \int_0^{\pi} f(x) \frac{\sin \frac{2n+1}{2} x}{\sin \frac{x}{2}} dx \tag{22.57}$$

it follows that

$$\lim_{N \to \infty} \frac{1}{N} \sum_0^N |s_n(0)|^2 = 0. \tag{22.58}$$

In the proof of this theorem, it is easy to see that we may replace the conclusion (22.58) which we are to establish, by the equivalent

$$\lim_{H \to \infty} \frac{1}{H} \int_0^{H} \left| \frac{2}{\pi} \int_0^{\pi} f(x) \frac{\sin ux}{x} dx \right|^2 du = 0. \tag{22.59}$$

By a Tauberian theorem of the type of Theorem XI, already established in a previous paper of the author,[58] it follows that (22.59) is completely equivalent to

$$\lim_{\varepsilon \to 0} \frac{2}{\pi} \int_0^{\infty} \frac{\sin^2 \varepsilon u}{\varepsilon u^2} \left| \frac{2}{\pi} \int_0^{\pi} f(x) \frac{\sin ux}{x} dx \right|^2 du. \tag{22.60}$$

On the assumption that $f(x) = 0$ if $x > \pi$, we have by the Plancherel theorem

$$\begin{aligned}
&\frac{2}{\pi} \int_0^{\infty} \frac{\sin^2 \varepsilon u}{\varepsilon u^2} \left| \frac{2}{\pi} \int_0^{\pi} f(x) \frac{\sin ux}{x} dx \right|^2 du \\
&= \frac{2}{\pi \varepsilon} \int_0^{\infty} \left| \frac{2}{\pi} \int_0^{\pi} f(x) \frac{\sin \varepsilon u \sin ux}{ux} dx \right|^2 du \\
&= \frac{2}{\pi \varepsilon} \int_0^{\infty} \left| \frac{2}{\pi} \int_0^{\pi} \frac{\sin \varepsilon u \sin ux}{u} d \int_x^{\infty} \frac{f(\xi)}{\xi} d\xi \right|^2 du \\
&= \frac{2}{\pi \varepsilon} \int_0^{\infty} \left| \frac{2}{\pi} \int_0^{\infty} \sin \varepsilon u \cos ux \, dx \int_x^{\infty} \frac{f(\xi)}{\xi} d\xi \right|^2 du \\
&= \frac{2}{\pi \varepsilon} \int_0^{\infty} \left| \frac{1}{\pi} \int_0^{\infty} (\sin u(x+\varepsilon) - \sin u(x-\varepsilon)) \, dx \int_x^{\infty} \frac{f(\xi)}{\xi} d\xi \right|^2 du \qquad (22.61) \\
&= \frac{2}{\pi \varepsilon} \int_0^{\infty} \left| \frac{1}{2\pi} \int_{-\infty}^{\infty} (\sin u(x+\varepsilon) - \sin u(x-\varepsilon)) \, dx \int_{|x|}^{\infty} \frac{f(\xi)}{\xi} d\xi \right|^2 du \\
&= \frac{2}{\pi \varepsilon} \int_0^{\infty} \left| \frac{1}{2\pi} \int_{-\infty}^{\infty} \sin ux \, dx \int_{|x-\varepsilon|}^{|x+\varepsilon|} \frac{f(\xi)}{\xi} d\xi \right|^2 du \\
&= \frac{1}{\pi^2 \varepsilon} \int_0^{\infty} \left| \int_{|x-\varepsilon|}^{|x+\varepsilon|} \frac{f(\xi)}{\xi} d\xi \right|^2 dx.
\end{aligned}$$

Thus (22.59) and (22.58) are equivalent to

$$\lim_{\varepsilon\to 0}\frac{1}{\varepsilon}\int_0^\infty\left|\int_{|x-\varepsilon|}^{|x+\varepsilon|}\frac{f(\xi)}{\xi}\,d\xi\right|^2 dx = 0. \tag{22.62}$$

By the Schwarz inequality,

$$\begin{aligned}\frac{1}{\varepsilon}\left|\int_{|x-\varepsilon|}^{|x+\varepsilon|}\frac{f(\xi)}{\xi}\,d\xi\right|^2 &\leqq \frac{1}{\varepsilon}\int_{|x-\varepsilon|}^{|x+\varepsilon|}|f(\xi)|^2\,d\xi\int_{|x-\varepsilon|}^{|x+\varepsilon|}\frac{d\xi}{\xi^2}\\ &= \left(\frac{1}{|x-\varepsilon|}-\frac{1}{|x+\varepsilon|}\right)\frac{1}{\varepsilon}\int_{|x-\varepsilon|}^{|x+\varepsilon|}|f(\xi)|^2\,d\xi.\end{aligned} \tag{22.63}$$

Thus by the boundedness of (22.55) we have uniformly in ε for all sufficiently large values of N:

$$\begin{aligned}\frac{1}{\varepsilon}\int_{N\varepsilon}^\infty\left|\int_{|x-\varepsilon|}^{|x+\varepsilon|}\frac{f(\xi)}{\xi}\,d\xi\right|^2 dx &\leqq \int_{N\varepsilon}^\infty \frac{2\varepsilon}{x^2-\varepsilon^2}\,d\int_0^x\frac{d\eta}{\varepsilon}\int_{|\eta-\varepsilon|}^{|\eta+\varepsilon|}|f(\xi)|^2\,d\xi\\ &\leqq \int_{N\varepsilon}^\infty\frac{4\varepsilon x\,dx}{(x^2-\varepsilon^2)^2}\int_0^x\frac{d\eta}{\varepsilon}\int_{|\eta-\varepsilon|}^{|\eta+\varepsilon|}|f(\xi)|^2\,d\xi\\ &=\frac{1}{2}\int_{N\varepsilon}^\infty\frac{4\varepsilon x\,dx}{(x^2-\varepsilon^2)^2}\int_{-x-\varepsilon}^{x+\varepsilon}|f(\xi)|^2\,d\xi\int_{\xi-\varepsilon}^{\xi+\varepsilon}\frac{d\eta}{\varepsilon}\\ &\leqq \text{const.}\int_{N\varepsilon}^\infty\frac{4\varepsilon x^2\,dx}{(x^2-\varepsilon^2)^2}\\ &=\text{const.}\int_N^\infty\frac{4x^2\,dx}{(x^2-1)^2}.\end{aligned} \tag{22.64}$$

This can be made as small as we wish by taking N large enough. Moreover,

$$\begin{aligned}\frac{1}{\varepsilon}\int_0^{N\varepsilon}\left|\int_{|x-\varepsilon|}^{|x+\varepsilon|}\frac{f(\xi)}{\xi}\,d\xi\right|^2 &= \frac{1}{\varepsilon}\int_0^{N\varepsilon}\left|\int_{|x-\varepsilon|}^{|x+\varepsilon|}\frac{1}{\xi}\,d\int_0^\xi f(\eta)\,d\eta\right|^2 dx\\ &=\frac{1}{\varepsilon}\int_0^{N\varepsilon}\left|\left[\frac{1}{\xi}\int_0^\xi f(\eta)\,d\eta\right]_{|x-\varepsilon|}^{|x+\varepsilon|}+\int_{|x-\varepsilon|}^{|x+\varepsilon|}\frac{d\xi}{\xi^2}\int_0^\xi f(\eta)\,d\eta\right|^2 dx\\ &\leqq\frac{1}{\varepsilon}\int_0^{N\varepsilon}\left|A\left(1+\log\frac{|x+\varepsilon|}{|x-\varepsilon|}\right)\right|^2 dx\\ &=A^2\int_0^N\left(1+\log\frac{|x+1|}{|x-1|}\right)^2 dx,\end{aligned} \tag{22.65}$$

where

$$A>\left|\frac{1}{\xi}\int_0^\xi f(\eta)\,d\eta\right| \qquad (0<\xi\leqq|N+1|\varepsilon) \tag{22.651}$$

By (22.56), we may make A as small as we wish for a given N by taking ε small enough. Now let N be so large that for all ε,

$$\frac{1}{\varepsilon}\int_{N\varepsilon}^{\infty}\left|\int_{|x-\varepsilon|}^{|x+\varepsilon|}\frac{f(\xi)}{\xi}\,d\xi\right|^2 dx < \varepsilon_1/2, \tag{22.66}$$

and then let ε be so small that

$$\frac{1}{\varepsilon}\int_{0}^{N\varepsilon}\left|\int_{|x-\varepsilon|}^{|x+\varepsilon|}\frac{f(\xi)}{\xi}\,d\xi\right|^2 dx < \varepsilon_1/2. \tag{22.67}$$

It will then follow that for this and all smaller values of

$$\frac{1}{\varepsilon}\int_{0}^{\infty}\left|\int_{|x-\varepsilon|}^{|x+\varepsilon|}\frac{f(\xi)}{\xi}\,d\xi\right|^2 dx < \varepsilon_1. \tag{22.68}$$

Thus (22.62) follows, and Theorem XXV is established.

23. Generalized types of summability. The subject-matter of the last section leads us to interesting reflections on the notion of summability itself. The ordinary processes for summing a series are linear processes: that is, they consist in replacing the partial sums of a series by linear combinations of partial sums, and then investigating the ordinary limits of these linear combinations. This much of linearity must always remain in a definition of summability, that if two summable series are added term by term, the corresponding sums are also added, in the sense that the new sum-series will be summable to the sum of the sums corresponding to the individual series.

It is possible, however, to consider summability from another standpoint, from which linearity is not so obvious an attribute of the process. We may confine our attention to series, or rather to their sequences of partial sums, which are summable to zero. Thus a method of summability will sum to A the series whose partial sums are:

$$s_1, s_2, \cdots, s_n, \cdots \tag{23.01}$$

if it will sum to zero the series whose partial sums are:

$$s_1 - A, s_2 - A, \cdots, s_n - A, \cdots. \tag{23.02}$$

We may thus center our concept of summability about the notion of *null-sequence* or *null-function*. Whatever definition we choose for the class of such sequences or functions, this class should be closed additively, in the sense that the sum of any two members of the class must belong to the class. *It is not however essential that a member of the class should be characterized by the vanishing of some particular linear transform of that member.* In this sense, we are introducing a non-linear theory of summability.[62]

It is clearly a desideratum of a definition of a null-function that every function tending to a zero limit at infinity should be null. This requirement at once makes the corresponding definition of summability consistent with convergence and inclusive of it in scope. It is not satisfied by definitions of the type that assert that f is null if

$$\int_{-\infty}^{\infty} |f(x)|^p \, dx < \infty \tag{23.03}$$

but is satisfied by the definition that f is null if

$$\lim_{B\to\infty} \frac{1}{B} \int_{-B}^{B} |f(x)|^2 \, dx = 0. \tag{23.04}$$

We have already seen that if $s(u)$ is defined as in (22.23), this last statement is equivalent to the assertion that

$$\lim_{\varepsilon\to 0} \frac{1}{2\varepsilon} \int_{-\infty}^{\infty} |s(u+\varepsilon) - s(u-\varepsilon)|^2 \, du = 0. \tag{23.05}$$

This suggests an even looser definition of a null-function, according to which a function $f(x)$ is null if

$$\frac{1}{2B} \int_{-B}^{B} |f(x)|^2 \, dx \tag{23.06}$$

is bounded, and if

$$\lim_{\varepsilon\to 0} \frac{1}{2\varepsilon} \int_{-A}^{A} |s(u+\varepsilon) - s(u-\varepsilon)|^2 \, du = 0 \tag{23.07}$$

for all finite A, or even if (23.07) is true for *some* A greater than 0. The first of these definitions will make $f(x)$ null if (23.06) holds and

$$\int_{-\infty}^{\infty} K(x-\xi) f(\xi) \, d\xi \tag{23.08}$$

is null, provided that

$$\int_{-\infty}^{\infty} (1+|x|)^{3+\varepsilon} \, |K_1(x)|^2 \, dx < \infty \tag{23.09}$$

and

$$\int_{-\infty}^{\infty} K(x) \, e^{iux} \, dx \neq 0, \qquad [-\infty < u < \infty]. \tag{23.10}$$

---

[62] As far as the author knows, strong summability of various sorts is the only example of a summability process of this sort now in the literature.

The second definition will have the same property, provided only (23.10) holds for all values of u in some neighborhood of 0. All this follows from the theory developed in the last section.

These two last definitions of summability therefore fit in well with the ordinary linear definitions of summability, which define the generalized limit of a function as an expression of the form

$$\lim_{x\to\infty} \frac{\int_{-\infty}^{\infty} K(x, \xi) f(\xi)\, d\xi}{\int_{-\infty}^{\infty} K(x, \xi)\, d\xi} \tag{23.11}$$

on an appropriate scale of measurement. They have the great advantage of not involving any reference to a particular kernel $K(x, \xi)$. They are, however, restricted to functions for which (23.06) is bounded.

If we confine our attention to functions $f(x)$ vanishing for negative arguments, this difficulty may be overcome in its turn. We now put

$$\sigma(z) = \frac{1}{2\pi} \int_{-\infty}^{\infty} f(x)\, e^{zx}\, dx \tag{23.12}$$

throughout the half-plane $\Re(z) > u_0$: Then, by analytic continuation if necessary, $\sigma(i\nu)$ may be defined in some cases on a section of the real axis containing the origin. We put

$$s(u) = \int_0^u \sigma(iw)\, dw \tag{23.121}$$

giving σ its boundary value along the axis of imaginaries. We now define nullity as in (23.06) and (23.07). If for all z with real part between $-\varepsilon$ and $u_0 + \varepsilon$

$$\int_{-\infty}^{\infty} |K(x)|\, e^{zx}\, dx < \infty, \tag{23.13}$$

we have

$$\frac{1}{2\pi} \int_{-\infty}^{\infty} e^{zx}\, dx \int_{-\infty}^{\infty} K(x-\xi) f(\xi)\, d\xi = \sigma(z) \int_{-\infty}^{\infty} K(\xi)\, e^{z\xi}\, d\xi, \tag{23.14}$$

so that if $\int_{-\infty}^{\infty} K(\xi)\, e^{z\xi}\, d\xi$ is free from zeros and singularities in the left half plane, $f(x)$ and $\int_{-\infty}^{\infty} K(x-\xi) f(\xi)\, d\xi$ are null or not null simultaneously. With an appropriate definition of "analytic continuation", zeros and singularities not at the origin are of no concern, provided the function

$\int_{-\infty}^{\infty} K(\xi)\, e^{z\xi}\, d\xi$ is representible on a single-sheeted Riemann surface in the left half-plane.

It will be noted that all the definitions here given of the nullity of $f(x)$ depend on the scale chosen for x. If for all large x,

$$\frac{y'(x)}{y(A)} < \text{const. for } x < A \tag{23.15}$$

and

$$\frac{1}{2A} \int_{-A}^{A} |f(y(x))|^2\, dx \to 0 \tag{23.16}$$

then it at once follows that

$$\frac{1}{2B} \int_{-B}^{B} |f(x)|^2\, dx \to 0. \tag{23.17}$$

A similar study of the effect of a change of scale on nullity of other types would be of interest.

24. **Some unsolved problems.** In a piece of work of the ambitious length of the present, it is perhaps worth while to point out to the reader promising future directions of research. The following remarks may therefore not be amiss.

(1) The closure of the set of translations of a given function has been investigated in class L_1 and in class L_2. The methods of proof have been widely different in the two cases, but both results may be stated in a single formulation, that the set of translations of $f(x)$ is closed in the appropriate class when and only when the Fourier transform of $f(x)$, when properly defined and chosen, has no zeros. This formulation continues to constitute a reasonable proposition for L_p, where p is intermediate between 1 and 2, or even exterior to this interval. Is this proposition true? Certainly, for neither of the special cases already given is the method of proof extensible without serious modification. My own suspicion is that the general theorem is at least true for $1 \leq p \leq 2$.

(2) Obviously the power of Tauberian theorems in number theory has not been exhausted. Is there any Tauberian theorem which will reach from one complex ordinate to another, and enable us to handle the more refined forms of the prime number theorem?

(3) In particular, can we make a direct study of the closure of the set of all polynomials in the functions

$$e^{-\lambda(x-\xi)} \frac{d}{dx} \frac{e^{x-\xi}}{e^{e^{x-\xi}}-1}, \tag{24.01}$$

and thus attack directly the problem of the zeros of the zeta function in the critical strip?

(4) In section 22, the conditions to which $K_1(x)$ and $K_2(x)$ are subjected in Theorem XXIV are probably needlessly stringent. What is the best possible theorem in this connection?

(5) The "quasi-Tauberian" theorem of section 21 has not yet been exploited to the full. In particular, in the discussion of the relations between Riesz summability, Cesàro summability, Hölder summability, and the like, it is extremely desirable to have theorems of this type in which the kernels are not of the form $K(x-y)$, but are in some sense *nearly* of this form. Here the cruder theorems, depending solely on the use of dominant functions, are probably not difficult to elicit. On the other hand, the more refined ones will almost certainly require the full armament of Carleman's theory of singular integral equations and of the modern von Neumann-Stone[63] calculus of operators. Indeed, the time will soon come when the entire Tauberian theory must be reconsidered from the point of view of this calculus.

BIBLIOGRAPHY.

The present list of memoirs, while it cannot in the nature of things be complete, represents an attempt to bring together in one place as much as possible of the literature of Tauberian theorems. Up to 1915, we have been able to draw on the bibliography of the Cambridge tract by Hardy and M. Riesz on The General Theory of Dirichlet Series, and up to about 1924, on Smail's History and Synopsis of the Theory of Summable Infinite Processes. The bibliography also contains certain memoirs mentioned in the notes which are not of themselves Tauberian.

K. Ananda-Rau.

(1) A note on a theorem of Mr. Hardy's. Proc. Lond. Math. Soc., (2) 17 (1918), 334–336.

(2) On Lambert's series. Proc. Lond. Math. Soc., (2) 19 (1921), 1–20.

(3) On the relation between the convergence of a series and its summability by Cesàro's means. Jour. Indian Math. Soc., 15 (1923–5), 264–268.

(4) On Dirichlet's series with positive coefficients. Rendiconti di Palermo, 54 (1930), 455–462.

S. Bochner and G. H. Hardy.

(1) Note on two theorems of Norbert Wiener. Jour. Lond. Math. Soc., 1 (1926), 240–242.

E. Bortolotti.

(1) Sulle condizioni di applicabilità e di coerenza dei processi di sommazione asintotica di algoritmi infiniti. Rend. del R. Ac. di Bologna, 31 (1926–7), 33–41.

[63] Von Neumann (1); Stone (1).

S. Bosanquet.

(1) The summability of Fourier series. Math. Gaz., 15 (1931), 293–296.
(2) On the summability of Fourier series. Proc. Lond. Math. Soc., (2) 31 (1930), 144–164.

L. S. Bosanquet and E. H. Linfoot.

(1) On the zero order summability of Fourier series. Jour. Lond. Math. Soc., 6 (1931), 117–126.
(2) Generalized means and the summability of Fourier series. Quarterly Journal (Oxford series), 2 (1931), 207–229.

H. E. Bray.

(1) Elementary properties of the Stieltjes integral. Ann. of Math., 20 (1919), 176–186.

T. J. I'a. Bromwich.

(1) An introduction to the theory of infinite series. London 1908. Cf. especially p. 251
(2) On the limits of certain infinite series and integrals. Math. Ann. 65, (1908), 350–369.

Anna Caldarera.

(1) Su talune estenzioni dei criteri di convergenza di Hardy e Landau. Note e memorie di mat. (Catania), 2 (1923), 77–98.

T. Carleman.

(1) A theorem concerning Fourier series. Proc. Lond. Math. Soc., (2) 21 (1923), 483–492.

H. S. Carslaw.

(1) Introduction to the theory of Fourier series and integrals. New edition London 1921. Cf. especially pp. 150–154, 234–243.

S. Chapman.

(1) On non-integral orders of summability of series and integrals. Proc. Lond. Math. Soc., (2) 9 (1910–11), 369–409. Cf. p. 374, second note.

M. Cipolla.

(1) Criteri di convergenza riducibili a quello di Hardy-Landau. Naples Rendiconti (3a), 27 (1921), 28–37.

P. J. Daniell.

(1) A general form of integral. Ann. of Math., 19 (1917), 279–294.

P. Dienes.

(1) Sur la sommabilité de la série de Taylor. C. R., 153 (1911), 802–805.

G. Doetsch.

(1) Ein Konvergenzkriterium für Integrale. Math. Ann., 82 (1920), 68–82.
(2) Die Integrodifferentialgleichungen vom Faltungstypus. Math. Ann., 89 (1923), 192–207.
(3) Sätze von Tauberschem Charakter im Gebiet der Laplace- und Stieltjes Transformation. Berliner Berichte, Phys.-Math. Klasse (1930), 144–157.
(4) Über den Zusammenhang zwischen Abelscher und Borelscher Summabilität. Math. Ann., 104 (1931), 403–414.

P. Fatou.

(1) Séries trigonométriques et séries de Taylor. Acta Math., 30 (1906), 335–400. (Thèse, Paris 1907.)

L. Fejér.

(1) Fourierreihe und Potenzreihe. Monatshefte f. Math., 28 (1917), 64–76.

M. Fekete.

(1) Viszgálatok a Fourier sorokrol. Math. és Termész. Ért., 34 (1916).

M. Fujiwara.

(1) Über die Verallgemeinerung des Tauberschen Satzes auf Doppelreihen. Science reports of the Tôhoku national university, Sendai, Japan, 8 (1919), 43–50.
(2) Über summierbare Reihen und Integrale. Tôhoku math. jour., 15 (1919), 323–329.
(3) Ein Satz über die Borelsche Summation. Tôhoku math. jour., 17 (1920), 339–343.

J. Hadamard.

(1) Sur la distribution des zéros de la fonction $\zeta(s)$ et ses consequences arithmétiques. Bull. Soc. Math. France, 24 (1896), 199–220.

G. H. Hardy.

(1) Researches in the theory of divergent series and divergent integrals. Quarterly journal, 35 (1904), 22–66.
(2) Theorems relating to the summability and convergence of slowly oscillating series. Proc. Lond. Math. Soc., (2) 8 (1909), 301–320.
(3) On the multiplication of Dirichlet series. Proc. Lond. Math. Soc., (2) 10 (1912), 396–405.
(4) On the summability of Fourier's series. Proc. Lond. Math. Soc., (2) 12 (1913), 365–372.
(5) An extension of a theorem on oscillating series. Proc. Lond. Math. Soc., (2), 12 (1913), 174–180.
(6) Note on Lambert's series. Proc. Lond. Math. Soc., (2) 13 (1913), 192–198.
(7) The application of Abel's method of summation to Dirichlet series. Quarterly journal, 47 (1916), 176–192.
(8) The second theorem of consistency for summable series. Proc. Lond. Math. Soc., (2) 15 (1916), 72–88.
(9) On certain criteria for the convergence of the Fourier series of a continuous function. Messenger of mathematics, 49 (1920), 149–155.
(10) The summability of a Fourier series by logarithmic means. Quarterly Journal (Oxford series), 2 (1931), 107–113.

G. H. Hardy and J. E. Littlewood.

(1) Contributions to the arithmetic theory of series. Proc. Lond. Math. Soc., (2) 11 (1913), 411–478.
(2) The relations between Borel's and Cesàro's methods of summation. Proc. Lond. Math. Soc., (2) 11 (1913), 1–16.
(3) Sur la série de Fourier d'une fonction à carré sommable. C. R. April 28, 1913.
(4) Tauberian theorems concerning power series and Dirichlet's series whose coefficients are positive. Proc. Lond. Math. Soc., (2) 13 (1913), 174–191.
(5) Some theorems concerning Dirichlet's series. Mess. math., 43 (1914), 134–147.
(6) Theorems concerning the summability of series by Borel's exponential method. Rendiconti di Palermo, 41 (1916), 36–53.
(7) Sur la convergence des séries de Fourier et des séries de Taylor. C. R. Dec. 24, 1917.
(8) The Riemann zeta-function and the theory of the distribution of primes. Acta Math., 41 (1918), 119–196.
(9) On the Fourier series of a bounded function. Proc. Lond. Math. Soc., (2) 17 (1918), xiii–xv. (Abstract).
(10) Abel's theorem and its converse. Proc. Lond. Math. Soc., (2) 18 (1920), 205–235.

(11) On a Tauberian theorem for Lambert's series, and some fundamental theorems in the analytic theory of numbers. Proc. Lond. Math. Soc., (2) 19 (1921), 21-29.
(12) Solution of the Cesàro summability problem for power-series and Fourier series. Math. Ztschr., 19 (1923), 67–96.
(13) Abel's theorem and its converse (II). Proc. Lond. Math. Soc., (2) 22 (1924), 254–269.
(14) The allied series of a Fourier series. Proc. Lond. Math. Soc., 24 (1925), 211–246.
(15) The strong summability of Fourier series. Proc. Lond. Math. Soc., (2) 26 (1926), 273–286.
(16) Elementary theorems concerning power series with positive coefficients and moment-constants of positive functions. Crelle, 157 (1926), 141–158.
(17) A further note on the converse of Abel's theorem. Proc. Lond. Math. Soc., (2) 25 (1926), 219–236.
(18) Notes on the theory of series (VII): on Young's convergence criterion for Fourier series. Proc. Lond. Math. Soc., (2) 28 (1928) 301–311.
(19) Notes on the theory of series (XI): on Tauberian theorems. Proc. Lond. Math. Soc., (2) 30 (1929), 23–37.
(20) Notes on the theory of series (III): on the summability of the Fourier series of a nearly continuous function. Proc. Camb. Phil. Soc., 23 (1927), 681–684.
(21) The equivalence of certain integral means. Proc. Lond. Math. Soc. records, (2) 22 (1924), xl–xliii.

G. H. Hardy and M. Riesz.

(1) The general theory of Dirichlet series. Cambridge Tract in Mathematics and Mathematical Physics, No. 18. Cambridge 1915.

G. H. Hardy and E. C. Titchmarsh.

(1) Self-reciprocal functions. Quarterly journal (Oxford series), 1 (1930), 196–232.

W. A. Hurwitz.

(1) A trivial Tauberian theorem. Bull. Am. Math. Soc., 32 (1926), 77–82.

S. Ikehara.

(1) An extension of Landau's theorem in the analytic theory of numbers. Journal of Math. and Phys. of the Mass. Inst. of Technology, 10 (1931), 1–12.

S. Izumi.

(1) A generalization of Tauber's theorem. Proc. Imperial Acad. Japan, 5 (1929), 57-59.

M. Jacob.

(1) Über den Eindeutigkeitssatz in der Theorie der verallgemeinerten trigonometrischen Integrale. Math. Ann., 100 (1926–7), 278–294.
(2) Über ein Theorem von Bochner-Hardy-Wiener. Jour. Lond. Math. Soc., 3 (1928), 182–187.

J. Karamata.

(1) Sur le mode de croissance regulière des fonctions. Mathematica (Cluj, Roumania), 4 (1930), 38–53.
(2) Über die Hardy-Littlewoodschen Umkehrungen des Abelschen Stetigkeitssatzes. Math. Zeitschr., 32 (1930), 319–320.
(3) Neuer Beweis und Verallgemeinerung einiger Tauberian-Sätze. Math. Zeitschr., 33 (1931), 294–300.
(4) Neuer Beweis und Verallgemeinerung der Tauberschen Sätze, welche die Laplacesche und Stieltjessche Transformation betreffen. Crelle, 164 (1931), 27–40.

A. Kienast.

(1) Extension to other series of Abel's and Tauber's theorems on power series. Proc. Lond. Math. Soc., (2) 25 (1926), 45–52.

K. Knopp.

(1) Neuere Untersuchungen in der Theorie der divergenten Reihen. Jahresber. d. deutschen Mathematikervereinigung, 32 (1923), 43–67.

(2) Theorie und Anwendung der unendlichen Reihen. Berlin, Springer, 1922. Chap. XIII.

E. Landau.

(1) Über die Konvergenz einiger Klassen von unendlichen Reihen am Rande des Konvergenzgebietes. Monatsh. f. Math. und Phys., 18 (1907), 8–28.

(2) Handbuch der Verteilung der Primzahlen. Leipzig 1909, 2v.

(3) Über die Bedeutung einiger neuerer Grenzwertsätze der Herren Hardy und Axer. Prace mat.-Fiz., 21 (1910), 97–177.

(4) Über einen Satz des Herrn Littlewood. Rendiconti di Palermo, 35 (1913), 265–276.

(5) Ein neues Konvergenzkriterium für Integrale. Münch. Sitzber., 43 (1913), 461–467.

(6) Darstellung und Begründung einiger neuerer Ergebnisse der Funktionentheorie. Berlin 1916, 2nd ed., 1930.

(7) Sobre los números primos en progressión aritmética. Rev. Mat. Hispano-Americana, 4 (1923), 1–16, 33–44.

(8) Vorlesungen über Zahlentheorie. Leipzig 1927, 3v.

H. Lebesgue.

(1) Recherches sur la convergence des séries de Fourier. Math. Ann., 61 (1905), 251–280.

P. Lévy.

(1) Sur les conditions d'application et sur la régularité des procédés de sommation des séries divergentes. Bull. Soc. Math. de France, 54 (1926), 1–25.

S. B. Littauer.

(1) On a theorem of Jacob. Jour. Lond. Math. Soc., 4 (1929), 226–231.

(2) A new Tauberian theorem with application to the summability of Fourier series and integrals. Jour. Math. Phys. Mass. Insti. Tech., 8 (1928), 216–234.

J. E. Littlewood.

(1) On the converse of Abel's theorem on power series. Proc. Lond. Math. Soc., (2) 9 (1910), 434–448.

H. J. Mellin.

(1) Die Theorie der asymptotischen Reihen vom Standpunkte der reziproken Funktionen und Integrale. Ann. Ac. Sci. Fennicae, (A) 18 (1922), No. 4, 108 p.

G. Mignosi.

(1) Inversione d'un teorema sul rapporto delle medie (Cp) di due serie. Naples Rendiconti (3a), 27 (1921), 17–28.

L. Neder.

(1) Über Taubersche Bedingungen. Proc. Lond. Math. Soc., (2) 23 (1925), 172–184.

J. von Neumann.

(1) Allgemeine Eigenwerttheorie Hermitescher Funktionaloperatoren. Math. Ann., 102 (1930), 49–131.

R. E. A. C. Paley.

(1) On the Cesàro summability of Fourier series and allied series. Proc. Camb. Phil. Soc., 26 (1930), 173-203.

M. Plancherel.

(1) Contribution à l'étude de la répresentation d'une fonction arbitraire par des intégrales définies. Rendiconti di Palermo, 30 (1910), 289–335.

A. Pringsheim.

(1) Über das Verhalten von Potenzreihen auf dem Konvergenzkreise. Münch. Ber., 30 (1900), 43–100.

(2) Über die Divergenz gewisser Potenzreihen an der Konvergenzgrenze. M. Ber., 31 (1901), 505–524.

(3) Über eine Konvergenzbedingung für unendliche Reihen, die durch iterierte Mittelbildung reduzibel sind. M. Ber., 50 (1920), 275-284.

M. Riesz.

(1) Une methode de sommation équivalente à la methode des moyennes arithmétiques. C. R., 12 June 1911.

(2) Über einen Satz des Herrn Fatou. Crelle, 140 (1911) 89 99.

R. Schmidt.

(1) Über das Borelsche Summierungsverfahren. Schriften der Königsberger gelehrten Gesellschaft, 1 (1925), 202–256.

(2) Über divergente Folgen und lineare Mittelbildungen. M. Ztsch., 22 (1925), 89–152.

W. Schnee.

(1) Über Dirichletsche Reihen. Rend. di Palermo, 27 (1909), 87–116.

J. Schur.

(1) Über lineare Transformationen in der Theorie der unendlichen Reihen. Crelle, 151 (1921), 79–111.

L. L. Smail.

(1) History and synopsis of the theory of summable infinite processes. U. of Oregon publ., 2 (Feb. 1925), No. 8.

M. H. Stone.

(1) Linear transformations in Hilbert space III. Operational methods and group theory. Proc. Mat. Acad. Sci., 16 (1930), 172–175.

O. G. Sutton.

(1) On a theorem of Carleman. Proc. Lond. Math. Soc., (2) 23 (1925), xlvii–li.

O. Szász.

(1) Verallgemeinerung und neuer Beweis einiger Sätze Tauberischer Art. Münch. Ber., (1929), 325–340.

(2) Über einen Satz von Hardy und Littlewood. Berliner Berichte 1930, 4p.

(3) Über die Approximation stetiger Funktionen durch gegebene Funktionenfolgen. Math. Ann., 104 (1930), 155–160.

(4) Abel hatványsortételével kapcsolatos újabb vizsgálatokról. Mat. és Phys. Lapok., 36 (1929), 10–22.

(5) Über Sätze Tauberscher Art. Jahresb. d. Deutschen Mathematikervereinigung, 39 (1930), 28–31.

(6) Über einige Sätze von Hardy und Littlewood. Göttinger Nachrichten (1930), Fachgruppe I, 315–333.

(7) Verallgemeinerung eines Littlewoodschen Satzes über Potenzreihen. J. L. M. S., 3 (1930), 254–262.

(8) Über Dirichletsche Reihen an der Konvergenzgrenze. Rend. del Cong. Math. di Bologna, (1928), 269-276.

A. Tauber.

(1) Ein Satz aus der Theorie der unendlichen Reihen. Monatshefte f. Math., 8 (1897), 273–277.

E. C. Titchmarsh.

(1) A theorem on infinite products. Jour. Lond. Math. Soc., 1 (1926), 35–37.
(2) On integral functions with real negative zeros. Proc. Lond. Math. Soc., (2) 26 (1927), 185–200.
(3) A contribution to the theory of Fourier transforms. Proc. Lond. Math. Soc., (2) 23 (1924), 279–289.

C. de la Vallée Poussin.

(1) Récherches analytiques sur la théorie des nombres premiers. Première partie. Ann. Soc. Sci. Brux., 20: 2 (1896), 183–256.

S. Verblunsky.

(1) The convergence of singular integrals. Proc. Camb. Phil. Soc., 26 (1930), 312–322.
(2) On summable trigonometric series. Proc. Lond. Math. Soc., (2) 31 (1930), 370–386.
(3) On the limit of a function at a point. Proc. Lond. Math. Soc., (2) 32 (1931), 163–200.
(4) A uniqueness theorem for trigonometric series. Quarterly Journal (Oxford series), (2) (1931), 81–90.

T. Vijayaraghavan.

(1) A theorem concerning the summability of a series by Borel's method. Proc. Lond. Math. Soc., (2) 27 (1928), 316–326.
(2) A Tauberian theorem. Jour. Lond. Math. Soc., 1 (1926), 113–230.

N. Wiener.

(1) The harmonic analysis of irregular motion. Jour. Math. and Phys. Mass. Inst. Tech., 5 (1925), 99–122.
(2) Une methode nouvelle pour la démonstration des théorèmes de M. Tauber. C. R., 184 (1927), 793–795.
(3) On a theorem of Bochner and Hardy. Jour. Lond. Math. Soc., 2 (1927), 118–123.
(4) A new method in Tauberian theorems. Jour. Math. Phys. M. I. T., 7 (1928), 161–184.
(5) A type of Tauberian theorem applying to Fourier series. Proc. Lond. Math. Soc., (2) 30 (1929), 1–8.
(6) The spectrum of an arbitrary function. Proc. Lond. Math. Soc., (2) 27 (1928), 487–496.
(7) Generalized harmonic analysis. Acta. Math., 55 (1930), 117–258.
(8) The operational calculus. Math. Ann., 95 (1926), 557–584.

A. Wintner.

(1) Untersuchungen über Funktionen großer Zahlen. (Erste Mitteilung) M. Ztschr., 28 (1928), 416–429.

W. H. Young.

(1) On the determination of the summability of a function by means of its Fourier constants. Proc. Lond. Math. Soc., (2) 12 (1912), 71–88.

Über eine Klasse singulärer Integralgleichungen.

Von Prof. N. Wiener
in Cambridge (Mass.)
und Dr. E. Hopf
in Berlin-Dahlem.

(Vorgelegt von Hrn. Bieberbach am 5. November 1931 [s. oben S. 605].)

Im folgenden wird die homogene lineare Integralgleichung

$$f(x) = \int_0^\infty K(x-y) f(y)\, dy \tag{1}$$

aufgelöst, wo die Kernfunktion $K(x)$ für große $|x|$ wie eine Exponentialfunktion von $|x|$ klein wird. Der einfachste Spezialfall ist die Gleichung von Lalesco, $K(x) = e^{-|x|}$. Von besonderer Wichtigkeit für die Astronomie ist der Fall $K(x) = \frac{1}{2} Ei(|x|) = \frac{1}{2}\int_{|x|}^{\infty} \frac{e^{-t}}{t} dt$, die Gleichung von Milne. Die Lösung ergibt hier die Temperaturverteilung in einer Sternatmosphäre im Strahlungsgleichgewicht[1]. Mit Hilfe der Theorie der Laplace-Transformationen und elementarfunktionentheoretischen Betrachtungen gewinnen wir die Hauptlösungen von (1), d. h. alle Lösungen $f(x)$, die für große x höchstens wie eine im Vergleich zum Kern schwächere Exponentialfunktion groß werden. Die Lösungen können durch explizite Integralformeln dargestellt werden.

Die analoge Integralgleichung

$$f(x) = \int_{-\infty}^{+\infty} K(x-y) f(y)\, dy \tag{2}$$

ist viel leichter diskutierbar, ihre Lösungen sind im wesentlichen Exponentialfunktionen. Ist $u = u^*$ eine n-fache Nullstelle der Funktion

$$1 - \int_{-\infty}^{+\infty} K(t) e^{ut} dt,$$

so ist

$$Q(x) e^{-u^* x}$$

[1] Vgl. etwa E. Hopf, Mathematisches zur Strahlungsgleichgewichtstheorie der Fixsternatmosphären. Math. Zeitschr. 33 (1931), S. 109.

Reprinted from *Sitzber. Deutsch. Akad. Wiss. Berlin, Kl. Math. Phys. Tech.*, 1931, pp. 696–706. (Courtesy of the Akademie-Verlag.)

eine Lösung von (2), unter Q ein beliebiges Polynom von höchstens $(n-1)$-tem Grade verstanden (vorausgesetzt, daß die Integrale sinnvoll sind). Die Verwandtschaft von (1) und (2) äußert sich darin, daß die Lösungen von (1) sich für große x asymptotisch ebenso verhalten wie gewisse Lösungen von (2), eine Tatsache, die schon bei den beiden erwähnten Beispielen hervorgetreten war.

Auflösung der Integralgleichung (1). Die Kernfunktion $K(x)$ sei reell und bis auf endlich viele Stellen stetig. Die Funktion $K(x)e^{s|x|}$ sei für mindestens ein positives s im Intervalle $-\infty < x < \infty$ quadratisch integrierbar. Es bedeutet dann keine Einschränkung, wenn wir annehmen, daß

$$\int_{-\infty}^{+\infty} (K(x)e^{s|x|})^2\,dx$$

für alle $s < 1$ konvergiert. Hieraus folgt, daß auch

$$\int_{-\infty}^{+\infty} |K(x)|\,e^{s|x|}\,dx$$

für $s < 1$ konvergiert, wie die SCHWARZsche Ungleichung lehrt, wenn man den Integranden in der Form $\left(K(x)e^{\frac{s+1}{2}|x|}\right)\cdot e^{-\frac{1-s}{2}|x|}$ schreibt.

Im folgenden werden die Lösungen von (1) betrachtet, für welche

$$f(x) = O(e^{\alpha x}) \tag{3}$$

mit festem, sonst aber beliebigen $\alpha < 1$ gilt.

Um (1) unter diesen Voraussetzungen aufzulösen, schreiben wir (1) zunächst in der Form

$$g(x) = f(x) - \int_{-\infty}^{+\infty} K(x-y)f(y)\,dy \tag{4}$$

mit

$$f(x) = 0,\ x < 0; \qquad g(x) = 0,\ x > 0, \tag{5}$$

wobei $g(x)$ für $x < 0$ durch die rechte Seite in (4) definiert ist. Wir führen nun die LAPLACE-Transformierten

$$\phi(u) = \int_{-\infty}^{+\infty} f(x)e^{ux}\,dx,\quad \gamma(u) = \int_{-\infty}^{+\infty} g(x)e^{ux}\,dx,\quad \varkappa(u) = \int_{-\infty}^{+\infty} K(x)e^{ux}\,dx \tag{6}$$

von f, g, K ein, unter

$$u = s + it$$

eine komplexe Variable verstanden. $\dfrac{\phi(s+it)}{\sqrt{2\pi}}$, $\dfrac{\gamma(s+it)}{\sqrt{2\pi}}$, $\dfrac{\varkappa(s+it)}{\sqrt{2\pi}}$ sind als Funktionen von t also die FOURIER-Transformierten von $f(x)e^{sx}$, $g(x)e^{sx}$, $K(x)e^{sx}$.

Wegen (5) ist

$$\text{(6a)}\qquad \phi(u)=\int_0^\infty f(x)e^{ux}dx\,,\quad \gamma(u)=\int_{-\infty}^0 g(x)e^{ux}dx\,.$$

Wegen (3) konvergiert das Integral (6a) für $\phi(u)$ absolut, solange $s<-\alpha$ ist. Somit gilt das

Lemma 1. $\phi(u)$ ist in der Halbebene $s<-\alpha$ regulär und in jeder Teilhalbebene beschränkt.

Wegen (4) und (5) ist für $x<0$ mit Rücksicht auf (3)

$$|g(x)|\leqq\int_0^\infty|K(x-y)|\,|f(y)|\,dy<\text{const}\int_0^\infty|K(x-y)|\,e^{\alpha y}\,dy\,.$$

Für $\alpha<\lambda<1$ ist ferner

$$\int_0^\infty|K(x-y)|\,e^{\alpha y}\,dy<\int_0^\infty|K(x-y)|\,e^{\lambda y}\,dy=e^{\lambda x}\int_{-x}^\infty|K(-y)|\,e^{\lambda y}\,dy\,,$$

also wegen $x<0$

$$|g(x)|<e^{\lambda x}\int_0^\infty|K(-y)|\,e^{\lambda y}\,dy\,;$$

somit $g(x)=O(e^{-\lambda|x|})$ für beliebiges $\lambda<1$. Das LAPLACE-Integral (6a) für $\gamma(u)$ konvergiert daher für $s>-1$ absolut, und es gilt das

Lemma 2. $\gamma(u)$ ist in der Halbebene $s>-1$ regulär und in jeder Teilhalbebene beschränkt.

Daß die Regularitäts-Halbebenen von ϕ und γ zusammen die u-Ebene ausfüllen und einen Streifen $(-1<s<-\alpha)$ gemeinsam haben, ist der Hauptpunkt der folgenden Betrachtungen. Unter den Voraussetzungen über $K(x)$ konvergiert das LAPLACE-Integral von $K(x)$ absolut für $|s|<1$, und $K(u)$ ist in diesem Streifen regulär. Wendet man nun die LAPLACE-Transformation auf die Gleichung (4) an, so resultiert

$$\begin{aligned}\gamma(u)&=\phi(u)-\int_{-\infty}^{+\infty}e^{ux}\,dx\int_{-\infty}^{+\infty}K(x-y)f(y)\,dy\\&=\phi(u)-\int_{-\infty}^{+\infty}f(y)\,dy\int_{-\infty}^{+\infty}K(x-y)\,e^{ux}\,dx\\&=\phi(u)-\int_{-\infty}^{+\infty}f(y)\,e^{uy}\,dy\int_{-\infty}^{+\infty}K(t)\,e^{ut}\,dt^{1},\end{aligned}$$

also

$$\text{(7)}\qquad \gamma(u)=\phi(u)\big(1-\varkappa(u)\big)\,.$$

[1] Dies ist die wohlbekannte Eigenschaft der LAPLACE-Transformation, eine Faltung in ein Produkt von Transformierten überzuführen. Vgl. G. DOETSCH, Die Integrodifferentialgleichungen vom Faltungstypus. Math. Annalen 89 (1923), S. 192.

Die Umkehrung der Integrationsreihenfolge ist dabei $(-1 < s = \Re(u) < -\alpha)$ gewiß erlaubt, da die Integrale absolut konvergieren. Zur Diskussion von (7) benötigen wir das

Lemma 3. In jedem Streifen $|s| \leqq \beta$ $(\beta < 1)$ besitzt die Funktion $1 - \varkappa(u)$ höchstens endlich viele Nullstellen. Bezeichnet man dieselben mit $u_1, u_2, \ldots, u_m$, so läßt sich $1 - \varkappa(u)$ im Streifen $|s| \leqq \beta$ in der Form

$$(8) \qquad 1 - \varkappa(u) = \frac{\sigma_+(u)}{\sigma_-(u)} \prod_1^m (u - u_\nu)$$

darstellen, wo $\sigma_+(u)$ in der Halbebene $s \geqq -\beta$, $\sigma_-(u)$ in der Halbebene $s \leqq +\beta$ regulär und nullstellenfrei ist. Die Beträge

$$\left|\sigma_+(u) u^{k+\frac{m}{2}}\right|, \quad \left|\sigma_-(u) u^{k-\frac{m}{2}}\right|$$

sind für genügend großes $|u|$ in den erwähnten Halbebenen zwischen positiven Schranken gelegen; k ist hierbei eine durch $K(x)$ wohlbestimmte ganze Zahl. Für einen symmetrischen Kern, $K = K(|x|)$, ist $\varkappa(u) = \varkappa(-u)$, m gerade, $m = 2n$, und $k = 0$.

Beweis. Es ist $\left(x = \xi + \frac{\pi}{t}\right)$

$$\varkappa(u) = \int_{-\infty}^{+\infty} K(x) e^{sx} e^{itx} dx = -\int_{-\infty}^{+\infty} K\left(\xi + \frac{\pi}{t}\right) e^{s\left(\xi + \frac{\pi}{t}\right)} e^{it\xi} d\xi,$$

somit, wenn im zweiten Integral ξ wieder durch x ersetzt wird und das erste Integral zum zweiten addiert wird,

$$2\varkappa(u) = \int_{-\infty}^{+\infty} \left\{K(x) e^{sx} - K\left(x + \frac{\pi}{t}\right) e^{s\left(x + \frac{\pi}{t}\right)}\right\} e^{itx} dx.$$

Eine einfache Zerspaltung lehrt, daß der Betrag des Integranden nicht größer als $(|s| < 1)$

$$e^{\frac{\pi}{|t|}} \left| K\left(x + \frac{\pi}{t}\right) - K(x) \right| e^{sx} + \left| e^{s\frac{\pi}{t}} - 1 \right| K(x) e^{sx}$$

ist. Somit ist für $|s| \leqq s_0 < 1$

$$2|\varkappa(u)| \leqq e^{\frac{\pi}{|t|}} \int_{-\infty}^{+\infty} \left| K\left(x + \frac{\pi}{t}\right) - K(x) \right| e^{s_0|x|} dx + \left| e^{s\frac{\pi}{t}} - 1 \right| \int_{-\infty}^{+\infty} K(x) e^{s_0|x|} dx,$$

woraus mit Rücksicht auf die Voraussetzungen über K leicht

$$(9) \qquad \varkappa(s + it) \to 0 \quad \text{für} \quad |t| \to \infty$$

folgt, und zwar gleichmäßig im Streifen $|s| \leqq s_0$. Die Behauptung über die Nullstellen von $1 - \varkappa(u)$ folgt hieraus unmittelbar. Ferner ist

$$\frac{\varkappa(s+it)}{\sqrt{2\pi}}$$

als Funktion von t die FOURIER-Transformierte von $K(x)e^{sx}$, einer nach Voraussetzung quadratisch integrablen Funktion von x. Nach einem bekannten Satze von PLANCHEREL[1] ist somit $|\varkappa(s+it)|$ quadratisch summierbar in $-\infty < t < +\infty$.

Wir setzen nun

$$\tau(u) = \left(1 - \varkappa(u)\right) \frac{(u^2-1)^{\frac{m}{2}}}{\prod\limits_1^m (u - u_\nu)} \left(\frac{u+1}{u-1}\right)^k, \tag{10}$$

wo k eine noch zu bestimmende ganze Zahl ist und unter $(u^2-1)^{\frac{m}{2}}$ derjenige in $|s| < 1$ eindeutige Zweig zu verstehen ist, der sich für große $|u|$ wie u^m verhält. Es sei $\beta < \beta' < 1$, β' jedoch so gewählt, daß der etwas größere Streifen $|s| \leqq \beta'$ keine neuen Nullstellen von $1 - \varkappa(u)$ enthält. $\tau(u)$ ist dann regulär und nullstellenfrei in $|s| \leqq \beta'$, und wir haben

$$\tau(u) \to 1 \quad \text{für} \quad |t| \to \infty \tag{11}$$

gleichmäßig für $|s| \leqq \beta'$. Wir betrachten den Zuwachs von $\log \tau(s+it)$, wenn t von $-\infty$ nach $+\infty$ läuft. Er ist wegen (11) ein ganzzahliges Vielfaches von 2π und kann durch passende Wahl von k in (10) zu Null gemacht werden. Nachdem k so bestimmt ist, betrachten wir denjenigen im Streifen $|s| \leqq \beta'$ eindeutigen Zweig von $\log \tau(u)$, für welchen $\log \tau(u)$ für $t \to -\infty$ nach Null konvergiert. Dann gilt $\log \tau(u) \to 0$ auch für $t \to +\infty$, somit

$$\log \tau(u) \to 0 \quad \text{für} \quad |t| \to \infty$$

gleichzeitig im Streifen $|s| \leqq \beta'$. Fernerhin ist $\tau(u)$ für große $|t|$ von der Form $\left(1 - \varkappa(u)\right)\left(1 + O\left(\frac{1}{|u|}\right)\right)$, somit ist auch $|\log \tau(u)|$ als Funktion von t quadratisch summierbar. Daher gilt die CAUCHYsche Integralformel

$$\log \tau(u) = \log \tau_+(u) - \log \tau_-(u),$$

$$\left\{ \log \tau_+(u) = -\frac{1}{2\pi i} \int\limits_{-\beta' - i\infty}^{-\beta' + i\infty} \frac{\log \tau(v)}{v - u}\, dv, \quad \log \tau_-(u) = -\frac{1}{2\pi i} \int\limits_{\beta' - i\infty}^{\beta' + i\infty} \frac{\log \tau(v)}{v - u}\, dv \right. \tag{12}$$

für $-\beta' < s = \Re(u) < \beta'$.

---

[1] Vgl. N. WIENER, Generalized harmonic analysis. Acta Math. 55 (1930), S. 117, insbes. S. 120—126.

Nun ist nach der SCHWARZschen Ungleichung

$$|\log \tau_-(u)|^2 \leqq \frac{1}{4\pi^2} \int_{\beta'-i\infty}^{\beta'+i\infty} |\log \tau(v)|^2 |dv| \cdot \int_{\beta'-i\infty}^{\beta'+i\infty} \frac{|dv|}{|u-v|^2},$$

und $\log \tau_-(u)$ ist für $s \leqq \beta < \beta'$ regulär und beschränkt. Ebenso ist $\log \tau_+(u)$ für $s \geqq -\beta > -\beta'$ regulär und beschränkt. Setzen wir endlich

$$\sigma_+(u) = \tau_+(u)(u+1)^{-k-\frac{m}{2}}, \quad \sigma_-(u) = \tau_-(u)(u-1)^{-k+\frac{m}{2}},$$

so folgt aus (10) und (12) die Formel (8) mit allen behaupteten Eigenschaften von σ_+ und σ_-.

Für symmetrische Kerne ist offenbar $\varkappa(u) = \varkappa(-u)$ und die Anzahl der Wurzeln von $1-\varkappa(u)=0$ im Streifen $|s| \leqq \beta$ ist gerade, $m = 2n$. Wegen der Realität von $K(x)$ ist $\overline{\varkappa(u)} = \varkappa(\bar{u})$ und für imaginäre u, $\bar{u} = -u$, ist $\overline{\varkappa(u)} = \varkappa(-u) = \varkappa(u)$, d. h. $\varkappa(u)$ reell. Setzt man in (10) $k = 0$, so ist auch $\tau(u)$ reell für imaginäre u. Wegen $\tau(u) \neq 0$ ist daher der Zuwachs von $\log \tau(u)$ längs der imaginären u-Achse gleich Null. Es wird noch einmal bemerkt, daß die Funktionen σ_+ und σ_- aus der Kernfunktion $K(x)$ durch explizite Integralformeln bestimmt sind.

Die Gleichung (7) ist nun leicht lösbar. $u_1, u_2, \cdots, u_m$ seien alle im Streifen $|s| \leqq \alpha$ gelegenen Nullstellen von $1-\varkappa(u)$, wobei α das α von (3) bedeutet. Wir bestimmen β mit $\alpha < \beta < 1$ so, daß im Streifen $|s| \leqq \beta$ keine neuen Nullstellen liegen, und wenden Lemma 3 mit diesem β an. (7) und (8) ergibt dann

$$\frac{\gamma(u)}{\sigma_+(u)} = \frac{\phi(u)}{\sigma_-(u)} \prod_1^m (u-u_\nu). \tag{13}$$

Nach Lemma 2 und 3 ist die linke Seite für $s \geqq -\beta$ regular und für große $|u|$ gleich $O\left(|u|^{k+\frac{m}{2}}\right)$. Nach Lemma 1 und 3 ist die rechte Seite für $s \leqq -\beta$ regulär und für große $|u|$ ebenfalls gleich $O\left(|k|^{k+\frac{m}{2}}\right)$. Somit definiert (13) eine ganze Funktion, die für große $|u|$ gleich $O\left(|u|^{k+\frac{m}{2}}\right)$, somit ein Polynom vom Grade $\leqq \left[k+\frac{m}{2}\right]$ ist. Wir beschränken uns im folgenden auf die Diskussion eines symmetrischen reellen Kernes,

$$K = K(|x|).$$

Dann ist $m = 2n$, $k = 0$, und (13) definiert also ein Polynom höchstens n-ten Grades. Es kann jedoch nicht von n-tem Grade sein, da nach Lemma (3) sonst $|\phi(u)|$ für große $|u|$ oberhalb einer positiven Schranke liegen müßte, während nach PLANCHERELS Theorem $|\phi(s+it)|$ als Funktion von t quadratisch integrierbar sein muß. Also ist

$$\phi(u) = \sigma_-(u) \frac{P_{n-1}(u)}{\prod_1^{2n}(u-u_\nu)}, \quad \gamma(u) = \sigma_+(u) P_{n-1}(u), \tag{14}$$

unter P_{n-1} ein beliebiges Polynom von höchstens $(n-1)$-tem Grade verstanden.

Wir müssen nun noch umgekehrt zeigen, daß die durch (14) explizit gegebenen Funktionen $\phi(u)$, $\gamma(u)$ wirklichen Lösungen $f(x)$, $g(x)$ von (4), (5) entsprechen. $\phi(u)$ ist für $s \leqq \beta$ bis auf Pole in $u = u_1, \ldots, u_{2n}$ regulär und für große $|u|$ gleich $O\left(\frac{1}{|u|}\right)$. $\gamma(u)$ ist für $s \geqq -\beta$ regulär und für große $|u|$ gleich $O\left(\frac{1}{|u|}\right)$. Mit Rücksicht auf PLANCHERELS Theorem haben somit die Integrale in den MELLINschen Umkehrformeln

$$f(x) = \frac{1}{2\pi i} \int_{-\beta-i\infty}^{-\beta+i\infty} \phi(u) e^{-ux} du, \quad g(x) = \frac{1}{2\pi i} \int_{-\beta-i\infty}^{-\beta+i\infty} \gamma(u) e^{-ux} du \tag{15}$$

einen wohlbestimmten Sinn, wenn man sie als Grenzwerte en moyenne auffaßt. Im Integral (15) für $f(x)$ kann offenbar die Integrationsabszisse beliebig weit nach links verlegt werden. Somit ist

$$f(x) e^{-\lambda x} = \frac{1}{2\pi} \int_{-\infty}^{+\infty} \phi(-\lambda+it) e^{-ixt} dt, \quad \lambda \geqq \beta,$$

und $\frac{\phi(-\lambda+it)}{\sqrt{2\pi}}$ ist umgekehrt die FOURIER-Transformierte von $f(x)e^{-\lambda x}$. Wegen der Gleichheit der Absolut-Quadratintegrale von Funktion und transformierter Funktion (3) ist

$$\int_{-\infty}^{+\infty} |f(x)|^2 e^{-2\lambda x} dx = \frac{1}{2\pi} \int_{-\infty}^{+\infty} |\phi(-\lambda+it)|^2 dt$$

für beliebig große λ. Da die rechte Seite für $\lambda \geqq \beta$ beschränkt ist, muß offenbar $f(x)$ für $x < 0$ bis auf eine Nullmenge verschwinden. Analog ergibt sich $g(x) = 0$ für $x > 0$ bis auf eine Nullmenge, d. h. (5). Wir zeigen nun, daß (4) erfüllt ist, d. h. daß die Funktion

$$\delta(x) = e^{-\beta x} g(x) - e^{-\beta x} f(x) + \int_{-\infty}^{+\infty} e^{-\beta x} K(x-y) f(y) dy \tag{16}$$

verschwindet (zunächst bis auf eine Nullmenge). Nun sind wegen (15) $\frac{1}{\sqrt{2\pi}}\phi(-\beta+it)$, $\frac{1}{\sqrt{2\pi}}\gamma(-\beta+it)$ umgekehrt die FOURIER-Transformierten von $f(x)e^{-\beta x}$, $g(x)e^{-\beta x}$. Das Integral in (16) hat die Form

$$(16a)\quad \int_{+-}^{+\infty} G(-y)F(y)dy\,,\quad G(y) = e^{-\beta(x+y)}K(x+y)\,,\quad F(y) = e^{-\beta y}f(y)\,,$$

wobei $|F|$ und $|G|$ gewiß quadratisch integrierbar sind. Nach der Vollständigkeitsrelation für FOURIER-Integrale (3) ist das Integral in (16a) gleich dem Integral über das Produkt der FOURIER-Transformierten von $G(y)$ und $F(y)$. Da offenbar $\frac{1}{\sqrt{2\pi}}e^{-itx}\varkappa(-\beta+it)$ als Funktion von t die FOURIER-Transformierte von $G(y)$ ist und $\frac{1}{\sqrt{2\pi}}\phi(-\beta+it)$ die Transformierte von $F(y)$ war, folgt

$$\int_{-\infty}^{+\infty} e^{-\beta x}K(x-y)f(y)dy = \frac{1}{2\pi}\int_{-\infty}^{+\infty}\phi(-\beta+it)\varkappa(-\beta+it)e^{-itx}dt\,.$$

Umgekehrt besagt dies, daß $\frac{1}{\sqrt{2\pi}}\phi(-\beta+it)\varkappa(-\beta+it)$ die FOURIER-Transformierte des Integrals in (16) ist. Daher ist die FOURIER-Transformierte von $\delta(x)$ gleich $\frac{1}{\sqrt{2\pi}}(\gamma-\phi+\phi\varkappa)\equiv 0$, und $\delta(x)$ verschwindet bis auf eine Nullmenge. Die durch (14) und (15) definierte Funktion $f(x)$ befriedigt also die Integralgleichung (1) fast überall. Schließlich zeigt eine geläufige Schlußweise nachträglich, daß aus (1) und der quadratischen Summierbarkeit von $|f(x)|e^{-\beta x}$ die Stetigkeit und Beschränktheit von $f(x)e^{-\beta x}$ folgt; diese Funktion ist gleich dem Integral (16a), das nach der SCHWARZschen Ungleichung gewiß eine beschränkte Funktion von x ist.

Asymptotisches Verhalten der Lösungen von (1). Wir beschränken uns wieder auf symmetrische Kerne, $K = K(|x|)$. Wir haben noch das Wachstum der durch (14) und (15) definierten Lösungen $f(x)$ von (1) zu untersuchen. Im Integral (15) für $f(x)$ kann nun die Integrationsabszisse von $-\beta$ nach $+\beta$ verlegt werden; der Wert des Integrales ändert sich dabei um die negative Summe der Residuen von $\phi(u)e^{-ux}$ in $|\Re(u)|\leqq\alpha$. Ein solches Residuum ist wegen (14) von der Form

$$-Q(x)e^{-u^*x}\,,$$

wo u^* eine in jenem Streifen gelegene Nullstelle von $1-\varkappa(u)$ und $Q(x)$ ein Polynom bedeutet, dessen Grad kleiner als die Vielfachheit der Nullstelle ist.

Somit ist

$$(17)\quad \begin{cases} f(x) = f_0(x) + r(x), \\ f_0(x) = \sum Q(x)e^{-u^*x}, \quad r(x) = \dfrac{1}{2\pi i}\displaystyle\int_{\beta-i\infty}^{\beta+i\infty} \phi(u)e^{-ux}du, \end{cases}$$

wo die Summe sich auf alle Wurzeln u^* in $|\Re(u)| \leqq \alpha$ bezieht. $f_0(x)$ ist eine Lösung der Integralgleichung (2), genügt also der Gleichung

$$(18)\qquad f_0(x) = \int_0^\infty K(x-y)f_0(y)\,dy + \int_{-\infty}^0 K(x-y)f_0(y)\,dy.$$

Nun ist gewiß $f_0(x) = O(e^{\beta|x|})$, woraus mit Rücksicht auf die Voraussetzungen über K leicht

$$(19)\qquad \int_{-\infty}^0 K(x-y)f_0(y)\,dy = O(e^{-\beta x}), \qquad x > 0$$

folgt. Nach (17) ist ferner $r(x)e^{\beta x}$ die Fourier-Transformierte von $\frac{1}{\sqrt{2\pi}}\phi(\beta - it)$, somit quadratisch integrierbar. Wegen (17), (18), (19) ist

$$r(x) = \int_0^\infty K(x-y)\,r(y)\,dy + O(e^{-\beta x}),$$

woraus durch Anwendung der Schwarzschen Ungleichung auf das Integral — der Integrand ist $(K(x-y)e^{-\beta y})\cdot(r(y)e^{\beta y})$ — leicht $r(x) = O(e^{-\beta x})$ folgt. Es gilt somit unter den zugrunde gelegten Voraussetzungen über K das

Theorem. Ist $2n$ die immer endliche Anzahl der in ihrer Vielfachheit gezählten Nullstellen der Funktion

$$1 - \int_{-\infty}^{+\infty} K(x)e^{ux}dx \qquad K(x) = K(|x|)$$

im Streifen $|\Re(u)| \leqq \alpha < 1$, so ist die Maximalzahl der linear unabhängigen Lösungen von (1) mit $f(x) = O(e^{(\alpha+\delta)x})$ und beliebigem $\delta > 0$ genau gleich n. Die Lösungen haben die Form

$$f(x) = \sum Q(x)e^{-u^*x} + O(e^{-\beta x}),\text{[1]}$$

wobei u^* eine der obigen Nullstellen bedeutet und $Q(x)$ ein Polynom ist, dessen Grad kleiner als die Vielfachheit von u^* ist; β ist eine Zahl mit $\alpha < \beta < 1$ derart, daß die Streifen $\alpha < \Re(u) \leqq \beta$, $-\alpha > \Re(u) \geqq -\beta$ keine Nullstellen enthalten.

Als weitere Besonderheit des symmetrischen Kernes sei erwähnt, daß unter den im Hauptteil der Lösungen auftretenden Termen immer einer

[1] (3) braucht nicht erfüllt zu sein, da auf dem Rande des Streifens $|\Re(u)| \leqq \alpha$ eine mehrfache Nullstelle liegen kann, und daher ein Term von der Größenordnung $x^l e^{\alpha x}$, $l < 0$, vorkommen kann.

mit $\Re(u^*) \leqq 0$ vorkommen muß, d. h. daß es keine Lösung von (1) mit $f(x) \to 0$, $x \to \infty$ gibt; denn wegen (14) und $\sigma_-(u) \neq 0$ hat $\phi(u)$ mindestens einen Pol, der nicht in der rechten Halbebene liegt.

Beispiele. Bei der *Lalescoschen Gleichung* ist $K(x) = \lambda e^{-|x|}$, soweit $\varkappa(u) = \dfrac{2\lambda}{1-u^2}$. Die Wurzeln von $1 - \varkappa = 0$ sind $u = \pm\sqrt{1-2\lambda}$, und es ist $n = 1$. Die Darstellung (8) ergibt sich direkt,

$$\sigma_+(u) = \frac{1}{u+1}.$$

$\sigma_-(u) = u - 1$

Für $\lambda \leqq 0$ gibt es keine Lösungen mit der Eigenschaft (3) und irgendeinem $\alpha < 1$. Für $\lambda > 0$ gibt es im wesentlichen nur eine Lösung. Wegen (14) ist

$$\phi(u) = \frac{u-1}{u^2 - 1 + 2\lambda}.$$

Wegen (17) ist weiter

$$f(x) = \frac{e^{\sqrt{1-2\lambda}\,x} + e^{-\sqrt{1-2\lambda}\,x}}{2} + \frac{e^{\sqrt{1-2\lambda}\,x} - e^{-\sqrt{1-2\lambda}\,x}}{2\sqrt{1-2\lambda}} + \frac{1}{2\pi i}\int\limits_{\beta - i\infty}^{\beta + i\infty} \phi(u)\, e^{-ux}\, du.$$

Das Integral muß hier verschwinden, da die Integrationsabszisse beliebig weit nach rechts gelegt werden kann.

Als weiteres, komplizierteres Beispiel sei die *Milnesche Gleichung* behandelt, $K(x) = \dfrac{1}{2}\displaystyle\int\limits_{|x|}^{\infty} \frac{e^{-t}}{t}\, dt$. Hier ist

$$\begin{aligned}
\varkappa(u) &= \frac{1}{2}\int\limits_{x=0}^{\infty}\int\limits_{t=x}^{\infty} \frac{e^{-t}}{t}\,(e^{ux} + e^{-ux})\, dt\, dx \\
&= \frac{1}{2}\int\limits_{t=0}^{\infty}\int\limits_{x=0}^{t} \frac{e^{-t}}{t}\,(e^{ux} + e^{-ux})\, dx\, dt \\
&= \frac{1}{2u}\int\limits_{0}^{\infty} \frac{e^{-(1-u)t} - e^{-(1+u)t}}{t}\, dt \\
&= \frac{1}{2u}\int\limits_{0}^{\infty} dt \int\limits_{1-u}^{1+u} e^{-wt}\, dw = \frac{1}{2u}\int\limits_{1-u}^{1+u} \frac{dw}{w} = \frac{1}{2u}\log\frac{1+u}{1-u},
\end{aligned}$$

wo derjenige im Streifen $|\Re(u)| < 1$ reguläre Zweig von log genommen ist. der für $u = 0$ verschwindet. $u = 0$ ist zweifache Wurzel von $1 - \varkappa = 0$, und sonst sind, wie man sich leicht überlegt, im Streifen $|\Re(u)| < 1$ keine

60*

weiteren Wurzeln vorhanden. Es gibt daher nur eine Lösung mit der Eigenschaft (3); sie verhält sich für große x wie eine lineare Funktion

$$f(x) = x + a + O(e^{-(1-\delta)x}) \tag{20}$$

mit beliebigem $\delta > 0$. Es ist uns in diesem Falle nicht gelungen, aus den allgemeinen Integraldarstellungen $f(x)$ in einfacher Gestalt zu gewinnen. Vermutlich ist sie eine neue Transzendente.

Im allgemeinen Falle ist die Voraussetzung (3) über die Lösungen von (1) durchaus natürlich, da $1-\varkappa$ unendlich viele Nullstellen im Streifen $|\Re(u)| < 1$ haben kann. Wenn jedoch $1-\varkappa$ nur endlich viele Nullstellen hat, wie in den beiden obigen Beispielen, kann man die Einseitigkeit der gewonnenen Lösungen auch unter geringeren Voraussetzungen als (3) beweisen, wie hier nicht näher ausgeführt sein möge.

Positiver Kern[1]. Es sei $K = K(|x|) > 0$ bis auf endlich viele Stellen. Ferner sei

$$\varkappa(0) \leqq 1, \quad \varkappa(1) > 1.$$

Wir zeigen, daß es dann unter den Lösungen von (1) eine für $x \geqq 0$ positive gibt. Unter diesen Voraussetzungen wächst nämlich $\varkappa(u) = \int_0^\infty K(x)(e^{ux} + e^{-ux})\,dx$ monoton, wenn u von 0 durch reelle Werte nach 1 läuft, und die Gleichung $1-\varkappa = 0$ hat im Streifen $|\Re(u)| < 1$ genau zwei reelle Wurzeln $\pm u^*$; $u^* \geqq 0$. Es gibt nun eine Lösung von (1) mit $f(x) = O(e^{u^* x})$ für $u^* > 0$, und $f(x) = O(x)$ im Falle $u^* = 0$, d. h. $\varkappa(0) = 1$. In beiden Fällen ist diese Lösung gewiß von einer Stelle ab positiv. Würde sie überhaupt Werte $\leqq 0$ annehmen, so müßte sie ihre untere Grenze irgendwo annehmen, etwa bei $x = x_0$. Wegen $K > 0$ ist daher

$$f(x_0) = \int_0^\infty K(x_0 - y) f(y)\,dy \geqq f(x_0) \int_0^\infty K(x_0 - y)\,dy,$$

oder

$$f(x_0)\left(1 - \int_0^\infty K(x_0 - y)\,dy\right) \geqq 0,$$

wo das Gleichheitszeichen nur für $f \equiv f(x_0)$ gelten kann. Somit ist $f(x_0) = 0$ unmöglich. $f(x_0) < 0$ ist ebenfalls unmöglich, da das Integral in der Klammer kleiner als $\int_{-\infty}^{+\infty} K(t)\,dt = \varkappa(0) \leqq 1$ ist. Daher ist $f(x) > 0$ für $x \geqq 0$. Die Lösung (20) der Milneschen Gleichung ist z. B. positiv.

[1] Vgl. auch E. Hopf, Über lineare Integralgleichungen mit positivem Kern. Sitzungsber. der Berliner Akad. d. Wiss. 1928. XVIII.

Ausgegeben am 28. Januar 1932.

THE HOMOGENEOUS CHAOS.

By Norbert Wiener.

---

1. Introduction. Physical need for theory. Statistical mechanics may be defined as the application of the concepts of Lebesgue integration to mechanics. Historically, this is perhaps putting the cart before the horse. Statistical mechanics developed through the entire latter half of the nineteenth century before the Lebesgue integral was discovered. Nevertheless, it developed without an adequate armory of concepts and mathematical technique, which is only now in the process of development at the hands of the modern school of students of integral theory.

In the more primitive forms of statistical mechanics, the integration or summation was taken over the manifold particles of a single homogeneous dynamical system, as in the case of the perfect gas. In its more mature form, due to Gibbs, the integration is performed over a parameter of distribution, numerical or not, serving to label the constituent systems of a dynamical ensemble, evolving under identical laws of force, but differing in their initial conditions. Nevertheless, the study of the mode in which this parameter of distribution enters into the individual systems of the ensemble does not seem to have received much explicit study. The parameter of distribution is essentially a parameter of integration only. As such, questions of dimensionality are indifferent to it, and it may be replaced by a numerical variable α with the

Reprinted from *Amer. J. Math.*, *60*, 1938, pp. 897–936. (Courtesy of the *American Journal of Mathematics*, The Johns Hopkins University.)

range $(0, 1)$. Any transformation leaving invariant the probability properties of the ensemble as a whole is then represented by a measure-preserving transformation of the interval $(0, 1)$ into itself.

Among the simplest and most important ensembles of physics are those which have a spatially homogeneous character. Among these are the homogeneous gas, the homogeneous liquid, the homogeneous state of turbulence. In these, while the individual systems may not be invariant under a change of origin, or, in other words, under the translation of space by a vector, the ensemble as a whole is invariant, and the individual systems are merely permuted without change of probability. From what we have said, the translations of space thus generate an Abelian group of equi-measure transformations of the parameter of distribution.

One-dimensional groups of equi-measure transformations have become well known to the mathematicians during the past decade, as they lie at the root of Birkhoff's famous ergodic theorem.[1] This theorem asserts that if we have given a set S of finite measure, an integrable function $f(P)$ on S, and a one-parameter Abelian group T^λ of equi-measure transformations of S into itself, such that

$$T^\lambda \cdot T^\mu = T^{\lambda+\mu} \qquad (-\infty < \lambda < \infty, \; -\infty < \mu < \infty), \tag{1}$$

then for all points P on S except those of a set of zero measure, and provided certain conditions of measurability are satisfied,

$$\lim_{A\to\infty} \frac{1}{A} \int_0^A f(T^\lambda P)\, d\lambda \tag{2}$$

will exist. Under certain more stringent conditions, known as metric transitivity, we shall have

$$\lim_{A\to\infty} \frac{1}{A} \int_0^A f(T^\lambda P)\, d\lambda = \int_S f(P)\, dV_P \tag{3}$$

almost everywhere. The ergodic theorem thus translates averages over an infinite range, taken with respect to λ, into averages over the set S of finite measure. Even without metric transitivity, the ergodic theorem translates the distribution theory of λ averages into the theory of S averages.

In the most familiar applications of the ergodic theorem, S is taken to be a spatial set, and the parameter λ is identified with the time. The theorem thus becomes a way of translating time averages into space averages, in a

---

[1] Cf. Eberhard Hopf, "Ergodentheorie," *Ergebnisse der Mathematik und ihrer Grenzgebiete*, vol. 5. See particularly § 14, where further references to the literature are given.

manner which was postulated by Gibbs without rigorous justification, and which forms the entire basis of his methods. Strictly speaking, the space averages are generally in phase-space rather than in the ordinary geometrical space of three dimensions. There is no reason, however, why the parameter λ should be confined to one taking on values of the time for its arguments, nor even why it should be a one-dimensional variable. We are thus driven to formulate and prove a multidimensional analogue of the classical Birkhoff theorem.

In the ordinary Birkhoff theorem, the transformations T^{λ} are taken to be one-one point transformations. Now, the ergodic theorem belongs fundamentally to the abstract theory of the Lebesgue integral, and in this theory, individual points play no rôle. In the study of chaos, individual values of the parameter of integration are equally unnatural as an object of study, and it becomes desirable to recast the ergodic theorem into a true Lebesgue form. This we do in paragraph 2.

Of all the forms of chaos occurring in physics, there is only one class which has been studied with anything approaching completeness. This is the class of types of chaos connected with the theory of the Brownian motion. In this one-dimensional theory, there is a simple and powerful algorithm of phase averages, which the ergodic theorem readily converts into a theory of averages over the transformation group. This theory is easily generalized to spaces of a higher dimensionality, without any very fundamental alterations. We shall show that there is a certain sense in which these types of chaos are central in the theory, and allow us to approximate to all types.

Physical theories of chaos, such as that of turbulence, or of the statistical theory of a gas or a liquid, may or may not be theories of equilibrium. In the general case, the statistical state of a chaotic system, subject to the laws of dynamics, will be a function of the time. The laws of dynamics produce a continuous transformation group, in which the chaos remains a chaos, but changes its character. This is at least the case in those systems which can continue to exist indefinitely in time without some catastrophe which essentially changes their dynamic character. The study, for example, of the development of a state of turbulence, depends on an existence theory which avoids the possibility of such a catastrophe. We shall close this paper by certain very general considerations concerning the demands of an existence theory of this sort.

2. Definition. Types of Chaos. A *continuous homogeneous chaos* in n dimensions is a scalar or vector valued measurable function $\rho(x_1, \cdots, x_n; \alpha)$ of $x_1, \cdots, x_n; \alpha$, in which $x_1, \cdots, x_n$ assume all real values, while α ranges over $(0, 1)$; and in which the set of values of α for which

$$\rho(x_1 + y_1, \cdots, x_n + y_n; \alpha) \text{ belongs to } S, \tag{4}$$

if it has a measure for any set of values of $y_1, \cdots, y_n$, has the same measure for any other set of values. In this paper, we shall confine our attention to scalar chaoses. A continuous homogeneous chaos is said to be *metrically transitive,* if whenever the sets of values of α for which $\rho(x_1, \cdots, x_n; \alpha)$ belongs to S and to S_1, respectively, have measures M and M_1, the set of values of α for which simultaneously

$$\rho(x_1, \cdots, x_n; \alpha) \text{ belongs to } S$$

and

$$\rho(x_1 + y_1, \cdots, x_n + y_n; \alpha) \text{ belongs to } S_1$$

has a measure which tends to M_1 as $y_1^2 + \cdots + y_n^2 \to \infty$.

If ρ is integrable, it determines the additive set-function

$$\mathfrak{F}(\Sigma; \alpha) = \int \cdots_{\Sigma} \int \rho(x_1, \cdots, x_n; \alpha) dx_1 \cdots dx_n. \tag{5}$$

On the other hand, not every additive set-function may be so defined. This suggests a more general definition of a *homogeneous chaos,* in which the chaos is defined to be a function $\mathfrak{F}(\Sigma; \alpha)$, where α ranges over $(0, 1)$ and Σ belongs to some additively closed set Ξ of measurable sets of points in n-space. We suppose that if Σ and Σ_1 do not overlap,

$$\mathfrak{F}(\Sigma + \Sigma_1; \alpha) = \mathfrak{F}(\Sigma; \alpha) + \mathfrak{F}(\Sigma_1; \alpha). \tag{6}$$

We now define the new point-set $\Sigma(y_1, \cdots, y_n)$ by the assertion

$$\Sigma(y_1, \cdots, y_n) \text{ contains } x_1 + y_1, \cdots, x_n + y_n \text{ when and only when } \Sigma \text{ contains } x_1, \cdots, x_n. \tag{7}$$

This leads to the definition of the additive set-function $\mathfrak{F}_{y_1, \ldots, y_n}(\Sigma; \alpha)$ by

$$\mathfrak{F}_{y_1, \ldots, y_n}(\Sigma; \alpha) = \mathfrak{F}(\Sigma(y_1, \cdots, y_n); \alpha). \tag{8}$$

If, then, for all classes S of real numbers,

(9) Measure of set of α's for which $\mathfrak{F}_{y_1, \ldots, y_n}(\Sigma; \alpha)$ belongs to class S

is independent of $y_1, \cdots, y_n$, in the sense that if it exists for one set of these numbers, it exists for all sets, and has the same value, and if it is measurable in $y_1, \cdots, y_n$, we shall call $\mathfrak{F}$ a homogeneous chaos. The notion of metrical transitivity is generalized in the obvious way, replacing $\rho(x_1, \cdots, x_n; \alpha)$ by $\mathfrak{F}(\Sigma; \alpha)$, and $\rho(x_1 + y_1, \cdots, x_n + y_n; \alpha)$ by $\mathfrak{F}_{y_1, \ldots, y_n}(\Sigma; \alpha)$.

The theorem which we wish to prove is the following:

THEOREM I. *Let $\mathfrak{F}(\Sigma;\alpha)$ be a homogeneous chaos. Let the functional*

$$\Phi\{\mathfrak{F}(\Sigma;\alpha)\} = g(\alpha) \tag{10}$$

be a measurable function of α, such that $\int_0^1 | g(\alpha) \log^+ | g(\alpha)| | d\alpha$ is finite. Then for almost all values of α,

$$\lim_{r\to\infty} \frac{1}{V(r)} \int \cdots_R \int \Phi\{\mathfrak{F}_{y_1,\ldots,y_n}(\Sigma;\alpha)\} dy_1 \cdots dy_n \tag{11}$$

exists, where R is the interior of the sphere

$$y_1^2 + y_2^2 + \cdots + y_n^2 = r^2 \tag{12}$$

and $V(r)$ is its volume. If in addition, $\mathfrak{F}(\Sigma;\alpha)$ is metrically transitive,

$$\lim_{r\to\infty} \frac{1}{V(r)} \int \cdots_R \int \Phi\{\mathfrak{F}_{y_1,\ldots,y_n}(\Sigma;\alpha)\} dy_1 \cdots dy_n = \int_0^1 \Phi\{\mathfrak{F}(\Sigma;\beta)\} d\beta \tag{13}$$

for almost all values of α.

3. Classical ergodic theorem. Lebesgue form. Theorem I is manifestly a theorem of the ergodic type. Let it be noticed, however, that we nowhere assume that the transformation of α given by $\beta = T\alpha$ when

$$\mathfrak{F}_{y_1,\ldots,y_n}(\Sigma;\beta) = \mathfrak{F}(\Sigma;\alpha) \tag{14}$$

is one-one. This should not be surprising, as the ergodic theorem is fundamentally one concerning the Lebesgue integral, and in the theory of the Lebesgue integral, individual points play no rôle.

Nevertheless, in the usual formulation of the ergodic theorem, the expression $f(T^\lambda P)$ enters in an essential way. Can we give this a meaning without introducing the individual transform of an individual point?

The clue to this lies in the definition of the Lebesgue integral itself. If $f(P)$ is to be integrated over a region S, we divide S into the regions $S_{a,b}(f)$, defined by the condition that over such a region,

$$a < f(P) \leq b. \tag{15}$$

We now write

$$\int_S f(P) dV_P = \lim_{\epsilon\to 0} \sum_{-\infty}^{\infty} n\,\epsilon\, m(S_{(n-1)\epsilon,\, n\epsilon}(f)). \tag{16}$$

The condition that $f(P)$ be integrable thus implies the condition that it be measurable, or that all the sets $S_{a,b}(f)$ be measurable.

Now, if T is a measure-preserving transformation on S, the sets $TS_{a,b}(f)$

will all be measurable, and will have, respectively, the same measures as the sets $S_{a,b}(f)$. We shall define the function $f(TP) = g(P)$ by the conditions

$$TS_{a,b}(f) = S_{a,b}(g). \tag{17}$$

If T conserves relations of inclusion of sets, up to sets of zero measure, this function will clearly be defined up to a set of values of P of zero measure, and we shall have

$$\int_S f(TP)\,dV_P = \int_S f(P)\,dV_P. \tag{18}$$

We may thus formulate the original or discrete case of the Birkhoff ergodic theorem, as follows: *Let S be a set of points of finite measure. Let T be a transformation of all measurable sub-sets of S into measurable sub-sets of S, which conserves measure, and the relation between two sets, that one contains the other except at most for a set of zero measure. Then except for a set of points P of zero measure,*

$$\lim_{N\to\infty} \frac{1}{N+1} \sum_0^N f(T^n P) \tag{19}$$

will exist.

The continuous analogue of this theorem needs to be formulated in a somewhat more restricted form, owing to the need of providing for the integrability of the functions concerned. It reads: *Let S be a set of points of finite measure. Let T^λ be a group of transformations fulfilling the conditions we have laid down for T in the discrete case just mentioned. Let $T^\lambda P$ be measurable in the product space of λ and of P. Then, except for a set of points P of zero measure,*

$$\lim_{A\to\infty} \frac{1}{A} \int_0^A f(T^\lambda P)\,d\lambda \tag{20}$$

will exist.

In the proofs of Birkhoff's ergodic theorem, as given by Khintchine and Hopf, no actual use is made of the fact that the transformation T is one-one, and the proofs extend to our theorem as stated here, without any change. The restriction of measurability, or something to take its place, is really necessary for the correct formulation of Khintchine's statement of the ergodic theorem, as Rademacher, von Neumann, and others have already pointed out.[2]

With the aid of the proper Lebesgue formulation of the ergodic theorem, the one-dimensional case of Theorem I follows at once. Actually more follows, as it is only necessary that g belong to L, instead of to the logarithmic class

[2] Cf. J. v. Neumann, *Annals of Mathematics*, 2, vol. 33, p. 589, note 11.

with which we replace it. To prove Theorem I in its full generality, we must establish a multidimensional ergodic theorem.

4. Dominated ergodic theorem. Multidimensional ergodic theorem. As a lemma to the multidimensional ergodic theorem, we first wish to establish the fact that if the function $f(P)$ in the ergodic theorem satisfies the condition

$$\int_S |f(P)| \log^+ |f(P)| \, dV_P < \infty, \tag{21}$$

then the expressions (19) and (20) not only exist, but the limits in question will be approached under the domination of a summable function of P. We shall prove this in the discrete case, for the sake of simplicity, but the result goes over without difficulty to the continuous case.

Let T be an equimeasure transformation of the set S of finite measure into itself, in the generalized sense of the last paragraph, and let W be a measurable sub-set of S, with the characteristic function $W(P)$. Let U be the set of all points P for which some $T^{-j}P$ belongs to W. Let $i(P)$, when P belongs to W, be defined as the smallest positive number n such that $T^{-n}P$ belongs to W, and let us call it the *index* of P. Every point of W, except for a set of measure 0, will have a finite index, since if we write W_∞ for the set of points without a finite index, no two sets $T^m W_\infty$ and $T^n W_\infty$ can overlap, while they all have the same measure and their sum has a finite measure. Thus except for a set of zero measure, we may divide W into the sets W_p, each consisting of the points of W of index p. It is easy to show that if $1 \leq p < \infty$, $1 \leq p' < \infty$, $0 \leq j < p$, $0 \leq j' < p'$, the sets $T^{-j}W_p$ and $T^{-j'}W_{p'}$ can not overlap over a set of positive measure unless $j = j'$, $p = p'$. Similarly, the sets $T^{-p}W_p$ and $T^{-p'}W_{p'}$ can not overlap over sets of positive measure, unless $p = p'$, and represent a dissection of W, except for a set of zero measure. Let us put W_{pq} for the logical product of W_p and T^pW_q; W_{pqr} for the logical product of W_{pq} and $T^{p+q}W_r$; and so on. Then if k is fixed, the sets $T^{-j}W_{p_1p_2\ldots p_k}$ cover U (except for sets of zero measure) once as j goes from 0 to p_1, once as it goes from p_1 to $p_1 + p_2$, and so on; making just k times between 0, inclusive, and $p_1 + p_2 + \cdots + p_k$, exclusive. Thus if S_K is the set of all the points in all the $W_{p_1p_2\ldots p_k}$ for which $p_1 + p_2 + \cdots + p_k = K$, the total measure of all the S_K's for $2^{N+1} \geq K \geq 2^N$ can not exceed $km(S)/2^N$.

Now let P lie in $T^{-j}W_{p_1p_2\ldots p_k}$, where $0 \leq j < p_1 + p_2 + \cdots + p_k$. Let us consider the sequence of numbers a_j, where if $p_1 + \cdots + p_l \leq j < p_1 + \cdots + p_{l+1}$, a_j is the greatest of the numbers

$$\frac{1}{j - p_1 - \cdots - p_l + 1}, \quad \frac{2}{j - p_1 - \cdots - p_{l-1} + 1},$$
$$\frac{3}{j - p_1 - \cdots - p_{l-2} + 1}, \ldots, \frac{l+1}{j+1}.$$

Then a_j will be the largest of the numbers

$$W(P), \frac{W(P)+W(TP)}{2}, \cdots, \frac{W(P)+W(TP)+\cdots+W(T^jP)}{j+1}.$$

The sum $\sum_0^k a_j$, for a fixed K, will have its maximum value when $p_1 = p_2 = \cdots = p_{k-1} = 1$; $p_k = K + 1 - k$, when it will be $k + \sum_{k+1}^{k} k/j \leq k(1 + \log K/k)$. In this case the sequence of the a_j's will be

$$\underbrace{1, 1, \cdots, 1,}_{k \text{ times}} \frac{k}{k+1}, \frac{k}{k+2}, \cdots, \frac{k}{K}.$$

This remark will be an easy consequence of the following fact: let us consider the sequence

$$(22) \qquad \cdots, \frac{\lambda}{n-1}, \frac{\lambda}{n}, 1, \frac{1}{2}, \cdots, \frac{1}{\left[\frac{n}{\lambda}\right]}, \frac{\lambda+1}{\left[\frac{n}{\lambda}\right]+n+1}, \cdots$$

and the modified sequence

$$(23) \qquad \cdots, \frac{\lambda}{n-1}, 1, \frac{1}{2}, \cdots,$$
$$\frac{1}{\left[\frac{n-1}{\lambda}\right]}, \frac{\lambda+1}{\left[\frac{n-1}{\lambda}\right]+n}, \frac{\lambda+1}{\left[\frac{n-1}{\lambda}\right]+n+1}, \cdots$$

where of course

$$\frac{\lambda+1}{n+\left[\frac{n}{\lambda}\right]} \leq \frac{1}{\left[\frac{n}{\lambda}\right]}; \quad \frac{\lambda+1}{n+\left[\frac{n}{\lambda}\right]+1} > \frac{1}{\left[\frac{n}{\lambda}\right]+1};$$
$$\frac{\lambda+1}{n-1+\left[\frac{n-1}{\lambda}\right]} \leq \frac{1}{\left[\frac{n-1}{\lambda}\right]}; \quad \frac{\lambda+1}{n+\left[\frac{n-1}{\lambda}\right]} > \frac{1}{\left[\frac{n-1}{\lambda}\right]+1}.$$

Apart from the arrangement, the terms of (22) will be the same as the terms of (23), except that in (23), $\dfrac{\lambda+1}{\left[\frac{n-1}{\lambda}\right]+n}$ replaces λ/n. Now,

$$\frac{\lambda+1}{\left[\frac{n-1}{\lambda}\right]+n} > \frac{\lambda}{n},$$

so that the sum of the terms in (23) is greater than that in (22).

Since the transforms of a given set have the same measure as the set, and the sets $T^{-j}S_K$ cover U exactly k times, we have

$$\frac{1}{k}\int_S \Big\{ W(P) + \max\left(W(P), \frac{W(P)+W(TP)}{2}\right) + \max\left(W(P), \frac{W(P)+W(TP)}{2}, \frac{W(P)+W(TP)+W(T^2P)}{3}\right) + \cdots + \max\left(W(P), \frac{W(P)+W(TP)}{2}, \cdots, \frac{W(P)+\cdots+W(T^{k-1}P)}{k}\right)\Big\} dV_P$$

$$\leq \sum_{K=0}^{\infty} m(S_K)\left(1+\log\frac{K}{k}\right)$$

$$\leq m(W)\left(1+\log\frac{2^N}{k}\right) + \sum_{j=1}^{\infty}\frac{km(S)}{2^{N+j}}\left(1+\log\frac{2^{N+j}}{k}\right)$$

$$\leq \left(1+\log\frac{2^N}{k}\right)\Big\{ m(W) + \text{const.}\ \frac{km(S)}{2^N}\Big\},$$

the constant being absolute. If we now put

$$N = \left[\log\frac{km(S)}{m(W)} \Big/ \log 2\right],$$

this last expression is dominated by

$$\text{const.}\ m(W)\left(1+\log\frac{m(S)}{m(W)}\right).$$

Since the constant is independent of k, we see that the Cesàro average of

$$\max\left(W(P), \frac{W(P)+W(TP)}{2}, \cdots, \frac{W(P)+\cdots+W(T^{k-1}P)}{k}\right)$$

is dominated by the same term. Thus

$$\int_S \max\left(W(P), \cdots, \frac{W(P)+\cdots+W(T^{k-1}P)}{k}\right) dV_P \leq \text{const.}\ m(W)\left(1+\log\frac{m(S)}{m(W)}\right),$$

and it follows by monotone convergence that there exists a function $W^*(P)$ such that

$$\int_S W^*(P)\,dV_P \leq \text{const.}\ m(W)\left(1+\log\frac{m(S)}{m(W)}\right), \tag{24}$$

and for all positive m,

$$\frac{W(P)+\cdots+W(T^mP)}{m+1} \leq W^*(P). \tag{25}$$

Now let $f(P)$ be a function such that

$$(21)\qquad \int_S |f(P)| \log^+ |f(P)|\, dV_P < \infty.$$

Let $W^{(N)}$ be the set of points such that

$$(26)\qquad 2^N \leq |f(P)| \leq 2^{N+1},$$

and let $f^{(N)}(P)$ be the function equal to $f(P)$ over this set of points, and 0 elsewhere. Then there exists a function $f^{(N)*}(P)$, such that

$$(27)\qquad \frac{f^{(N)}(P) + \cdots + f^{(N)}(T^m P)}{m+1} \leq f^{(N)*}(P) \qquad (m = 0, 1, 2, \cdots),$$

and

$$(28)\qquad \int_S f^{(N)*}(P)\, dV_P \leq \text{const. } m(W^{(N)}) \left(1 + \log^+ \frac{m(S)}{m(W^{(N)})}\right) 2^N.$$

Hence if

$$(29)\qquad \sum_{-\infty}^{\infty} m(W^{(N)}) \left(1 + \log^+ \frac{m(S)}{m(W^{(N)})}\right) 2^N < \infty$$

and

$$(30)\qquad f^*(P) = \sum_{-\infty}^{\infty} f^{(N)*}(P),$$

then

$$(31)\qquad \frac{f(P) + \cdots + f(T^m P)}{m+1} \leq f^*(P) \qquad (m = 0, 1, 2, \cdots),$$

and

$$(32)\qquad \int_S f^*(P)\, dV_P \leq \text{const. } \sum m(W^{(N)}) \left(1 + \log^+ \frac{m(S)}{m(W^{(N)})}\right) 2^N.$$

However,

$$(33)\qquad \sum_{-\infty}^{\infty} m(W^{(N)}) \left(1 + \log^+ \frac{m(W^{(N)})}{m(S)}\right) 2^N$$

$$\leq \sum_{-\infty}^{\infty} \int_{W^{(N)}} |f(P)|\, dV_P \left(1 + \log^+ \frac{2^{N+1} m(S)}{\int_S |f(P)|\, dV_P}\right)$$

$$\leq \int_S |f(P)| \left(1 + \log^+ \frac{2\,|f(P)|\, m(S)}{\int_S |f(Q)\, dV_Q}\right) dV_P$$

$$\leq \int_S |f(P)|\, dV_P \left(1 + \log^+ \frac{2m(S)}{\int_S |f(P)|\, dV_P}\right)$$

$$+ \int_S |f(P)| \log^+ |f(P)|\, dV_P.$$

Thus $\int_S f^*(P)dV_P$ has an upper bound which is less than a function of $\int_S |f(P)|\,dV_P$ and $\int_S |f(P)| \log^+ |f(P)|\,dV_P$, tending to 0 as they both tend to 0. This establishes our theorem of the existence of a uniform dominant.

There is a sense in which (21) *is a best possible condition.* That is, if

$$\psi(x) = o(\log^+ x), \tag{34}$$

the condition

$$\int_S (\psi(|f(P)|)\,|f(P)|\,dV_P < \infty \tag{35}$$

is not sufficient for the existence of a uniform dominant. For let S be a set of measure 1, subdivided into mutually exclusive sets S_n, of measures respectively 2^{-n}. Let S_n be divided into mutually exclusive sets $S_{n,1}, \cdots, S_{n,\nu_n}$, all of equal measures. Let T transform $S_{n,k}$ into $S_{n,(k+1)}$ $(k < \nu^n)$, and S_{n,ν_n} into $S_{n,1}$. Let $f(P)$ be defined by

$$f(P) = a_n > 0 \text{ on } S_{n_1}, \quad (n = 1, 2, \cdots); \quad f(P) = 0 \text{ elsewhere.} \tag{36}$$

Then the smallest possible uniform dominant of

$$\frac{1}{N+1} \sum_0^N f(T^n P) \tag{37}$$

is

$$f^*(P) = \frac{a_n}{k} \text{ on } S_{n,k}, \tag{38}$$

and we have

$$\int_S f^*(P)dV_P = \Sigma \frac{a_n \Omega(\log \nu_n) 2^{-n}}{\nu_n}. \tag{39}$$

Thus if

$$\nu_n = 2^{2n} a_n, \qquad a_n = \Omega(n!), \tag{40}$$

the function $f^*(P)$ will belong to L if and only if

$$\infty > \Sigma \frac{a_n \Omega(\log a_n) 2^{-n}}{\nu_n} > \text{const.} \int_S f(P) \log^+ f(P) dV_P. \tag{41}$$

While we have proved the dominated ergodic theorem merely as a lemma for the multidimensional ergodic theorem, the theorem, and more particularly, the method by which we have proved it, have very considerable independent interest. We may use these methods to deduce von Neumann's mean ergodic theorem from the Birkhoff theorem; or vice versa, we may deduce the Birkhoff theorem, at least in the case of a function satisfying (21), from the von

Neumann theorem. These facts however are not relevant to the frame of the present paper, and will be published elsewhere.

We shall now proceed to the proof of the multidimensional ergodic theorem, which we shall establish in the two-dimensional case, although the method is independent of the number of dimensions. Let $T_1^\lambda T_2^\mu$ be a two-dimensional Abelian group of transformations of the set S (of measure 1) into itself, in the sense in which we have used this term in paragraph 3. Let $T_1^\lambda T_2^\mu P$ be measurable in λ, μ, and P. We now introduce a new variable x, ranging over $(0, 1)$, and form the product space Σ of P and x. We introduce the one-parameter group of transformations of this space, T^ρ, by putting

$$T^\rho(P, x) = (T_1^{\rho \cos 2\pi x} T_2^{\rho \sin 2\pi x} P, x) \tag{42}$$

The expressions $T^\rho(P, x)$ will be measurable in ρ and (P, x), and the transformations T^ρ will all preserve measure on Σ. Thus by the ergodic theorem, for almost all points (P, x) of Σ, if $f(P) = f(P, x)$ belongs to L, the limit

$$\lim_{A\to\infty} \frac{1}{A} \int_0^A f(T^\rho(P, x))\, d\rho = \lim_{A\to\infty} \frac{1}{A} \int_0^A f(T_1^{\rho \cos 2\pi x} T_2^{\rho \sin 2\pi x} P)\, d\rho \tag{43}$$

will exist. If condition (26) is satisfied, it will follow by dominated convergence that the limit

$$\lim_{A\to\infty} \frac{1}{A} \int_0^A d\rho \int_0^1 f(T_1^{\rho \cos 2\pi x} T_2^{\rho \sin 2\pi x} P)\, dx \tag{44}$$

will exist for almost all points (P, x), and hence for almost all points P.

For the moment, let us assume that $f(P)$ is non-negative. Then there is a Tauberian theorem, due to the author,[3] which establishes that the expression (44) is equivalent to

$$\lim_{A\to\infty} \frac{1}{\pi A^2} \int_0^A \rho\, d\rho \int_0^{2\pi} f(T_1^{\rho \cos \theta} T_2^{\rho \sin \theta} P)\, d\theta. \tag{45}$$

The only point of importance which we must establish in order to justify this Tauberian theorem is that

$$\int_0^1 \rho^{1+iu}\, d\rho = \frac{1}{2 + iu} \neq 0 \tag{46}$$

for real values of u. Since every function $f(P)$ satisfying (21) is the difference of two non-negative functions satisfying this condition, (45) is established in the general case.

It will be observed that we have established our multidimensional ergodic

[3] N. Wiener, "Tauberian theorems," *Annals of Mathematics*, 2, vol. 33, p. 28.

theorem on the basis of assumption (21), and not on that of the weaker assumption that $f(P)$ belongs to L. What the actual state of affairs may be, we do not know. At any rate, all attempts to arrive at a direct analogue of the Khintchine proof for one dimension have broken down. The one-dimensional proof makes essential use of the fact that the difference of two intervals is always an interval, while the difference between two spheres is not always a sphere.

The precise statement of the multidimensional ergodic theorem is the following: *Let S be a set of points of measure* 1, *and let $T_1^{\lambda_1}T_2^{\lambda_2}\cdots T_n^{\lambda_n}$ be an Abelian group of equimeasure transformations of S into itself, in the sense of paragraph* 3. *Let $T_1^{\lambda_1}\cdots T_n^{\lambda_n}P$ be measurable in $\lambda_1,\cdots,\lambda_n$; P. Let R be the set of values of $\lambda_1,\cdots,\lambda_n$ for which*

$$\lambda_1^2+\lambda_2^2+\cdots+\lambda_n^2\leq r^2 \tag{47}$$

and let $V(r)$ be its volume. Let $f(P)$ satisfy the condition (21). *Then for almost all values of P,*

$$\lim_{r\to\infty}\frac{1}{V(r)}\int\cdots_R\int f(T_1^{\lambda_1}\cdots T_n^{\lambda_n}P)\,d\lambda_1\cdots d\lambda_n \tag{48}$$

exists.

That part of Theorem I which does not concern metric transitivity is an immediate corollary.

5. Metric transitivity. Space and Phase Averages in a Chaos. If the function $f(P)$ is positive, clearly

$$\begin{aligned}\tag{49} &\int\cdots\int_{\lambda_1^2+\cdots+\lambda_n^2\leq r^2-\mu_1^2-\cdots-\mu_n^2} f(T_1^{\lambda}\cdots T_n^{\lambda}P)\,d\lambda_1\cdots d\lambda_n\\ &\leq\int\cdots\int_{\lambda_1^2+\cdots+\lambda_n^2\leq r^2} f(T_1^{\lambda_1}\cdots T_n^{\lambda_n}T_1^{\mu_1}\cdots T_n^{\mu_n}P)\,d\lambda_1\cdots d\lambda_n\\ &\leq\int\cdots\int_{\lambda_1^2+\cdots+\lambda_n^2\leq r^2+\mu_1^2+\cdots+\mu_n^2} f(T_1^{\lambda_1}\cdots T_n^{\lambda_n}P)\,d\lambda_1\cdots d\lambda_n.\end{aligned}$$

Hence

$$\begin{aligned}\tag{50} \lim_{r\to\infty}\frac{1}{V(r)}\int\cdots_R\int f(T_1^{\lambda_1}\cdots T_n^{\lambda_n}T_1^{\mu_1}\cdots T_n^{\mu_n}P)\,d\lambda_1\cdots d\lambda_n\\ =\lim_{r\to\infty}\frac{1}{V(r)}\int\cdots_R\int f(T_1^{\lambda_1}\cdots T_n^{\lambda_n}P)\,d\lambda_1\cdots d\lambda_n,\end{aligned}$$

and expression (48) has the same value for P and all its transforms under the group $T_1^{\lambda_1}\cdots T_n^{\lambda_n}$. The condition of positivity is clearly superfluous. Thus in case expression (48) does not almost everywhere assume a single

9

value, there will be two classes S_1 and S_2 of elements of S, each of positive measure, and each invariant under all the transformations $T_1^{\lambda_1} \cdots T_n^{\lambda_n}$.

A condition which will manifestly exclude such a contingency is that if S_1 and S_2 are two sub-sets of S of positive measure, and ϵ is a positive quantity, there always exists a transformation $T = T_1^{\lambda_1} \cdots T_n^{\lambda_n}$, such that

$$\left| \frac{mS_1(TS_2)}{mS_2} - mS_1 \right| < \epsilon. \tag{51}$$

From this it will immediately follow that if a chaos is metrically transitive in the sense of paragraph 3, the group of transformations of the α space generated by translations of the chaos will have the property we have just stated, and under the assumptions of Theorem I,

$$\lim_{r\to\infty} \frac{1}{V(r)} \int \cdots_R \cdots \int \Phi\{\mathfrak{F}_{y_1, \ldots, y_n}(\Sigma;\alpha)\} dy_1 \cdots dy_n \tag{52}$$

will exist and have the same value for almost all values of α.

If almost everywhere

$$\lim_{r\to\infty} \frac{1}{V(r)} \int \cdots_R \cdots \int \Phi\{\mathfrak{F}_{y_1, \ldots, y_n}(\Sigma;\alpha)\} dy_1 \cdots dy_n = A, \tag{53}$$

and

$$\frac{1}{V(r)} \int \cdots_R \cdots \int \Phi\{\mathfrak{F}_{y_1, \ldots, y_n}(\Sigma;\alpha)\} dy_1 \cdots dy_n < g(\alpha) \tag{54}$$

where $g(\alpha)$ belongs to L, then by dominated convergence,

$$\begin{aligned} A &= \lim_{r\to\infty} \frac{1}{V(r)} \int \cdots_R \cdots \int dy_1 \cdots dy_n \int_0^1 d\alpha \Phi\{\mathfrak{F}_{y_1, \ldots, y_n}(\Sigma;\alpha)\} \\ &= \int_0^1 \Phi\{\mathfrak{F}(\Sigma;\alpha)\} d\alpha. \end{aligned} \tag{55}$$

That is, the average of $\Phi\{\mathfrak{F}(\Sigma;\alpha)\}$, taken over the finite phase space of α, is almost everywhere the same as the average of $\Phi\{\mathfrak{F}_{y_1, \ldots, y_n}(\Sigma;\alpha)\}$ taken over the infinite group space of points $y_1, \cdots, y_n$. This completes the establishment of Theorem I, and gives us a real basis for the study of the homogeneous chaos.[4]

6. Pure one-dimensional chaos. The simplest type of pure chaos is that which has already been treated by the author in connection with the Brownian motion. However, as we wish to generalize this theory to a multi-

[4] The material of this chapter, in the one-dimensional case, has been discussed by the author with Professor Eberhard Hopf several years ago, and he wishes to thank Professor Hopf for suggestions which have contributed to his present point of view.

plicity of dimensions, instead of referring to existing articles on the subject, we shall present it in a form which emphasizes its essential independence of dimensionality.

The type of chaos which we shall consider is that in which the expression $\mathfrak{F}(\Sigma;\alpha)$ has a distribution in α dependent only on the measure of the set Σ; and in which, if Σ_1 and Σ_2 do not overlap, the distributions of $\mathfrak{F}(\Sigma_1,\alpha)$ and $\mathfrak{F}(\Sigma_2,\alpha)$ are independent, in the sense that if $\phi(x,y)$ is a measurable function, and either side of the equation has a sense,

$$(56)\qquad \int_0^1\int_0^1 \phi(\mathfrak{F}(\Sigma_1;\alpha),\mathfrak{F}(\Sigma_2,\beta))\,d\alpha d\beta = \int_0^1 \phi(\mathfrak{F}(\Sigma_1,\alpha),\mathfrak{F}(\Sigma_2,\alpha))\,d\alpha.$$

We assume a similar independence when n non-overlapping sets $\Sigma_1,\cdots,\Sigma_n$ are concerned. It is by no means intuitively certain that such a type of chaos exists. In establishing its existence, we encounter a difficulty belonging to many branches of the theory of the Lebesgue integral. The fundamental theorem of Lebesgue assures us of the possibility of adding the measures of a denumerable assemblage of measurable sets, to get the measure of their sum, if they do not overlap. Accordingly, behind any effective realization of the theory of Lebesgue integration, there is always a certain denumerable family of sets in the background, such that all measurable sets may be approximated by denumerable combinations of these. This family is not unique, but without the possibility of finding it, there is no Lebesgue theory.

On the other hand, a theory of measure suitable for the description of a chaos must yield the measure of any assemblage of functions arising from a given measurable assemblage by a translational change of origin. This set of assemblages is essentially non-denumerable. Any attempt to introduce the notion of measure in a way which is invariant under translational changes of origin, without the introduction of some more restricted set of measurable sets, which does not possess this invariance, will fail to establish those essential postulates of the Lebesgue integral which deal with denumerable sets of points. There is no way of avoiding the introduction of constructional devices which seem to restrict the invariance of the theory, although once the theory is obtained it may be established in its full invariance.

Accordingly, we shall start our theory of randomness with a division of space, whether of one dimension or of more, into a denumerable assemblage of sub-sets. In one dimension, this division may be that into those intervals whose coördinates are terminating binary numbers, and in more dimensions, into those parallelepipeds with edges parallel to the axes and with terminating binary coördinates for the corner points. We then wish to find a self-consistent distribution-function for the mass in such a region, dependent only on the volume, and independent for non-overlapping regions.

This problem does not admit of a unique solution, although the solution becomes essentially unique if we adjoin suitable auxiliary conditions. Among these conditions, for example, is the hypothesis that the distribution is symmetric, as between positive and negative values, has a finite mean square, and that the measure of the set of α's for which $\mathfrak{F}(S; \alpha) > A$ is a continuous function of A.[5] Without going into such considerations, we shall assume directly that the measure of the set of instances in which the value of $\mathfrak{F}(S; \alpha)$ in a region S of measure M lies between a and $b > a$, is

$$(57) \qquad \frac{1}{\sqrt{2\pi M}} \int_a^b \exp\left(-\frac{u^2}{2M}\right) du.$$

The formula

$$(58) \qquad \frac{1}{\sqrt{2\pi M_1 M_2}} \int_{-\infty}^{\infty} \exp\left(-\frac{u^2}{2M_1} - \frac{(v-u)^2}{2M_2}\right) du = \frac{1}{\sqrt{M_1 + M_2}} \exp\left(-\frac{v^2}{2(M_1 + M_2)}\right)$$

shows the consistency of this assumption.

The distributions of mass among the sets of our denumerable assemblage may be mapped on the line segment $0 \leq \alpha \leq 1$, in such a way that the measure of the set of instances in which a certain contingency holds will go into a set of values of α of the same measure. This statement needs a certain amount of elucidation. To begin with, the only sets of instances whose measures we know are those determined by

$$(59) \qquad \begin{array}{c} a_1 \leq \mathfrak{F}(S_1; \alpha) \leq b_1 \\ a_2 \leq \mathfrak{F}(S_2; \alpha) \leq b_2 \\ \cdots\cdots\cdots \\ a_n \leq \mathfrak{F}(S_n; \alpha) \leq b_n; \end{array}$$

where $S_1, S_2, \cdots, S_n$ are to be found among our denumerable set of subdivisions of space. However, once we have established a correspondence between the measures of these specific sets of contingencies and their corresponding sets of values of α, we may use the measure of any measurable set of values of α to define the measure of its corresponding set of contingencies.

The correspondence between sets of contingencies and points on the line $(0, 1)$ is made by determining a hierarchy of sets of contingencies

$$(60) \qquad \begin{array}{c} a_1^{(m,n)} \leq \mathfrak{F}(S_1^{(m,n)}; \alpha) \leq b_1^{(m,n)} \\ \cdots\cdots\cdots \\ a_\nu^{(m,n)} \leq \mathfrak{F}(S_\nu^{(m,n)}; \alpha) \leq b_\nu^{(m,n)}. \end{array}$$

---

[5] Cf. the recent investigations of Cramér and P. Lévy.

Let us call such a contingency $C_{m,n}$. If m is fixed, let all the contingencies $C_{m,n}$ $(n=1,2,\cdots)$ be mutually exclusive, and let them be finite in number. Let us be able to write

$$C_{m,n}=\sum_{k=1}^{N} C_{m+1,n_k}. \tag{61}$$

If S' is one of our denumerable sets of regions of space, let every $C_{m,n}$ with a sufficiently large index m be included in a class determined by a set of conditions concerning the mass on S' alone, and restricting it to a set of values lying in an interval (c,d), corresponding to an integral of form (57) and of arbitrarily small value. (Here d may be ∞, or c may be $-\infty$.) Let us put $C_{1,1}$ for the entire class of all possible contingencies, and let us represent it by the entire interval $(0,1)$. Let us assume that $C_{m,n}$ has been mapped into an interval of length corresponding to its probability, in accordance with (57), and let this interval be divided in order of the sequence of their n_k's into intervals corresponding respectively to the component C_{m+1,n_k}'s, and of the same measure. Except for a set of points of zero measure, every point of the segment $(0,1)$ of α will then be determined uniquely by the sequence of the intervals containing it and corresponding to the contingencies $C_{m,n}$ for successive values of m. This sequence will then determine uniquely (except in a set of cases corresponding to a set of values of α of zero measure) the value of $\mathfrak{F}(S_n;\alpha)$ for every one of our original denumerable set of sets S_n.

So far, everything that we have said has been independent of dimensionality. We now proceed to something belonging specifically to the one-dimensional case. If the original sets S_n are the sets of intervals with binary end-points, of such a form that they may be written in the binary scale

$$(d_1d_2\cdots d_k\cdot d_{k+1}\cdots d_l,\; d_1\cdots d_k\cdot d_{k+1}\cdots d_l+2^{-l-1}) \tag{62}$$

where $d_1,\cdots,d_l$ are digits which are either 0 or 1, then any interval whatever of length not exceeding $2^{-\mu}$ and lying in (a,b) (where a and b are integers) may be written as the sum of not more than two of the $(b-a)2^{\mu+2}$ intervals of form (62) lying in (a,b) and of length $2^{-\mu-1}$, not more than two of the $(b-a)2^{\mu+2}$ intervals of length $2^{-\mu-2}$, and so on. The probability that the value of $|\mathfrak{F}(S_n;\alpha)|$ should exceed A, or in other words, the measure of the set of α's for which it exceeds A, is

$$\frac{2}{\sqrt{2\pi m(S_n)}}\int_A^\infty \exp\left(-\frac{u^2}{2m(S_n)}\right)du=o\{\exp(-Am(S_n)^{-\frac{1}{2}})\}. \tag{63}$$

Now let us consider the total probability that the value of $|\mathfrak{F}(S_n;\alpha)|$ should exceed $2^{-(\mu+1)(\frac{1}{2}-\epsilon)}$ for any one of the $(b-a)2^{\mu+1}$ intervals of length

$2^{-\mu-1}$, or $2^{-(\mu+2)(\frac{1}{2}-\epsilon)}$ for any one of the $(b-a)2^{\mu+2}$ intervals of length $2^{-\mu-2}$, or so on. This probability can not exceed

$$\sum_{k=1}^{\infty}(b-a)2^{\mu+k}\,o(\exp(-2^{(\mu+k)\epsilon}))=o(2^{\mu}\exp(-2^{(\mu+1)\epsilon})). \tag{64}$$

On the other hand, the sum of $|\mathfrak{F}(S_n;\alpha)|$ for all the $2+2+\cdots$ intervals must in any other case be equal to or less than

$$2\sum_{k=1}^{\infty}2^{-(\mu+k)(\frac{1}{2}-\epsilon)}=O(2^{-\mu(\frac{1}{2}-\epsilon)}). \tag{65}$$

Thus there is a certain sense in which over a finite interval, and except for a set of values of α of arbitrarily small positive measure, the total mass in a sub-interval of length $\leq 2^{-\mu}$ tends uniformly to 0 with $2^{-\mu}$. On this basis, we may extend the functional $\mathfrak{F}(S;\alpha)$ to all intervals S. It is already defined for all intervals with terminating binary end-points. If (c,d) is any interval whatever, let $c_1, c_2, \cdots$ be a sequence of terminating binary numbers approaching c, and let $d_1, d_2, \cdots$ be a similar sequence approaching d. Then except for a fixed set of values of α of arbitrarily small measure,

$$\lim_{m,n\to\infty}|\mathfrak{F}((c_n,d_n)+\alpha)-\mathfrak{F}((c_m,d_m);\alpha)|=0, \tag{66}$$

and we may put

$$\mathfrak{F}((c,d);\alpha)=\lim_{m\to\infty}\mathfrak{F}((c_m,d_m);\alpha). \tag{67}$$

Formula (57), and the fact that $\mathfrak{F}(S_1;\alpha)$ and $\mathfrak{F}(S_2;\alpha)$ vary independently for non-overlapping intervals S_1 and S_2, will be left untouched by this extension.

Thus if Σ is an interval and T^{λ} a translation through an amount λ, we can define $\mathfrak{F}(T^{\lambda}\Sigma;\alpha)$, and it will be equally continuous in λ over any finite range of λ except for a set of values of α of arbitrarily small measure. From this it follows at once that it is measurable in λ and α together. Furthermore, we shall have

(68) Measure of set of α's for which $\mathfrak{F}(T^{\lambda}\Sigma;\alpha)$ belongs to C =
Measure of set of α's for which $\mathfrak{F}(\Sigma;\alpha)$ belongs to C.

Thus $\mathfrak{F}(\Sigma;\alpha)$ is a homogeneous chaos. We shall call it *the pure chaos.*

If $\Phi\{\mathfrak{F}(\Sigma;\alpha)\}$ is a functional dependent on the values of $\mathfrak{F}(\Sigma;\alpha)$ for a finite number of intervals Σ_n, then if λ is so great that none of the intervals Σ_n overlaps any translated interval $T^{\lambda}\Sigma_m$, $\Phi\{\mathfrak{F}(\Sigma;\alpha)\}$ will have a distribution entirely independent of $\Phi\{\mathfrak{F}(T^{\lambda}\Sigma;\alpha)\}$. As every measurable functional may be approached in the L sense by such a functional, we see at once that $\mathfrak{F}(\Sigma;\alpha)$ is metrically transitive.[6]

---

[6] Except that the method of treatment has been adapted to the needs of § 7, the

7. Pure multidimensional chaos. In order to avoid notational complexity, we shall not treat the general multidimensional case explicitly, but shall treat the two-dimensional case by a method which will go over directly to the most general multidimensional case. If our initial sets S_n are the rectangles with terminating binary coördinates for their corners and sides parallel to the axes, and we replace (a, b) by the square with opposite vertices (p, q) and $(p+r, q+r)$, an argument of exactly the same sort as that which we have used in the last paragraph will show that except in a set of cases of total probability not exceeding

$$(69) \qquad \text{const.} \sum_{k=1}^{\infty} \sum_{l=1}^{\infty} 2^{\mu+k} 2^{l}\, o(\exp(-2^{(\mu+k+l)\epsilon})) = o(2^{\mu} \exp(-2^{(\mu+1)\epsilon}))$$

the sum of $|\mathfrak{F}(S_n; \alpha)|$ for a denumerable set of binary rectangles with base $\leq 2^{-\mu}$, of the form (62), and adding up to make a vertical interval lying in the square (p, q), $(p+r, q+r)$, must be equal to or less than

$$(65) \qquad 2 \sum_{k=1}^{\infty} 2^{-(\mu+k)(\frac{1}{2}-\epsilon)} = O(2^{-\mu(\frac{1}{2}-\epsilon)}).$$

If we now add this expression up for all the base intervals of type (62) necessary to exhaust a horizontal interval of magnitude not exceeding $2^{-\mu}$, we shall again obtain an expression of the form (65). It hence follows that if we take the total mass on the coördinate rectangles within a given square, this will tend to zero uniformly with their area, except for a set of values of α of arbitrarily small measure. From this point the two-dimensional argument, and indeed the general multidimensional argument, follows exactly the same lines as the one-dimensional argument. It is only necessary to note that if

$$a_n \to a, \quad b_n \to b, \quad c_n \to c, \quad d_n \to d$$

then the rectangles (a_m, b_m), (c_m, d_m) and (a_n, b_n), (c_n, d_n) differ at most by four rectangles of small area.[7]

From this point on, we shall write $\mathfrak{P}(S; \alpha)$ for a pure chaos, whether in one or in more dimensions.

8. Phase averages in a pure chaos. If $f(P)$ is a measurable step-function, the definition of

$$(70) \qquad \int f(P)\, d_P \mathfrak{P}(S; \alpha)$$

---

results of this section have previously been demonstrated by the author. (*Proceedings of the London Mathematical Society*, 2, vol. 22 (1924), pp. 454-467).

[7] Here we represent a rectangle by giving two opposite corners.

is obvious, for it reduces to the finite sum

$$\sum_{1}^{N} f_n \mathfrak{P}(S_n;\alpha) \tag{71}$$

where f_n are the N values assumed by $f(P)$, and S_n respectively are the sets over which these values are assumed. Let us notice that

$$\begin{aligned}
\int_0^1 d\alpha \left| \int f(P)\,d_P\mathfrak{P}(S;\alpha)\right|^2 &= \sum_{m=1}^{N}\sum_{n=1}^{N}\int_0^1 d\alpha f_m \bar{f}_n \mathfrak{P}(S_m;\alpha)\mathfrak{P}(S_n;\alpha) \\
&= \sum_{1}^{N} |f_n|^2 \int_0^1 (\mathfrak{P}(S_n;\alpha))^2 d\alpha \\
&= \sum_{1}^{N} |f_n|^2 \frac{1}{\sqrt{2\pi m(S_n)}}\int_{-\infty}^{\infty} u^2 \exp\left(-\frac{u^2}{2m(S_n)}\right) du \\
&= \sum_{1}^{N} |f_n|^2 m(S_n)\frac{1}{\sqrt{2\pi}}\int_{-\infty}^{\infty} u^2 e^{-(u^2/2)}du \\
&= \sum_{1}^{N} |f_n|^2 m(S_n) = \int |f(P)|^2 dV_P,
\end{aligned} \tag{72}$$

the integral being taken over the whole of space. In other words, the transformation from $f(P)$ as a function of P, to $\int f(P)\,d_P\mathfrak{P}(S;\alpha)$ as a function of α, retains distance in Hilbert space.[8] Such a transformation, by virtue of the Riesz-Fischer theorem, may always be extended by making limits in the mean correspond to limits in the mean. Thus both in the one-dimensional and in the many-dimensional case, we may *define*

$$\int f(P)\,d_P\mathfrak{P}(S;\alpha) = \underset{n\to\infty}{\text{l. i. m.}} \int f_n(P)\,d_P\mathfrak{P}(S;\alpha) \tag{73}$$

where $f(P)$ is a function belonging to L^2, and the sequence $f_1(P), f_2(P), \cdots$ is a sequence of step-functions converging in the mean to $f(P)$ over the whole of space. The definition will be unambiguous, except for a set of values of α of zero measure.

If S is any measurable set, we have

$$\begin{aligned}
\int_0^1 d\alpha\{\mathfrak{P}(S;\alpha)\}^n &= \frac{1}{\sqrt{2\pi m(S)}}\int_{-\infty}^{\infty} u^n \exp\left(-\frac{u^2}{2m(S)}\right) du \\
&= (m(S))^{n/2}\frac{1}{\sqrt{2\pi}}\int_{-\infty}^{\infty} u^n e^{-(u^2/2)} du \\
&\begin{cases} = 0 \text{ if } n \text{ is odd} \\ = (m(S))^{n/2}(n-1)(n-3)\cdots 1 \text{ if } n \text{ is even.}\end{cases}
\end{aligned} \tag{74}$$

---

[8] Cf. Paley, Wiener, and Zygmund, *Mathematische Zeitschrift*, vol. 37 (1933), pp. 647-668.

This represents $(m(S))^{n/2}$, multiplied by the number of distinct ways of representing n objects as a set of pairs. Remembering that if $S_1, S_2, \cdots, S_{2n}$ are non-overlapping, their distributions are independent, we see that if the sets $\Sigma_1, \Sigma_2, \cdots, \Sigma_{2n}$ are either totally non-overlapping, or else such that when two overlap, they coincide, we have [9]

$$(75) \qquad \int_0^1 \mathfrak{P}(\Sigma_1;\alpha)\cdots\mathfrak{P}(\Sigma_n;\alpha)\,d\alpha = \Sigma\Pi\int_0^1 \mathfrak{P}(\Sigma_j;\alpha)\,\mathfrak{P}(\Sigma_k;\alpha)\,d\alpha,$$

where the product sign indicates that the $2n$ terms are divided into n sets of pairs, j and k, and that these factors are multiplied together, while the addition is over all the partitions of $1,\cdots,2n$ into pairs. If $2n$ is replaced by $2n+1$, the integral in (75) of course vanishes.

Since $\mathfrak{P}(S;\alpha)$ is a linear functional of sets of points, and since both sides of (75) are linear with respect to each $\mathfrak{P}(\Sigma_k;\alpha)$ separately, (75) still holds when $\Sigma_1, \Sigma_2, \cdots, \Sigma_{2n}$ can be reduced to sums of sets which either coincide or do not overlap, and hence holds for all measurable sets.

Now let $f(P_1,\cdots,P_n)$ be a measurable step-function: that is, a function taking only a finite set of finite values, each over a set of values $P_1,\cdots,P_n$ which is a product-set of measurable sets in each variable P_k. Clearly we may define

$$(76) \qquad \int\cdots\int f(P_1,\cdots,P_n)\,d_{P_1}\mathfrak{P}(S;\alpha)\cdots d_{P_n}\mathfrak{P}(S;\alpha)$$

in a way quite analogous to that in which we have defined (70), and we shall have

$$(77) \qquad \int_0^1 d\alpha\int\cdots\int f(P_1,\cdots,P_n)\,d_{P_1}\mathfrak{P}(S;\alpha)\cdots d_{P_n}\mathfrak{P}(S;\alpha)$$
$$= \Sigma\int\cdots\int f(P_1,P_1,P_2,P_2,\cdots,P_n,P_n)\,dV_{P_1}\cdots dV_{P_n}$$

where the summation is carried out for all possible divisions of the $2n$ P's into pairs. Similarly in the odd case

$$(78) \qquad \int_0^1 d\alpha\int\cdots\int f(p_1,\cdots,P_{2n+1})\,d_{P_1}\mathfrak{P}(S;\alpha)\cdots d_{P_n}\mathfrak{P}(S;\alpha) = 0.$$

We may apply (77) to give a meaning to

$$(79) \qquad \int_0^1 d\alpha\,\left|\int\cdots\int f(P_1,\cdots,P_n)\,d_{P_1}\mathfrak{P}(S;\alpha)\cdots d_{P_n}\mathfrak{P}(S;\alpha)\right|^2.$$

If $f(P_1,\cdots,P_n)$ is a measurable step-function, and

[9] Cf. Paley, Wiener, and Zygmund, *loc. cit.*, formula (2.05).

$$(80) \qquad |f(P_1,\cdots,P_n)| \leq |f_1(P_1)\cdots f_n(P_n)|\,; \qquad \int |f_k(P)|^2 dV_P \leq A \qquad (k=1,2,\cdots)$$

we shall have

$$(81) \qquad \int_0^1 d\alpha \left| \int \cdots \int f(P_1,\cdots P_n)\, d_{P_1}\mathfrak{P}(S;\alpha)\cdots d_{P_n}\mathfrak{P}(S;\alpha) \right|^2 \leq A^n(2n-1)(2n-3)\cdots 1.$$

If now $f(P_1,\cdots,P_n)$ is an integrable function satisfying (80), but not necessarily a step-function, let

$$(82) \qquad f(\nu;P_1,\cdots,P_n) = \frac{1}{\nu}\,\mathrm{sgn}\, f(P_1,\cdots,P_n)\,[\nu f(P_1,\cdots,P_n)\,\mathrm{sgn}\, f(P_1,\cdots,P_n)].$$

Clearly almost everywhere

$$(83) \qquad \int \cdots \int f(\nu;P_1,\cdots,P_n)\, d_{P_1}\mathfrak{P}(S;\alpha)\cdots d_{P_n}\mathfrak{P}(S;\alpha) \leq \prod_1^n \int f_k(P)\, d_P\mathfrak{P}(S;\alpha)$$

and

$$(84) \qquad \overline{\lim_{\mu,\nu\to\infty}} \Big| \int \cdots \int f(\mu;P_1,\cdots,P_n)\, d_{P_1}\mathfrak{P}(S;\alpha)\cdots d_{P_n}\mathfrak{P}(S;\alpha) - \int \cdots \int f(\nu;P_1,\cdots,P_n)\, d_{P_1}\mathfrak{P}(S;\alpha)\, d_{P_2}\mathfrak{P}(S;\alpha) \Big|$$
$$\leq \prod_1^n \int f_k(P)\, d_P\mathfrak{P}(S;\alpha) \left[\epsilon + \sum_1^n \left\{ \frac{\int_R f_k(P)\, d_P\mathfrak{P}(S;\alpha)}{\int f_k(P)\, d_P\mathfrak{P}(S;\alpha)} \right\} \right]$$

where R represents the exterior of a sphere of arbitrarily large volume. Let it be noted that both the numerator and the denominator of this fraction have Gaussian distributions, but that the mean square value of the numerator is arbitrarily small. Thus except for a set of values of α of arbitrarily small measure, the right side of expression (84) is arbitrarily small, so that we may write

$$(85) \qquad \lim_{\mu,\nu\to\infty} \Big\{ \int \cdots \int f(\mu;P_1,\cdots,P_n)\, d_{P_1}\mathfrak{P}(S;\alpha)\cdots d_{P_n}\mathfrak{P}(S;\alpha) - \int \cdots \int f(\nu;P_1,\cdots,P_n)\, d_{P_1}\mathfrak{P}(S;\alpha)\cdots d_{P_n}\mathfrak{P}(S;\alpha) \Big\} = 0.$$

Thus by dominated convergence,

$$(86) \qquad \lim_{\mu\to\infty} \int \cdots \int f(\mu;P_1,\cdots,P_n)\, d_{P_1}\mathfrak{P}(S;\alpha)\cdots d_{P_n}\mathfrak{P}(S;\alpha)$$

exists for almost all values of α, and we may write it by definition

$$(87) \qquad \int \cdots \int f(P_1, \cdots, P_n)\, d_{P_1}\mathfrak{P}(S;\alpha) \cdots d_{P_n}\mathfrak{P}(S;\alpha).$$

This will clearly be unique, except for a set of values of α of zero measure. There will then be no difficulty in checking (77), (78), and (81).

9. Forms of chaos derivable from a pure chaos. Let us assume that $f(P)$ belongs to L^2, or that $f(P_1, \cdots, P_n)$ is a measurable function satisfying (80). Let us write $\widehat{PQ}$ for the vector in n-space connecting the points P and Q. Then the function

$$(88) \qquad \int \cdots \int f(\widehat{PP_1}, \cdots, \widehat{PP_n})\, d_{P_1}\mathfrak{P}(S;\alpha) \cdots d_{P_n}\mathfrak{P}(S;\alpha) = F(P;\alpha)$$

is a metrically transitive differentiable chaos. This results from the fact that $\mathfrak{P}(S;\alpha)$ is a metrically transitive chaos, and that a translation of P generates a similar translation of all the points P_k. The sum of a finite number of functions of the type (88) is also a metrically transitive differentiable chaos. To show that $F(P;\alpha)$ is measurable in P and α simultaneously, we merely repeat the argument of (83)–(86) with both P and α as variables.

We shall call a chaos such as (88) a *polynomial chaos homogeneously of the n-th degree,* and a sum of such chaoses a *polynomial chaos* of the degree of its highest term. In this connection, we shall treat a constant as a chaos homogeneously of degree zero.

By the multidimensional ergodic theorem, if Φ is a functional such that

$$(89) \qquad \int_0^1 |\Phi(F(P;\alpha))| \log^+ |\Phi(F(P;\alpha))|\, d\alpha < \infty$$

we shall have

$$(90) \qquad \lim_{r\to\infty} \frac{1}{V(r)} \int_R \Phi(F(P;\alpha))\, dV_P = \int_0^1 \Phi(F(P;\alpha))\, d\alpha$$

for almost all values of α. Since the distribution of $F(P;\alpha)$ is dominated by the product of a finite number of independent Gaussian distributions, we even have

$$(91) \qquad \int_0^1 |F(P;\alpha)|^n d\alpha < \infty$$

for all positive integral values of n. In a wide class of cases this enables us to establish a relation of the type of (89).

In formula (90), we have an algorithm for the computation of the right-hand side. For example, if

$$F(P;\alpha) = \int f(\widehat{PP_1})\, d_{P_1}\mathfrak{P}(S;\alpha), \tag{92}$$

and $P + Q$ is the vector sum of P and Q, we have for almost all α,

$$\lim_{r\to\infty} \frac{1}{V(r)} \int_R F(P+Q;\alpha)\bar{F}(Q;\alpha)\, dV_Q = \int f(P+Q)\bar{f}(Q)\, dV_Q, \tag{93}$$

the integral being taken over the whole of space; if

$$F(P;\alpha) = \int \cdots \int f(\widehat{PP_1}, \widehat{PP_2})\, d_{P_1}\mathfrak{P}(S;\alpha)\, d_{P_2}\mathfrak{P}(S;\alpha), \tag{94}$$

we have almost always

$$\begin{aligned} \lim_{r\to\infty} \frac{1}{V(r)} \int_R & F(P+Q;\alpha)\bar{F}(Q;\alpha)\, dV_Q \\ &= \left| \int f(Q,Q)\, dV_Q \right|^2 + \int\int f(P+Q, P+M)\bar{f}(Q,M)\, dV_Q dV_M \\ &\quad + \int\int f(P+Q,P+M)\bar{f}(M,Q)\, dV_Q dV_M; \end{aligned} \tag{95}$$

and if

$$F(P;\alpha) = \int\int\int f(\widehat{PP_1}, \widehat{PP_2}, \widehat{PP_3})\, d_{P_1}\mathfrak{P}(S;\alpha)\, d_{P_2}\mathfrak{P}(S;\alpha)\, d_{P_3}\mathfrak{P}(S;\alpha), \tag{96}$$

we have almost everywhere

$$\begin{aligned} \lim_{r\to\infty} \frac{1}{V(r)} \int_R & F(P+Q;\alpha)\bar{F}(Q;\alpha)\, dV_Q \\ = \int\int\int & \{ f(Q,Q,P+M)\bar{f}(M,S,S) \\ & + f(Q,Q,P+M)\bar{f}(S,M,S) + f(Q,Q,P+M)\bar{f}(S,S,M) \\ & + f(Q,P+M,Q)\bar{f}(M,S,S) + f(Q,P+M,Q)\bar{f}(S,M,S) \\ & + f(Q,P+M,Q)\bar{f}(S,S,M) + f(P+M,Q,Q)\bar{f}(M,S,S) \\ & + f(P+M,Q,Q)\bar{f}(S,M,S) + f(P+M,Q,Q)\bar{f}(S,S,M) \\ & + f(P+Q,P+M,P+S)\bar{f}(Q,M,S) \\ & + f(P+Q,P+M,P+S)\bar{f}(Q,S,M) \\ & + f(P+Q,P+M,P+S)\bar{f}(M,Q,S) \\ & + f(P+Q,P+M,P+S)\bar{f}(M,S,Q) \\ & + f(P+Q,P+M,P+S)\bar{f}(S,Q,M) \\ & + f(P+Q,P+M,P+S)\bar{f}(S,M,Q) \}\, dV_Q dV_M dV_S. \end{aligned} \tag{97}$$

We have similar results in the non-homogeneous case. Thus if

$$\begin{aligned} F(P;\alpha) = A &+ \int f(\widehat{PP_1})\, d_{P_1}\mathfrak{P}(S;\alpha) \\ &+ \int\int g(\widehat{PP_1}, \widehat{PP_2})\, d_{P_1}\mathfrak{P}(S;\alpha)\, d_{P_2}\mathfrak{P}(S;\alpha), \end{aligned} \tag{98}$$

we have almost everywhere

$$
(99) \qquad \lim_{r\to\infty} \frac{1}{V(r)} \int_R F(P+Q;\alpha)\bar{F}(Q;\alpha)dV_Q
$$
$$
= A\int \bar{g}(Q,Q)dV_Q + \bar{A}\int g(Q,Q)dV_Q
$$
$$
+ \int f(P+Q)\bar{f}(Q)dV_Q + \left|\int g(Q,Q)dV_Q\right|^2
$$
$$
+ \int\int g(P+Q,P+M)\bar{g}(Q,M)dV_Q dV_M
$$
$$
+ \int\int g(P+Q,P+M)\bar{g}(M,Q)dV_Q dV_M.
$$

10. Chaos theory and spectra.[10] The function

$$
(100) \qquad \lim_{r\to\infty} \frac{1}{V(r)} \int_R F(P+Q)\bar{F}(Q)dV_Q = G(P)
$$

occupies a central position in the theory of harmonic analysis. If it exists and is continuous for every value of P, the function $F(P)$ is said to have an n-dimensional spectrum. To define this spectrum, we put

$$
(101) \qquad F_r(P) = \begin{cases} F(P) \text{ on } R; \\ 0 \text{ elsewhere.} \end{cases}
$$

It is then easy to show by an argument involving considerations like those of (49) that if

$$
(102) \qquad G_r(P) = \frac{1}{V(r)} \int_\infty F_r(P+Q)\bar{F}_r(Q)dV_Q,
$$

the integral being taken over the whole of space, then we have

$$
(103) \qquad G(P) = \lim_{r\to\infty} G_r(P).
$$

Since, if O is the point with zero coördinates, by the Schwarz inequality,

$$
(104) \qquad |G(P)| \leq G(O),
$$

the limit in (103) is approached boundedly.

If now we put

$$
(105) \qquad \phi_r(U) = (2\pi)^{-(n/2)} V(r)^{-\frac{1}{2}} \operatorname*{l.i.m.}_{s\to\infty} \int_S F_r(P)e^{iU\cdot P}dV_P
$$

where S is the interior of a sphere of radius s about the origin, the n-fold Parseval theorem will give us

---

[10] Cf. N. Wiener, "Generalized harmonic analysis," *Acta Mathematica*, vol. 55 (1930).

$$(106) \qquad |\phi_r(U)|^2 = (2\pi)^{-n} \int_\infty G_r(P) e^{iU \cdot P} dV_P.$$

If $M(U)$ is a function with an absolutely integrable Fourier transform, we shall have

$$(107) \quad \int_{-\infty}^{\infty} |\phi_r(U)|^2 M(U) dV_U = (2\pi)^{-n} \int_\infty G_r(P) dV_P \int_\infty M(U) e^{iU \cdot P} dV_U,$$

and hence

$$(108) \quad \lim_{r \to \infty} \int_{-\infty}^{\infty} |\phi_r(U)|^2 M(U) dV_U = (2\pi)^{-n} \int G(P) dV_P \int_\infty M(U) e^{iU \cdot P} dV_U$$

which will always exist. Let us put

$$(109) \qquad \mathcal{H}\{M(U)\} = (2\pi)^{-n} \int_\infty G(P) dV_P \int_\infty M(U) e^{iU \cdot P} dV_U.$$

If S is any set of points of finite measure, and $S(P)$ is its characteristic function, let us put

$$(110) \qquad \overline{\mathcal{H}}(S) = \underset{M(U) \geq S(U)}{\text{l. u. b.}} \mathcal{H}(M(U)),$$

and

$$(111) \qquad \underline{\mathcal{H}}(S) = \underset{M(U) \leq S(U)}{\text{g. l. b.}} \mathcal{H}(M(U)).$$

If $\underline{\mathcal{H}}(S)$ and $\overline{\mathcal{H}}(S)$ have the same value, we shall write it $\mathcal{H}(S)$, and shall call it the *spectral mass* of F on S. It will be a non-negative additive set-function of S, and may be regarded as determining the spectrum of F.

If $f(P_1, \cdots, P_n)$ satisfies (80) and $F(P; \alpha)$ is defined as in (88), we know that for any given P,

$$(112) \quad G(P; \alpha) = \lim_{r \to \infty} \frac{1}{V(r)} \int_R F(P+Q; \alpha) \bar{F}(Q; \alpha) = \int_0^1 F(P, \beta) \bar{F}(0, \beta) d\beta$$

for almost all values of α. This alone is not enough to assure that $F(\mathrm{P}, \alpha)$ has a spectrum for almost all values of α, as the sum of a non-denumerable set of sets of zero measure is not necessarily of zero measure. On the other hand, except for a set of values of α of zero measure, $G(P, \alpha)$ exists for all points P with rational coördinates.

We may even extend this result, and assert that if

$$(113) \qquad F_\theta(P; \alpha) = \frac{1}{V(\theta)} \int_{\text{length of } \widehat{PS} \leq \theta} F(S; \alpha) dV_S$$

and

$$(114) \qquad G_\theta(P; \alpha) = \lim_{r \to \infty} \frac{1}{V(r)} \int_R F_\theta(P+Q; \alpha) \bar{F}_\theta(Q; \alpha) dV_Q,$$

then except for a set of values of α of zero measure, $G_\theta(P;\alpha)$ exists for all points P with rational coördinates and all rational parameters θ, and it is easily proved that for almost all values of α, as θ tends to 0 through rational values,

$$\lim_{\theta\to 0}\lim_{r\to\infty}\frac{1}{V(r)}\int_R |F_\theta(Q;\alpha)-F(Q;\alpha)|^2\,dV_Q=0. \tag{115}$$

Now, by the Schwarz inequality,

$$\begin{aligned}(116)\quad &\left|\frac{1}{V(r)}\int_R F_\theta(P+Q;\alpha)\bar{F}_\theta(Q;\alpha)dV_Q\right.\\ &\left.-\frac{1}{V(r)}\int_R F_\theta(P_1+Q;\alpha)\bar{F}_\theta(Q;\alpha)dV_Q\right|\\ &\leq\left\{G_\theta(O;\alpha)\left(\frac{1}{V(r)}\int_R |F_\theta(P+Q;\alpha)-F_\theta(P_1+Q;\alpha)|^2\,dV_Q\right)\right\}^{\frac{1}{2}}\\ &\leq\left\{G_\theta(O;\alpha)\left(\frac{1}{V(r)}\int_R dV_Q\left(\frac{1}{V(\theta)}\left[\int_{|\widehat{P+Q,S}|\leq\theta}-\int_{|\widehat{P+Q,S}|\leq\theta}\right]|F(S;\alpha|^2\,dV_S\right.\right.\right.\\ &\left.\left.\left.\times\left(\frac{1}{V(\theta)}\left[\int_{|\widehat{P+Q,S}|\leq\theta}-\int_{|\widehat{P_1+Q,S}|\leq\theta}\right]dV_S\right)\right\}^{\frac{1}{2}}\\ &\leq G_\theta(O;\alpha)O(|\widehat{PP_1}|^{\frac{1}{2}}).\end{aligned}$$

It thus follows that if (114) exists for a given θ and all P's with rational coördinates, it exists for that θ and all real P's whatever. We may readily show that

$$G_\theta(O;\alpha)\leq G(O;\alpha). \tag{117}$$

By another use of the Schwarz inequality,

$$\begin{aligned}(118)\quad &\left|\frac{1}{V(r)}\int_R F_\theta(P+Q;\alpha)\bar{F}_\theta(Q;\alpha)dV_Q\right.\\ &\qquad\left.-\frac{1}{V(r)}\int_R F(P+Q;\alpha)\bar{F}(Q;\alpha)dV_Q\right|\\ &\leq\frac{1}{V(r)}\int_R |F_\theta(P+Q;\alpha)-F(P+Q;\alpha)|\;|F_\theta(Q;\alpha)|\,dV_Q\\ &\quad+\frac{1}{V(r)}\int_R |F(P+Q;\alpha)|\;|F_\theta(Q;\alpha)-F(Q;\alpha)|\,dV_Q\\ &\leq\{G(O;\alpha)\frac{1}{V(r)}\int_R |F_\theta(P+Q;\alpha)-F(P+Q;\alpha)|^2\,dV_Q\}^{\frac{1}{2}}\\ &\quad+\{G(O;\alpha)\frac{1}{V(r)}\int_R |F_\theta(Q;\alpha)-F(Q;\alpha)|^2\,dV_Q\}^{\frac{1}{2}}.\end{aligned}$$

Combining (115) and (118), we see that except for a set of values of α of zero measure, we have for all P,

$$G(P;\alpha)=\lim_{\theta\to 0}G_\theta(P;\alpha). \tag{119}$$

We thus have an adequate basis for spectrum theory. This will extend, not merely to functions $F(P, \alpha)$ defined as in (88), but to finite sums of such functions. It will even extend to the case of any differentiable chaos $F(P; \alpha)$, for which $F(P+Q; \alpha)\bar{F}(Q; \alpha)$ is an integrable function of α, and for which (115) holds. For a metrically transitive chaos, this latter will be true if

$$(120) \qquad \lim_{\theta \to 0} \int_0^1 | F_\theta(P; \alpha) - F(P; \alpha) |^2 \, d\alpha = 0.$$

Under this assumption, we have proved that $F(P)$ has a spectrum, and the same spectrum, for all values of α.

This enables us to answer a question which has been put several times, as to whether there is any relation between the spectrum of a chaos and the distribution of its values. There is no unique relation of the sort. The function

$$(121) \qquad \int g(P+Q)\bar{g}(Q)\, dV_Q,$$

where g belongs to L^2, may be so chosen as to represent any Fourier transform of a positive function of L, and if $f(P, Q, M)$ is a bounded step-function, the right-hand side of (97) will clearly be the Fourier transform of a positive function of L. In particular, let $f(P_1, P_2, P_3) = f(P_1)f(P_2)f(P_3)$,

$$(122) \qquad F_1(P; \alpha) = \int f_1(\widehat{PP_1})\, d_{P_1}\mathfrak{P}(S; \alpha)$$

and choose $f_1(Q)$ in such a way that

$$(123) \qquad \int f_1(P+Q)\bar{f}_1(Q)\, dV_Q = \text{right-hand side of (97).}$$

Then

$$(124) \quad \int_0^1 (F_1(P; \alpha))^{2n} d\alpha - \left(\int_\infty | f_1(P) |^2 \, dV_P\right)^n (2n-1)(2n-3) \cdots 1$$

and if $F(P, \alpha)$ is defined as in (96),

$$(125) \quad \int_0^1 (F(P; \alpha))^{2n} d\alpha = \left(\int | f(P) |^2 dV_P\right)^{3n} (6n-1)(6n-3) \cdots 1)$$

so that for all but at most one value of n,

$$(126) \qquad \int_0^1 (F_1(P; \alpha)^{2n} d\alpha \neq \int_0^1 (F(P; \alpha))^{2n} d\alpha$$

and we obtain in F and F_1 two chaoses with identical spectra but different distribution functions. On the other hand, if

$$(127) \qquad \int_\infty | f_1(P) |^2 \, dV_P = \int_\infty | f_2(P) |^2 \, dV_P,$$

the chaoses

$$\int f_1(\widehat{PP_1})\, d_{P_1}\mathfrak{P}(S;\alpha) \tag{128}$$

and

$$\int f_2(\widehat{PP_1})\, d_{P_1}\mathfrak{P}(S;\alpha) \tag{129}$$

will have the same distribution functions, but may have very different spectra.

11. The discrete chaos.[11] Let us now divide the whole of Euclidean n-space dichotomously into sets $S_{m,n}$, such that every two sets S_{m_1,n_1} and S_{m_2,n_2} have the same measure, and that each $S_{m,n}$ is made up of exactly two non-overlapping sets $S_{m+1,k}$. Let us divide all these sets into two categories, "occupied," and "empty." Let us require that the probability that a set be empty depend only on its measure, and that the probability that two non-overlapping sets be empty be the product of the probabilities that each be empty. Let us assume that both empty and occupied sets exist. Let every set contained in an empty set be empty, while if a set be occupied, let at least one-half always be occupied. We thus get an infinite class of schedules of emptiness and occupiedness, and methods analogous to those of paragraph 6 may be used to map the class of these schedules in an almost everywhere one-one way on the line $(0,1)$ of the variable α, in such a way that the set of schedules for which a given finite number of regions are empty or occupied will have a probability equal to the measure of the corresponding set of values of α.

By the independence assumption, the probability that a given set $S_{m,n}$ be empty must be of the form $e^{-Am(S_{m,n})}$. If $S_{m,n}$ is divided into the 2^ν intervals $S_{m+\nu,n_1}, \cdots, S_{m+\nu,n_{2^\nu}}$ at the ν-th stage of sub-division, the probability that just one is occupied and the rest are empty is

$$2^\nu(1 - \exp(-Am(S_{m,n})/2^\nu) \exp\left(-\frac{2^\nu - 1}{2^\nu} Am(S_{m,n})\right). \tag{130}$$

This contingency at the $\nu + 1$-st stage is a sub-case of this contingency at the ν-th stage. If we interpret probability to mean the same thing as the measure of the corresponding set of α's, then by monotone convergence, the probability that at every stage, all but one of the subdivisions of $S_{m,n}$ are empty, while the remaining one is occupied, will be the limit of (130), or

$$Am(S_{m,n})\exp(-Am(S_{m,n})). \tag{131}$$

[11] The ideas of this paragraph are related to discussions the author has had with Professor von Neumann, and the main theorem is equivalent to one enunciated by the latter.

10

Such a series of stages of subdivision will have as its occupied regions exactly those which contain a given point.

The probability that the occupied regions are exactly those which contain two points is the probability that each half of $S_{m,n}$ contain exactly one point, plus the probability that one-half is empty, and that in the occupied half, each quarter will contain exactly one point, plus and so on. This will be

$$
\begin{aligned}
(132) \quad & \left\{ \frac{Am(S_{m,n})}{2} \exp\left(- \frac{Am(S_{m,n})}{2} \right) \right\}^2 \\
& + 2 \exp\left(- \frac{Am(S_{m,n})}{2} \right) \left\{ \frac{Am(S_{m,n})}{2} \exp\left(- \frac{Am(S_{m,n})}{4} \right)^2 \right. \\
& + \cdots = \exp(- Am(S_{m,n}) m(S_{m,n}))^2 (\tfrac{1}{4} + \tfrac{1}{8} + \quad) \\
& = \frac{(Am(S_{m,n}))^2}{2} \exp(- Am(S_{m,n})).
\end{aligned}
$$

If the probability that the occupied regions are exactly those containing $k - 1$ points is

$$
\frac{(Am(S_{m,n}))^{k-1}}{(k-1)!} \exp(- Am(S_{m,n}))
$$

then a similar argument will show that the probability that the occupied regions are exactly those containing k points will be

$$
\begin{aligned}
(133) \quad & \sum_{j=1}^{k-1} \frac{1}{j!} \frac{1}{(k-j)!} (Am(S_{m,n}))^k \exp(- Am(S_{m,n})) \left(1 + \frac{1}{2^{k-1}} + \frac{1}{4^{k-1}} + \cdots \right) \\
& = \frac{1}{k!} (2^k - 2) \left(\frac{1}{1 - \frac{1}{2^{k-1}}} \right) (Am(S_{m,n}))^k \exp(- Am(S_{m,n})) \\
& = \frac{1}{k!} (Am(S_{m,n}))^k \exp(- Am(S_{m,n})).
\end{aligned}
$$

Thus by mathematical induction, the probability that the occupied regions are exactly those containing k points will be

$$
\frac{1}{k!} (Am(S_{m,n}))^k \exp(- Am(S_{m,n}))
$$

and the sum of this for all values of k will be

$$
(134) \qquad \sum_{0}^{\infty} \frac{1}{k!} (Am(S_{m,n}))^k \exp(- Am(S_{m,n})) = 1.
$$

In other words, except for a set of contingencies of probability zero, the occupied regions will be exactly those containing a given finite number of points.

We may proceed at once from the fact that the probability that a set S_1 contains exactly k points is

$$\frac{1}{k!}(Am(S_1))^k e^{-Am(S_1)}$$

while the probability that the non-overlapping set S_2 contains exactly k points is

$$\frac{1}{k!}(Am(S_2))^k e^{-Am(S_2)}$$

to the fact that the probability that the set S_1+S_2 contains exactly k points is

$$(135)\qquad \sum_0^k \frac{1}{j!}\frac{1}{(k-j)!}(Am(S_1))^j(Am(S_2))^{k-j}e^{-Am(S_1+S_2)} = \frac{1}{k!}(Am(S_1+S_2))^k e^{-Am(S_1+S_2)}.$$

From this, by monotone convergence, it follows at once that the probability that any set S which is the sum of a denumerable set of our fundamental regions $S_{m,n}$ should contain exactly k points is

$$\frac{1}{k!}(Am(S))^k e^{-Am(S)}.$$

It is then easy to prove this for all measurable sets S.

We are now in a position to prove that the additive functional $\mathfrak{D}(S;\alpha)$, onsisting in the number of points in the region S on the basis of the schedule corresponding to α, is a homogeneous metrically transitive chaos. The rôle which continuity filled in paragraphs 6 and 7, of allowing us to show that $\mathfrak{F}_{y_1,\ldots,y_n}(S;\alpha)$ was measurable in $y_1,\cdots,y_n$ and α, is now filled by the fact that the probability that any of the points in a region lie within a very small distance of the boundary, is for any Jordan region the probability that a small region be occupied, and is small. The metric transitivity of the chaos results as before from the independence of the distribution in non-overlapping regions.

The discrete or Poisson chaos which we have thus defined is the chaos of an infinite random shot pattern, or the chaos of the gas molecules in a perfect gas in statistical equilibrium according to the old Maxwell statistical mechanics. It also has important applications to the study of polycrystalline aggregates, and to similar physical problems.

Two important formulae are

$$(136)\qquad \int_0^1 \mathfrak{D}(S;\alpha)\,d\alpha = e^{-Am(S)}\sum_1^\infty \frac{k}{k!}(Am(S))^k = Am(S),$$

and

$$(137)\quad \int_0^1 (\mathfrak{D}(S;\alpha))^2 d\alpha = e^{-Am(S)} \sum_1^\infty \frac{k^2}{k!}(Am(S))^k = (Am(S))^2 + Am(S).$$

Let it be noted that if we define

$$(138)\quad \int f(P)\,d_P\mathfrak{D}(S;\alpha)$$

for a measurable step-function $f(P)$ as in (70), by

$$(139)\quad \sum_1^N f_n \mathfrak{D}(S_n;\alpha),$$

(72) is replaced by

$$(140)\quad \int_0^1 d\alpha \left| \int f(P)\,d_P\mathfrak{D}(S;\alpha) - A\int_\infty f(Q)\,dV_Q \right|^2$$
$$= \sum_{m=1}^N \sum_{n=1}^N \int_0^1 d\alpha\, f_m \bar{f}_n \mathfrak{D}(S_m;\alpha)\mathfrak{D}(S_n;\alpha)$$
$$-2\mathfrak{R}\{\bar{A}\sum_{m=1}^\infty \int_0^1 d\alpha\, f_m \mathfrak{D}(S_m;\alpha)\int_\infty f(Q)\,dV_Q + \left|A\int_\infty f(Q)\,dV_Q\right|^2\}$$
$$= \sum_{m=1}^N |f_m|^2 Am(S_m)$$
$$= A\int_\infty |f(P)|^2\, dV_P.$$

Thus the transformation from $f(P)$ as a function of P, to

$$(141)\quad \int f(P)\,d_P\mathfrak{D}(S;\alpha) - A\int_\infty f(Q)\,dV_Q$$

as a function of α, retains distance in Hilbert space, apart from a constant factor, and if $f(P)$ belongs to L and L^2 simultaneously, and $\{f_n(P)\}$ is a sequence of step-functions converging in the mean both in the L sense and in the L^2 sense to $f(P)$, we may *define*

$$(142)\quad \int f(P)\,d_P\mathfrak{D}(S;\alpha) = A\int_\infty f(Q)\,dV_Q$$
$$+ \underset{n\to\infty}{\text{l.i.m.}}\left(\int f_n(P)\,d_P\mathfrak{D}(S;\alpha) - A\int_\infty f_n(Q)\,dV_Q\right).$$

As in the case of (73), this definition is substantially unique. We may prove the analogue of (93) in exactly the same way as (93) itself, and shall obtain

$$(143)\quad \lim_{r\to\infty}\frac{1}{V(r)}\int_R \left\{ \int_\infty (f(P+\widehat{Q)M};\alpha)\,d_M\mathfrak{D}(S;\alpha) - A\int_\infty f(M)\,dV_M \right\}$$
$$\times \left\{ \int_\infty \bar{f}(\widehat{QM};\alpha)\,d_M\mathfrak{D}(S;\alpha) - A\int_\infty \bar{f}(M)\,dV_M \right\} dV_Q$$
$$= A^2 \int_\infty f(P+Q)\bar{f}(Q)\,dV_Q.$$

As we may see by appealing to the theory of spectra, one interpretation of this in the one-dimensional case is the following: *If a linear resonator be set into motion by a haphazard series of impulses forming a Poisson chaos, the effect, apart from that of a constant uniform stream of impulses, will have the same power spectrum as the energy spectrum of the response of the resonator to a single impulse.*

12. The weak approximation theorem for the polynomial chaos. We wish to show that the chaoses of paragraph 9 are in some sense everywhere dense in the class of all metrically transitive homogeneous chaoses. We shall show that if $\mathfrak{F}(S;\alpha)$ is any homogeneous chaos in n dimensions, there is a sequence $\mathfrak{F}_k(S;\alpha)$ of polynomial chaoses as defined in paragraph 9, such that if $S_1, \cdots, S_\nu$ is any finite assemblage of bounded measurable sets in n-space selected from among a denumerable set, and

$$(144) \qquad \int_0^1 |\mathfrak{F}(S_\lambda;\alpha)|^\mu \, d\alpha < \infty \qquad (\lambda = 1, 2, \cdots, \nu)$$

is finite, then

$$(145) \qquad \int_0^1 \mathfrak{F}(S_1;\alpha) \cdots \mathfrak{F}(S_\nu;\alpha)\, d\alpha = \lim_{n\to\infty} \int_0^1 \mathfrak{F}_n(S_1;\alpha) \cdots \mathfrak{F}_n(S_\nu;\alpha)\, d\alpha.$$

We first make use of the fact that if the probability that a quantity u be greater in absolute value than A, be less than

$$(146) \qquad \frac{2}{\sqrt{2\pi B}} \int_A^\infty e^{-(u^2/2B)}\, du,$$

then if $\psi(u)$ is any even measurable function bounded over $(-\infty, \infty)$, we may find a polynomial $\psi_\epsilon(u)$, such that the mean value of

$$(147) \qquad |\psi(u) - \psi_\epsilon(u)|^n,$$

which will be

$$(148) \qquad \frac{1}{\sqrt{2\pi B}} \int_{-\infty}^\infty |\psi(u) - \psi_\epsilon(u)|^n \, e^{-(u^2/2B)}\, du,$$

is less than ϵ. Since it is well known that if $\phi(u)$ is a continuous function vanishing outside a finite interval, and

$$(149) \qquad \sum_1^\infty A_n H_n(u)\, e^{-(u^2/2)}$$

is the series for $\phi(u)$ in Hermite functions, then we have uniformly

$$\phi(u) = \lim_{t\to 1-0} \sum_{1}^{\infty} A_n t^n H_n(u) e^{-(u^2/2)}, \tag{150}$$

to establish the existence of $\psi_\epsilon(u)$, we need only prove it in the case in which

$$\psi(u) = u^k e^{-Cu^2} \tag{151}$$

for an arbitrarily small value of C: as for example for $C = 1/4n_1B$. We shall then have

$$\left| \psi(u) - u^k \sum_{0}^{N} \frac{(cu^2)^k}{k!} \right| < |u|^k \sum_{0}^{\infty} \frac{(cu^2)^k}{k!} = |u|^k e^{u^2/4n_1B}, \tag{152}$$

so that by dominated convergence, and if we take N large enough, we may make

$$\frac{1}{\sqrt{2\pi B}} \int_{-\infty}^{\infty} |\psi(u) - \psi_\epsilon(u)|^n e^{-(u^2/2B)} du < \epsilon \qquad (n \leq n_1). \tag{153}$$

Now let

$$\psi_K(u) = \begin{cases} 0 & (|u| < K); \\ 1 & (|u| \geq K); \end{cases} \tag{154}$$

and let us put

$$\mathcal{G}(P;\alpha) = \frac{1}{V(r)} \mathcal{P}(S;\alpha) \quad (S = \text{interior of } |\widehat{QP}| \leq r). \tag{155}$$

The chaos

$$\mathcal{L}(P;\alpha) = \psi_K(\mathcal{G}(P;\alpha)) \tag{156}$$

may then be approximated by polynomial chaoses in such a way as to approximate simultaneously to all polynomials in $\mathcal{L}(P;\alpha)$ by corresponding polynomials in the approximating chaoses. Since the distribution of the values of $\mathcal{G}(P;\alpha)$ will be Gaussian, with a root mean square value proportional to a power of r, and $\mathcal{G}(P;\alpha)$ will be independent in spheres of radius η about two points P_1 and P_2 more remote from each other than $2r + 2\eta$, it follows that if we take K to be large enough, we may make the probability that $\mathcal{L}(P;\alpha)$ differs from 0 between two spheres of radii respectively $r + \eta$ and H about a given point where it differs from 0, as small as we wish.

We now form the new chaos

$$\int_{|\widehat{PQ}| < x} \mathcal{L}(Q;\alpha) dV_Q, \tag{157}$$

which we may also approximate, with all its polynomial functionals, by a sequence of polynomial chaoses. The use of polynomial approximations tending boundedly to a step function over a finite range will show us that this is also true of the chaos determined by

$$\psi_\gamma\Big(\int_{|\widehat{PQ}|<x} \mathcal{L}(Q;\alpha)\,dV_Q = \mathcal{M}(P;\alpha). \tag{158}$$

By a proper choice of the parameters, this can be made to have arbitrarily nearly all its mass uniformly distributed over regions arbitrarily near to arbitrarily small spheres, all arbitrarily remote from one another, except in an arbitrarily small fraction of the cases. We then form

$$\frac{1}{(2\pi k)^{n/2}}\int_\infty (\mathcal{M}(Q;\alpha)+\delta)\exp\left(-\frac{|\widehat{PQ}|^2}{2k}\right)dV_Q = \mathcal{N}(P;\alpha) \tag{159}$$

where δ is taken to be very small. This chaos again, as far as all its polynomial functionals are concerned, will be approximable by polynomial chaoses. Since it is bounded away from 0 and ∞, and since over such a range the function $1/x$ may be approximated uniformly by polynomials, it follows that in our sense,

$$1/\mathcal{N}(P;\alpha) \tag{160}$$

is approximable by polynomial chaoses.

If $\varpi(P)$ is any measurable function for which arbitrarily high moments are always finite, it is easy to show that

$$\frac{1}{(2\pi k)^{n/2}}\int_\infty \varpi(Q)(\mathcal{M}(Q;\alpha)+\delta)\exp\left(-\frac{|\widehat{PQ}|^2}{2k}\right)dV_Q = \mathcal{W}(P;\alpha) \tag{161}$$

is approximable by polynomial chaoses. Multiplying expressions (160) and (161), it follows that

$$\mathcal{W}(P;\alpha)/\mathcal{N}(P;\alpha) = \mathcal{U}(P;\alpha) \tag{162}$$

is approximable by polynomial chaoses.

If A is a large enough constant, depending on the choice of the constant ϵ, we have

$$\begin{aligned} &\frac{1}{(2\pi k)^{n/2}}\int_{|P|>A}\exp\left(-\frac{|P|^2}{2k}\right)dV_P \\ &\qquad = \int_A^\infty x^n e^{-(x^2/2k)}\,dx\Big/\int_0^\infty x^n e^{-(x^2/2k)}\,dx \\ &\qquad < \frac{1}{(2\pi k)^{n/2}}\exp\left(-\frac{(A-\epsilon)^2}{2k}\right). \end{aligned} \tag{163}$$

Thus by the proper choice of the parameters of $\mathcal{M}(P;\alpha)$, if we take k small enough and then δ small enough, the chaos (162) will consist as nearly as we wish, from the distribution standpoint, of an infinite assemblage of convex cells of great minimum dimension, in each of which the function $\varpi(P)$ is repeated, with the origin moved to some point remote from the boundary.

Now let $\mathfrak{F}(S;\alpha)$ be a metrically transitive homogeneous chaos. Let us form

$$(164) \qquad \mathfrak{F}(r;S;\alpha) = \frac{1}{V(r)} \int_R \mathfrak{F}_{x_1,\ldots,x_n}(S;\alpha)\,dx_1 \cdots dx_n.$$

Clearly by the fundamental theorem of the calculus, over any finite region in $(x_1,\cdots,x_n)$, we shall have for almost all points and almost all values of α,

$$(165) \qquad \mathfrak{F}(S;\alpha) = \lim_{r\to 0} \mathfrak{F}(r;S;\alpha);$$

and if (144) holds, it is easy to show that

$$(166) \qquad \int_0^1 |\mathfrak{F}(r;S;\alpha)|^n d\alpha < \text{const.}$$

From this it follows that

$$(167) \qquad \lim_{r\to 0} \int_0^1 |\mathfrak{F}(r;S;\alpha) - \mathfrak{F}(S;\alpha)|^n d\alpha = 0$$

and by the ergodic theorem, except for a set of values of α of zero measure, as r tends to 0 through a denumerable set of values,

$$(168) \qquad \lim_{r\to 0} \frac{1}{V(r)} \int_R |\mathfrak{F}_{x_1,\ldots,x_n}(r;S;\alpha) - \mathfrak{F}_{x_1,\ldots,x_n}(S;\alpha)|^n dx_1 \cdots dx_n = 0.$$

With this result as an aid, enabling us to show that the distribution of $\mathfrak{F}(S;\alpha)$ is only slightly affected by averaging within a small sphere with a given radius, or even within any small region near enough to a small sphere with a given radius, we may proceed as in (161) and (162) and form the chaos

$$(169) \qquad \mathfrak{F}_k(S;\alpha) = \frac{1}{\mathcal{N}(\mathcal{P};\alpha)(2\pi k)^{n/2}} \int_\infty \mathfrak{F}_{x_1,\ldots,x_n}(S;\beta) \times (\mathcal{M}(x_1,\cdots,x_n;\alpha)+\delta) \exp \frac{\left(-\sum_1^n x_j^2\right)}{2k}\, dx_1 \cdots dx_n.$$

For almost all β, in each of the large cells of this chaos, (169) will have as nearly as we wish the same distribution as some $\mathfrak{F}_{x_1, \ldots, x_n}(S;\alpha)$, where $(x_1, \cdots, x_n)$ lies in the interior of the cell, remote from the boundary. These cells may so be determined that except for those filling an arbitrarily small proportion of space, all are convex regions with a minimum dimension greater than some given quantity.

To establish (143), it only remains to show that the average of a quantity depending on a chaos over a large cell tends to the same limit as its average over a large sphere. To show this, we only need to duplicate the argument of paragraph 4, where we prove the multidimensional ergodic theorem, for large pyramids with the origin as a corner, instead of for large spheres about the origin. We may take the shapes and orientations of these pyramids to form a denumerable assemblage, from which we may pick a finite assemblage which will allow us to approach as closely as we want to any cell for which the ratio of the maximum to the minimum distance from the origin within it does not exceed a given amount. It is possible to show that by discarding cells whose measure is an arbitrarily small fraction of the measure of all space, the remaining cells will have this property.

13. The physical problem. The transformation of a chaos. The statistical theory of a homogeneous medium, such as a gas or liquid, or a field of turbulence, deals with the problem, given the statistical configuration and velocity distribution of the medium at a given initial time, and the dynamical laws to which it is subject, to determine the configuration at any future time, with respect to its statistical parameters. This of course is not a problem in the first instance of the history of the individual system, but of the entire ensemble, although in proper cases it is possible to show that almost all systems of the ensemble do actually share the same history, as far as certain specified statistical parameters are concerned.

The dynamical transformations of a homogeneous system have the very important properties, that they are independent of any choice of origin in time or in space. Leaving the time variable out of it, for the moment, the simplest space transformations of a homogeneous chaos $\mathfrak{F}(S;\alpha)$ which have this property are the polynomial transformations which turn it into

$$
\begin{aligned}
&K_0 + \int K_1(x_1 - y_1, x_2 - y_2, \cdots, x_n - y_n)\mathfrak{F}_{y_1, \ldots, y_n}(S;\alpha)\,dy_1 \cdots dy_n \\
(170)\qquad &+ \cdots \\
&+ \int \cdots \int K_\nu(x_1 - y_1^{(1)}, \cdots, x_n - y_n^{(1)}, \cdots, x_1 - y_1^{(\nu)}, \cdots, \\
&\quad x_n - y_n^{(\nu)})\mathfrak{F}_{y_1^{(1)}, \ldots, y_n^{(1)}}(S;\alpha) \cdots \\
&\qquad\qquad \mathfrak{F}_{y_1^{(\nu)}, \ldots, y_n^{(\nu)}}(S;\alpha)\,dy_1^{(1)} \cdots dy_n^{(\nu)}.
\end{aligned}
$$

These are a sub-class of the general class of polynomial transformations

$$
\begin{aligned}
&K_0 + \int K_1(x_1, \cdots, x_n; y_1, \cdots, y_n) \mathfrak{F}_{y_1, \ldots, y_n}(S; \alpha)\, dy_1 \cdots dy_n \\
(171) \qquad &+ \cdots \\
&+ \int \cdots \int K_\nu(x_1, \cdots, x_n; y_1^{(1)}, \cdots, y_n^{(1)}; \cdots; y_1^{(\nu)}, \cdots, y_n^{(\nu)}) \\
&\quad \times \mathfrak{F}_{y_1^{(1)}, \ldots, y_n^{(1)}}(S; \alpha) \cdots \mathfrak{F}_{y_1^{(\nu)}, \ldots, y_n^{(\nu)}}(S; \alpha)\, dy_1^{(1)} \cdots dy_n^{(\nu)}.
\end{aligned}
$$

If a transformation of type (171) is invariant with respect to position in space, it must belong to class (170). On the other hand, in space of a finite number of dimensions and in any of the ordinary spaces of an infinite number of dimensions, polynomials are a closed set of functions, and hence every transformation may be approximated by a transformation of type (171).

A polynomial transformation such as (170) of a polynomial chaos yields a polynomial chaos. If then we can approximate to the state of a dynamical system at time 0 by a polynomial chaos, and approximate to the transformation which yields its status at time t by a polynomial transformation, we shall obtain for its state at time t, the approximation of another polynomial chaos. The theory of approximation developed in the last section will enable us to show this.

On the other hand, the transformation of a dynamical system induced by its own development is infinitely subdivisible in the time, and except in the case of linear transformations, this is not a property of polynomial transformations. Furthermore, when these transformations are non-linear, they are quite commonly not infinitely continuable in time. For example, let us consider the differential equation

$$
(172) \qquad \frac{\partial u}{\partial t} + u \frac{\partial u}{\partial x} = 0.
$$

This corresponds to the history of a space-distribution of velocity transferred by particles moving with that velocity. Its solutions are determined by the equation

$$
(173) \qquad u(x, t) = u(x - tu(x, t), 0),
$$

or if ψ is the inverse function of $u(x, 0)$,

$$
(174) \qquad x - tu(x, t) = \psi(u(x, t)).
$$

Manifestly, if two particles with different velocities are allowed to move long enough to allow their space-time paths to cross, $u(x, t)$ will cease to exist as a single-valued function. This will always be the case for *some* value of t if $u(x, 0)$ is not constant, and for almost all values of t and α if it is a polynomial chaos.

By Lagrange's formula, (174) may be inverted into

$$u(x, t) = \sum_{0}^{\infty} \frac{(-t)^n}{n!} \frac{\partial^n}{\partial x^n} \{(u(x, 0))^n\}. \tag{175}$$

In a somewhat generalized sense, the partial sums of this formally represent polynomial transformations of the initial conditions. However, it is only for a very special sort of bounded initial function, and for a finite value of the time, that they converge. It is only in this restricted sense that the polynomial transformation represents a true approximation to that given by the differential equation.

It will be seen that the useful application of the theory of chaos to the study of particular dynamical chaoses involves a very careful study of the existence theories of the particular problems. In many cases, such as that of turbulence, the demands of chaos theory go considerably beyond the best knowledge of the present day. The difficulty is often both mathematical and physical. The mathematical theory may lead inevitably to a catastrophe beyond which there is no continuation, either because it is not the adequate presentation of the physical facts; or because after the catastrophe the physical system continues to develop in a manner not adequately provided for in a mathematical formulation which is adequate up to the occurrence of the catastrophe; or lastly, because the catastrophe does really occur physically, and the system really has no subsequent history. The hydrodynamical investigations required in the case of turbulence are directly in the spirit of the work of Oseen and Leray, but must be carried much further.

The study of the history of a mechanical chaos will then proceed as follows: we first determine the transformation of the initial conditions generated by the dynamics of the ensemble. We then determine under what assumptions the initial conditions admit of this transformation for either a finite or an infinite interval of time. Then we approximate to the transformation for a given range of values of the time by a polynomial transformation. Then, having regard to a definition of distance between two functions determined by the transformation, we approximate to the initial chaos by a polynomial chaos. Next we apply the polynomial transformation to the polynomial chaos, and obtain an approximating polynomial chaos at time t. Finally, we apply our algorithm of the pure chaos to determine the averages of the statistical parameters of this chaos, and express these as functions of the time.

The results of such an investigation belong to a little-studied branch of statistical mechanics: the statistical mechanics of systems not in equilibrium. To study the classical, equilibrium theory of statistical mechanics by the methods of chaos theory is not easy. As yet we lack a method of representing all forms of homogeneous chaos, which will tell us by inspection when two differ merely by an equimeasure transformation of the parameter of distribution. In certain cases, in which the equilibrium is stable, the study of the history of a system with an arbitrary initial chaos will yield us for large values of t an approximation to equilibrium, but this will often fail to be so, particularly in the case of differentiable chaoses, or the only equilibrium may be that in which the chaos reduces to a constant.

MASSACHUSETTS INSTITUTE OF TECHNOLOGY.

THE ERGODIC THEOREM

BY NORBERT WIENER

1. Ergodic theory has its roots in statistical mechanics. Both in the older Maxwell theory and in the later theory of Gibbs, it is necessary to make some sort of logical transition between the average behavior of all dynamical systems of a given family or ensemble, and the historical average behavior of a single system. This transition was not carried out with any rigor until the theory of Lebesgue measure had been developed. The fundamental theorems are those due to Koopman, von Neumann, and Carleman, on the one hand, and to Birkhoff, on the other. A careful account of them and their proofs is to be found in Eberhard Hopf's *Ergodentheorie* (Berlin, 1937) in the series *Ergebnisse der Mathematik und ihrer Grenzgebiete.* The bibliography of that monograph is so complete that it relieves me from all need of furnishing one on my own account. It is important to point out, however, that Birkhoff's first paper (Proc. Nat. Acad. Sci. (1931)) contains Theorem IV of this paper.

The fundamental theorems of ergodic theory, which emerged in the epoch 1931–32, are recognized as theorems pertaining in the first instance to the abstract theory of the Lebesgue integral. Much of this paper will be devoted to new proofs of these known results and to their proper orientation mutually and with respect to the rest of analysis. They have several variant forms, but perhaps the simplest form of the von Neumann theorem reads:

THEOREM I. *Let S be a measurable set of points of finite measure. Let T be a transformation of S into itself, which transforms every measurable subset of S into a set of equal measure, and whose inverse has the same property. Let $f(P)$ be a function defined over S and of Lebesgue class L^2. Then there exists a function $f_1(P)$, also belonging to L^2 and such that*

$$\lim_{N\to\infty} \int_S \left| f_1(P) - \frac{1}{N+1} \sum_{n=0}^{N} f(T^n P) \right|^2 dV_P = 0. \tag{1.01}$$

Birkhoff's theorem reads:

THEOREM II. *Let S and T be as in Theorem* I. *Let $f(P)$ be a function defined over S and of Lebesgue class L. Then, except for a set of points P of zero measure,*

$$f_1(P) = \lim_{N\to\infty} \frac{1}{N+1} \sum_{n=0}^{N} f(T^n P) \tag{1.02}$$

will exist and belong to L.

These theorems have continuous analogues. These are, respectively:

Received December 7, 1938; an invited address before the American Mathematical Society at its meeting held April 7–8 in conjunction with the centennial celebration of Duke University.

Reprinted from *Duke Math. J., 5,* 1939, pp. 1–18. (Courtesy of the *Duke Mathematical Journal.*)

THEOREM I′. *Let S be as in the hypothesis of Theorem* I. *Let T^λ satisfy the conditions given for T in the hypothesis of Theorem* I, *for every real value of λ, and let*

$$T^\lambda(T^\mu P) = T^{\lambda+\mu}P. \tag{1.03}$$

Let $f(P)$ be a function defined over S and of Lebesgue class L^2, and let $f(T^\lambda P)$ be measurable in the product space of λ and P. Then there exists a function $f_1(P)$, also belonging to L^2 and such that

$$\lim_{N\to\infty} \int_S \left| f_1(P) - \frac{1}{N}\int_0^N f(T^\lambda P)\,d\lambda \right|^2 dV_P = 0. \tag{1.04}$$

THEOREM II′. *Let S and T^λ be as in the hypothesis of Theorem* I′. *Let $f(P)$ be a function defined over S and of Lebesgue class L, and let $f(T^\lambda P)$ be measurable in the product space of λ and P. Then, except for a set of points P of zero measure,*

$$f_1(P) = \lim_{N\to\infty} \frac{1}{N}\int_0^N f(T^\lambda P)\,d\lambda \tag{1.05}$$

will exist as a function in L.

Theorem II′ has a very close formal analogy to the fundamental theorem of the calculus. One form of the latter reads:

THEOREM III. *Let $f(x)$ belong to L. Then, for all values of x with the exception of a set of zero measure,*

$$\lim_{\epsilon\to 0} \frac{1}{\epsilon}\int_0^\epsilon f(x+\lambda)\,d\lambda = f(x). \tag{1.06}$$

This analogy becomes even closer if we substitute for Theorem III

THEOREM III′. *On the hypothesis of Theorem* II′, *except for a set of points P of zero measure,*

$$f(P) = \lim_{\epsilon\to 0} \frac{1}{\epsilon}\int_0^\epsilon f(T^\lambda P)\,d\lambda. \tag{1.07}$$

It will be seen that the only difference between the pattern of Theorem II′ and Theorem III′ is that in the first case $N \to \infty$, while in the second case $\epsilon \to 0$. This suggests the existence of a central theorem in which we suppose neither the one nor the other, but deal with $\frac{1}{A}\int_0^A f(T^\lambda P)\,d\lambda$ for all values of A. Birkhoff in fact has proved a theorem of this type.[1] A sharper form of it is the following:

THEOREM IV. *On the hypothesis of Theorem* II, *let $f(P) \geqq 0$ on S and let*

$$f^*(P) = \operatorname*{l.u.b.}_{0<A<\infty} \frac{1}{A+1}\sum_{n=0}^{A} f(T^n P); \tag{1.08}$$

[1] Cf. G. D. Birkhoff, Proc. Nat. Acad. Sci., vol. 17(1931), pp. 650–660; also N. Wiener, *The homogeneous chaos*, Amer. Jour. Math., vol. 60(1938), p. 907.

or on the hypothesis of Theorem II′, *let*

$$f^*(P) = \underset{0<A<\infty}{\text{l.u.b.}} \frac{1}{A} \int_0^A f(T^\lambda P)\, d\lambda. \tag{1.09}$$

In either case, if $\alpha > 0$, *the measure of the set of points* P *for which* $f^*(P) \geqq \alpha$ *does not exceed*

$$\frac{1}{\alpha} \int_S f(P)\, dV_P. \tag{1.10}$$

It also does not exceed

$$\frac{2}{\alpha} \int_{f(P) \geqq \frac{1}{2}\alpha} f(P)\, dV_P. \tag{1.11}$$

The importance of Theorem IV has been much neglected in the subsequent literature and can be properly appreciated only by a direct reference to Birkhoff's own work. As a corollary of Theorem IV, we have

THEOREM V. *If we define* $f^*(P)$ *as in Theorem* IV, *then if* $f(P)$ *belongs to* L^p $(p > 1)$, *so does* $f^*(P)$; *while if*

$$\int_S f(P) \log^+ f(P) dV_P < \infty, \tag{1.12}$$

then $f^*(P)$ *belongs to* L.

This last theorem is intimately connected with an inequality of Hardy and Littlewood.[2]

All of the theorems so far quoted have analogues in which the group T^λ is replaced by an Abelian group with more than one generator. Theorem I becomes

THEOREM I″. *Let* S *be a measurable set of points of finite measure. Let* $T_1^{\lambda_1} T_2^{\lambda_2} \cdots T_n^{\lambda_n}$ *satisfy the conditions given for* T *in the hypothesis of Theorem* I, *for every set of real values of* $(\lambda_1, \cdots, \lambda_n)$, *and let*

$$T_1^{\lambda_1} T_2^{\lambda_2} \cdots T_n^{\lambda_n} (T_1^{\mu_1} \cdots T_n^{\mu_n} P) = T_1^{\lambda_1+\mu_1} \cdots T_n^{\lambda_n+\mu_n} P. \tag{1.13}$$

Let $f(P)$ *be a function defined over* S *and of Lebesgue class* L^2, *and let* $f(T_1^{\lambda_1} \cdots T_n^{\lambda_n} P)$ *be measurable in the product space of* $(\lambda_1, \cdots, \lambda_n)$ *and* P. *Then there exists a function* $f_1(P)$, *also belonging to* L^2 *and such that*[3]

$$\lim_{\Lambda\to\infty} \int_S \left| f_1(P) - \frac{1}{V(\Lambda)} \int_{\lambda_1^2+\cdots+\lambda_n^2 \leqq \Lambda^2} \cdots \int f(T_1^{\lambda_1} \cdots T_n^{\lambda_n} P)\, d\lambda_1 \cdots d\lambda_n \right|^2 dV_P = 0. \tag{1.14}$$

[2] Cf. Hardy, Littlewood, and Pólya, *Inequalities*, Cambridge, 1934.

[3] Here we use $V(r)$ for the volume of a sphere of radius r in n-space.

Theorem II becomes

THEOREM II″. *Let S and $T_1^{\lambda_1} \cdots T_n^{\lambda_n}$ be as in the hypothesis of Theorem* I″, *and let $f(P)$ be a function defined over S and of Lebesgue class L. Let $f(T_1^{\lambda_1} \cdots T_n^{\lambda_n} P)$ be measurable in the product space of $(\lambda_1, \cdots, \lambda_n)$ and P. Then, except for a set of points P of zero measure,*

$$f_1(P) = \lim_{\Lambda \to \infty} \frac{1}{V(\Lambda)} \int \cdots \int_{\lambda_1^2 + \cdots + \lambda_n^2 \leq \Lambda^2} f(T_1^{\lambda_1} \cdots T_n^{\lambda_n} P)\, d\lambda_1 \cdots d\lambda_n \tag{1.15}$$

will exist.

Theorem III′ will become

THEOREM III″. *On the hypothesis of Theorem* III′, *except for a set of points P of zero measure,*

$$f(P) = \lim_{\Lambda \to 0} \frac{1}{V(\Lambda)} \int \cdots \int_{\lambda_1^2 + \cdots + \lambda_n^2 \leq \Lambda^2} f(T_1^{\lambda_1} \cdots T_n^{\lambda_n} P)\, d\lambda_1 \cdots d\lambda_n . \tag{1.16}$$

Theorem IV will become

THEOREM IV′. *On the hypothesis of Theorem* II″, *let*

$$f^*(P) = \operatorname*{l.u.b.}_{0<\Lambda<\infty} \frac{1}{V(\Lambda)} \int \cdots \int_{\lambda_1^2 + \cdots + \lambda_n^2 \leq \Lambda^2} f(T_1^{\lambda_1} \cdots T_n^{\lambda_n} P)\, d\lambda_1 \cdots d\lambda_n . \tag{1.17}$$

If $\alpha > 0$, the measure of the set of points P for which $f^*(P) \geqq \alpha$ does not exceed

$$\frac{B_n}{\alpha} \int_S f(P)\, dV_P , \tag{1.18}$$

where B_n depends only on the number n of dimensions. It also does not exceed

$$\frac{2B_n}{\alpha} \int_{f(P) \geqq \frac{1}{2}\alpha} f(P) dV_P . \tag{1.19}$$

The generalization of Theorem V has verbally the same statement as Theorem V itself, with the exception that the definition of $f^*(P)$ is made as in Theorem IV′ instead of as in Theorem IV. We shall call this theorem *Theorem* V′.

The order of proof will be the following: we shall first establish the theorems of the I type.[4] Then we give a completely autonomous proof of the theorems of the IV type. Theorems of the III and the V type result from theorems of the IV type alone, while theorems of the II type result from the combination of theorems of the I and the IV type.

[4] Theorem I has been established in a very direct way in a recent unpublished paper by F. Riesz, and Theorem I″ by Dunford.

2. In the proof of theorems of the I type, the following lemmas are of great service:

LEMMA A. *Let $\phi_k(x)$ belong to L^2 for $1 \leqq k \leqq n$. Let*

$$\int |\phi_k(x)|^2 \, dx = A \tag{2.01}$$

for all admissible values of k. Then

$$\int \left| \frac{1}{n} \sum_1^n \phi_k(x) \right|^2 dx \leqq A. \tag{2.02}$$

LEMMA A′. *Let $\phi(x, y)$ be measurable in x and y, and let*

$$\int |\phi(x, y)|^2 \, dx = A \qquad \textit{for all } y. \tag{2.03}$$

Let $\psi(y)$ belong to L. Then $\int \phi(x, y)\psi(y) \, dy$ belongs to L^2, and

$$\int \left| \int \phi(x, y)\psi(y) \, dy \right|^2 dx \leqq \left[\int |\psi(y)| \, dy \right]^2 A. \tag{2.04}$$

LEMMA B. *Let $\phi_k(x)$ be a set of functions of L^2 for $1 \leqq k \leqq n$. Then there exists a value of k, say k_1, such that*

$$\int \left| \phi_{k_1}(x) - \frac{1}{n} \sum_{k=1}^n \phi_k(x) \right|^2 dx \leqq \int |\phi_{k_1}(x)|^2 \, dx - \int \left| \frac{1}{n} \sum_{k=1}^n \phi_k(x) \right|^2 dx. \tag{2.05}$$

LEMMA B′. *Let $\phi(x, y)$ be measurable in x and y, and let $\int |\phi(x, y)|^2 \, dx$ be uniformly bounded in y. Let $\psi(y)$ be a non-negative function of class L such that $\int \psi(y) \, dy = 1$. Then there exists a value of y, say y_1, such that*

$$\int \left| \phi(x, y_1) - \int \phi(x, y)\psi(y) \, dy \right|^2 dx \leqq \int |\phi(x, y_1)|^2 \, dx - \int \left| \int \phi(x, y)\psi(y) \, dy \right|^2 dx. \tag{2.06}$$

In all these lemmas, the limits of integration when left blank may be given any suitable values.

To establish Lemma A, we notice that

$$\begin{aligned} \int \left| \frac{1}{n} \sum_1^n \phi_k(x) \right|^2 dx &= \frac{1}{n^2} \sum_{j=1}^n \sum_{k=1}^n \int \phi_j(x) \overline{\phi_k(x)} \, dx \\ &\leqq \frac{1}{n^2} \sum_{j=1}^n \sum_{k=1}^n \left\{ \int |\phi_j(x)|^2 \, dx \int |\phi_k(x)|^2 \, dx \right\}^{\frac{1}{2}} = A. \end{aligned} \tag{2.07}$$

As to Lemma B, we may obviously choose k_1 so that

$$R\int \overline{\phi_{k_1}(x)}\left(\frac{1}{n}\sum_{k=1}^{n}\phi_k(x)\right)dx \geqq \int \left|\frac{1}{n}\sum_{k=1}^{n}\phi_k(x)\right|^2 dx. \tag{2.08}$$

Otherwise we should have

$$R\frac{1}{n}\sum_{j=1}^{n}\int \overline{\phi_j(x)}\left(\frac{1}{n}\sum_{k=1}^{n}\phi_k(x)\right)dx < \int \left|\frac{1}{n}\sum_{k=1}^{n}\phi_k(x)\right|^2 dx, \tag{2.09}$$

and this is a contradiction. Then

$$\begin{aligned}\int \left|\phi_{k_1}(x) - \frac{1}{n}\sum_{k=1}^{n}\phi_k(x)\right|^2 dx & \\ &= \int |\phi_{k_1}(x)|^2 dx - 2R\int \overline{\phi_{k_1}(x)}\left(\frac{1}{n}\sum_{k=1}^{n}\phi_k(x)\right)dx \\ &\quad + \int \left|\frac{1}{n}\sum_{k=1}^{n}\phi_k(x)\right|^2 dx \\ &\leqq \int |\phi_{k_1}(x)|^2 dx - \int \left|\frac{1}{n}\sum_{k=1}^{n}\phi_k(x)\right|^2 dx.\end{aligned} \tag{2.10}$$

Lemmas A′ and B′ are proved in an exactly analogous manner.

We now turn to the proof of Theorem I. Let us put

$$f_{(k)}(P) = \frac{1}{2^k}\sum_{j=0}^{2^k-1} f(T^jP). \tag{2.11}$$

We then have for $k > m$

$$f_{(k)}(P) = \frac{1}{2^{k-m}}\sum_{j=0}^{2^{k-m}-1} f_{(m)}(T^{2^m j}P). \tag{2.12}$$

Thus by Lemma A, the numbers $\int_S |f_{(k)}(P)|^2\, dV_P$ form a decreasing sequence of positive terms, so that

$$F = \lim_{k\to\infty}\int_S |f_{(k)}(P)|^2\, dV_P \tag{2.13}$$

exists and is non-negative. By Lemma B, if $k > m > 0$, there exists an integer j in the range $0,\ 2^{k-m} - 1$ such that

$$\begin{aligned}\int_S |f_{(m)}(T^{2^m j}P) - f_{(k)}(P)|^2\, dV_P &\leqq \int_S |f_{(m)}(T^{2^m j}P)|^2\, dV_P - \int_S |f_{(k)}(P)|^2\, dV_P \\ &\leqq \int_S |f_{(m)}(P)|^2\, dV_P - F,\end{aligned} \tag{2.14}$$

since T preserves measure. Also

$$(2.15)\qquad \int_S |f_{(m)}(T^\nu P) - f_{(k)}(T^{\nu-2m_j} P)|^2 dV_P \leqq \int_S |f_{(m)}(P)|^2 dV_P - F.$$

Now let m_μ be so large that

$$(2.16)\qquad \int_S |f_{(m_\mu)}(P)|^2 dV_P < 2^{-\mu} + F.$$

It will follow that there exists a sequence of numbers ν_μ such that

$$(2.17)\qquad \int_S |f_{(m_\mu)}(T^{\nu_\mu} P) - f_{(m_{\mu+1})}(T^{\nu_{\mu+1}} P)|^2 dV_P < 2^{1-\mu}.$$

Then if $\mu > \lambda$,

$$(2.18)\qquad \int_S |f_{(m_\mu)}(T^{\nu_\mu} P) - f_{(m_\lambda)}(T^{\nu_\lambda} P)|^2 dV_P \leqq \frac{2^{\frac{1}{2}(1-\lambda)}}{2^{\frac{1}{2}} - 1}.$$

Thus, by the Riesz-Fischer theorem, there exists a function $f_1(P)$ such that

$$(2.19)\qquad \lim_{\mu\to\infty} \int_S |f_1(P) - f_{(m_\mu)}(T^{\nu_\mu} P)|^2 dV_P = 0.$$

Again,

$$(2.20)\qquad \lim_{\mu\to\infty} \int_S |f_1(TP) - f_{(m_\mu)}(T^{\nu_\mu+1} P)|^2 dV_P = 0.$$

Now,

$$\int_S |f_{(m_\mu)}(T^{\nu_\mu} P) - f_{(m_\mu)}(T^{\nu_\mu+1} P)|^2 dV_P$$

$$(2.21)\qquad = \int_S |2^{-m_\mu} f(T^{\nu_\mu} P) - 2^{-m_\mu} f(T^{\nu_\mu+2^{m_\mu}} P)|^2 dV_P$$

$$\leqq 2^{2-2m_\mu} \int_S |f(P)|^2 dV_P.$$

Thus

$$(2.22)\qquad \lim_{\mu\to\infty} \int_S |f_{(m_\mu)}(T^{\nu_\mu} P) - f_{(m_\mu)}(T^{\nu_\mu+1} P)|^2 dV_P = 0,$$

and

$$(2.23)\qquad \int_S |f_1(P) - f_1(TP)|^2 dV_P = 0.$$

Hence for all n

$$(2.24)\qquad f_1(P) = f_1(TP) = f_1(T^n P).$$

Now let us put

$$(2.25)\qquad f_2(P) = f(P) - f_1(P).$$

Then there exists a sequence $\{m_\mu\}$ such that

$$\lim_{\mu\to\infty}\int_S |f_{2(m_\mu)}(P)|^2\,dV_P = 0. \tag{2.26}$$

However, as in (2.13),

$$\lim_{k\to\infty}\int_S |f_{2(k)}(P)|^2\,dV_P \tag{2.27}$$

exists. Thus

$$\lim_{k\to\infty}\int_S |f_{(k)}(P) - f_1(P)|^2\,dV_P = 0. \tag{2.28}$$

Now, let the integer N have the development

$$N = 2^\nu + \sum_{\mu=0}^{\nu-1} a_\mu 2^\mu \qquad (a_\mu = 0 \text{ or } 1;\, a_\nu = 1) \tag{2.29}$$

in the binary scale. Then

$$\begin{aligned}&\int_S \left| f_1(P) - \frac{1}{N}\sum_{k=0}^{N-1} f(T^k P)\right|^2 dV_P \\ &\qquad \leq \left\{\sum_{\mu=0}^{\nu} \frac{a_\mu 2^\mu}{N}\left\{\int |f_1(P) - f_{(\mu)}(P)|^2\,dV_P\right\}^{\frac{1}{2}}\right\}^2.\end{aligned} \tag{2.30}$$

This establishes (1.01) and Theorem I.

Theorem I′ may be established in a like manner. We shall proceed to the more general case of Theorem I″. Here we shall prove in the first instance that there exists a function $f_1(P)$ of L^2 such that

$$\begin{aligned}\lim_{\Lambda\to\infty}\int_S \Bigg| f_1(P) - \frac{1}{(2\pi\Lambda)^{\frac{1}{2}n}}\int_{-\infty}^{\infty}\cdots\int_{-\infty}^{\infty} f(T_1^{\lambda_1}\cdots T_n^{\lambda_n}P) \\ \cdot\exp\left(-\sum_1^n \frac{\lambda_k^2}{2\Lambda}\right)d\lambda_1\cdots d\lambda_n\Bigg|^2 dV_P = 0.\end{aligned} \tag{2.31}$$

Let us notice that

$$\begin{aligned}&\frac{1}{[2\pi(\Lambda_1+\Lambda_2)]^{\frac{1}{2}n}}\int_{-\infty}^{\infty}\cdots\int_{-\infty}^{\infty} f(T_1^{\lambda_1}\cdots T_n^{\lambda_n}P)\exp\left(-\sum_1^n \frac{\lambda_k^2}{2(\Lambda_1+\Lambda_2)}\right)d\lambda_1 \\ &\cdots d\lambda_n = \frac{1}{(2\pi\Lambda_1)^{\frac{1}{2}n}}\int_{-\infty}^{\infty}\cdots\int_{-\infty}^{\infty}\exp\left(-\sum_1^n\frac{\lambda_k^2}{2\Lambda_1}\right)d\lambda_1\cdots d\lambda_n\,\frac{1}{(2\pi\Lambda_2)^{\frac{1}{2}n}} \\ &\qquad\cdot\int_{-\infty}^{\infty}\cdots\int_{-\infty}^{\infty} f(T_1^{\lambda_1+\mu_1}\cdots T_n^{\lambda_n+\mu_n}P)\exp\left(-\sum_1^n\frac{\mu_k^2}{2\Lambda_2}\right)d\mu_1\cdots d\mu_n.\end{aligned} \tag{2.32}$$

By Lemma A′,

$$\int_S\left|\frac{1}{(2\pi\Lambda)^{\frac{1}{2}n}}\int_{-\infty}^{\infty}\cdots\int_{-\infty}^{\infty} f(T_1^{\lambda_1}\cdots T_n^{\lambda_n}P)\exp\left(-\sum_1^n\frac{\lambda_k^2}{2\Lambda}\right)d\lambda_1\cdots d\lambda_n\right|^2 dV_P \tag{2.33}$$

is monotone decreasing in Λ and has a non-negative limit F. Thus by Lemma B′, if $\Lambda > \Lambda_1 > 0$, there exists a set $(\mu_1, \cdots, \mu_n)$ such that

$$\begin{aligned}\int_S \Bigg| \frac{1}{(2\pi\Lambda)^{\frac{1}{2}n}} \int_{-\infty}^{\infty} \cdots \int_{-\infty}^{\infty} f(T_1^{\lambda_1} \cdots T_n^{\lambda_n} P) \exp\left(-\sum_1^n \frac{\lambda_k^2}{2\Lambda}\right) d\lambda_1 \cdots d\lambda_n \\ - \frac{1}{(2\pi\Lambda_1)^{\frac{1}{2}n}} \int_{-\infty}^{\infty} \cdots \int_{-\infty}^{\infty} f(T_1^{\lambda_1+\mu_1} \cdots T_n^{\lambda_n+\mu_n} P) \exp\left(-\sum_1^n \frac{\lambda_k^2}{2\Lambda_1}\right) d\lambda_1 \\ \cdots d\lambda_n \Bigg|^2 dV_P \leqq \int_S \Bigg| \frac{1}{(2\pi\Lambda_1)^{\frac{1}{2}n}} \int_{-\infty}^{\infty} \cdots \int_{-\infty}^{\infty} f(T_1^{\lambda_1} \cdots T_n^{\lambda_n} P) \\ \cdot \exp\left(-\sum_1^n \frac{\lambda_k^2}{2\Lambda_1}\right) d\lambda_1 \cdots d\lambda_n \Bigg|^2 dV_P - B.\end{aligned} \tag{2.34}$$

As in the proof of (2.19), we may choose the vectors $(\mu_{\nu_1}, \cdots, \mu_{\nu_n})$ and the numbers Λ_ν in such a way that the sequence

$$\frac{1}{(2\pi\Lambda_\nu)^{\frac{1}{2}n}} \int_{-\infty}^{\infty} \cdots \int_{-\infty}^{\infty} f(T_1^{\lambda_1+\mu_{\nu_1}} \cdots T_n^{\lambda_n+\mu_{\nu_n}} P) \exp\left(-\sum_1^n \frac{\lambda_k^2}{2\Lambda_\nu}\right) d\lambda_1 \cdots d\lambda_n \tag{2.35}$$

converges in the mean in the L^2 sense to a function $f_1(P)$.

The argument corresponding to (2.21)–(2.23) proceeds as follows. We have

$$\begin{aligned}\lim_{\Lambda_\nu \to \infty} \frac{1}{(2\pi\Lambda_\nu)^{\frac{1}{2}n}} \int_{-\infty}^{\infty} \cdots \int_{-\infty}^{\infty} \Bigg| \exp\left(-\sum_1^n \frac{(\lambda_k + \mu_{\nu_k} + \mu_k)^2}{2\Lambda_\nu}\right) \\ - \exp\left(-\sum_1^n \frac{(\lambda_k + \mu_{\nu_k})^2}{2\Lambda_\nu}\right) \Bigg| d\lambda_1 \cdots d\lambda_n = 0.\end{aligned} \tag{2.36}$$

Again,

$$\begin{aligned}\frac{1}{(2\pi\Lambda_\nu)^{\frac{1}{2}n}} \int_{-\infty}^{\infty} \cdots \int_{-\infty}^{\infty} \{f(T_1^{\lambda_1+\mu_{\nu_1}+\mu_1} \cdots T_n^{\lambda_n+\mu_{\nu_n}+\mu_n} P) \\ - f(T_1^{\lambda_1+\mu_{\nu_1}} \cdots T_n^{\lambda_n+\mu_{\nu_n}} P)\} \exp\left(-\sum_1^n \frac{\lambda_k^2}{2\Lambda_\nu}\right) d\lambda_1 \cdots d\lambda_n \\ = \frac{1}{(2\pi\Lambda_\nu)^{\frac{1}{2}n}} \int_{-\infty}^{\infty} \cdots \int_{\infty}^{\infty} f(T_1^{\lambda_1} \cdots T_n^{\lambda_n} P) \left\{\exp\left(-\sum_1^n \frac{(\lambda_k + \mu_{\nu_k} + \mu_k)^2}{2\Lambda_\nu}\right) \\ - \exp\left(-\sum_1^n \frac{(\lambda_k + \mu_{\nu_k})^2}{2\Lambda_\nu}\right)\right\} d\lambda_1 \cdots d\lambda_n.\end{aligned} \tag{2.37}$$

Thus by Lemma A′,

$$\begin{aligned}\lim_{\Lambda_\nu \to \infty} \int_S \Bigg| \frac{1}{(2\pi\Lambda_\nu)^{\frac{1}{2}n}} \int_{-\infty}^{\infty} \cdots \int_{-\infty}^{\infty} f(T_1^{\lambda_1+\mu_{\nu_1}+\mu_1} \cdots T_n^{\lambda_n+\mu_{\nu_n}+\mu_n} P) \\ \cdot \exp\left(-\sum_1^n \frac{\lambda_k^2}{2\Lambda_\nu}\right) d\lambda_1 \cdots d\lambda_n \\ - \frac{1}{(2\pi\Lambda_\nu)^{\frac{1}{2}n}} \int_{-\infty}^{\infty} \cdots \int_{-\infty}^{\infty} f(T_1^{\lambda_1+\mu_{\nu_1}} \cdots T_n^{\lambda_n+\mu_{\nu_n}} P) \\ \cdot \exp\left(-\sum_1^n \frac{\lambda_k^2}{2\Lambda_\nu}\right) d\lambda_1 \cdots d\lambda_n \Bigg|^2 dV_P = 0.\end{aligned} \tag{2.38}$$

The argument of (2.23)–(2.28) is duplicated without substantial change.

To proceed from (2.31) to (1.14), let us notice that if $S_\Lambda(\lambda_1, \cdots, \lambda_n)$ is the characteristic function of the sphere $\lambda_1^2 + \lambda_2^2 + \cdots + \lambda_n^2 \leqq \Lambda^2$, we have

$$\lim_{\Lambda\to\infty} \frac{1}{V(\Lambda)} \int_{-\infty}^{\infty} \cdots \int_{-\infty}^{\infty} \Big| S_\Lambda(\lambda_1, \cdots, \lambda_n) - \frac{1}{(2\pi\Lambda_1)^{\frac{1}{2}n}} \int_{-\infty}^{\infty} \cdots \int_{-\infty}^{\infty} S_\Lambda(\lambda_1 + \mu_1, \cdots, \lambda_n + \mu_n) \cdot \exp\left(-\sum_1^n \frac{\mu_k^2}{2\Lambda_1}\right) d\mu_1 \cdots d\mu_n \Big|^2 d\lambda_1 \cdots d\lambda_n = 0. \tag{2.39}$$

Thus, by Lemma A′,

$$\lim_{\Lambda\to\infty} \int_S dV_P \Big| \int_{-\infty}^{\infty} \cdots \int_{-\infty}^{\infty} \{f(T_1^{\lambda_1} \cdots T_n^{\lambda_n} P) - f_1(T_1^{\lambda_1} \cdots T_n^{\lambda_n} P)\} \frac{1}{V(\Lambda)} \Big\{ S_\Lambda(\lambda_1, \cdots, \lambda_n) - \frac{1}{(2\pi\Lambda_1)^{\frac{1}{2}n}} \int_{-\infty}^{\infty} \cdots \int_{-\infty}^{\infty} S_\Lambda(\lambda_1 + \mu_1, \cdots, \lambda_n + \mu_n) \cdot \exp\left(-\sum_1^n \frac{\mu_k^2}{2\Lambda_1}\right) d\mu_1 \cdots d\mu_n \Big\} d\lambda_1 \cdots d\lambda_n \Big|^2 = 0. \tag{2.40}$$

Hence by (2.31)

$$\overline{\lim_{\Lambda\to\infty}} \int_S \Big| f_1(P) - \frac{1}{V(\Lambda)} \underset{\lambda_1^2+\cdots+\lambda_n^2 \leqq \Lambda^2}{\int \cdots \int} f(T_1^{\lambda_1} \cdots T_n^{\lambda_n} P)\, d\lambda_1 \cdots d\lambda_n \Big|^2 dV_P \leqq \overline{\lim_{\Lambda_1\to\infty}} \int_S \Big| f_1(P) - \frac{1}{(2\pi\Lambda_1)^{\frac{1}{2}n}} \int_{-\infty}^{\infty} \cdots \int_{-\infty}^{\infty} f(T_1^{\lambda_1} \cdots T_n^{\lambda_n} P) \cdot \exp\left(-\sum_1^n \frac{\lambda_k^2}{2\Lambda_1}\right) d\lambda_1 \cdots d\lambda_n \Big|^2 dV_P = 0. \tag{2.41}$$

Theorems I, I′, and I″ deal with functions of the class L^2. From these, similar theorems arise concerning functions of the class L. If in the hypothesis of any of these theorems we assume $f(P)$ only to belong to L, we shall show that (1.01) may be replaced by

$$\lim_{N\to\infty} \int_S \Big| f_1(P) - \frac{1}{N+1} \sum_{n=0}^{N} f(T^n P) \Big| dV_P = 0; \tag{2.42}$$

(1.04) by

$$\lim_{N\to\infty} \int_S \Big| f_1(P) - \frac{1}{N} \int_0^N f(T^\lambda P)\, d\lambda \Big| dV_P = 0; \tag{2.43}$$

and (1.14) by

$$\lim_{\Lambda\to\infty}\int_S \left| f_1(P) - \frac{1}{V(\Lambda)} \int_{\lambda_1^2+\cdots+\lambda_n^2 \leqq \Lambda^2} \cdots \int f(T_1^{\lambda_1} \cdots T_n^{\lambda_n} P)\, d\lambda_1 \cdots d\lambda_n \right| dV_P = 0. \tag{2.44}$$

As all the proofs are exactly alike, we shall establish (2.42). We may approximate in the L sense to $f(P)$ with any desired degree of accuracy by a bounded function $g(P)$, which obviously belongs to L^2. Next we form $g_1(P)$ as we have formed $f_1(P)$ in the L^2 case. By the Schwarz inequality,

$$\lim_{N\to\infty}\int_S \left| g_1(P) - \frac{1}{N+1}\sum_{n=0}^{N} g(T^n P)\right| dV_P = 0. \tag{2.45}$$

Again

$$\int_S \left| \frac{1}{N+1}\sum_0^N f(T^n P) - \frac{1}{N+1}\sum_0^N g(T^n P)\right| dV_P \leqq \int_S |f(P) - g(P)|\, dV_P. \tag{2.46}$$

Thus

$$\overline{\lim_{M,N\to\infty}}\int_S \left| \frac{1}{N+1}\sum_0^N f(T^n P) - \frac{1}{M+1}\sum_0^N f(T^n P)\right| dV_P \leqq 2\int_S |f(P) - g(P)|\, dV_P. \tag{2.47}$$

Since $g(P)$ converges in the mean in the L sense to $f(P)$, the right side of (2.47) tends to 0, so that we may replace it by 0. If we now use the L form of the Riesz-Fischer theorem, (2.42) follows at once.

3. We now turn to the proofs of Theorems IV and IV′. For Theorem IV′ we need a lemma of the Vitali type:[5]

LEMMA C′. *Let S be a set of points of finite outer measure in n-space. Let each point P of S be the center of a sphere $\Sigma(P)$. Then there exists a finite set of spheres $\Sigma(P)$, non-overlapping, and of total volume exceeding $A_n m_e(S)$, where A_n is a positive absolute constant depending only on the number n of dimensions.*

To establish this lemma, let us notice that it is trivially true unless the radii of the spheres $\Sigma(P)$ have a finite upper bound. If they have such a bound, let it be B. Let S_ν be that subset of S for which the spheres $\Sigma(P)$ have radii $\leqq 2^{1-\nu}B$ and $>2^{-\nu}B$. Let $S_1' = S_1$, and let us define by mathematical induction S_ν' as that part of S_ν which is more remote than $5B$ from any point of S_1', more remote than $5B\cdot 2^{-1}$ from any point of S_2', and so on, being more remote than $5B\cdot 2^{-\nu+2}$ from any point of $S_{\nu-1}'$.

Let us lay upon S_1 a mesh of cubes with side $5B$, and if any cube of this mesh contain a point of S_1, let us pick out one of these points, P_m. Then the

[5] Cf. E. W. Hobson, *The Theory of Functions of a Real Variable*, vol. 1, Cambridge, 1927, §136. Cf. also Titchmarsh, *Theory of Functions*, §11.41.

sum of the volumes of the spheres $\Sigma(P_m)$ will be more than a fixed fraction of the sum of the meshes of the cubes corresponding to them, and hence more than a fixed fraction of the volumes of these cubes, together with all adjacent cubes. None of these spheres can overlap a sphere whose center is not in an adjacent cube, and hence each can overlap at most a fixed finite number of spheres. Thus if we enumerate these spheres in any order, and discard in turn every sphere overlapping a sphere which we have not already discarded, we shall ultimately obtain a finite set of non-overlapping spheres whose total volume will exceed a fixed submultiple A'_n of the volume of all the cubes containing points of S_1, together with all adjacent cubes. These cubes contain every point of S_1, together with all the points which are in any S_ν, but fail to be included in S'_ν by virtue of their vicinity to points of S_1.

We now cover S'_2 with a mesh of half the linear magnitude, and proceed as above. We shall find a finite set of spheres with centers at points of S'_2, obviously not overlapping any sphere of the previous set, not overlapping one another, and with a total volume exceeding A'_n multiplied by the total volume of all the cubes containing S'_2, together with all adjacent cubes. These cubes together with the similar set obtained from S_1 contain S_1, S_2, and all the points in later S_ν, but are excluded from the corresponding S'_ν on account of their propinquity to S_1 or S'_2.

By continuing this process, we obtain a denumerable set of spheres $\Sigma(P)$, exceeding in volume $A'_n m_e(S)$ and non-overlapping. We may clearly take a finite number of these with total volume exceeding $A_n m_e(S)$, if $A_n < A'_n$.

In the one-dimensional case, Lemma C′ is replaced by the somewhat more precise

LEMMA C. *Let S be a set of points of finite measure on the line. Let every point P of S be the right-hand terminus of an interval $I(P)$. Then there exists a finite subset of intervals $I(P)$, non-overlapping, and of total length exceeding $(1 - \epsilon)m_e(S)$, where ϵ is any positive number.*

This lemma is due to Sierpinski (see footnote 5) and is completely elementary.

From Lemma C and Lemma C′, respectively, we may deduce

LEMMA D. *Let $f(x)$ belong to L on the infinite line. Then the set S_α of values of x for which*

$$\underset{0<\Lambda<\infty}{\text{l.u.b.}} \frac{1}{\Lambda} \int_{x-\Lambda}^{x} |f(\xi)|\, d\xi > \alpha > 0 \tag{3.01}$$

does not exceed $\frac{1}{\alpha} \int_{-\infty}^{\infty} |f(\xi)|\, d\xi$ *in measure.*

LEMMA D′. *Let $f(x_1, \cdots, x_n)$ belong to L in infinite n-space. Then the set S_α of points $(y_1, \cdots, y_n)$ for which*

$$\underset{0<\Lambda<\infty}{\text{l.u.b.}} \frac{1}{V(\Lambda)} \int \cdots \int_{x_1^2+\cdots+x_n^2 \leq \Lambda^2} |f(x_1 + y_1, \cdots, x_n + y_n)|\, dx_1 \cdots dx_n > \alpha > 0 \tag{3.02}$$

does not exceed

$$\frac{1}{\alpha A_n}\int_{-\infty}^{\infty}\cdots\int_{-\infty}^{\infty}|f(x_1,\cdots,x_n)|\,dx_1\cdots dx_n \tag{3.03}$$

in measure.

The proof is the same in both cases. In Lemma D, we can pick out a finite set of non-overlapping intervals, of total length exceeding $(1-\epsilon)m(S_\alpha)$ and such that the integral of $f(x)$ over each interval exceeds α multiplied by its length. Thus

$$\alpha(1-\epsilon)m(S_\alpha) \leqq \int_\Sigma |f(\xi)|\,d\xi, \tag{3.04}$$

where Σ is a finite set of non-overlapping intervals.

The discrete analogue of Lemma D is

LEMMA E. *Let* $\sum_{-\infty}^{\infty}|A_\nu|<\infty$. *Then the number of values of* ν *for which*

$$\underset{0<\mu<\infty}{\text{l.u.b.}}\ \frac{1}{\mu}\sum_{k=\nu-\mu+1}^{k=\nu}|A_k|>\alpha>0 \tag{3.05}$$

does not exceed $\frac{1}{\alpha}\sum_{-\infty}^{\infty}|A_k|$.

Here the proof goes exactly as in Lemma D. The discrete equivalent of Lemma C is trivial.

We proceed to the proof of Theorem IV. It is clear that, for a particular P, the number of values of ν on $(0, N)$ for which

$$\underset{0<\mu\leqq\nu}{\text{l.u.b.}}\ \frac{1}{\mu}\sum_{k=\nu-\mu+1}^{k=\nu}f(T^{-k}P)>\alpha>0 \tag{3.06}$$

does not exceed

$$\frac{1}{\alpha}\sum_0^N f(T^{-k}P). \tag{3.07}$$

If we now integrate over S with respect to P, we see that the sum of the measures of the sets of values of P for which

$$\underset{0<\mu\leqq\nu}{\text{l.u.b.}}\ \frac{1}{\mu}\sum_{k=\nu-\mu+1}^{k=\nu}f(T^{-k}P)>\alpha>0 \qquad (\nu=1,2,\cdots,N) \tag{3.08}$$

will not exceed

$$\frac{N+1}{\alpha}\int_S f(P)\,dV_P. \tag{3.09}$$

Since $T^{-\nu}$ is an equimeasure transformation, we get

$$\frac{1}{N}\sum_{\nu=1}^{N}\Big(\text{measure of set of values of } P \text{ for which } \underset{0<\mu\leqq\nu}{\text{l.u.b.}}\ \frac{1}{\mu}\sum_{k=0}^{k=\mu-1}f(T^kP)>\alpha>0\Big)\leqq\frac{1}{\alpha}\int_S f(P)\,dV_P. \tag{3.10}$$

We now use the simple lemma that if B_ν is an increasing sequence for which

$$\frac{1}{N+1}\sum_{\nu=1}^{N} B_\nu \leqq B \qquad \text{for all } N, \tag{3.11}$$

then

$$B_\nu \leqq B \qquad \text{for all } \nu. \tag{3.12}$$

Thus for all values of ν,

$$\begin{aligned}&\left(\text{measure of set of values of } P \text{ for which } \underset{0<\mu\leqq\nu}{\text{l.u.b.}}\ \frac{1}{\mu}\sum_{k=0}^{\mu-1} f(T^k P) > \alpha > 0\right) \\ &\qquad\qquad \leqq \frac{1}{\alpha}\int_S f(P)\,dV_P,\end{aligned} \tag{3.13}$$

which gives us

$$\begin{aligned}&\Big(\text{measure of set of all values of } P \text{ for which} \\ &\qquad \underset{0<\mu<\infty}{\text{l.u.b.}}\ \frac{1}{\mu}\sum_{k=0}^{\mu-1} f(T^k P) > \alpha > 0\Big) \leqq \frac{1}{\alpha}\int_S f(P)\,dV_P.\end{aligned} \tag{3.14}$$

Let us now turn to Theorem IV′. In Lemma D′, let

$$f(x_1, \cdots, x_n) = \begin{cases} f(T_1^{x_1} \cdots T_n^{x_n} P) & \text{if } x_1^2 + \cdots + x_n^2 \leqq \Lambda^2; \\ 0 & \text{otherwise.} \end{cases} \tag{3.15}$$

Then if we integrate with respect to P over S, we see that in the product space of $(\mu_1, \cdots, \mu_n)$ and P, the measure of the set of points for which

$$\underset{0<\Lambda<\infty}{\text{l.u.b.}}\ \frac{1}{V(\Lambda)} \underset{\substack{\lambda_1^2+\cdots+\lambda_n^2\leqq\Lambda^2 \\ (\lambda_1+\mu_1)^2+\cdots+(\lambda_n+\mu_n)^2\leqq\Lambda^2}}{\int\cdots\int} f(T_1^{\lambda_1+\mu_1}\cdots T_n^{\lambda_n+\mu_n}P)\,d\lambda_1\cdots d\lambda_n > \alpha > 0 \tag{3.16}$$

does not exceed

$$\frac{1}{\alpha A_n}\int_S dV_P \underset{\lambda_1^2+\cdots+\lambda_n^2\leqq\Lambda^2}{\int\cdots\int} f(T_1^{\lambda_1}\cdots T_n^{\lambda_n}P)\,d\lambda_1\cdots d\lambda_n = \frac{V(\Lambda)}{\alpha A_n}\int_S f(P)\,dV_P. \tag{3.17}$$

Since $T_1^{\mu_1} \cdots T_n^{\mu_n}$ is an equimeasure transformation, we may replace the measure of the set for which (3.16) holds by

$$\begin{aligned}&\int\cdots\int d\mu_1\cdots d\mu_n \Big\{\text{measure of the set of values of } P \text{ for which} \\ &\underset{0<\Lambda<\infty}{\text{l.u.b.}}\ \frac{1}{V(\Lambda)} \underset{\substack{\lambda_1^2+\cdots+\lambda_n^2\leqq\Lambda^2 \\ (\lambda_1+\mu_1)^2+\cdots+(\lambda_n+\mu_n)^2\leqq\Lambda^2}}{\int\cdots\int} f(T_1^{\lambda_1}\cdots T_n^{\lambda_n}P)\,d\lambda_1\cdots d\lambda_n > \alpha > 0\Big\}.\end{aligned} \tag{3.18}$$

A fortiori,

$$\int_{\mu_1^2+\cdots+\mu_n^2 \leqq \frac{1}{4}\Lambda^2} \cdots \int d\mu_1 \cdots d\mu_n \Big\{ \text{measure of set of values of } P \text{ for which}$$

$$\text{l.u.b.}_{0<\Lambda\leqq\frac{1}{2}\Lambda} \frac{1}{V(\Lambda)} \int_{\lambda_1^2+\cdots+\lambda_n^2\leqq\Lambda^2} \cdots \int f(T_1^{\lambda_1} \cdots T_n^{\lambda_n} P)\, d\lambda_1 \cdots d\lambda_n > \alpha > 0 \Big\} \tag{3.19}$$

$$\leqq \frac{V(\Lambda)}{\alpha A_n} \int_S f(P)\ dV_P.$$

Thus

$$\Big\{ \text{measure of set of values of } P \text{ for which}$$

$$\text{l.u.b.}_{0<\Lambda\leqq\frac{1}{2}\Lambda} \frac{1}{V(\Lambda)} \int_{\lambda_1^2+\cdots+\lambda_n^2\leqq\Lambda^2} \cdots \int f(T_1^{\lambda_1} \cdots T_n^{\lambda_n} P)\, d\lambda_1 \cdots d\lambda_n > \alpha > 0 \Big\} \tag{3.20}$$

$$\leqq \frac{2^n}{\alpha A_n} \int_S f(P)\, dV_P.$$

If we now put $B_n = A_n 2^{-n}$, Theorem IV′ follows at once.

Theorems IV and IV′ have the secondary conclusions (1.11) and (1.19), respectively. To obtain these, let us introduce the function $h(P)$ which is $f(P)$ when this exceeds $\frac{1}{2}\alpha$ and is 0 otherwise. Clearly

$$f^*(P) \leqq h^*(P) + \tfrac{1}{2}\alpha. \tag{3.21}$$

Thus in Theorem IV

$$\begin{aligned} &\{\text{measure of set of values of } P \text{ for which } f^*(P) > \alpha\} \\ &\qquad \leqq \{\text{measure of set of values of } P \text{ for which } h^*(P) > \tfrac{1}{2}\alpha\} \\ &\qquad \leqq \frac{2}{\alpha} \int_S h(P)\, dV_P = \frac{2}{\alpha} \int_{f(P)\geqq\frac{1}{2}\alpha} f(P)\, dV_P. \end{aligned} \tag{3.22}$$

This yields (1.11), and (1.19) may be obtained in the identical manner. These are similar to certain results of Hardy and Littlewood, but are somewhat weaker.

Let us now turn to the proof of Theorem II. By (2.42), we may choose N so large that

$$\int_S \left| f_1(P) - \frac{1}{N+1} \sum_{n=0}^{N} f(T^n P) \right| dV_P < \epsilon^2. \tag{3.23}$$

Then by Theorem IV, except for a set of values of P of measure not exceeding ϵ, we shall have

$$\text{l.u.b.}_{0<A<\infty} \left| \frac{1}{A+1} \sum_{n=0}^{A} \left[f_1(T^n P) - \frac{1}{N+1} \sum_{m=0}^{N} f(T^{m+n} P) \right] \right| < \epsilon. \tag{3.24}$$

We have, for large A,

$$\begin{aligned}(3.25)\qquad \frac{1}{A+1}\sum_{N+1}^{A} f(T^n P) &\leqq \frac{1}{A+1}\sum_{n=0}^{A}\frac{1}{N+1}\sum_{m=0}^{N} f(T^{m+n}P)\\ &\leqq \frac{1}{A+1}\sum_{0}^{A+N} f(T^n P).\end{aligned}$$

Thus, except for a set of values of P of measure not exceeding ϵ, we have

$$(3.26)\qquad \overline{\lim_{A\to\infty}}\left| f_1(P) - \frac{1}{A+1}\sum_{n=0}^{A} f(T^n P)\right| < \epsilon.$$

Since ϵ can be given consecutively values from a convergent series, we see that, except over a set whose measure does not exceed the remainder of this series, we have uniformly

$$(3.27)\qquad f_1(P) = \lim_{A\to\infty}\frac{1}{A+1}\sum_{n=0}^{A} f(T^n P).$$

This establishes the Birkhoff ergodic theorem. The parallel Theorems II′ and II″ are proved in exactly the same way.

Theorem III and its n-dimensional analogue, which is the particular case of Theorem III″ which we obtain when the group reduces to the n-dimensional translation group, may be proved on the basis of Theorems IV and IV′, or even on the basis of the Lemmas D and D′. Every function of class L is the sum of a continuous bounded function and a function of small L-norm. Let $f(x)$ be the sum of $\phi(x)$ and $\psi(x)$, where $\phi(x)$ is continuous and $\int_{-\infty}^{\infty} |\psi(x)|\,dx < \frac{1}{2}\epsilon^2$. Then

$$\begin{aligned}(3.28)\qquad \overline{\lim_{\eta\to 0}}\left| f(x) - \frac{1}{\eta}\int_x^{x+\eta} f(\xi)\,d\xi\right| &= \overline{\lim_{\eta\to 0}}\left|\psi(x) - \frac{1}{\eta}\int_x^{x+\eta}\psi(\xi)\,d\xi\right|\\ &\leqq |\psi(x)| + \psi^*(x),\end{aligned}$$

where

$$(3.29)\qquad \psi^*(x) = \underset{0<\eta<\infty}{\text{l.u.b.}}\ \frac{1}{\eta}\int_x^{x+\eta} |\psi(\xi)|\,d\xi.$$

By Lemma D, the set of points for which $\psi^*(x) > \epsilon$ has a measure not exceeding $\frac{1}{2}\epsilon$, and the same is clearly true of $|\psi(x)|$. Thus, except for a set of points of measure not exceeding ϵ, we have

$$(3.30)\qquad \overline{\lim_{\eta\to 0}}\left| f(x) - \frac{1}{\eta}\int_x^{x+\eta} f(\xi)\,d\xi\right| \leqq \epsilon.$$

The proof now proceeds as in Theorem II. Theorem III″, as far as the translation group is concerned, is proved in the same way.

To prove Theorem III′ and the general case of Theorem III″, let us note that, for a fixed P, the set of values of λ for which we fail to have

$$f(T^{\lambda}P) = \lim_{\eta\to 0}\frac{1}{\eta}\int_{\lambda}^{\lambda+\eta} f(T^{\xi}P)\,d\xi, \tag{3.31}$$

or the set of values of $(\lambda_1, \cdots, \lambda_n)$ for which we fail to have

$$f(T_1^{\lambda_1}\cdots T_n^{\lambda_n}P) = \lim_{\Lambda\to 0}\frac{1}{V(\Lambda)}\underset{\mu_1^2+\cdots+\mu_n^2\leqq\Lambda^2}{\int\cdots\int} f(T_1^{\lambda_1+\mu_1}\cdots T_n^{\lambda_n+\mu_n}P)\,d\mu_1\cdots d\mu_n \tag{3.32}$$

are sets of zero measure. Thus in the product space of these spaces and P, the corresponding sets of values are of zero measure. That is, for almost all values of λ or $(\lambda_1, \cdots, \lambda_n)$, the set of points P for which we fail to have (3.31) or (3.32) valid is of zero measure. Therefore, for at least one of these values, the set is of zero measure. The result follows by an equimeasure transformation.

We finally come to Theorem V. Let $M(x)$ be the measure of the set of points over which $f(P) > x$, and let $M^*(x)$ be the measure of the set of points over which $f^*(P) \geqq x$. We shall have

$$\begin{aligned} \int_S s(f(P))\,dV_P &= -\int_0^\infty s(x)\,dM(x);\\ \int_S s(f^*(P))\,dV_P &= -\int_0^\infty s(x)\,dM^*(x). \end{aligned} \tag{3.33}$$

By Theorem IV, we have formally for $\nu > 0$

$$\begin{aligned} \int_0^\infty m^*(x)x^\nu\,dx &\leqq -\text{const.}\int_0^\infty x^{\nu-1}\,dx\int_x^\infty y\,dM(y)\\ &= -\text{const.}\int_0^\infty y\,dm(y)\int_0^y x^{\nu-1}\,dx\\ &= -\text{const.}\int_0^\infty y^{\nu+1}\,dM(y). \end{aligned} \tag{3.34}$$

Thus if $\int_0^\infty y^{\nu+1}\,dM(y)$ is bounded, it will follow that

$$0 = \lim_{\xi\to\infty}\int_\xi^{2\xi} m^*(x)x^\nu\,dx = \lim_{\xi\to\infty} M^*(2\xi)\xi^{\nu+1}. \tag{3.35}$$

We may hence integrate by parts in (3.34) and obtain

$$-\int_0^\infty x^{\nu+1}\,dm^*(x) \leqq -\text{const.}\int_0^\infty y^{\nu+1}\,dM(y). \tag{3.36}$$

Similarly,

$$\text{(3.37)} \quad \begin{aligned} \int_1^\infty m^*(x)\,dx &\leqq -\text{const.} \int_1^\infty y\,dM(y) \int_1^y \frac{dx}{x} \\ &= -\text{const.} \int_1^\infty y \log y\,dM(y). \end{aligned}$$

As before, if $\int f(P) \log^+ f(P)\, dV_P$ is bounded,

$$\text{(3.38)} \quad -\int_1^\infty x\,dm^*(x) \leqq -\text{const.} \int_1^\infty y \log y\,dM(y).$$

These theorems, apart from the less accurate constants, are the same as theorems given by Hardy, Littlewood, and Pólya (loc. cit.).

MASSACHUSETTS INSTITUTE OF TECHNOLOGY.

ENTROPY AND INFORMATION

BY

NORBERT WIENER

It has already been ascertained by Shannon and Wiener that amount of information is a quantity corresponding to negative entropy. Wiener has already suggested in his book *Cybernetics* that this is relevant to the action of a possible Maxwell demon. The Maxwell demon is able to lower the entropy of the mechanical output compared with the mechanical input because the demon itself possesses a negative entropy of information. This negative entropy must give information of the momentum and position of particles approaching the gateway operated by the demon in such a way that when the particle collides with the gateway, the gateway is either open or locked shut, so that no absorption of energy appears. This involves a mode of transmission more rapid than the motion of the particles to be separated, and a mode which is probably light. This light itself has a negative entropy which decreases whenever the frequency of the light is lowered by a Compton effect or similar cause.

Thus the Maxwell demon gives at least one way for comparing entropy of light with mechanical entropy. Since the information of the Maxwell demon includes both position and momentum it can be only imperfect information in quantum theory. This whole point of view suggests that one place to look for Maxwell demons may be in the phenomena of photosynthesis. It also suggests that absorption and emission of light may be an essential part of enzyme action in other forms of catalysis. Since there has been at least a suggestion that enzyme action and similar biological catalytic phenomena operate across an inert membrane, this light may give at least a partial explanation of the phenomena observed. At present the absence of an adequate entropic theory of radiation and in general of an adequate quantum theory of radiation able to take care of the particle-like as well as the wave-like aspects of light make a more precise verification of the opinions suggested here an impossibility.

MASSACHUSETTS INSTITUTE OF TECHNOLOGY,
CAMBRIDGE, MASS.

PROBLEMS OF SENSORY PROSTHESIS

NORBERT WIENER

This is an age in which a great deal is being done to help the handicapped. There is an improvement in artificial limbs for the maimed, in new sorts of spectacles, and in hearing aids. All this is returning to useful life a considerable part of those who would in other generations have been hopelessly handicapped. Even the two greatest handicaps from which humanity suffers—blindness and deafness—are the object of powerful and determined attacks with the object that where these defects cannot be removed or corrected by medical means, these victims may still be allowed to take, as nearly as possible, a normal part in life.

Notice that these two afflictions are sensory defects, that is, defects in the reception of impressions from the outer universe and from other human beings. From the standpoint of the outer universe, blindness is overwhelmingly the greater of the two losses. The sense of hearing, as brought out by such investigations as have been made by the Bell Telephone Company in their study of speech, and as further evidenced by the size of the speech areas in the brain, is a sense with vastly less variety than the sense of sight. On the other hand, normal communication between man and man goes by mouth and ear much more than by all other channels, and the social and emotional damage done by deafness is disproportionately great when one compares it with the social loss of the blind. The typical emotional picture of the deaf man is that of a reserved, self-contained, neurotic personality, whereas the typical picture of the blind man who has achieved any degree whatever of equilibrium is that of a euphorically confident and rather cocky personality.

In order to make any effort to replace the lost sense, either of the blind or of the deaf, it is necessary for us to make a rather accurate measure of what they have lost. It is then necessary for us to see if there are any relatively unused channels into the human nervous system which are capable of supplying the whole or any considerable part of what is lost. These are problems of physiology and communication engineering, and have a most important mathematical side.

To bring out this mathematical side adequately, I wish to refer to

---

The twenty-third Josiah Willard Gibbs Lecture, delivered at New York City, December 28, 1949, under the auspices of the American Mathematical Society; received by the editors April 10, 1950.

communication mechanisms which are more accessible to us in their whole than the human nervous system. A very interesting type of machine which is engaged in continual conversation with its surroundings is the remotely controlled power station. It is not unusual for water power to be available under just those circumstances which make the site highly unsuitable for human habitation. Under these conditions, a power station may be built without any resident engineer or operative. The place of such a resident operative is taken in part by a visiting operative who packs in at long intervals to replace and repair vital parts, but who is much more effectively replaced by automatic operation and long distance control.

In the automatic operation of a power station, one meets much the same problems which one finds in the operation of a railroad switching and signal system. There are certain things which must not be done at all costs. No generator should be connected with the bus bars until it is brought up to speed, and is running in the proper phase relations with the others. A penstock must not be thrown suddenly open in the lower end, or it will collapse under atmospheric pressure; and, on the other hand, it must not be subjected to the shock of too sudden turning on of water, or it will burst. These contingencies must be blocked in the same way as the conjunction of a closed switch and an open signal must be blocked in railroad practice.

These are relatively simple undertakings. As stated, they seem scarcely deep enough to have a mathematical form. However, behind them is a principle which is well susceptible to mathematical treatment. The principle, namely, is that the performance of a piece of apparatus should be fed back to it as information on which it is to operate.

What is more interesting to us mathematically at present, however, is not these purely local devices which modify the nature of the terminal apparatus in the generating station, and may be considered as part of the operation itself, but rather the messages which go between the load dispatcher and the station. Such messages must exist. A little cloud passes over the city to which the power is sent, and thousands of lights go on in the offices, shops and homes. This immediately demands the sending of more power; and this in turn means the starting of new turbines, the switching in of new generators, and a thousand and one other operations. Let there be an accident in the remote generating station, and even if this accident can be compensated for temporarily by automatic machinery, a message must go out to the plant engineer to send in a troubleshooter to restore the plant to maximum efficiency. In other words, the automatic substation must listen, and it must speak. Since, however, its ears and its mouth have been made by engineers for these specific purposes, and not installed by a complicated embryological process, we know far better what to expect of them than we do of human mouths and ears.

It may be remarked in passing that it is quite common practice to transmit these messages to and from the power station either over ordinary telephone lines and similar facilities, or over a carrier sent over the power line itself. In either case, they are generally coded messages like those of teletypewriters. If we wish to go to the trouble to make them vocal, all of the resources of the telephone and phonograph are at our disposal, but there is generally no particular point in this rather precious imitation of human function.

What the casual listener may miss in this account is the fact that the operation of such a system involves several stages of linguistic translation. The message from the dispatcher to the station is or should be committed to the line in the form best suitable for line transmission. On the other hand, it must appear in the station, not in the form of a coded sequence of dots and dashes, but in the form of a series of openings of switches, turnings on of valves, measurements of the speeds of motors or generators, and so on. The language of command must be translated into the language of action.

To understand this, we must realize what information is, and how it can be measured. When I send a series of dots and dashes, the problem of the information that it carries cannot be answered in terms of the message I send, taken by itself apart from all other messages. A single click may be able to do a large number of things, such as opening a sluice, connecting a generator to the bus bars, or replacing a burnt out electronic valve; but it cannot do these things *differentially.* It is only when I know the number of separate alternatives which must be transmitted and combined that I can determine how many different messages I must be able to send. It then appears that the significance of a signal is determined not only by that signal itself, but by the whole set of signals I might have sent in place of the actual one. In other words, significance is a property belonging to ensembles.

Another property belonging to larger and smaller ensembles is that of probability, or what is its equivalent, their Lebesgue measure. I do not intend to go into the whole question of the theoretical basis of probability, but shall only state that, under certain rather wide conditions, a smaller set of signals has a probability when taken to be part of a larger set, and that this probability is a number between 0 and 1. Now, the probability that two independent events should happen simultaneously is the *product* of their individual probabilities. On the other hand, if information is to have a significance even remotely resembling its ordinary meaning, the amount of information given by two independent signals should be the *sum* of the amount of information which they convey separately. Under these conditions, since multiplication of probabilities corresponds to addition of information, the amount of information given by a signal can only be a constant multiple of the logarithm of its probability. Moreover, since the information is positive when the probability is less than one, the ratio between the logarithm of the probability and the amount of information must be negative.

The notion of the logarithms of probability is a familiar one for statistical mechanics. The logarithm of a probability is called *entropy*. In an ordinary statistical system, the entropy can increase spontaneously, but can never so decrease. In the transmission of information, we may lose information, but can never gain it, except from a new source. Thus in the translation which occurs between the message sent into the automatic station, and the operation of that station, there may be a loss of information, but there cannot be a gain. Under ideal circumstances, we attempt a coding which will correspond as nearly as possible to the actual operation of the power station, so that we are not sending in unused material, as we should for example if we were to transmit our orders by voice. Nevertheless, it is hardly to be expected that our code is so perfect that we lose no information.

Let us go from this problem to the problem of transmission of a message by voice or by eye from person to person. There is in the first instance a limitation of the rate at which information can be transmitted. It depends on the system of repeatable and recognizable signals which can be made with our vocal organs or our pen, or whatever other means we use of active communication, and which will be accepted by the air, or whatever other channel of transmission we use. I say recognizable, but I have already introduced an ambiguity thereby. Recognizable how, and by what? If I allow the recipient the full choice of any detecting and analyzing instrument which he may choose to demand, I get a certain rate of transmission of information, which is characteristic of the sender and of the means of transmission but which is not characteristic of the reception of this message by the ordinary listener or viewer. It is information of this type with which the communication engineer has been chiefly concerned in the past, as it is by all means the simplest to handle. However, when this information penetrates through the ear or the eye into those parts of the human nervous system which are specifically concerned with sound or sight, it undergoes a translation by a mechanism of which we know something, but not everything. As a matter of fact, it undergoes two or more translations. The first of these represent what is necessary to penetrate into the human sense organ, and through that sense organ to those parts of the brain which are in more or less permanent connection with it. The later translations interpret this superficial sensory representation into meaning, and give to what is received a more or less permanent place in memory and understanding. In the case of sound, the first translation is called *phonetic* and the later translations are termed *semantic.*

The problems of the loss of information in the two human translations, phonetic and semantic, are not at present susceptible to a direct investigation through the structure of the nervous system. They are however susceptible to an investigation which is empirical and experimental. Work of this sort has been done by the Bell Telephone Company, with special reference to the recognizability of sound as speech. The original measures of information, which were given more or less crudely, referred to the bands of frequency necessary for the formation of intelligible speech; and in other words were almost more measures of what the outside line and the air could carry, than of what the ear could translate. The reconstruction of intelligible speech from clicks and buzzes, by way of wave filters and resonators and other apparatus of the sort, shows that the loss of information between the ear and the part of the nervous system belonging to the ear is overwhelmingly great, and that a crude imitation of speech involving less than 1/10 and perhaps even 1/100 of the information carried over a telephone line, is enough for adequate conversation.

It was in view of this work that Dr. Wiesner of the Electronics Laboratory of MIT discussed with me the question as to whether anything could be done for a totally deaf man, to replace the missing sense of hearing in its most important employment, that of the recognition of conversation. The question was put to me by Dr. Wiesner, and the two of us were independently led to the same answer: namely, that the recognizable part of speech was so scanty, that it was not beyond our hope to replace it by the sense of touch backed up with adequate equipment. In other words, it appeared to us that it is perfectly possible to make the transition between sound in the outside air, and the semantic recognition of speech, by an artificial phonetic stage, making use of touch, and supplemented by adequate electrical tools. Having come to this conclusion, we started the detailed design of such apparatus. We took speech through an ordinary microphone.

not dissimilar to that used as a receiver for the usual hearing aids, and quite adequate to take in all that is intelligible in speech. We then split the band received into a number of frequency bands, making use of wave filters. As we intended to use the hand as a receiver, five seemed the natural number of bands to take, because we could then carry them to the five fingers. However, we are not yet finally decided that five is the right number and may experiment on supplementing it by additional stations in the palm of the hand, so as to bring the number up to six or seven. The different bands we have so far taken to be of equal logarithmic length, or length when measured in terms of the number of keys of the piano corresponding to them. Whether this is the best mode of speech, we are not yet sure. It is a subject for further experiment.

At any rate, after receiving and separating these bands we rectify them, which means, nearly, that we replace the time-patterns of that band by its envelope. These rectified messages are then used to modulate an equal number of vibrators at frequencies of something like 100 cycles each. This is because the skin is very insensitive, both to nearby constant messages and to messages of excessively high frequency. These modulated messages are then carried either to electrical mechanical vibrators placed on the five fingers, or to electrodes designed for the direct stimulation of the nerve endings in the skin.

From the standpoint of lightness of weight, it would seem that a direct electrical stimulation is desirable. However, there are several difficulties to be overcome. The variable resistances of the skin, and the fact that current rather than voltage is what stimulates touch, both indicate that it is desirable to use constant current electrodes, which can be properly designed. However, even with these, the margin between the lower threshold of feeling and the lower threshold of pain is so small that our apparatus must work over a very narrow range, and our present models need considerable modification.

The mechanical vibrator, on the other hand, tends to be undesirable, because its current demand is at present so great as to require an apparatus which is still too heavy to be comfortably portable. We believe that whether we finally arrive at mechanical or electrical stimulators, the problem of making the apparatus portable, and indeed of comparable weight with that actually used in existing hearing aids, is not too difficult to be solved, and is more one of development than anything else.

Let me first state how our results stood when we left off work early in the summer. This work was undertaken by Dr. Leon Levine, with the cooperation of Dr. Alex Bavelas of our Psychology Department in matters which concerned techniques of teaching. It was desirable to see if there was in fact a sufficient basis of recognizable and distinguishable syllables in speech for us to go ahead. In the beginning, we did not experiment with the deaf, but with hearing people whose ears were filled with such a jumble of artificial noise that they were quite completely deaf for the purposes of the experiment. We then gave them a sequence of syllables by the machine accompanied with a visual pattern display of these syllables. After they had learned these, we subjected them to a series of trial syllables selected from the sequence; and we recorded the successes and failures. The best run we had was a series of 80 selections among 12 syllables with only six errors.

It has been our idea from the beginning that the learning process necessary for achieving proficiency with the new method of receiving speech must be something quite long and arduous, and we have not wished to involve any deaf mutes in an experiment which might give them a false hope, or at any rate a bad learning technique. However, we did make a certain number of experiments with a blind deaf mute, who came to us with his brother. The two had learned to communicate together by speech, which was recognized by the deaf mute when he placed his hands on his brother's larynx. The deaf mute could himself enunciate a few words, but the enunciation was very breathy and bad. After he had been put in the apparatus, he was for the first time in his life able to compare his brother's speech with his own, particularly in the enunciation of his own name and that of his brother. Within a matter of minutes, and certainly of hours, the improvement in his enunciation was not merely noticeable, but overwhelming. It was perfectly clear to us that our apparatus is an adequate way for a deaf mute, even for a blind deaf mute, to monitor his own speech.

I want to say at this point that soon after we had begun our work, we recognized very definitely that it was closely related to the visual speech developed by the Bell Telephone Company. I also wish to say that the patents which the Bell Telephone Company has based on this work were so phrased as to cover the work we had done, although our methods had not been actually employed experimentally by the Bell workers. The advantage which we believe our methods possess over visual speech, in addition to the obvious one that they can be used by blind deaf mutes, are that we have been working for portability in the apparatus. It was our conviction before we started, and it has been very greatly confirmed by our results, that the apparatus which we finally achieve should be completely portable, and should not deprive one of the concurrent use of such an important sense as vision. It is only by this portability that the speaker's words can be continually monitored, so that he is never allowed to fall into the faulty speech of the congenital deaf mute. We hope that, with the use of this apparatus, the speech-training of the deaf mute can be begun in early childhood, and that although it will be arduous training, it may not at that period be too different from that training which hearing children receive in learning to speak.

It will be a curious study to see what these hearing aids will do for the musical sense. They were not designed for this purpose, and I do not expect any very great results. Still, people like Helen Keller have reached a surprising degree of proficiency in recognizing musical vibrations with their fingers, and it is scarcely to be expected that the results with these machines will be worse than with the bare fingers. All in all, however, the social function of hearing is so nearly exclusively a matter of the reception of speech, that a person who can follow speech in the basis of sound carried by the air, and can do this with reasonable proficiency, can scarcely be considered socially deaf.

It is a little unfortunate that we have to use so valuable a part of the body as the hand for a receiver. We intend to use the left hand alone, and even as far as this is concerned, we intend to hold the receivers in a flexible glove, so that a gross use of the grasp by the hand is not incompatible with the simultaneous reception of speech. The trouble with the other sensitive receiving parts of the body is that the best of them lie around the mouth which must be left without impediment to participate in conversation.

It is thus conceivable at least to us that deafness as a considerable social problem might even be eliminated, except for those not mentally able to face the stringent training needed, or for those with serious concomitant physical defects. We have no such expectation for the replacement of vision.

There is reason to believe that the total amount of information received at a given time by a normal eye is of the order of 50 to 100 times what is received in the same time by a normal ear. In this, the type of measurement of vision as given by the oculist's ratio is deceptive. Vision which is 50/100 means ability to recognize details at 50 feet which would ordinarily be recognized at 100. However, the number of details varies as the *square* of the distance, so that from this point of view 30/100 vision should give about a quarter of the amount of possible information which normal vision gives. On this basis, if the whole auditory region of the cortex were turned over to the purposes of vision, we should expect about 10/100 vision. This is not blindness, but very poor vision indeed. Touch is an even worse surrogate for vision. Under these circumstances, it scarcely seems worth while to try to make a universal apparatus to replace vision for all purposes.

There are two directions in which progress should be expected. One of these is a reading apparatus, which will translate the pages of an ordinary book into a sound or touch pattern recognizable by the blind. The other is an aid whereby the blind man may go around more freely out of doors and in unfamiliar rooms. In the construction of an apparatus of this latter sort, there is a limitation below as well as a limitation above. Any apparatus must be better than what the blind man can do for himself by using such odds and ends of sensation as auditory echo, air temperature, air pressure of the forehead, and so forth.

Finally, besides the problem of the blind and the deaf, the problem of the maimed is also one with a profound sensory aspect which is often neglected. The better an artificial limb is from the standpoint of variability of performance and flexibility, the more information as to its position and as to the strains on it which the cripple needs if he is to walk securely. Similar remarks apply to artificial hands. There is a very considerable future in the art of applying strain and pressure gauges to artificial limbs, in such a way as to furnish the cripple with better sensory monitoring than he can obtain with the aid of anything but the skin of his stump. What the cripple has lost is not only motion, but sensation as well. He is an ataxic, as well as a paralytic, and anything that can be done to furnish him with sensory feedback channels other than those which he normally possesses is likely to make a great contribution to his effectiveness and to his well-being.

MASSACHUSETTS INSTITUTE OF TECHNOLOGY

HOMEOSTASIS IN THE INDIVIDUAL AND SOCIETY

BY

NORBERT WIENER

Massachusetts Institute of Technology

One of the great ideas introduced into medicine by Claude Bernet is that of homeostasis—that is of a mechanism in the living organism which maintains the internal environment at a level at which healthy life is possible. Our blood pressure, the rate of our pulse, the rate of our breathing, the action of our kidneys, are all determined by homeostatic mechanisms which generally function so well that we do not notice them and, when they fail to function, bring on acute disorders such as fever, dyspnoea, tachycardia, uraemia, and the like.

A generation ago, it would have been difficult to find outside of the human body adequate analogues of these homeostatic mechanisms. One indeed was well-known, that of the governor in the steam-engine, and this special mechanism has been the theme of a classical paper by Clarke Maxwell in the 1860's. But the fact that the steam-engine governor was but one of a multitude of mechanisms, which we now class as the feedback type, regulating not only the speed, but the position, or the temperature, or some other property of physical systems, had not come to general consciousness. In these feedback systems an error of performance is returned to the system as a partial basis of its future performance, and in this way swings up and swings down are corrected when they are begun, without waiting for them to take on large proportions. At present, one of the cheaper means for introducing such a correction and for performing the combinations which make it possible, is the use of the vacuum tube.

Feedback mechanisms in general increase the uniformity of performance of a system whatever the load may be. Strictly speaking, this needs a severe qualification. If the load should be excessive and the amount of feedback demanded by the load also excessive, the feedback mechanism will unstabilize performance rather than stabilize it. Such a system will embark on an ever wilder series of oscillations until it breaks down, or at any rate the fundamental laws of the performance change.

In my book *Cybernetics* (John Wiley & Sons, The Technology Press, Hermann et Cie., 1948) I have given a parallel discussion of living and mechanical feedback mechanisms. I wish to call to the attention of the reader the existence of feedback mechanisms which contain both a living and mechanical part. There has been a recent invention[1] at the

[1] R. Bickford, *EEG Journal,* May, 1950.

Reprinted from *J. Franklin Inst., 251,* 1951, pp. 65–68. (Courtesy of the *Journal of the Franklin Institute.*)

Mayo Clinic of a mechanical anaesthetist, which regulates the depth of anaesthesia of an animal or a human by an injection device which injects barbiturates into the veins, or ether into the breathing mask, and the injection is measured in accordance with the depth of anaesthesia of the animal as made manifest through its electroencephalogram, which is directly interpreted into a mechanical basis for the injection. This principle is in the long run manifestly the right way to regulate anaesthesia, and it amounts to an artificial chain of homeostasis combining elements in the body and elements outside. If the validity of this method of homeostasis be granted in the one case of anaesthesia, we may scarcely expect anaesthesia to be the sum total of its usefulness. There are many conditions of great peril which are the result of a breakdown of homeostasis.

A very simple one, and also a very unusual one is that of curare poisoning. If there were a way for the degree of paralysis of the muscular endplate to be scanned for muscular performance, it would be an ideal thing to treat curare poisoning by a respirator whose intensity of action was regulated correctly by observing the paralysis of these endplates; or perhaps a better method would be to use an apparatus reading continuously the CO_2 concentration in the blood. Again, the medication of the heart is notably difficult, and requires great skill and judgment. It may well be in the future that when the signs of an incipient coronary catastrophe are present, we may learn of some drug whose dosage may be determined by a properly designed mechanical reading of the electrocardiogram.

Frankly, without stepping beyond the limitations of the layman in judging in what cases such a treatment is possible and in what cases it is not, I am very sanguine about the future possibilities of feedback-regulated therapy.

Besides these vital phenomena of homeostasis, there are a considerable number which belong to posture, and to the living phenomena of voluntary performance rather than to the maintenance of a livable internal environment. Your one-legged man finds that his artificial leg gives him less information concerning the state of his joints than he really needs to walk with the greatest freedom. I am not the only person who has suggested to various Government agencies that it is highly desirable to treat the problem of artificial limbs as essentially a problem of feedback, by attaching to the joints strain gauges, recording by vibrators on the normal skin of the amputee. But like my other colleagues who have maintained these ideas, I have not found my voice echoed in official quarters. I may be permitted to suspect that the misfortunes of the maimed constitute a sort of vested interest.

Besides the importance of the feedback in motor deficiencies, sensory deficiencies also have an important feedback element. It is foolish to separate deafness as inability to hear, from dumbness as inability to speak. Either or both constitute an inability to use sound waves as an effective two-sided tool of conversation. The deaf patient has no adequate way of monitoring his own voice, so that it shall accord with the standards of hearing people. If then he is given an adequate way of turning sound waves into sensory experience, this monitoring and the feedback it gives rise to will enable him to use his own voice so that it conforms to the canons of an adequate conversational tool. This has been our object in a series of experiments at MIT, in which we have tried to secure a tactile reception of sound waves adequate for the understanding of speech.

The blind man's cane is a feedback. It enables him to create disturbances by which he can scan the world around him. The design of better pieces of apparatus of this sort promises to mitigate greatly the lot of the blind, even as a similar ability almost assures us of the possibility to mitigate the lot of the deaf.

There are, however, regions and important regions in which we are all deaf and blind. The horror of walking through an atom-bombed city is that one may kill oneself without hurting oneself. The Geiger counter is a great tool to prevent this, if we have the presence of mind to look at it, but a Geiger counter alone will never have the compelling force of the searing pain of ordinary low-temperature fire. Any adequate Geiger counter to be carried by those whose occupation may at any time expose them to the peril of nuclear radiation, should not merely follow the radiation in a visible scale or by audible clicks. When the radioactive level goes beyond a certain point, it should hurt the person carrying it definitively and sharply. Similarly, if we could only make the incipient stages of cancer or of heart disease desperately painful, we should go far to eliminate them from the list of killers. They are painful enough indeed when they are thoroughly established, but this is generally not until the cancer has metastasized into large regions of vital tissue, or until the life of the coronary patient hangs on, on one or two patent arteries. Thus with all the power that we now possess, the body corporeal does not possess an adequate homeostasis for all its needs. Even less does the body politic.

The body politic is not without homeostasis, or at least the intention of having homeostasis. This is what we mean, for example, by calling the Constitution of the United States "a Constitution of checks and balances." It is our intention that the administration serve as a protection against any inordinate assumption of power on the part of the Legislature, and that the Legislature watch jealously over any incursions of the Executive into its own particular realm, while the Judiciary should observe the action of both and see that they conform to national traditions. I have said, "national traditions." There is the rub. On the one hand, we do not wish to be bound forever to decisions made in the past, under what were perhaps irrelevant circumstances.

On the other hand, national traditions will not function without a certain minimum of historical consciousness and sagacity on the part of the population at large. We are very sensitive about a policy which will lead to great unemployment or to great poverty, provided we may expect this policy to come to ripeness within our own time. On the other hand, as the history of the slogan, "Peace in our time," has shown, we are not very sensitive to a future catastrophe, however certain we are that it will arise to plague our children when we are laid in our graves. The future is terribly unreal to the majority of us. And how shall it not be, when the past is equally unreal to us?

The continuity with the past which belongs to the older settled regions of Europe and to a lesser but a very real extent to the older settled regions of the United States, is dependent not only on an acquaintance with written history, but with the continual presence of the houses, the roads, the farms and the cities established by past generations, and with a familiarity with the mode of life which developed them. When a Yankee basketmaker will show you in his shed the tools which his great grandfather forged from bog iron and which he learned to use after the custom of the Indian to split the annual rings of the red ash and make his splints, he will do so with a guileless sense of the contemporaneity of the past, which is very far removed from the pride of the New England aristocrat in his genealogy. His past lies in his barn with its bins, its tools and its baskets; and if his thoughts carry him back to a time when the roads were just emerging out of the Indian trails, and the mode of life of the white man was subject to continual borrowing from that of the red man, his thoughts will also carry him forward to at least some crude consideration of the time when his children's children will plow his land and make his baskets. To such a man, the possibility of a stabilizing mechanism is present, and he will have a real reaction to it, for otherwise his house will become an empty cellar-hole, and his land the possession of a stranger. This is for him a death beyond death, in a very real and present sense.

However, in the great cities, or in the esoteric hot house civilization of Southern California, where a man's parents lie in the soil of Iowa or Nebraska, and a man's neighbors are pursuing their purposes in life without any reference to his own, it is useless to ask that he should regard his own grandchildren as anything but slightly modified sorts of strangers. The span of social memory which is needed for the homeostatic action of an historical sense is too great for transients and squatters. Yet without this span, the regions which have been built out of the desert and the burnt hillsides in a magnificient defiance of nature, represent far too much of a defiance of nature to possess the elements necessary for their continued existence. To respect the future, we must be aware of the past; and if the regions where that awareness of the past is real have shrunken down to a pin point on our vast map, then so much the worse for us and our children, and our children's children.

A Factorization of Positive Hermitian Matrices

N. WIENER & E. J. AKUTOWICZ

Communicated by J. L. WALSH

It is our purpose to give an independent proof of the matrix-analogue of a classical theorem of G. SZEGÖ[1] which runs as follows:

If $F(\theta)$, $0 \leqq \theta \leqq 2\pi$, is a non-negative function of class L^1, and if

$$-\infty < \int_0^{2\pi} \log F(\theta)\, d\theta,$$

then there exists a complex-valued function $G(\theta)$ of class L^2, the Fourier series of which is of power series type,

$$G(\theta) \sim \sum_0^{\infty} g_n e^{in\theta},$$

and such that

$$F(\theta) = |G(\theta)|^2$$

for almost all θ.

The generalization of this result, to be proved in this paper, is the following

Theorem. *If $H(\theta) = (h_{jk}(\theta))$ is an Hermitian positive-definite matrix-valued function defined on $0 \leqq \theta \leqq 2\pi$ such that*

$$h_{jk}(\theta) \quad \text{belongs to} \quad L^1(0, 2\pi) \qquad (j, k = 1, \cdots, q)$$

and

$$-\infty < \int_0^{2\pi} \log \det H(\theta)\, d\theta,$$

then $H(\theta)$ can be factored as

$$H(\theta) = N(\theta)N(\theta)^*,$$

[1] G. SZEGÖ, *Orthogonal Polynomials*, A.M.S. Colloq. Publ., 1939, §10.2.

Reprinted from *J. Math. & Mech.*, *8*, 1959, pp. 111–120. (Courtesy of the *Journal of Mathematics and Mechanics*, Indiana University.)

where $N(\theta) = (n_{jk}(\theta))$ *is such that*

$$n_{jk}(\theta) \text{ belongs to } L^2(0, 2\pi) \qquad (j, k = 1, \cdots, q)$$

and each $n_{jk}(\theta)$ *has a Fourier series of power series type (without negative frequencies).*

Although the essential idea of the present argument derives from the theory of linear least-squares prediction of multiple time series, no use will be made of the theory of stochastic processes in what follows.[2]

The theorem of SZEGÖ is precisely the special case of the general theorem for a 1×1 matrix, and will constitute the first step in the inductive proof of the general theorem. Therefore the present argument concerns the step from a $q - 1 \times q - 1$ matrix to a $q \times q$ matrix.

1. A generalization of Hadamard's determinant inequality. We shall use the following notation only in this section in order to state a determinantal inequality. For any $q \times q$ matrix $H \equiv (h_{jk})$ with complex coefficients, the symbol

$$H\begin{pmatrix} j_1 j_2 & \cdots & j_p \\ k_1 k_2 & \cdots & k_p \end{pmatrix}$$

stands for the determinant

$$\begin{vmatrix} h_{j_1k_1} & h_{j_2k_1} & \cdots & h_{j_pk_1} \\ h_{j_1k_2} & h_{j_2k_2} & \cdots & h_{j_pk_2} \\ \vdots & \vdots & & \vdots \\ h_{j_1k_p} & h_{j_2k_p} & \cdots & h_{j_pk_p} \end{vmatrix}.$$

If H is a $q \times q$ *Hermitian positive-definite matrix, then for* $p < q$

$$H\begin{pmatrix} 1 & 2 & \cdots & q \\ 1 & 2 & \cdots & q \end{pmatrix} \leqq H\begin{pmatrix} 1 & 2 & \cdots & p \\ 1 & 2 & \cdots & p \end{pmatrix} H\begin{pmatrix} p+1 & \cdots & q \\ p+1 & \cdots & q \end{pmatrix}. \tag{1.1}$$

This implies for an Hermitian positive-definite matrix-valued function $H(\theta) = (h_{jk}(\theta))$, $0 \leqq \theta \leqq 2\pi$, that

$$\det H(\theta) \leqq h_{11}(\theta) \cdots h_{qq}(\theta),$$

[2] WIENER & MASANI, *Acta Math.* 98 (1957), p. 148, have obtained the same factorization theorem as a corollary of certain facts in the time-domain analysis of stationary multivariate stochastic processes. Their deduction does not show clearly how the integrability of the logarithm of the determinant of the given matrix implies its factorability. A difficulty, mentioned by WIENER & MASANI (*loc. cit.* p. 112) and having its origin in the possibility that the components of the vector process span subspaces "inclined at zero angle," does not in fact arise as shown in the present note.

HELSON & LOWDENSLAGER, *Acta Math.* 99 (1958), have also proved this theorem by still different methods.

and, hence, if

$$-\infty < \int_0^{2\pi} \log \det H(\theta)\, d\theta,$$

that

$$-\infty < \int_0^{2\pi} \log h_{qq}(\theta)\, d\theta,$$

and therefore

$$-\infty < \int_0^{2\pi} \log H\begin{pmatrix} 1 & 2 & \cdots & q-1 \\ 1 & 2 & \cdots & q-1 \end{pmatrix} d\theta. \tag{1.2}$$

(1.2) will be made use of directly.

2. Preliminary reductions. Suppose $H(\theta) = (h_{jk}(\theta))$ is a given Hermitian positive-definite $q \times q$ matrix function defined on $0 \leqq \theta \leqq 2\pi$ and such that

$$h_{jk}(\theta) \quad \textit{belongs to} \quad L^1(0, 2\pi) \tag{2.1}$$

and

$$-\infty < \int_0^{2\pi} \log \det H(\theta)\, d\theta. \tag{2.2}$$

Conditions (2.1) and (2.2) imply

$$-\infty < \int_0^{2\pi} \log h_{jj}(\theta)\, d\theta < \infty, \qquad j = 1, \cdots, q.$$

Therefore SZEGÖ's theorem implies $h_{jj}(\theta) = |g_j(\theta)|^2$, where g_j belongs to $L^2(0, 2\pi)$ and has Fourier series with only non-negative frequencies. Therefore

$$H(\theta) = \begin{bmatrix} g_1 & 0 & \cdots & 0 \\ 0 & g_2 & & \vdots \\ \vdots & & \ddots & \\ 0 & & \cdots & g_q \end{bmatrix} \begin{bmatrix} 1 & h_{12}/g_1\bar{g}_2 & \cdots & h_{1q}/g_1\bar{g}_q \\ h_{21}/\bar{g}_1 g_2 & 1 & & \\ \vdots & & \ddots & \\ h_{q1}/\bar{g}_1 g_q & & \cdots & 1 \end{bmatrix} \begin{bmatrix} \bar{g}_1 & 0 & \cdots & 0 \\ 0 & \bar{g}_2 & & \vdots \\ \vdots & & \ddots & \\ 0 & & \cdots & \bar{g}_q \end{bmatrix} \tag{2.3}$$

for almost all θ.[3] Writing, for $1 < j,\ k < q$,

$$\begin{aligned} f_{jk}(\theta) &= h_{jk}(\theta)/(g_j(\theta)\bar{g}_k(\theta)) && \text{if} \quad j < k, \\ f_{jk}(\theta) &= 1 && \text{if} \quad j = k, \\ f_{jk}(\theta) &= \bar{f}_{kj}(\theta) && \text{if} \quad j > k, \end{aligned}$$

it follows that the matrix $F(\theta) = (f_{jk}(\theta))$ is an Hermitian positive-definite

[3] We shall usually omit the qualification "almost all θ."

matrix for almost all θ. Hence every principal minor is non-negative, and in particular $|f_{ik}(\theta)| \leqq 1$.

Thus the problem is reduced[4] to factoring an Hermitian positive matrix function $F(\theta) = (f_{ik}(\theta))$, such that $f_{ii}(\theta) \equiv 1$ and

$$-\infty < \int_0^{2\pi} \log \det F(\theta)\, d\theta, \tag{2.4}$$

into a product $M(\theta)M(\theta)^*$ where $M(\theta) = (m_{ik}(\theta))$ is such that

$$m_{ik}(\theta) \quad \text{is} \quad \text{bounded} \tag{2.5}$$

and

(2.6) $m_{ik}(\theta)$ has only non-negative frequencies in its Fourier series.

3. The problem further reduced to factoring a simpler matrix. In this section we shall apply the induction assumption according to which for $q > 1$ any $q - 1 \times q - 1$ matrix satisfying the conditions of the last paragraph of §2 can be factored as described there.

Now let $F(\theta)$ be a $q \times q$ matrix,

$$F(\theta) = \begin{bmatrix} 1 & f_{12}(\theta) & \cdots & f_{1q}(\theta) \\ f_{21}(\theta) & 1 & \cdots & f_{2q}(\theta) \\ \vdots & & & \vdots \\ f_{q1}(\theta) & & \cdots & 1 \end{bmatrix},$$

which is Hermitian and positive-definite and such that (2.4) holds. Put

$$F_0(\theta) = \begin{bmatrix} 1 & f_{12}(\theta) & \cdots & f_{1,q-1}(\theta) \\ f_{21}(\theta) & 1 & \cdots & f_{2,q-1}(\theta) \\ \vdots & & & \vdots \\ f_{q-1,1}(\theta) & & \cdots & 1 \end{bmatrix}.$$

---

[4] It may be useful to recall the

Lemma. *If $\phi(\theta)$ belongs to $L(0, 2\pi)$ and $\psi(\theta)$ is bounded and both have Fourier series without negative frequencies then their product has a Fourier series without negative frequencies.*

According to FEJÉR and LEBESGUE there exist polynomials P_N of degree $\leqq N$ in $e^{i\theta}$ such that, for almost all θ, $P_N(e^{i\theta}) \to \psi(\theta)$ boundedly. Therefore, by dominated convergence,

$$\int_0^{2\pi} \phi \cdot P_N \cdot e^{in\theta}\, d\theta \to \int_0^{2\pi} \phi \cdot \psi \cdot e^{in\theta}\, d\theta.$$

But, for $n = 1, 2, \cdots$, for all N, the left hand side is null, since ϕ is of power series type. Hence

$$\int_0^{2\pi} \phi \cdot \psi \cdot e^{in\theta}\, d\theta = 0, \qquad n = 1, 2, \cdots,$$

as required.

By the generalized Hadamard inequality of §1, (1.2),

$$-\infty < \int_0^{2\pi} \log \det F_0(\theta)\, d\theta, \tag{3.1}$$

and hence, by our inductive hypothesis,

$$F_0(\theta) = M_0(\theta) \cdot M_0(\theta)^*, \tag{3.2}$$

where $M_0(\theta) = (\mu_{jk}(\theta))$ is such that $\mu_{jk}(\theta)$ is bounded and has only non-negative frequencies appearing in its Fourier series. Consider now the matrix equation

$$F(\theta) = \begin{pmatrix} \mu_{11} & \cdots & \mu_{1,q-1} & 0 \\ \mu_{21} & \cdots & \mu_{2,q-1} & 0 \\ \vdots & & \vdots & \vdots \\ \mu_{q-1,1} & \cdots & \mu_{q-1,q-1} & 0 \\ 0 & \cdots & 0 & 1 \end{pmatrix} \begin{pmatrix} 1 & 0 & \cdots & 0 & \gamma_1 \\ 0 & 1 & \cdots & 0 & \gamma_2 \\ \vdots & & & & \vdots \\ 0 & & \cdots & 1 & \gamma_{q-1} \\ \bar{\gamma}_1 & & \cdots & \bar{\gamma}_{q-1} & 1 \end{pmatrix} \begin{pmatrix} \bar{\mu}_{11} & \cdots & \bar{\mu}_{q-1,1} & 0 \\ \vdots & & \vdots & \vdots \\ \bar{\mu}_{1,q-1} & \cdots & \bar{\mu}_{q-1,q-1} & 0 \\ 0 & \cdots & 0 & 1 \end{pmatrix}$$

to determine $\gamma_1(\theta), \cdots, \gamma_{q-1}(\theta)$. This is equivalent to (3.2) together with the following linear system for the γ_k's:

$$\sum_{k=1}^{q-1} \mu_{jk}\gamma_k = f_{jq}, \qquad j = 1, \cdots, q-1.$$

Since $\det M_0(\theta) = (\det F_0(\theta))^{\frac{1}{2}}$ and since $F_0(\theta)$ satisfies (3.1) this system has a unique solution $\gamma_1(\theta), \cdots, \gamma_{q-1}(\theta)$ for almost all θ. Writing the above matrix equation as $QCQ^* = F$ we have that $C = Q^{-1}F(Q^{-1})^*$. But F is positive. Hence C is positive. Hence

$$|\gamma_j(\theta)| \leqq 1, \qquad j = 1, \cdots, q-1.$$

Therefore the problem is further reduced to factoring the Hermitian positive matrix

$$\Gamma(\theta) = \begin{pmatrix} 1 & 0 & \cdots & 0 & \gamma_1 \\ 0 & 1 & & & \\ \vdots & & & \vdots & \vdots \\ 0 & & & 1 & \gamma_{q-1} \\ \bar{\gamma}_1 & & \cdots & \bar{\gamma}_{q-1} & 1 \end{pmatrix} \tag{3.3}$$

where

$$\det \Gamma(\theta) = 1 - \sum_{k=1}^{q-1} |\gamma_k(\theta)|^2$$

satisfies the condition

$$-\infty < \int_0^{2\pi} \log \det \Gamma(\theta)\, d\theta. \tag{3.4}$$

That det $\Gamma(\theta)$ has this property follows because every matrix that has been introduced as a factor of the original $H(\theta)$ by our reductions has a determinant whose absolute value has this property.

4. The Hilbert space $\mathcal{L}_\Gamma^2$. Since the matrix $\Gamma(\theta)$ of (3.3) is Hermitian and non-negative, an inner product can be introduced into the set of (equivalence classes of) all vectors

$$\mathbf{f}(\theta) = (f_1(\theta), \cdots, f_q(\theta)), \tag{4.1}$$

where the $f_i(\theta)$ are complex-valued, Lebesgue measurable functions on $0 \leqq \theta \leqq 2\pi$, the $\mathbf{f}(\theta)$'s being such that

$$\int_0^{2\pi} \mathbf{f}(\theta)\Gamma(\theta)\mathbf{f}(\theta)^* \, d\theta < \infty. \tag{4.2}$$

The inner product for such $\mathbf{f}$, $\mathbf{g}$ is by definition

$$(\mathbf{f}, \mathbf{g})_\Gamma = \int_0^{2\pi} \mathbf{f}(\theta)\Gamma(\theta)\mathbf{g}(\theta)^* \, d\theta.$$

In this way we obtain a separable pre-Hilbert space, whose completion will be denoted $\mathcal{L}_\Gamma^2$.

It will be convenient to have a certain lower bound for the norm (4.2). For fixed θ let U be a unitary matrix diagonalizing Γ:

$$U^*\Gamma U = \Lambda = \begin{pmatrix} \lambda_1 & & \\ & \ddots & \\ & & \lambda_q \end{pmatrix}.$$

Then, putting $\mathbf{y} = \mathbf{x}U$,

$$\begin{aligned} \mathbf{x}\Gamma\mathbf{x}^* = \mathbf{y}\Lambda\mathbf{y}^* &= \lambda_1 |y_1|^2 + \cdots + \lambda_q |y_q|^2 \\ &= \frac{\lambda_1\lambda_2 \cdots \lambda_q}{\lambda_2 \cdots \lambda_q} |y_1|^2 + \cdots + \frac{\lambda_1\lambda_2 \cdots \lambda_q}{\lambda_1\lambda_2 \cdots \lambda_{q-1}} |y_q|^2 \\ &\geqq \frac{\det \Gamma}{\sigma_{q-1}} (|y_1|^2 + \cdots + |y_q|^2), \end{aligned}$$

where

$$\sigma_{q-1} = (\lambda_2\lambda_3 \cdots \lambda_q) + \cdots + (\lambda_1\lambda_2 \cdots \lambda_{q-1})$$

is the sum of all $q - 1 \times q - 1$ principal minors of Λ, and therefore of Γ. Hence, from the explicit form of Γ, for all θ, $\sigma_{q-1} \leqq q$. It follows that

$$\mathbf{x}\Gamma\mathbf{x}^* \geqq \frac{\det \Gamma}{q} \cdot \sum_1^q |x_i|^2$$

and

$$(4.3) \qquad ||\mathbf{x}||_{\Gamma}^{2} \geqq \frac{1}{q} \int_{0}^{2\pi} \sum_{1}^{q} |x_j|^2 \det \Gamma \, d\theta.$$[5]

5. Existence of a non-null innovation space. For non-negative integers n let $\mathbf{\Sigma}_n$ denote the closed linear span in $\mathfrak{L}_{\Gamma}^{2}$ of the set of vectors

$$\{e^{in_1\theta}\boldsymbol{\phi}_1, \cdots, e^{in_q\theta}\boldsymbol{\phi}_q \,|n_j \geqq n, \qquad j = 1, \cdots, q\},$$

where $\boldsymbol{\phi}_j$ has the constant function 1 in the jth place and 0's elsewhere.

Lemma 5.1. *Each element* $\boldsymbol{\phi}_1, \cdots, \boldsymbol{\phi}_q$ *has a non-null projection orthogonal to* $\mathbf{\Sigma}_1$.

Put $\mathbf{P} = (P_1, \cdots, P_q)$, where P_j's are polynomials in $e^{i\theta}$ without constant term. Such $\mathbf{P}$'s are dense in $\mathbf{\Sigma}_1$. On the other hand, according to the lower bound (4.3),

$$||\boldsymbol{\phi}_j - \mathbf{P}||^2 \geqq \frac{1}{q}\int_0^{2\pi} |1 - P_j|^2 \det \Gamma \, d\theta \geqq \frac{1}{q} \exp\left\{\frac{1}{2\pi}\int_0^{2\pi} \log \det \Gamma \, d\theta\right\} > 0,$$

where we have, in succession, taken a smaller lower bound, used a classical result[6] of Szegö on one-sided approximation and used the hypothesis that $\log \det \Gamma$ belongs to L^1. Therefore $\boldsymbol{\phi}_j$ does not belong to $\mathbf{\Sigma}_1$, $j = 1, \cdots, q$. *Q.e.d.*

Let $\boldsymbol{\psi}_1', \cdots, \boldsymbol{\psi}_q'$ be the projections of $\boldsymbol{\phi}_1, \cdots, \boldsymbol{\phi}_q$ into the orthogonal complement of $\mathbf{\Sigma}_1$. The vectors $\boldsymbol{\psi}_1', \cdots, \boldsymbol{\psi}_q'$ can be orthonormalized into r orthogonal unit vectors in $\mathfrak{L}_{\Gamma}^{2}$, to be denoted henceforth as $\boldsymbol{\psi}_1, \cdots, \boldsymbol{\psi}_r$, where $0 < r \leqq q$. Actually, $r = q$, as will be shown below.

6. Emptiness of the infinitely remote future. In order to obtain Fourier expansions for $\boldsymbol{\phi}_1, \cdots, \boldsymbol{\phi}_q$ in $\mathfrak{L}_{\Gamma}^{2}$ in terms of $e^{in_1\theta}\boldsymbol{\psi}_1, \cdots, e^{in_r\theta}\boldsymbol{\psi}_r$, where $n_1, \cdots, n_r$ run independently over all non-negative integers, the following lemma is essential.

Lemma 6.1. *If* S_N *denotes the closed linear span in* $\mathfrak{L}_{\Gamma}^{2}$ *of the vectors*

$$\{e^{in_1\theta}\boldsymbol{\phi}_1, \cdots, e^{in_q\theta}\boldsymbol{\phi}_q \,|n_j \geqq N, \qquad j = 1, \cdots, q\},$$

then

$$\bigcap_{N>0} \mathrm{S}_N = (0).$$

Suppose $\mathbf{f}$ belongs to $\bigcap_{N>0}\mathrm{S}_N$. It is permissible to take $\mathbf{f}$ to be of the form (4.1) $\mathbf{f} = (f_1, \cdots, f_q)$. According to Szegö's theorem, which can be applied in view of (3.4),

$$(6.1) \qquad \det \Gamma = \gamma_+\bar{\gamma}_+,$$

[5] We owe this inequality to Professor P. R. Masani of Bombay whom we should like to thank for several critical discussions on this paper, which resulted in essential improvements.

[6] G. Szegö, *loc. cit.*[1], pp. 293–294.

where γ_+ belongs to L^2 and has a Fourier series of power series type. We recall that $\gamma^+(\theta) \neq 0$ for almost all θ (Lebesgue measure). By (4.3) and (6.1) for any $\mathbf{g} = (g_1, \cdots, g_q)$ belonging to $\mathfrak{L}_\Gamma^2$,

$$||\mathbf{g}||_\Gamma^2 \geqq \frac{1}{q}\int_0^{2\pi} \sum_1^q |g_j\gamma_+|^2 \, d\theta.$$

Therefore $g_j\gamma_+$ belongs to L^2. Let us put P_{jN}^k for polynomials in $e^{iN\theta}$, $e^{i(N+1)\theta}$, $\cdots$, and $\mathbf{P}_N^k = (P_{1N}^k, \cdots, P_{qN}^k)$, these quantities being such that

$$\lim_{k\to\infty} ||\mathbf{f} - \mathbf{P}_N^k|| = 0, \qquad \text{for all } N. \tag{6.2}$$

On the other hand,

$$||\mathbf{f} - \mathbf{P}_N^k||^2 \geqq \sum_1^q \int_0^{2\pi} |f_j\gamma_+ \quad P_{jN}^k\gamma_+|^2 \, d\theta.$$

For N sufficiently large this lower bound is bounded away from zero uniformly in k (by Parseval's identity) unless $f_1 = 0, \cdots, f_q = 0$.

7. Certain one-sided developments. The orthogonal unit vectors $\psi_1, \cdots, \psi_r$ defined at the end of §5 generate a set $\mathbf{B}$ of vectors $\{e^{in_1\theta}\psi_1, \cdots, e^{in_r\theta}\psi_r\}$, where $n_1, n_2, \cdots, n_r$ run independently over all non-negative integers. The set $\mathbf{B}$ is evidently orthonormal in $\mathfrak{L}_\Gamma^2$. It is also a basis for $\mathbf{\Sigma}_0$. For if f belongs to $\mathbf{\Sigma}_0$, the difference

$$\mathbf{f} - \sum_{\psi \varepsilon \mathbf{B}} (\mathbf{f}, \psi)_\Gamma \psi$$

belongs to $\mathbf{S}_N$ for $N = 1, 2, \cdots$. By Lemma 6.1 this difference must be null. It follows at once that we can obtain Fourier expansions for ϕ_j that are convergent in $\mathfrak{L}_\Gamma^2$ to ϕ_j of the form

$$\phi_j = \sum_{\psi \varepsilon \mathbf{B}} (\phi_j, \psi)_\Gamma \psi, \qquad j = 1, \cdots, q,$$

or, more explicitly,

$$\phi_j = \psi_1 \sum_{n_1=0}^{\infty} a_{jn_1}^{(1)} e^{in_1\theta} + \cdots + \psi_r \sum_{n_r=0}^{\infty} a_{jn_r}^{(r)} e^{in_r\theta},$$

where

$$a_{j\nu}^{(k)} = (\phi_j, \psi_k e^{i\nu\theta})_\Gamma.$$

Put

$$M_{jk}(\theta) \sim \sum_{\nu=0}^{\infty} a_{j\nu}^{(k)} e^{i\nu\theta}, \qquad j = 1, \cdots, q, \qquad k = 1, \cdots, r,$$

so that M_{jk} belongs to $L^2(0, 2\pi)$, and

$$\phi_j = M_{j1}\psi_1 + \cdots + M_{jr}\psi_r, \qquad j = 1, \cdots, q.$$

Hence, for $1 \leqq j, k \leqq q$,

$$(\phi_k e^{in\theta}, \phi_j)_\Gamma = (\sum_J M_{kJ} e^{in\theta} \psi_J , \sum_L M_{jL} \psi_L)_\Gamma = \sum_{J,L} (M_{kJ} e^{in\theta} \psi_J , M_{jL} \psi_L)_\Gamma$$

$$= \sum_{J,L} \sum_{\nu=0}^{\infty} \sum_{\mu=0}^{\infty} a_{k\nu}^{(J)} \overline{a_{j\mu}^{(L)}} (e^{i(\nu+n)\theta} \psi_J , e^{i\mu\theta} \psi_L)_\Gamma$$

$$= \sum_{J,L} \sum_{\mu=\max(0,n)}^{\infty} a_{k,\mu-n}^{(J)} \overline{a_{j\mu}^{(L)}} (\psi_J , \psi_L)_\Gamma$$

$$= \sum_{J=1}^{r} \sum_{\mu=\max(0,n)}^{\infty} a_{k,\mu-n}^{(J)} \overline{a_{j\mu}^{(J)}}$$

$$= \frac{1}{2\pi} \int_0^{2\pi} e^{in\theta} \sum_{J=1}^{r} M_{jJ}(\theta) \overline{M_{kJ}(\theta)} \, d\theta, \qquad n = 0, \pm 1, \pm 2, \cdots .$$

On the other hand, $(\phi_k e^{in\theta}, \phi_j)$ can also be expressed in terms of the coefficients of the matrix $\Gamma(\theta)$:

$$(\phi_k e^{in\theta}, \phi_j)_\Gamma = \int_0^{2\pi} e^{in\theta} \delta_{jk} \, d\theta \qquad \text{if} \qquad 1 \leqq j, k \leqq q - 1,$$

$$(\phi_q e^{in\theta}, \phi_j)_\Gamma = \int_0^{2\pi} e^{in\theta} \overline{\gamma_j(\theta)} \, d\theta \qquad \text{if} \qquad 1 \leqq j \leqq q - 1,$$

$$(\phi_k e^{in\theta}, \phi_q)_\Gamma = \int_0^{2\pi} e^{in\theta} \gamma_k(\theta) \, d\theta \qquad \text{if} \qquad 1 \leqq k \leqq q - 1,$$

$$(\phi_q e^{in\theta}, \phi_q)_\Gamma = \int_0^{2\pi} e^{in\theta} \, d\theta, \qquad n = 0, \pm 1, \cdots .$$

By the uniqueness of Fourier coefficients in $L^1(0, 2\pi)$,

$$\begin{pmatrix} & & \gamma_1 \\ & & \vdots \\ I_{q-1} & \cdot & \vdots \\ & & \gamma_{q-1} \\ \bar{\gamma}_1 \; \cdots & \bar{\gamma}_{q-1} & 1 \end{pmatrix} = (M_{jk})(M_{jk})^*, \tag{7.1}$$

where I_{q-1} stands for the $q - 1$ dimensional unit matrix. That $M_{jk}(\theta)$ is bounded follows from the diagonal equations

$$\sum_{J=1}^{r} |M_{jJ}(\theta)|^2 = 1, \qquad j = 1, \cdots , q.$$

The matrix $(M_{jk}(\theta))$ has q rows and r ($\leqq q$) columns. Suppose $r < q$. Then if $(M_{jk}(\theta))$ is augmented by $q - r$ columns of zeros on the right and $(M_{jk}(\theta))^*$ is augmented by $q - r$ rows of zeros below, the product $(M_{jk}(\theta)(M_{jk}(\theta))^*$ remains unchanged. But the determinant of the augmented $(M_{ik}(\theta))$ is identically zero, whereas by (7.1)

$$|\det M_{ik}|^2 = \det \Gamma(\theta).$$

This contradiction proves that $r = q$, and completes the proof of the existence of a factorization of the required sort.

Massachusetts Institute of Technology
Cambridge, Massachusetts